ADVANCED TOPICS IN SCATTERING THEORY AND BIOMEDICAL ENGINEERING

Proceedings of the 9th International Workshop on
Mathematical Methods in Scattering Theory and Biomedical Engineering

ADVANCED TOPICS IN SCATTERING THEORY AND BIOMEDICAL ENGINEERING

Patras, Greece 9–11 October 2009

Editors

A. Charalambopoulos
University of Ioannina, Greece

D. I. Fotiadis
University of Ioannina, Greece

D. Polyzos
University of Patras, Greece

World Scientific

NEW JERSEY • LONDON • SINGAPORE • BEIJING • SHANGHAI • HONG KONG • TAIPEI • CHENNAI

Published by

World Scientific Publishing Co. Pte. Ltd.
5 Toh Tuck Link, Singapore 596224
USA office: 27 Warren Street, Suite 401-402, Hackensack, NJ 07601
UK office: 57 Shelton Street, Covent Garden, London WC2H 9HE

British Library Cataloguing-in-Publication Data
A catalogue record for this book is available from the British Library.

ADVANCED TOPICS IN SCATTERING THEORY AND BIOMEDICAL ENGINEERING
Proceedings of the 9th International Workshop on Mathematical Methods in Scattering Theory and Biomedical Engineering

ISBN-13 978-981-4322-02-7
ISBN-10 981-4322-02-4

Printed in Singapore by B & Jo Enterprise Pte Ltd

Preface

In the last couple of decades, Biomedical Engineering and applications of Wave Scattering Theory to bioengineering have attracted the attention of many scientists and research organizations worldwide. Biomedical Engineering is related to the use of engineering principles in biomedicine while wave scattering theory deals with wave propagation phenomena in non-homogeneous continuum systems. Those scientific fields are projected to undergo through a very fast growth during the next years. Thus, there is an even greater need to better understand the complex biological and natural underlying mechanisms.

In the early 1990s we created a working group on those two scientific fields having as common background applied mathematics. Initially, the workshops we have organized attracted the attention of some friends, being although very good researchers. The group was expanded and the workshop took another form, since many students and young researchers decided to participate and present their ideas. Now the scientific program of the workshop is based on plenary presentations and in depth discussions. We have invited distinguished scientists from abroad to present their work and provide a scientific forum for further development and research. New links among scientists have been created which helps to better understand problems, to develop new ideas and promote successful applications.

The workshop this year is dedicated to Prof. Ioannis Vardoulakis, an eminent scientist, who is known for his contribution in the modeling of geomaterials, geo-hazards and geotechnical processes. Prof. Vardoulakis was a member of the organizing committee of the workshop. He passed away beginning of September 2009. His contribution cannot be described in such a short preface. He has played a catalytic role in the development of both research and applications in mechanics. He was a great teacher, researcher and above all a great person.

In this volume of proceedings we have collected most of the papers presented in the 9th International Workshop on Applied Mathematics on Scattering Theory and Biomedical Engineering, held in Patras, from 9-11 October 2009. The papers are presented in sessions and cover applied mathematics, direct and inverse problems in scattering theory, enhanced elastic theories, scientific computing and biomedical engineering.

The Organizing Committee of the 9th workshop is grateful to Ms Maria Pikou and Mrs Vicky Papageorgiou for the organization of the workshop and the publishing of this volume of proceedings. We are also indebted to our universities for the financial support of the workshop. Our group will work for the organization of the next workshop and we remain always enthousiastic in the idea of having together great scientists and young researchers in a continuous processing of exchanging ideas and working together.

Ioannina, January 2010

Antonios Charalambopoulos, *University of Ioannina, Greece*
Dimitrios I. Fotiadis, *University of Ioannina, Greece*
Demosthenes Polyzos, *University of Patras, Greece*

Contents

Mathematical Methods in Scattering and Biomedical Engineering

THE DOUBLE ANALYTIC STRUCTURE THAT ALLOWS THE INTRODUCTION OF VECTOR ELLIPSOIDAL HARMONICS

G. DASSIOS

University of Patras and ICE/HT-FORTH, Greece

The ellipsoidal harmonics are the harmonics of the anisotropic space, just as the spherical harmonics are the harmonics of the isotropic space. The spherical harmonics were introduced in 1785 by Lagrange (zonal harmonics) and Laplace (tesseral harmonics). The ellipsoidal harmonics were introduced by Lamé in 1837. Hansen introduced vector spherical harmonics in 1935 as a means to solve antenna radiation problems. The vector ellipsoidal harmonics were introduced by the author and Tsampas in 2009. In contrast to all other cases, the theory of vector ellipsoidal harmonics demands the introduction of two different inner products to secure orthogonality. In this presentation we demonstrate this peculiarity of the ellipsoidal system which, to some extent, it is responsible for the delay of the introduction of vector ellipsoidal harmonics.

1. Introduction

Special functions were generated from the method of separation of variables for partial differential equations, a method that was initiated by Bernoulli (1753), Lagrange (1759), D'Alembert (1763) and was developed, to perhaps the central core of linear analysis, by Fourier in his celebrated 1822 paper on heat conduction. Separation of variables led to spectral theory for linear operators with physical applications that can hardly be compared with any other mathematical method in Physics. A quick glance to the mathematical basis of Vibration Theory, or Quantum Mechanics will make this statement obvious. Separation of variables and the eigenfunction expansions they support, provide the most effective way to solve analytically boundary value problems. But, life is seldom so generous, and in order to secure separability we need linearity of the governing operators, simple geometry and a lot of luck!

In three dimensions, separation of variables for a scalar partial differential equation leads to three families of ordinary differential equations which are connected by two separation constants. In almost all cases of physical interest the mathematical characteristics of these families, such as convergence, orthogonality, competence and so on, are concluded from the general theory of Sturm-Liouville systems. In fact, in the early 1950's, Eisenhart, Moon and Spencer attempted to find all coordinate systems that allow separation for the Laplace and the Helmholtz equations. They found that, among all the orthogonal curvilinear coordinate systems with first and second degree coordinate surfaces, there are only 11 systems that allow for splitting of the partial differential equation to three ordinary differential equations. Among these coordinate systems, the simpler one is the Cartesian and the most complicated one is the ellipsoidal.

If the partial differential operator acts on a vector field, which is decomposed in three scalar components with respect to some curvilinear system, then separation of variables seeks for the splitting of the vector equation into three scalar partial differential equations, each one involving only a single component of the decomposed field. Then, each one of these three scalar equations can be further separated into three ordinary differential equations and the theory of scalar equations is readily applied.

In the present work we will investigate what exactly happens in the case where this program is not applicable, as it is the case of ellipsoidal harmonics, and then we will try to find what is the best we can do with this peculiar system.

2. Vector Spherical Harmonics

The naive approach to solve a vector problem, for which the corresponding scalar problem is separable, is to expand every Cartesian component in terms of the corresponding scalar eigenfunctions and look for solutions in the form

$$\boldsymbol{U}(\boldsymbol{r}) = \sum_n \sum_m \boldsymbol{C}_{nm} u_{nm}(\boldsymbol{r}) \tag{1}$$

where

$$\boldsymbol{C}_{nm} = (C^1_{nm}, C^2_{nm}, C^3_{nm}) \tag{2}$$

denotes the vector of the unknown coefficients. If we apply the boundary conditions to the above form of the solution we will find that the evaluation of the coefficients $\boldsymbol{C}_{nm}$ is either very hard, or impossible. The reason for this difficulty is that we casted the vectorial character of the field $\boldsymbol{U}$ into the coefficients $\boldsymbol{C}_{nm}$ and not into the eigenfunctions, which carry information about the natural geometry of the problem. Therefore, the right approach is to introduce vector eigenfunctions that are compatible with the geometry of the boundaries [4], [5]. But this is not an easy task. In fact, vectorial eigenfunctions are known only for spherical and spheroidal geometries.

The vector spherical harmonics were introduced by Hansen [2]. The interior solid harmonics are given by

$$\boldsymbol{A}_{1n}^m(\boldsymbol{r}) = r^n \boldsymbol{r} Y_n^m(\hat{\boldsymbol{r}}) - \frac{1}{2n+3}\nabla(r^{n+2}Y_n^m(\hat{\boldsymbol{r}})) \tag{3}$$

$$\boldsymbol{A}_{2n}^m(\boldsymbol{r}) = \nabla \times \boldsymbol{A}_{1n}^m(\boldsymbol{r}) = \nabla \times (r^n \boldsymbol{r} Y_n^m(\hat{\boldsymbol{r}})) \tag{4}$$

$$\boldsymbol{A}_{3n}^m(\boldsymbol{r}) = \nabla \times \boldsymbol{A}_{2n}^m(\boldsymbol{r}) = \nabla\nabla \cdot \boldsymbol{A}_{1n}^m(\boldsymbol{r}) \tag{5}$$

and the exterior solid harmonics are given by

$$\boldsymbol{B}_{1n}^m(\boldsymbol{r}) = \boldsymbol{r}\frac{Y_n^m(\hat{\boldsymbol{r}})}{r^{n+1}} + \frac{1}{2n-1}\nabla\left(\frac{Y_n^m(\hat{\boldsymbol{r}})}{r^{n-1}}\right) \tag{6}$$

$$\boldsymbol{B}_{2n}^m(\boldsymbol{r}) = \nabla \times \boldsymbol{B}_{1n}^m(\boldsymbol{r}) = \nabla \times \left(\boldsymbol{r}\frac{Y_n^m(\hat{\boldsymbol{r}})}{r^{n+1}}\right) \tag{7}$$

$$\boldsymbol{B}_{3n}^m(\boldsymbol{r}) = \nabla \times \boldsymbol{B}_{2n}^m(\boldsymbol{r}) = \nabla\nabla \cdot \boldsymbol{B}_{1n}^m(\boldsymbol{r}) \tag{8}$$

where

$$Y_n^m(\hat{\boldsymbol{r}}) = C_n^m P_n^m(\cos\vartheta)e^{im\varphi} \tag{9}$$

denotes the complex normalized form of the scalar surface spherical harmonics, with $C_n^m$the normalization constants. The functions $\boldsymbol{A}_{in}^m$ and $\boldsymbol{B}_{in}^m$ solve the vector Laplace equation

$$\Delta\boldsymbol{F}(\boldsymbol{r}) = \nabla\nabla \cdot \boldsymbol{F}(\boldsymbol{r}) - \nabla \times \nabla \times \boldsymbol{F}(\boldsymbol{r}). \tag{10}$$

In applying boundary conditions, the important functions are the following surface vector harmonics

$$\boldsymbol{P}_n^m(\hat{\boldsymbol{r}}) = \hat{\boldsymbol{r}} Y_n^m(\hat{\boldsymbol{r}}) \tag{11}$$

$$\boldsymbol{B}_n^m(\hat{\boldsymbol{r}}) = \frac{1}{\sqrt{n(n+1)}}\boldsymbol{D}Y_n^m(\hat{\boldsymbol{r}}) \tag{12}$$

$$\boldsymbol{C}_n^m(\hat{\boldsymbol{r}}) = \frac{1}{\sqrt{n(n+1)}}(\boldsymbol{D}Y_n^m(\hat{\boldsymbol{r}})) \times \hat{\boldsymbol{r}} \tag{13}$$

where

$$\boldsymbol{D} = r\nabla = \hat{\boldsymbol{\vartheta}}\frac{\partial}{\partial\vartheta} + \frac{\hat{\boldsymbol{\varphi}}}{\sin\vartheta}\frac{\partial}{\partial\varphi}. \tag{14}$$

Observe that $\boldsymbol{P}_n^m$ are radial functions, while $\boldsymbol{B}_n^m$ and $\boldsymbol{C}_n^m$ are tangential functions. The set $\{\boldsymbol{P}_n^m, \boldsymbol{B}_n^m, \boldsymbol{C}_n^m\}_{n=0,m=-n}^{n=\infty,m=n}$ is a local orthogonal system.

Performing the indicated differentiations we obtain

$$\boldsymbol{A}_{1n}^m(\boldsymbol{r}) = \frac{\sqrt{n+1}}{2n+3} r^{n+1}[\sqrt{n+1}\boldsymbol{P}_n^m - \sqrt{n}\boldsymbol{B}_n^m] \tag{15}$$

$$\boldsymbol{A}_{2n}^m(\boldsymbol{r}) = \sqrt{n(n+1)} r^n \boldsymbol{C}_n^m \tag{16}$$

$$\boldsymbol{A}_{3n}^m(\boldsymbol{r}) = (n+1)\sqrt{n} r^{n-1}[\sqrt{n}\boldsymbol{P}_n^m + \sqrt{n+1}\boldsymbol{B}_n^m] \tag{17}$$

where $\boldsymbol{A}_{2n}^m$ are tangential for every n and m and $\boldsymbol{A}_{1n}^m$, $\boldsymbol{A}_{3n}^m$ are both perpendicular to $\boldsymbol{A}_{2n}^m$ for every n and m.

Any vector harmonic function inside a sphere has the expansion

$$\boldsymbol{F}(\boldsymbol{r}) = \sum_{n,m}[c_{1n}^m \boldsymbol{A}_{1n}^m(\boldsymbol{r}) + c_{2n}^m \boldsymbol{A}_{2n}^m(\boldsymbol{r}) + c_{3n}^m \boldsymbol{A}_{3n}^m(\boldsymbol{r})]. \tag{18}$$

For Dirichlet data on the boundary $r = a$, the function $\boldsymbol{F}$ is then written as

$$\boldsymbol{F}(a\hat{\boldsymbol{r}}) = \sum_{n,m}[p_n^m \boldsymbol{P}_n^m(\hat{\boldsymbol{r}}) + b_n^m \boldsymbol{B}_n^m(\hat{\boldsymbol{r}}) + c_n^m \boldsymbol{C}_n^m(\hat{\boldsymbol{r}})] \tag{19}$$

where p_n^m and b_n^m are linear combinations of c_{1n}^m and c_{3n}^m and c_n^m are proportional to c_{2n}^m. The orthogonality of $\boldsymbol{P}_n^m$, $\boldsymbol{B}_n^m$, $\boldsymbol{C}_n^m$ allows for the analytic evaluation of the unknown coefficients c_{1n}^m c_{2n}^m and c_{3n}^m. A similar approach solves the Neumann problem.

3. Vector Ellipsoidal Harmonics

The challenging question now is whether the above logic for the sphere can be extended to the case of the ellipsoidal system. We scrutinize the procedure that led to the definition of Vector Spherical Harmonics (VSH) for a general orthogonal curvilinear system (ξ_1, ξ_2, ξ_3) and a vector harmonic field

$$\boldsymbol{F}(\boldsymbol{\xi}) = f^1(\boldsymbol{\xi})\widehat{\boldsymbol{\xi}_1} + f^2(\boldsymbol{\xi})\widehat{\boldsymbol{\xi}_2} + f^3(\boldsymbol{\xi})\widehat{\boldsymbol{\xi}_3} \tag{20}$$

and we see that separability for the vector Laplacian is possible only when the following conditions among the metric coefficients h_{ξ_1}, h_{ξ_2}, h_{ξ_3} hold [5]

- $h_{\xi_1} = 1$
- $\frac{\partial}{\partial \xi_1} \frac{h_{\xi_2}}{h_{\xi_3}} = 0$
- $h_{\xi_2} h_{\xi_3}$ is proportional either to 1 or to ξ_1^2

It is easy to see that no one of these conditions holds for the ellipsoidal system. Hence, we can not introduce Vector Ellipsoidal Harmonics (VEH) which are as good as the VSH. But let's ignore this obstacle for the moment and proceed further, in order to find what is the best we can do with the ellipsoidal system [3].

The ellipsoidal coordinates (ρ, μ, ν) are connected to the Cartesian ones by the relations

$$x_1 = \frac{\rho\mu\nu}{h_2 h_3} \tag{21}$$

$$x_2 = \frac{\sqrt{\rho^2 - h_3^2}\sqrt{\mu^2 - h_3^2}\sqrt{h_3^2 - \nu^2}}{h_1 h_3} \tag{22}$$

$$x_3 = \frac{\sqrt{\rho^2 - h_2^2}\sqrt{h_2^2 - \mu^2}\sqrt{h_2^2 - \nu^2}}{h_1 h_2} \tag{23}$$

with

$$0 \leqslant \nu^2 \leqslant h_3^2 \leqslant \mu^2 \leqslant h_2^2 \leqslant \rho^2 < \infty \tag{24}$$

and

$$h_1^2 = a_2^2 - a_3^2, h_2^2 = a_1^2 - a_3^2, h_3^2 = a_1^2 - a_2^2 \tag{25}$$

are the semi-focal distances of the system. The ellipsoidal metric coefficients are

$$h_\rho^2 = \frac{(\rho^2 - \mu^2)(\rho^2 - \nu^2)}{(\rho^2 - h_3^2)(\rho^2 - h_2^2)} \tag{26}$$

$$h_\mu^2 = \frac{(\mu^2 - \rho^2)(\mu^2 - \nu^2)}{(\mu^2 - h_3^2)(\mu^2 - h_2^2)} \tag{27}$$

$$h_\nu^2 = \frac{(\nu^2 - \rho^2)(\nu^2 - \mu^2)}{(\nu^2 - h_3^2)(\nu^2 - h_2^2)} \tag{28}$$

from where it is obvious that none of the above conditions is satisfied, since all coefficients depend on all variables. Let

$$\mathbb{E}_n^m(\rho, \mu, \nu) = E_n^m(\rho) E_n^m(\mu) E_n^m(\nu) \tag{29}$$

$$\mathbb{F}_n^m(\rho, \mu, \nu) = F_n^m(\rho) E_n^m(\mu) E_n^m(\nu) \tag{30}$$

be the interior and exterior solid ellipsoidal harmonics, respectively, and let

$$S_n^m(\mu, \nu) = E_n^m(\mu) E_n^m(\nu) \tag{31}$$

be the corresponding surface ellipsoidal harmonics. As we mentioned earlier, the importance of the theory of vector spherical harmonics is that the the surface vector spherical harmonics $\boldsymbol{P}_n^m$, $\boldsymbol{B}_n^m$, $\boldsymbol{C}_n^m$ are complete and orthogonal over the unit sphere S^2. That allows for the calculation of the coefficients of an eigenexpansion in closed form.

For the spherical case, we choose one function in the normal direction $\hat{\boldsymbol{r}}$ and two tangential functions in the directions of ∇Y and $\hat{\boldsymbol{r}} \times \nabla Y$. Following a similar approach for the ellipsoid, we choose the normal direction $\hat{\boldsymbol{\rho}}$ and the two tangential directions ∇S and $\hat{\boldsymbol{\rho}} \times \nabla S$. Hence, we define the vector surface ellipsoidal harmonics as

$$\boldsymbol{R}_n^m(\mu, \nu) = d(\rho, \mu, \nu) \hat{\boldsymbol{\rho}} S_n^m(\mu, \nu) \tag{32}$$

$$\boldsymbol{D}_n^m(\mu, \nu) = f(\rho, \mu, \nu) \nabla S_n^m(\mu, \nu) \tag{33}$$

$$\boldsymbol{T}_n^m(\mu, \nu) = g(\rho, \mu, \nu) \hat{\boldsymbol{\rho}} \times \nabla S_n^m(\mu, \nu) \tag{34}$$

and specify the functions d, f and g in such a way as to secure as much orthogonality as possible. We know that complete orthogonality among these functions is not possible. So, let's try for the best possible, i.e. let's try to maximize the set of orthogonal relations among the functions (32)-(34).

The orthogonality relations we need to have are

$$\oint_{S_\rho} \boldsymbol{R}_n^m \cdot \boldsymbol{D}_{n'}^{m'} l_\rho ds_\rho = 0 \tag{35}$$

$$\oint_{S_\rho} \boldsymbol{D}_n^m \cdot \boldsymbol{T}_{n'}^{m'} l_\rho ds_\rho = 0 \tag{36}$$

$$\oint_{S_\rho} \boldsymbol{T}_n^m \cdot \boldsymbol{R}_{n'}^{m'} l_\rho ds_\rho = 0 \tag{37}$$

$$\oint_{S_\rho} \boldsymbol{R}_n^m \cdot \boldsymbol{R}_{n'}^{m'} l_\rho ds_\rho = R_n^m \delta_{nn'} \delta_{mm'} \tag{38}$$

$$\oint_{S_\rho} \boldsymbol{D}_n^m \cdot \boldsymbol{D}_{n'}^{m'} l_\rho ds_\rho = D_n^m \delta_{nn'} \delta_{mm'} \tag{39}$$

$$\oint_{S_\rho} \boldsymbol{T}_n^m \cdot \boldsymbol{T}_{n'}^{m'} l_\rho ds_\rho = T_n^m \delta_{nn'} \delta_{mm'}. \tag{40}$$

A long, detailed and tedious investigation leads to the choice

$$d(\rho,\mu,\nu) = 1 \tag{41}$$

$$f(\rho,\mu,\nu) = \rho\frac{\sqrt{\rho^2-\mu^2}\sqrt{\rho^2-\nu^2}}{\sqrt{\rho^2-h_3^2}\sqrt{\rho^2-h_2^2}} \tag{42}$$

$$g(\rho,\mu,\nu) = \rho \tag{43}$$

$$l_\rho(\mu,\nu) = \frac{1}{\sqrt{\rho^2-\mu^2}\sqrt{\rho^2-\nu^2}} \tag{44}$$

for which the orthogonality relations (35)-(39) are satisfied. Orthogonality (40) is managed if we replace the weighting function l_ρ by the function

$$\tilde{l}_\rho(\mu,\nu) = \frac{\sqrt{\rho^2-\mu^2}\sqrt{\rho^2-\nu^2}}{(\rho^2-h_3^2)(\rho^2-h_2^2)}. \tag{45}$$

Hence, complete orthogonality for the ellipsoidal system demands the use of two different inner products! For the relative proofs we refer to [1].

Note that, in contrast to the spherical system, where the surface harmonics $\{\boldsymbol{P}_n^m, \boldsymbol{B}_n^m, \boldsymbol{C}_n^m\}$ given by (11)-(13), are independent of the radial variable r, the ellipsoidal system $\{\boldsymbol{R}_n^m, \boldsymbol{D}_n^m, \boldsymbol{T}_n^m\}$ is $\rho-$dependent. In other words, in the case of the ellipsoid, the set of surface harmonics changes as we move from one ellipsoidal surface to another.

The completeness of the vector ellipsoidal harmonics, over the surface of any ellipsoid, is a consequence of the fact that every spherical harmonic is expandable in ellipsoidal harmonics and vice versa.

The expansion algorithm for a smooth vector field $\boldsymbol{F}$ defined on the ellipsoid $\rho =$ constant works as follows

$$\begin{aligned}\boldsymbol{F}(\mu,\nu;\rho) &= \frac{1}{h_\rho}\sum_{n=0}^{\infty}\sum_{m=1}^{2n+1} A_n^m(\rho)\boldsymbol{R}_n^m(\mu,\nu;\rho)\\ &+ \frac{1}{h_\rho}\sum_{n=1}^{\infty}\sum_{m=1}^{2n+1} B_n^m(\rho)\boldsymbol{D}_n^m(\mu,\nu;\rho)\\ &+ \frac{1}{h_\rho}\sum_{n=1}^{\infty}\sum_{m=1}^{2n+1} C_n^m(\rho)\boldsymbol{T}_n^m(\mu,\nu;\rho).\end{aligned} \tag{46}$$

Taking the inner product of (46) with $\boldsymbol{R}_{n'}^{m'}(\mu,\nu;\rho)$, with respect to the inner product defined by the weighting function $l_\rho(\mu,\nu)$ we obtain the coefficient $A_n^m(\rho)$. Taking the inner product of (46) with $\boldsymbol{D}_{n'}^{m'}(\mu,\nu;\rho)$, with respect to the inner product defined by the weighting function $\tilde{l}_\rho(\mu,\nu)$ we obtain the coefficient $B_n^m(\rho)$. Then the function

$$\begin{aligned}\boldsymbol{F}'(\mu,\nu;\rho) &= \boldsymbol{F}(\mu,\nu;\rho) \\ &\quad - \frac{1}{h_\rho}\sum_{n=0}^{\infty}\sum_{m=1}^{2n+1} A_n^m(\rho)\boldsymbol{R}_n^m(\mu,\nu;\rho) \\ &\quad - \frac{1}{h_\rho}\sum_{n=1}^{\infty}\sum_{m=1}^{2n+1} B_n^m(\rho)\boldsymbol{D}_n^m(\mu,\nu;\rho) \qquad (47)\end{aligned}$$

becomes known and has the expansion

$$\boldsymbol{F}'(\mu,\nu;\rho) = \frac{1}{h_\rho}\sum_{n=1}^{\infty}\sum_{m=1}^{2n+1} C_n^m(\rho)\boldsymbol{T}_n^m(\mu,\nu;\rho). \qquad (48)$$

Finally, taking the second inner product with $\boldsymbol{T}_{n'}^{m'}(\mu,\nu;\rho)$ we obtain the coefficients $C_n^m(\rho)$. Hence, the above expansion can be completely calculated.

4. Conclusions

It is not possible to introduce vector ellipsoidal harmonics which behave as nice as the vector spherical harmonics. Perhaps, this is the reason why the vector ellipsoidal harmonics were lacking from the literature for 74 years after the vector spherical harmonics were introduced. Here we define a complete set of vector ellipsoidal harmonics, that depend on the ellipsoid on which they are living, and that are orthogonal with respect to two analytic structures defined by two inner products having different weighting functions. Using these two inner products it is easy to calculate the coefficients of any vectorial eigenfunction expansion. Just as it is with the case of spherical geometry, one set of vector surface ellipsoidal harmonics is locally normal to the particular ellipsoid and the other two sets are tangential to the surface of the ellipsoid at this point. The crucial difference though is that, in general, the directions of the position vector and the corresponding normal do not coincide, and this is the source of many difficulties with the ellipsoidal system.

The introduction of Vector Spherical Harmonics in 1935, provided a significant freedom in solving boundary value problems in spherical geometry. The present introduction of vector ellipsoidal harmonics identifies an area of classical applied mathematics that is open to many theoretical and real life problems. For example, we know that every vector boundary value problem in ellipsoidal geometry that has been solved up to now involves a tremendous amount of calculations, since it was done the "wrong" (but the only possible) way, i.e. using scalar ellipsoidal harmonics. Now, it can be solved the "right" way by using vector eigenfunctions. It is of interest to see how much easier a vector problem can be solved using vector instead of scalar ellipsoidal harmonics. Even more important it is to investigate which vector problems that were impossible with the scalar eigenfunctions are tractable with the corresponding vector eigenfunctions. We hope that these questions will find some answers in the near future.

References

1. G. Dassios and M. Tsampas. Vector ellipsoidal harmonics and neuronal current decomposition in the brain. *(Inverse Problems and Imaging)*, 3:243–257, 2009.
2. W.W. Hansen. A new type of expansion in radiation problems. *Physical Review*, 47:139–143, 1935.
3. E.W. Hobson. *The Theory of Spherical and Ellipsoidal Harmonics.* Cambridge University Press, U.K., Cambridge, first edition, 1931.
4. P.M. Morse and H. Feshbach. *Methods of Theoretical Physics*, volume I. McGraw-Hill, New York, first edition, 1953.
5. P.M. Morse and H. Feshbach. *Methods of Theoretical Physics*, volume II. McGraw-Hill, New York, first edition, 1953.

THE INVERSE SCATTERING PROBLEM IN LINEAR ELASTICITY VIA A PAIR OF NONLINEAR INTEGRAL EQUATIONS

D. GINTIDES AND L. MIDRINOS

Department of Mathematics,
National Technical University of Athens,
Athens, Greece
E-mail: dgindi@math.ntua.gr , leomid@central.ntua.gr

We consider the inverse problem of determining the shape of a rigid scatterer inside a homogeneous and isotropic elastic medium. The problem is stated in R^2 and the wave field is assumed to be time-harmonic. The inverse boundary value problem consists of solving a pair of nonlinear and ill-posed integral equations for the unknown boundary derived from a single layer representation of the scattered field. A linearization of the equations is performed using Fréchet derivatives of the integral operators. The ill-posedness of the system leads to regularization via Tikhonov method. Numerical examples are given which illustrate the applicability of the method.

1. Introduction

If an elastic propagating wave illuminates a bounded rigid obstacle, an inclusion or a cavity then it is scattered in all directions. The scattered wave is separated into a longitudinal and a transversal wave. The direct problem is to find the scattered field from the knowledge of the geometrical and physical properties of the obstacle and the incident wave. The direct problem is linear and well posed provided that the obstacle is smooth. The corresponding integral equation theory is based on Noether's theory [8] since in general the integrals are singular. One important inverse problem is to recover the geometry of the obstacle from the knowledge of the two far field patterns, the incident field and the type of the boundary conditions. In this work we assume that the environment is a homogeneous and isotropic elastic medium and that the scatterer is rigid that is the displacement field satisfies Dirichlet boundary conditions. For this geometrical inverse problem we use the idea of a pair of nonlinear integral equations introduced by Kress and Rundell [7] for Laplace and Helmholtz equations.

In Sec. 2 we describe the direct scattering problem and the equivalent transformation in a uniquely solvable equation using the representation of the scattered field as a combination of a single and a double layer potential. Next, we provide the analytic form of the far fields which is derived from the integral representation of the scattered field. In Sec. 3 we first define the inverse problem. Then, we present a local uniqueness result based on the Faber-Krahn inequality which can be applied to deduce uniqueness if we have the a priori knowledge that the scatterer is bounded by a disk with known radius. A linearization of the nonlinear and ill-posed integral equations for the unknown boundary is performed using Fréchet derivatives of the integral operators. The ill-posedness of the system leads to regularization via Tikhonov method. Numerical examples are given in the last section.

2. Direct scattering problem

We consider the scattering of time-harmonic incident plane wave by a rigid scatterer within a two-dimensional isotropic homogeneous medium with Lamé constants μ and λ, satisfying $\lambda + \mu > 0$ and $\mu > 0$. The scatterer is assumed to be a bounded domain D with smooth boundary $\partial D = \Gamma$. u^{sc} denotes the scattered displacement field, $\omega > 0$ the circular frequency and ρ the density of the elastic medium. For simplicity we assume $\rho \equiv 1$ throughout. We assume that the scattered field $u^{sc} \in C^2(R^2 \backslash \overline{D}) \cup C(R^2 \backslash D)$ and satisfies the Navier equation

$$\mu \Delta u^{sc} + (\lambda + \mu) \nabla (\nabla \cdot u^{sc}) + \rho \omega^2 u^{sc} = 0 \tag{1}$$

and Dirichlet boundary condition

$$u^{sc} + u^{inc} = 0 \ \ on \ \ \Gamma \tag{2}$$

where u^{inc} is the incident plane wave. It is known [8] that any solution of (1) can be decomposed as: $u = u_p + u_s$ where u_p is the longitudinal wave and u_s is the transversal wave which are defined by

$$u_p := -\frac{1}{k_p^2} \nabla (\nabla \cdot u), \quad u_s := u - u_p$$

and the wavenumbers are given by $k_p^2 := \frac{\omega^2}{\lambda + 2\mu}, \quad k_s^2 := \frac{\omega^2}{\mu}$.

So, we have two types of plane waves: longitudinal or transversal waves. Our method works for any type of excitation. Both types due to the mode conversion on the boundary produce p and s scattered waves. The incident waves are of the form $u_p^{inc}(x) := d\, e^{ik_p d\cdot x}$ or $u_s^{inc}(x) := a\, e^{ik_s d\cdot x}$, where $d \cdot a = 0$, if we have p or s plane waves respectively.

To ensure uniqueness for the direct problem[8] u^{sc} is required to satisfy the Kupradze radiation condition

$$\lim_{r\to\infty} \sqrt{r}\left(\frac{\partial u_p}{\partial r} - ik_p u_p\right) = 0, \quad \lim_{r\to\infty} \sqrt{r}\left(\frac{\partial u_s}{\partial r} - ik_s u_s\right) = 0, \quad r = |x| \quad (3)$$

uniformly in all directions. Adopting an indirect approach to derive uniquely solvable integral equations, we will represent the scattered field as a combined double- and single- layer potential. First, we introduce some necessary tensors for the definition of the potentials. The fundamental solution for the Navier equation is given by

$$\Phi(x,y) := \frac{i}{4\mu} H_0^{(1)}(k_s r) I + \frac{i}{4\omega^2} \nabla_x \nabla_x^T \left[H_0^{(1)}(k_s r) - H_0^{(1)}(k_p r)\right] \quad (4)$$

where $r = |x-y|$ and $H_0^{(1)}$ is the Hankel function of order zero and of the first kind. A convenient form is:

$$\Phi(x,y) = \Phi_1(r) I + \Phi_2(r) J(x,y)$$

where I is the identity matrix and the functions Φ_1, Φ_2 and the matrix J are given by

$$\Phi_1(r) = \frac{i}{4\mu} H_0^{(1)}(k_s r) - \frac{i}{4\omega^2 r}\left[k_s H_1^{(1)}(k_s r) - k_p H_1^{(1)}(k_p r)\right],$$

$$\Phi_2(r) = \frac{i}{4\omega^2}\left[\frac{2k_s}{r} H_1^{(1)}(k_s r) - k_s^2 H_0^{(1)}(k_s r) - \frac{2k_p}{r} H_1^{(1)}(k_p r) + k_p^2 H_0^{(1)}(k_p r)\right]$$

$$J(x,y) = \frac{(x-y)(x-y)^T}{r^2}$$

with the Hankel function of order one $H_1^{(1)} = -H_0^{(1)'}$.

Also we need the following tensor:

$$[T_y \Phi(x,y)]^T = \frac{\Phi_1'(r)}{r} U_1(y,x)^T + \frac{\Phi_2'(r)}{r} J(x,y) U_1(y,x)^T + \frac{\Phi_2(r)}{r^2} U_2(y,x)^T \quad (5)$$

where prime denotes differentiation with respect to r and

$$U_1(x,y) := \lambda \nu(x)(x-y)^T + \mu(x-y)\nu(x)^T + \mu\nu(x)\cdot(x-y) I$$

$$U_2(x,y) := (\lambda+2\mu)\nu(x)(x-y)^T + \mu(x-y)\nu(x)^T + \mu\nu(x)\cdot(x-y)\left[I - 4J(x,y)\right]$$

where ν is the unit normal vector. We express the scattered field in the form

$$u^{sc}(x) = (D\varphi)(x) - i\eta(S\varphi)(x) \tag{6}$$

where $(S\varphi)(x) := \int_\Gamma \Phi(x,y)\varphi(y)ds(y)$ is the single layer potential and $(D\varphi)(x) := \int_\Gamma \left[T_y\Phi(x,y)\right]^T \varphi(y)ds(y)$ is the double layer potential. Applying the boundary properties of the potentials on (2) we infer that $\varphi(x)$ is the solution of the integral equation

$$\frac{1}{2}\varphi(x) + (D\varphi)(x) - i\eta(S\varphi)(x) = -u^{inc}(x)$$

$$\Rightarrow \left[I + 2D - 2i\eta S\right]\varphi = f \tag{7}$$

where $f = -2u^{inc}$. The integral equation (7) has exactly one solution and furthermore the solution depends continuously on f [3]. In [3] it is proved that uniqueness holds for square integrable densities on Γ. For the inverse problem is necessary to know the far-field patterns of the single and double layer potentials. From the asymptotics for the Hankel functions $H_0^{(1)}$ and $H_1^{(1)}$ we obtain that

$$u_p^\infty(\widehat{x}) = -\frac{e^{\frac{i\pi}{4}}i}{\lambda+2\mu}\sqrt{\frac{k_p}{8\pi}}\int_\Gamma J(\widehat{x})\left[F(\widehat{x},y) + \frac{\eta}{k_p}I\right]e^{-ik_p\widehat{x}\cdot y}\varphi(y)ds(y) \tag{8}$$

$$u_s^\infty(\widehat{x}) = -\frac{e^{\frac{i\pi}{4}}i}{\mu}\sqrt{\frac{k_s}{8\pi}}\int_\Gamma \left[I - J(\widehat{x})\right]\left[F(\widehat{x},y) + \frac{\eta}{k_s}I\right]e^{-ik_s\widehat{x}\cdot y}\varphi(y)ds(y) \tag{9}$$

where $\widehat{x} := \frac{x}{|x|}$, $J(\widehat{x}) = \widehat{x}\otimes\widehat{x}$ and $F(\widehat{x},y) = \lambda\widehat{x}\nu(y)^T + \mu\nu(y)\widehat{x}^T + \mu\nu(y)\cdot\widehat{x}I$.

3. Inverse Scattering Problem

The inverse elastic scattering problem that we want to solve is to determine the domain D from the knowledge of the pair of the far-field patterns $u^\infty(\widehat{x}) := (u_p^\infty(\widehat{x}), u_s^\infty(\widehat{x}))$ for all $\widehat{x} \in \Omega$, where Ω is the unit circle, for one incident direction d and one frequency ω. The general question about uniqueness for one wave is not solved yet but Schiffer's type theorems hold [1]. Unfortunately, Schiffer's type theorems assume the knowledge of the far fields for infinitely many incident waves and do not hold in our case. For one incident wave the problem is uniquely solvable under the assumption that the scatterers belong to a certain class of domains S, such that if $D_1, D_2 \in \mathcal{S}$, then

$$0 < E\left(D^*\right) < \frac{\mu\,\pi\,k_{01}^2}{\rho\,\omega^2}$$

and $\mathbb{R}^2\backslash(D_1 \cup D_2)$ is connected. In the above inequality, $D^* = D_1 \cup D_2 - \overline{D_1 \cap D_2}$, $E(D^*)$ is the area of D^* and k_{01} is the first zero of the Bessel function $J_0(k)$. The main argument about the difference set and the Faber Krahn inequality in acoustics is proposed in [4] and the extension to elasticity is given in [5]. We observe that the unknown boundary Γ and density h of the corresponding solutions u_p and u_s satisfy the two equations

$$Sh = -u^{inc}\big|_{\Gamma} \tag{10}$$

$$S^{\infty}h = u^{\infty} \tag{11}$$

where S^{∞} is the pair of far field patterns $(S_p^{\infty}(\widehat{x}), S_s^{\infty}(\widehat{x}))$ derived from the single layer representation of the scattered field and u^{∞} is the pair of the measured far field data. We can prove in the same way as in [6,7] that the following theorem holds:

Theorem 3.1. *The inverse obstacle scattering problem and the two-by-two system of integral equations* (10) - (11) *are equivalent.*

We choose the following parameterizations for the boundary

$$\Gamma = \{x(t) \ : t \in [0, 2\pi)\} \ \ and \ \ x(t) = r(t)(cost, sint)$$

where the radial function r is a 2π periodic smooth function. For solving this nonlinear system of integral equations we proceed to a linearization using the Fréchet derivatives of the operators in direction q with respect to r. The full linearization of the system (10) - (11) leads to

$$S(r,h) + S(r,\psi) + S^{'}(r,h;q) = -\left(u^{inc}(r) + u^{inc'}(r;q)\right) \tag{12}$$

$$S^{\infty}(r,h) + S^{\infty}(r,\psi) + S^{\infty'}(r,h;q) = u^{\infty}(\widehat{x}) \tag{13}$$

An iterative procedure based on (12) - (13) can be implemented directly as in [7]. Another method, with smaller computational cost, is for a given current approximation r first solve (11) for h and then, keeping h fixed, solve the linearized equation:

$$S(r,h) + S^{'}(r,h;q) = -\left(u^{inc}(r) + u^{inc'}(r;q)\right) \tag{14}$$

to obtain the update $r+q$ for the radial function. The *Fréchet* derivative $S^{'}(r,h;q)$ of $S(r,h)$ has the form

$$
\begin{aligned}
S^{'}(r,h;q)(t) = {} & \frac{i}{4\omega^2}\int_0^{2\pi}\left\{\frac{(x(t)-x(\tau))\cdot(\zeta(t)-\zeta(\tau))}{r}\left\{\left[-k_s^3H_1^{(1)}(k_sr)\right.\right.\right. \\
& +\frac{1}{r}\left.\left(k_s^2H_2^{(1)}(k_sr)-k_p^2H_2^{(1)}(k_pr)\right)\right]I+\left[k_s^3H_1^{(1)}(k_sr)\right. \\
& \left.\left.-k_p^3H_1^{(1)}(k_pr)+\frac{2}{r}\left(k_p^2H_2^{(1)}(k_pr)-k_s^2H_2^{(1)}(k_sr)\right)\right]J(x(t),x(\tau))\right\} \\
& +\left[\frac{2k_s}{r}H_1^{(1)}(k_sr)-k_s^2H_0^{(1)}(k_sr)-\frac{2k_p}{r}H_1^{(1)}(k_pr)+k_p^2H_0^{(1)}(k_pr)\right] \\
& \cdot\left[\frac{(x(t)-x(\tau))\,(\zeta(t)-\zeta(\tau))^T}{r^2}+\frac{(\zeta(t)-\zeta(\tau))\,(x(t)-x(\tau))^T}{r^2}\right. \\
& \left.\left.-2\frac{(\zeta(t)-\zeta(\tau))\cdot(x(t)-x(\tau))}{r^2}J(x(t),x(\tau))\right]\right\}h(x(\tau))d\tau,\quad t\in[0,2\pi]
\end{aligned}
$$

The kernel is smooth with diagonal term

$$\frac{x^{'}(t)\cdot\zeta^{'}(t)}{r}\left[-\frac{1}{2\mu\pi r}+\frac{1}{4w^2\pi r}\left(k_s^2-k_p^2\right)\right]I$$

where $\zeta(t):=q(t)(cost,sint)$ and

$$r:=r(t,\tau)=\begin{cases}|x(t)-x(\tau)| & \text{if } t\neq\tau\\ |x^{'}(t)| & \text{if } t=\tau\end{cases}$$

Analogously, we have

$$u_p^{inc'}(x;q)=d\;ik_pd\cdot\zeta(t)\,e^{ik_pd\cdot x(t)}$$

$$u_s^{inc'}(x;q)=a\;ik_sd\cdot\zeta(t)\,e^{ik_sd\cdot x(t)}$$

if the incident wave is p or s wave respectively.

4. Numerical Examples

In this section we present two numerical reconstructions which illustrate the effectiveness of the inversion method. We have used trigonometric polynomials of degree m and we have chosen $2n$ equidistant quadrature points to approximate the integral operators through trapezoidal rule. Special care for the singular integrals for the direct problem has been taken based on the work presented in [2]. Due to the ill - posedness of the system Tikhonov regularization was needed. We denote by l_0 and l_1 the regularization parameters for penalizing q and h respectively. In both examples the incident field is taken as the longitudinal plane wave with incident direction $d = (1, 0)$ and we chose $\eta = 1$, $\mu = 1.5$ and $\lambda = 2.5$ *Lamé* constants. In the first example we consider a peanut - shaped boundary with parametrization $x(t) = \sqrt{0.5cos^2t + 0.15sin^2t}\ (cost, sint)$. Reconstructions with $n = 64$ collocation points, $\omega = 1.5$ circular frequency, $m = 3$ Fourier coefficients, $l_1 = 10^{-4}$ and $l_0 = 10^{-3}$ are shown in Fig. 1. In the second example we consider a kite - shaped boundary with parametrization $x(t) = 0.45\,(cost + 0.65cos(2t) - 0.3, 1.5sint)$. Reconstructions with $n = 64$ collocation points, $\omega = 3.5$ circular frequency, $m = 9$ Fourier coefficients, $l_1 = 10^{-3}$ and $l_0 = 10^{-2}$ are shown in Fig. 2.

Acknowledgments

This research was supported by the grant PEBE 2007 of the National Technical University of Athens.

References

1. J. F. Ahner and G. C. Hsiao, SIAM J. Appl. Math. **31**, 677-685 (1976).
2. R. Chapko, R. Kress and L. Monch, IMA Journal of Numerical Analysis **20(4)**, 601-619 (2000).
3. G. Dassios and Z. Rigou, SIAM J. Appl. Math., **55** 1345-1361 (1995).
4. D. Gintides, Inverse Problems **21**, 1195-1205 (2005).
5. D. Gintides, Local uniqueness for the Dichlet problem in linear elasticity , (to appear).
6. O. Ivanyshyn and R. Kress, Mathematical Methods in Scattering Theory and Biomedical Engineering, 39-50 (2006).
7. R. Kress and W. Rundell, Inverse Problems **21**, 1207-1223 (2005).
8. V. D. Kupradze, *Potential Methods in the Theory of Elasticity.* Israel Program for Scientific Translations, Jerusalem 1965.

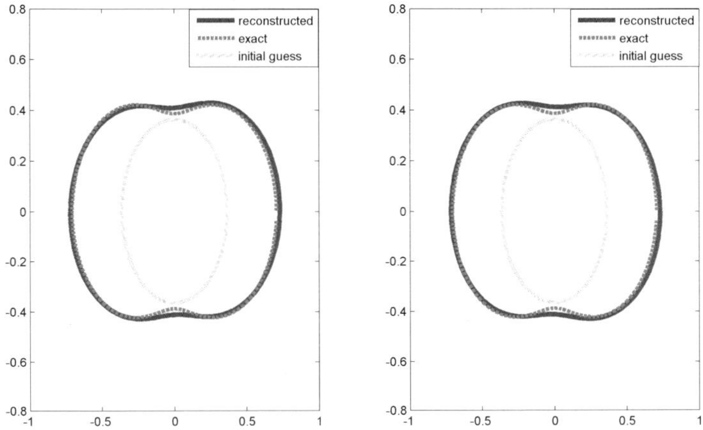

Figure 1. Reconstructions with six iterations, no noise (left) and 5% noise (right).

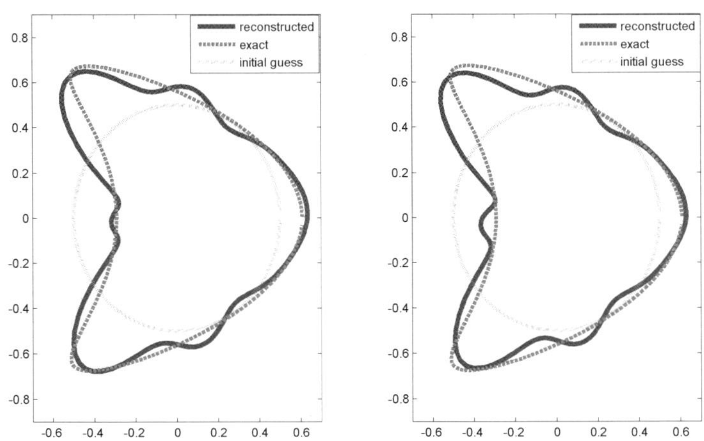

Figure 2. Reconstructions with four iterations, no noise (left) and 4% noise (right).

LOW–FREQUENCY MODELING OF THE INTERACTION OF A MAGNETIC DIPOLE AND TWO METALLIC SPHERICAL BODIES IN A CONDUCTIVE MEDIUM

PANAYIOTIS VAFEAS

Division of Applied Mathematics & Mechanics, Department of Engineering Sciences, University of Patras, GR – 265 04 Patras, Greece

DOMINIQUE LESSELIER

Département de Recherche en Electromagnétisme, Laboratoire des Signaux et Systèmes, UMR 8506 CNRS–SUPÉLEC–Univ. Paris Sud 11, 3, rue Joliot–Curie, FR – 91192 Gif–sur–Yvette, France

This work concerns the interaction of a time–harmonic magnetic dipole, arbitrarily orientated in the three–dimensional space, with two perfectly conducting spherical bodies embedded in a homogeneous conductive medium. For many practical applications involving two buried obstacles nearly located such as Earth's subsurface electromagnetic probing or other physical cases (e.g. geo–electromagnetics), the bispherical geometry might provide a very good approximation. The particular physics here concerns two solid impenetrable bodies under a magnetic dipole excitation, where the scattering boundary value problem is attacked via rigorous low–frequency expansions in terms of integral powers $(ik)^n$, $n \geq 0$, k being the complex wave number of the exterior medium, for the incident, scattered and total electric and magnetic fields. Our goal is to obtain the most important terms of the low–frequency expansions of the electromagnetic fields, that is the static (for $n = 0$) and the dynamic ($n = 1,2,3$) terms. In particular, for $n = 1$ there are no incident fields and thus no scattered ones, while for $n = 0$ the Rayleigh electromagnetic term is obtained in terms of infinite series. Emphasis is given on the calculation of the next two nontrivial terms (at $n = 2$ and at $n = 3$) of the aforementioned fields. Consequently, those are found in closed form from exact solutions of coupled (at $n = 2$, to the one at $n = 0$) or uncoupled (at $n = 3$) Laplace equations and they are given in compact fashion, as infinite series expansions for $n = 2,3$. Let us notice that the difficulty of the Poisson's that has to be solved for $n = 2$ is not minor and for this case we present only the methodology to obtain the solution. Finally, our analytical approach calls for the use of the well–known cut–off method in order to obtain properly closed solutions.

1. Introduction

The main field of application of the present research, especially in mineral exploration, is Earth's subsurface electromagnetic probing, where we are often faced with the problem to retrieve an anomaly of certain kind, such as metallic ores, from three–dimensional magnetic fields, which are measured along a borehole when a low–frequency [1] time–harmonic source is placed nearby, usually fixed at the surface of the Earth. In that way, interesting information concerning orientations, sizes, shapes, magnetic and electric parameters of the anomalies are recovered. However, this is not an easy task, since the inverse problem cannot be tackled in robust fashion unless proper models of the field interaction are available.

There are many computational codes available to that matter as illustrated early on in [2], whilst several complementary investigations are presented in [3]. Nevertheless, simple methodologies such as to obtain low–frequency expansions by the manipulation of Debye potentials are described in the classical book [4] and remain useful, while a recent and detailed contribution by [5] illustrates this type of approach. Even the low–frequency analytical expansions in the spherical case were not available until three years ago when the low–frequency solution for a perfectly conducting sphere in a conductive medium with dipolar excitation was presented [6]. Before then, results had been made available for a penetrable sphere illuminated by a dipolar magnetic field within a conductive host medium via a Mie series expansion [7]. On the other hand, in [8] the electric potential and field between two different spheres in an infinite medium supporting an external uniform electric field is calculated analytically by applying the Green's function method. In [9,10,11] or in more recent works [12,13] one can find useful reference results within the already ample library of scattering by simple shapes (e.g. spheroids and ellipsoids) utilizing analytical methods. Finally, two interesting papers [14,15] concern the low–frequency acoustic scattering using the bispherical coordinate system.

Here, the goal is to build a versatile set of mathematical tools in order to infer information on two unknown subsurface bodies, from the fields, which they scatter off when illuminated by a known primary source, let us say a magnetic dipole, operated at low frequency. In order to put those tools together, we work within the framework of the well–known low–frequency scattering theory [1], i.e., by expanding the electromagnetic fields (incident, as well as scattered) in positive integral powers of (ik), k being the complex wave number of the exterior medium at the operation frequency, and by appropriately calculating the vector fields at each n. High–contrast cases, for which the ratio between the conductivity of the bodies and the one of the embedding medium is high, are approximated via the introduction of perfectly impenetrable bodies, and the scattering problem is transformed into a succession, one at each n, of possibly coupled boundary value problems formulated according to second–

order Laplace's partial differential equations with proper perfectly reflecting boundary conditions, resulting in closed–form solutions.

Most of the literature (e.g., [6]) concern spherical obstacles, a fact that does not hold true for the case of general buried bodies. Despite the fact that the most general work concerns the ellipsoidal approximation [13,16], the simplest and important case is the spheroid [12,16]. However, in our case we face the problem of two different buried objects and thus we are obliged to use the best fitting system to the particular physics, i.e., the bispherical coordinate system [16], in order to obtain the most important terms of the low–frequency expansions of the electromagnetic fields, i.e., the static (for $n=0$) and the dynamic ($n=1,2,3$) terms. In particular, for $n=1$ there are no incident fields, while for $n=0$ the Rayleigh electromagnetic term is easily obtained in terms of infinite series. Emphasis is given on the calculation of the next two nontrivial terms (at $n=2$, where only the methodology is presented, and at $n=3$) of the magnetic and the electric fields. Those are found in closed form from exact solutions of coupled (at $n=2$, to the one at $n=0$) or uncoupled (at $n=3$) Laplace equations and they are given as infinite series expansions.

2. Mathematical Formulation

The particular physics concerns two different solid impenetrable bodies of surfaces S_1 and S_2 under a magnetic dipole excitation. Specifically, we are focused onto the case of two highly conductive spheres buried in a less conductive, homogeneous and isotropic, nonmagnetic, medium with conductivity σ^+ and permeability μ^+, where $k^+ \equiv k = \left(\omega\mu^+\sigma^+/2\right)^{1/2}(1+i)$ stands for the wave number, at low circular frequency ω, of the external space V^+. Those bodies are illuminated by a low–frequency localized vector source, which is modeled with little restriction as a magnetic dipole $\mathbf{m}$ with arbitrary location ($\mathbf{r}_0$) and orientation. Under the assumption of time harmonic dependence, the electromagnetic incident fields $\mathbf{H}^i$, $\mathbf{E}^i$ are scattered by the solid spheres, creating the corresponding scattered fields $\mathbf{H}^s$, $\mathbf{E}^s$, while the total magnetic and electric fields $\mathbf{H}^t$, $\mathbf{E}^t$ are the sums of those fields, i.e.,

$$\mathbf{H}^t = \mathbf{H}^i + \mathbf{H}^s \quad \text{and} \quad \mathbf{E}^t = \mathbf{E}^i + \mathbf{E}^s, \quad \mathbf{r} \in V^+ - \{\mathbf{r}_0\} \tag{1}$$

The boundary value problem is attacked via low frequency expansions in terms of integral powers of (ik),

$$\mathbf{H}^x = \sum_{n=0}^{\infty} \mathbf{H}_n^x (ik)^n \quad \text{and} \quad \mathbf{E}^x = \sum_{n=0}^{\infty} \mathbf{E}_n^x (ik)^n, \quad x = i,s,t \tag{2}$$

for all the electromagnetic fields, incident (i), scattered (s) and total (t).

As a consequence of the low–frequency assumption and harmonic time dependence, Maxwell equations, which connect the magnetic and the electric fields, are reduced as follows,

$$\sigma^{+}\nabla\times\mathbf{E}_{n}^{x}=-\mathbf{H}_{n-2}^{x},\quad n\geq 2\quad\text{and}\quad\nabla\times\mathbf{H}_{n}^{x}=\sigma^{+}\mathbf{E}_{n}^{x},\quad n\geq 0,\tag{3}$$

whereas $\nabla\cdot\mathbf{H}_{n}^{x}=\nabla\cdot\mathbf{E}_{n}^{x}=0$, $n\geq 0$ in $V^{+}-\{\mathbf{r}_{0}\}$ for every $x=i,s,t$.

Some extended algebraic calculations on the incident fields produced by the magnetic dipole lead us to the following low–frequency relations as powers of (ik), where at the end it is proved that the first four powers of the fields, that is the static term for $n=0$ and the dynamic terms for $n=1,2,3$ are sufficient enough to describe the problem. In detail,

$$\mathbf{H}^{i}=\left[\mathbf{H}_{0}^{i}+\mathbf{H}_{2}^{i}(ik)^{2}+\mathbf{H}_{3}^{i}(ik)^{3}\right]+\mathrm{O}\left((ik)^{4}\right),\quad\mathbf{E}^{i}=\left[\mathbf{E}_{2}^{i}(ik)^{2}\right]+\mathrm{O}\left((ik)^{4}\right),\tag{4}$$

where in view of the definitions $\mathbf{R}=\mathbf{r}-\mathbf{r}_{0}$ and $R=|\mathbf{r}-\mathbf{r}_{0}|$, we obtain

$$\mathbf{H}_{0}^{i}=\frac{\mathbf{m}}{4\pi}\cdot\left(\frac{3\mathbf{R}\otimes\mathbf{R}}{R^{2}}-\tilde{\mathbf{I}}\right)\frac{1}{R^{3}},\tag{5}$$

$$\mathbf{H}_{2}^{i}=-\frac{\mathbf{m}}{4\pi}\cdot\left(\frac{\mathbf{R}\otimes\mathbf{R}}{R^{2}}+\tilde{\mathbf{I}}\right)\frac{1}{2R}\quad\text{and}\quad\mathbf{E}_{2}^{i}=-\frac{\mathbf{m}}{4\pi\sigma^{+}}\times\frac{\mathbf{R}}{R^{3}},\tag{6}$$

$$\mathbf{H}_{3}^{i}=\frac{\mathbf{m}}{4\pi}\cdot\left(-\frac{2}{3}\tilde{\mathbf{I}}\right),\tag{7}$$

where the symbol "$\otimes$" denotes juxtaposition and $\tilde{\mathbf{I}}$ stands for the unit dyadic. We observe that for the incident magnetic field the dynamic term for $n=1$ is not present, while for the incident electric field the only term that survives is the dynamic term for $n=2$, reflecting exactly the same physical and mathematical treatment to the scattered fields (2) for $x=s$. Inserting the wave number k into those similar expressions like (4), some trivial analysis for $\mathbf{r}\in V^{+}-\{\mathbf{r}_{0}\}$ leads to

$$\mathbf{H}^{s}=\left[\mathbf{H}_{0}^{s}+\left(\omega\mu^{+}\sigma^{+}\right)\sqrt{\frac{\omega\mu^{+}\sigma^{+}}{2}}\mathbf{H}_{3}^{s}\right]+\left(\omega\mu^{+}\sigma^{+}\right)\left[\sqrt{\frac{\omega\mu^{+}\sigma^{+}}{2}}\mathbf{H}_{3}^{s}-\mathbf{H}_{2}^{s}\right]i+\mathrm{O}\left((ik)^{4}\right),\tag{8}$$

$$\mathbf{E}^{s}=\left[-\left(\omega\mu^{+}\sigma^{+}\right)\mathbf{E}_{2}^{s}\right]i+\mathrm{O}\left((ik)^{4}\right),\tag{9}$$

The electric field is purely imaginary, while the magnetic field is a complex number, indicating that the electromagnetic fields at $n=2$ are adequate for the full solutions, since the contribution of the $\mathbf{H}_{3}^{s}$ is of minor significance.

Straightforward calculations on Maxwell's equations (4) for $x=s$ result in a set of boundary value problems possibly coupled to one another, from the static one at $n=0$ to dynamic ones at higher values of n up to $n=3$, which are given in terms of the harmonic potentials Φ_0^s, $\mathbf{\Phi}_2^s$ and Φ_3^s via

$$\Delta\mathbf{H}_0^s=\mathbf{0} \;\Rightarrow\; \mathbf{H}_0^s=\nabla\Phi_0^s \quad \left(\nabla\cdot\mathbf{H}_0^s=0,\ \nabla\times\mathbf{H}_0^s=\mathbf{0}\right), \tag{10}$$

$$\Delta\mathbf{H}_2^s=\mathbf{H}_0^s \Rightarrow \mathbf{H}_2^s=\mathbf{\Phi}_2^s+\frac{1}{2}\left(\mathbf{r}\Phi_0^s\right) \text{ and } \sigma^+\mathbf{E}_2^s=\nabla\times\mathbf{H}_2^s \;\left(\nabla\cdot\mathbf{H}_2^s=\nabla\cdot\mathbf{E}_2^s=0\right), \tag{11}$$

$$\Delta\mathbf{H}_3^s=\mathbf{0} \;\Rightarrow\; \mathbf{H}_3^s=\nabla\Phi_3^s \quad \left(\nabla\cdot\mathbf{H}_3^s=0,\ \nabla\times\mathbf{H}_3^s=\mathbf{0}\right), \tag{12}$$

where $\mathbf{H}_0^s$, $\mathbf{H}_2^s$, $\mathbf{E}_2^s$ and $\mathbf{H}_3^s$ are to be calculated, while $\mathbf{H}_1^s=\mathbf{E}_0^s=\mathbf{E}_1^s=\mathbf{E}_3^s=\mathbf{0}$. This set of problems must be solved with the proper boundary conditions being applied on the surfaces for the total fields, i.e., cancellation of the normal component of the magnetic field and of the tangential component of the electric field, which in view of the outward unit normal vector $\hat{\boldsymbol{n}}$ on S_1 and S_2, yields

$$\hat{\boldsymbol{n}}\cdot\left(\mathbf{H}_n^i+\mathbf{H}_n^s\right)=0,\ n=0,2,3 \quad \text{and} \quad \hat{\boldsymbol{n}}\times\left(\mathbf{E}_2^i+\mathbf{E}_2^s\right)=\mathbf{0}, \quad \text{for} \quad \mathbf{r}\in S_1\cup S_2. \tag{13}$$

Hence, we have to solve the boundary value problems (10)–(13) with Neumann–type boundary conditions, considering bispherical coordinates that fit mostly to the physics of scattering by two metallic spherical objects of different radii.

3. Bispherical Harmonic Eigenfunctions and Eigenexpansions

Let us consider two different spheres of radii a_1 and a_2 with centers that are located a distance $d>a_1+a_2$ apart. The bispherical coordinate system [16] (ζ,θ,φ), $\zeta\in\mathbb{R}$, $\theta\in[0,\pi]$ and $\varphi\in[0,2\pi)$ that is able to identify the two given spheres with two specified values of one of the system variables is connected to the Cartesian system ($\mathbf{r}=(x_1,x_2,x_3)$ and $\mathbf{r}_0=(x_{10},x_{20},x_{30})$) through

$$x_1=\frac{c\sin\theta\cos\varphi}{\cosh\zeta-\cos\theta},\quad x_2=\frac{c\sin\theta\sin\varphi}{\cosh\zeta-\cos\theta},\quad x_3=\frac{c\sinh\zeta}{\cosh\zeta-\cos\theta}, \tag{14}$$

where $2c$ denotes the focii distance. If we demand that the sphere of radius a_1 correspond to the value $\zeta=-\zeta_1$ on S_1 and the sphere of radius a_2 correspond to the value $\zeta=\zeta_2$ on S_2 with $\zeta_1,\zeta_2>0$, then our problem (10)–(12) must be solved within the region $-\zeta_1\le\zeta\le\zeta_2$ (non–penetrable spheres) with boundary conditions (13) on $\zeta=-\zeta_1,\zeta_2$. A geometrical analysis provides us with

$$c=(2d)^{-1}\sqrt{d^2-(a_1-a_2)^2}\sqrt{d^2-(a_1+a_2)^2},\quad \zeta_i=\sinh^{-1}(c/a_i) \text{ for } i=1,2, \tag{15}$$

which calculate the unknowns c, ζ_1 and ζ_2 via the given values d, a_1 and a_2. The outward unit normal vector on the surface of each sphere is furnished by the formula via the Cartesian basis $\hat{\boldsymbol{x}}_\kappa$, $\kappa=1,2,3$,

$$\hat{\boldsymbol{n}} = \frac{|\sinh\zeta|\sin\theta\cos\varphi}{\cosh\zeta-\cos\theta}\hat{\boldsymbol{x}}_1 + \frac{|\sinh\zeta|\sin\theta\sin\varphi}{\cosh\zeta-\cos\theta}\hat{\boldsymbol{x}}_2 + \frac{|\sinh\zeta|(\cosh\zeta\cos\theta-1)}{\sinh\zeta(\cosh\zeta-\cos\theta)}\hat{\boldsymbol{x}}_3 \quad (16)$$

and the differential operator of gradient ∇, in this geometry, assumes the form

$$\nabla = \frac{\cosh\zeta-\cos\theta}{c}\left[\hat{\boldsymbol{\zeta}}\frac{\partial}{\partial\zeta} + \hat{\boldsymbol{\theta}}\frac{\partial}{\partial\theta} + \frac{\hat{\boldsymbol{\varphi}}}{\sin\theta}\frac{\partial}{\partial\varphi}\right], \quad (17)$$

where $\hat{\boldsymbol{\zeta}}$, $\hat{\boldsymbol{\theta}}$, $\hat{\boldsymbol{\varphi}}$ denote the coordinate vectors of the system [16], while $\hat{\boldsymbol{n}} = \hat{\boldsymbol{\zeta}}$ for $\zeta < 0$, $\hat{\boldsymbol{n}} = \hat{\boldsymbol{x}}_3$ for $\zeta = 0$ and $\hat{\boldsymbol{n}} = -\hat{\boldsymbol{\zeta}}$ for $\zeta > 0$.

We introduce the interior $u_\ell^{(i)ms}$ and the exterior $u_\ell^{(e)ms}$ harmonic eigenfunctions of degree ℓ ($\ell = 0,1,2,\ldots$) and of order m ($m = 0,1,2,\ldots,\ell$) in terms of the associated Legendre functions [17] P_ℓ^m of the first kind as

$$u_\ell^{(i)mq} = e^{\left(\ell+\frac{1}{2}\right)\zeta} P_\ell^m(\cos\theta) f^{mq}(\varphi) \quad \text{and} \quad u_\ell^{(e)mq} = e^{-\left(\ell+\frac{1}{2}\right)\zeta} P_\ell^m(\cos\theta) f^{mq}(\varphi), \quad (18)$$

respectively, which both are regular on the axis of symmetry, while

$$f^{mq}(\varphi) = \begin{cases} \cos m\varphi, & q = e \\ \sin m\varphi, & q = o \end{cases} \quad \text{for every} \quad \varphi \in [0, 2\pi), \quad (19)$$

with q denoting the even (e) or the odd (o) part of the eigenfunctions. Then, every harmonic function u in bispherical geometry is written as follows,

$$\Delta u = 0 \quad \Rightarrow \quad u = \sqrt{\cosh\zeta-\cos\theta}\sum_{\ell=0}^{\infty}\sum_{m=0}^{\ell}\sum_{q=e,o}\left[A_\ell^{(i)mq}u_\ell^{(i)mq} + A_\ell^{(e)mq}u_\ell^{(e)mq}\right], \quad (20)$$

where $A_\ell^{(i)mq}$ and $A_\ell^{(e)mq}$ are constant coefficients. In addition, for the region of observation, of interest here, which is for $\mathbf{r} < \mathbf{r}_0$, we utilize the relation

$$\frac{1}{R} \equiv \frac{1}{|\mathbf{r}-\mathbf{r}_0|} = \sqrt{\cosh\zeta-\cos\theta}\sum_{\ell=0}^{\infty}\sum_{m=0}^{\ell}\sum_{q=e,o}\left[\rho_\ell^{mq} e^{\left(\ell+\frac{1}{2}\right)\zeta} P_\ell^m(\cos\theta) f^{mq}(\varphi)\right], \quad (21)$$

where if $\varepsilon_m = 1$ for $m = 0$ and $\varepsilon_m = 2$ for $m \geq 1$, then

$$\rho_\ell^{mq} \equiv \rho_\ell^{mq}(\mathbf{r}_0) = \frac{\varepsilon_m}{c}\frac{(\ell-m)!}{(\ell+m)!}\sqrt{\cosh\zeta_0-\cos\theta_0}\, e^{-\left(\ell+\frac{1}{2}\right)\zeta_0} P_\ell^m(\cos\theta_0) f^{mq}(\varphi_0). \quad (22)$$

In order to solve (10)–(13), we use useful recurrence relations and identities [17], as well as the uniformly convergent expansion

$$\frac{1}{\sqrt{\cosh\zeta-\cos\theta}} = \sqrt{2}\sum_{\ell=0}^{\infty} e^{-\left(\ell+\frac{1}{2}\right)|\zeta|} P_\ell(\cos\theta). \quad (23)$$

4. Bispherical Low–Frequency Electromagnetic Fields

Our purpose is to derive closed analytical expressions for the non–vanishing scattered magnetic fields $\mathbf{H}_0^s$, $\mathbf{H}_2^s$ and $\mathbf{H}_3^s$, as well as for the scattered electric field $\mathbf{E}_2^s$, since $\mathbf{H}_1^s = \mathbf{E}_0^s = \mathbf{E}_1^s = \mathbf{E}_3^s = \mathbf{0}$. In order to achieve that, we must solve independently the problems (10) and (12) to obtain $\mathbf{H}_0^s$ and $\mathbf{H}_3^s$, respectively and then proceed to the problem (11) to obtain $\mathbf{H}_2^s$ and, thus, $\mathbf{E}_2^s$, which is much more complicated due to its coupling with (10). The proper boundary conditions on the surfaces $\zeta = -\zeta_1, \zeta_2$ of the metallic objects given by (13) fit the aforementioned boundary value problems and the expressions (5)–(7) for the incident fields are properly used, in view of (21) and (22). The position of the magnetic dipole $\mathbf{m}$ at $\mathbf{r} = \mathbf{r}_0$ is taken into consideration utilizing equation (14), whilst the harmonic potentials Φ_0^s, $\mathbf{\Phi}_2^s$ and Φ_3^s follow the previous analysis based on relations (18)–(20). Under this aspect and in view of relations (16), (17) and (23) we perform a number of long and tedious calculations to obtain the electromagnetic fields, by making extensive use of reference [17].

The scattered magnetic field $\mathbf{H}_3^s$ assumes the following form,

$$\mathbf{H}_3^s = \nabla\left\{\sqrt{\cosh\zeta - \cos\theta}\left[\sum_{\ell=1}^{\infty}\left(a_\ell^{(i)1} e^{\frac{2\ell+1}{2}\zeta} + a_\ell^{(e)1} e^{-\frac{2\ell+1}{2}\zeta}\right) P_\ell^1(\cos\theta)\cos\varphi \right.\right.$$
$$+\sum_{\ell=1}^{\infty}\left(a_\ell^{(i)2} e^{\frac{2\ell+1}{2}\zeta} + a_\ell^{(e)2} e^{-\frac{2\ell+1}{2}\zeta}\right) P_\ell^1(\cos\theta)\sin\varphi$$
$$\left.\left.+\sum_{\ell=0}^{\infty}\left(a_\ell^{(i)3} e^{\frac{2\ell+1}{2}\zeta} + a_\ell^{(e)3} e^{-\frac{2\ell+1}{2}\zeta}\right) P_\ell(\cos\theta)\right]\right\}. \qquad (24)$$

A more complicated analysis leads to the scattered field $\mathbf{H}_0^s$, that is

$$\mathbf{H}_0^s = \nabla\left[\sqrt{\cosh\zeta - \cos\theta}\sum_{\ell=0}^{\infty}\sum_{m=0}^{\ell}\sum_{q=e,o}\left(b_\ell^{(i)mq} e^{\frac{2\ell+1}{2}\zeta} + b_\ell^{(e)mq} e^{-\frac{2\ell+1}{2}\zeta}\right) P_\ell^m(\cos\theta) f^{mq}(\varphi)\right], \quad (25)$$

Finally, a very cumbersome manipulation of the boundary value problem (11) with (13) results into the scattered fields $\mathbf{H}_2^s$ and $\mathbf{E}_2^s$, which have the form

$$\frac{\mathbf{H}_2^s}{\sqrt{\cosh\zeta - \cos\theta}} = \sum_{\ell=0}^{\infty}\sum_{m=0}^{\ell}\sum_{q=e,o}\left[\left(\mathbf{h}_\ell^{(i)mq} + \frac{b_\ell^{(i)mq}\mathbf{r}}{2}\right) e^{\frac{2\ell+1}{2}\zeta} + \left(\mathbf{h}_\ell^{(e)mq} + \frac{b_\ell^{(e)mq}\mathbf{r}}{2}\right) e^{-\frac{2\ell+1}{2}\zeta}\right] P_\ell^m(\cos\theta) f^{mq}(\varphi) \quad (26)$$

and once the magnetic field $\mathbf{H}_2^s$ is known, the electric field $\mathbf{E}_2^s$ is calculated via (11), where $\mathbf{h}_\ell^{(i/e)mq} = c_\ell^{(i/e)mq}\hat{\mathbf{x}}_1 + d_\ell^{(i/e)mq}\hat{\mathbf{x}}_2 + e_\ell^{(i/e)mq}\hat{\mathbf{x}}_3$ for $\ell = 0,1,2,\ldots$, $m = 0,1,2,\ldots,\ell$, $q = e,o$.

The constants $a_\ell^{(i)\kappa}$, $a_\ell^{(e)\kappa}$ for $\ell \ge 0$ ($\kappa = 3$) and for $\ell \ge 1$ ($\kappa = 1,2$), the constants $b_\ell^{(i)mq}$, $b_\ell^{(e)mq}$ for $\ell = 0,1,2,\ldots$, $m = 0,1,2,\ldots,\ell$, $q = e,o$ and the components of the vector constants $\mathbf{h}_\ell^{(i/e)mq}$ for $\ell = 0,1,2,\ldots$, $m = 0,1,2,\ldots,\ell$,

$q = e, o$ satisfy infinite "six term recurrence" equations, which in matrix form can be written as $\tilde{\mathbf{A}}\boldsymbol{x} = \boldsymbol{b}$. The vector $\boldsymbol{x}$ contains the six sets of unknowns $\ell - 1$, ℓ, $\ell + 1$ for both interior (i) and exterior (e) cases and using the "cut–off" method we obtain the solution in the closed form (8) and (9).

References

1. G. Dassios and R.E. Kleinman, *Low Frequency Scattering*, Oxford University Press, Oxford (2000).
2. Z. Xiong and A.C. Tripp, "Electromagnetic Scattering of Large Structures in Layered Earths using Integral Equations", *Radio Science*, **30**, 921–929 (1995).
3. M.L. Oristaglio and B.R. Spies (eds.), *Three Dimensional Electromagnetics*, SEG, Tulsa (1999).
4. A.A. Kaufman and G.V. Keller, *Inductive Mining Prospecting*, Elsevier Science, New York (1985).
5. B. Bourgeois, K. Suignard and G. Perrusson, "Electric and Magnetic Dipoles for Geometric Interpretation of Three–Component Electromagnetic Data in Geophysics", *Inverse Problems*, **16**, 1225–1262 (2000).
6. P. Vafeas, G. Perrusson and D. Lesselier, "Low–Frequency Solution for a Perfectly Conducting Sphere in a Conductive Medium with Dipolar Excitation", *Progr. Electromagnetics Res.*, **49**, 87–111 (2004).
7. T.M. Habashy, R.W. Groom and B.R. Spies, "Beyond the Born and Rytov Approximations: A Nonlinear Approach to Electromagnetic Scattering", *J. Geophys. Res.*, **98**, 1759–1775 (1993).
8. P.C. Chaumet and J.P. Dufour, "Electric Potential and Field Between Two Different Spheres", *J. Electrostatics*, **43**, 145–159 (1998).
9. O.C. Ao, H. Braunisch, K. O'Neill and J.A. Kong, "Quasi–Magnetostatic solution for a Conductive and Permeable Spheroid with Arbitrary Excitation", *IEEE Trans. Geoscie. Remote Sensing*, **39**, 2689–2701 (2001).
10. J.J. Bowman, P.L. Uslenghi and T.B. Senior (eds.), *Electromagnetic and Acoustic Scattering by Simple Shapes*, North Holland, Amsterdam (1969).
11. V.K. Varadan and V.V. Varadan (eds.), *Acoustic, Electromagnetic and Elastic Wave Scattering. Low and High Frequency Asymptotics*, North Holland, Amsterdam (1987).
12. P. Vafeas, G. Perrusson and D. Lesselier, "Low–Frequency Scattering from Perfectly Conducting Spheroidal Bodies in a Conductive Medium with Magnetic Dipole Excitation", *Int. J. Eng. Sci.*, **47** (**3**), 372–390 (2009).
13. G. Perrusson, P. Vafeas and D. Lesselier, "Low–Frequency Dipolar Excitation of a Perfect Ellipsoidal Conductor", *Quart. Appl. Math.*, in press (2009).
14. A. Charalambopoulos, G. Dassios and M. Hadjinicolaou, "An Analytic Solution for Low–Frequency Scattering by Two Soft Spheres", *SIAM J. Appl. Math.*, **58** (**2**), 370–386 (1998).
15. G. Dassios, M. Hadjinicolaou, G. Kamvyssas and A.N. Kandili, "On the Polarizability Potential for Two Spheres", *Int. J. Eng. Sci.*, **44**, 1520–1533 (2006).
16. P. Moon and E. Spencer, *Field Theory Handbook*, Springer–Verlag, Berlin (1971).
17. E.W. Hobson, *The Theory of Spherical and Ellipsoidal Harmonics*, Chelsea Publishing Company, New York (1965).

SIZE EFFECTS IN THE MECHANICAL PROPERTIES OF CARBON NANOTUBES

K.I. TSERPES

Laboratory of Technology & Strength of Materials, Department of Mechanical Engineering & Aeronautics, Patras 26500, Greece

SP. PANTELAKIS

Laboratory of Technology & Strength of Materials, Department of Mechanical Engineering & Aeronautics, Patras 26500, Greece

We study the size effects in the mechanical properties of armchair and zigzag single-walled carbon nanotubes by employing an atomistic-based numerical method. The simulations are carried out using an 3D FE model representing the exact atomic lattice of the nanotubes. The C-C bonds are modeled using elastic beam elements. Under the assumption of linear behavior for the C-C bonds, Young's modulus and shear modulus are derived as functions of nanotube diameter while tensile, bending and torsional rigidities as functions of nanotube length. Results show that with increasing the nanotube diameter, the elastic moduli increase until they stabilize at a certain diameter. The same dependency was found between the bending rigiditys and nanotube length. Tensile rigidity was found to be independent. By simulating the non-linear behavior of the C-C bonds with the modified Morse interatomic potential, we also study the effect of nanotube length in the tensile strength of vacancy- and SW-defected carbon nanotubes. For the defect patterns considered a minor length effect was found indicating lack of interaction between aggregated defects.

1. Introduction

Carbon nanotubes (CNTs) are molecular-scale tubes of graphitic carbon discovered in 1991 by the Japanese scientist Iijima [1]. They possess excellent electronic, electrochemical, thermal and field-emission properties. As the C-C bond is one of the strongest chemical bond in nature, it is expected to lend them also extraordinary mechanical properties. Indeed, both experiments and analysis have revealed that they possess an elastic modulus higher than 1 TPa, a tensile strength in the range of 130 GPa, an elastic strain up to 5% and a failure strain up to 20%. Combining such mechanical properties and fiber-like structure, CNTs offer unique potential for reinforcing polymers either as replacements of conventional fibers or as fillers for enhancing the properties of the existed advanced composites. However, establishment of CNTs as reinforcements requires the determination of their precise mechanical properties and behavior.

For such a requirement to be met, the exact dimensions of CNTs must be known. Due to the very small size of CNTs, their exact dimensions cannot be measured. Therefore, a study of the size effect on the mechanical properties of CNTs is needed. On the other hand, the large deviation between experimentally measured and theoretically predicted strength of CNTs is possibly due to the large number of defects contained in the nanotubes which interact simulating the presence of large holes. Therefore, the size effect on the strength of defected CNTs and the possible interaction of defects must be also studied. Advancing from these two needs, we study in the present paper the size effects in the mechanical properties of armchair and zigzag CNTs.

2. FE analysis of CNTs

By treating CNTs as space-frame structures, their mechanical behavior can be analyzed using classical structural mechanics methods [2]. To accomplish this work, a 3D FE model of CNTs developed using the ANSYS commercial FE code, was used. For the modeling of the bonds, the 3D elastic BEAM4 ANSYS element is used. Fig.1 depicts how the hexagon, which is the constitutional element of CNTs nanostructure, is simulated as structural element of a space-frame. In the same way the entire nanotube lattice is simulated. The simulation leads to the correspondence between bond length and element length as well as between wall thickness and element thickness. By assuming a circular cross-sectional area for the element, corresponds to the element diameter.

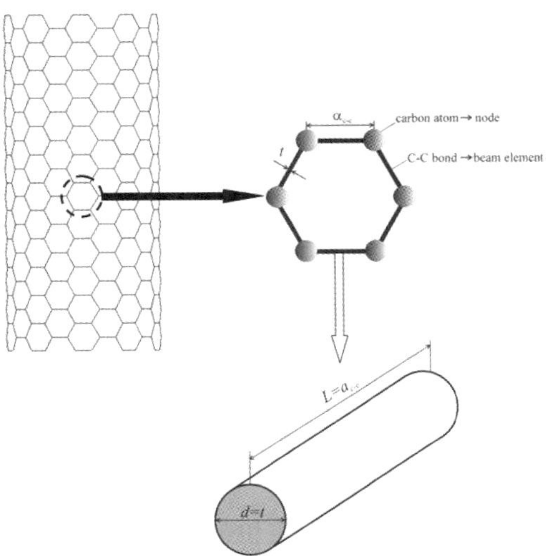

Figure 1. Simulation of a CNT as a space-frame structure.

3. Simulation of nanotube's behavior

In this work, both linear and non-linear behavior has been adopted for the nanotubes. These behaviors are dictated by specific assumptions regarding the behavior of the C-C bonds.

3.1. *Linear behavior of C-C bond*

In this case, the elastic moduli of the beam elements are derived using a linkage between molecular and structural mechanics [2]. According to this linkage, the following molecular mechanics expressions of bond stretching U_r, bond bending U_θ and combined torsion U_τ

$$U_r = \frac{1}{2}k_r(r - r_0)^2 = \frac{1}{2}k_r(\Delta r)^2 \tag{1}$$

$$U_\vartheta = \frac{1}{2}k_\vartheta(\vartheta - \vartheta_0)^2 = \frac{1}{2}k_r(\Delta\vartheta)^2 \tag{2}$$

$$U_\tau = U_\phi + U_\omega = \frac{1}{2}k_\tau(\Delta\phi)^2 \tag{3}$$

where k_r, k_θ, k_ω are the force constants and r, θ, τ, the corresponding deflections, are combined with the corresponding following classical mechanics expressions U_A, U_M and U_T

$$U_A = \frac{1}{2}\frac{EA}{L_n}(\Delta L_n)^2 \tag{4}$$

$$U_M = \frac{1}{2}\frac{EI}{L_n}(2\alpha)^2 \tag{5}$$

$$U_T = \frac{1}{2}\frac{GJ}{L_n}(\Delta\beta)^2 \tag{6}$$

where E and G are the bond Young's modulus and shear modulus, respectively, ΔL_n, 2α and $\Delta\beta$ are the corresponding macro-deflections, to get the following relations between beam rigidities and force constants

$$\frac{EA}{L_n} = k_r, \frac{EI}{L_n} = k_\vartheta, \frac{GJ}{L_n} = k_\tau \tag{7}$$

By assuming a circular cross-section with diameter d Eqs. (7) give

$$d = 4\sqrt{\frac{k_\theta}{k_r}}, \; E = \frac{k_r^2 L_n}{4\pi k_\theta}, \; G = \frac{k_r^2 k_\tau L_n}{8\pi k_\theta^2} \tag{8}$$

which fully define the beam element.

3.2. *Non-linear behavior of C-C bond*

The non-linear behavior of the C-C bond was simulated using the modified Morse interatomic potential [3]. According to this potential, the interatomic force between two atoms that are in distance r is given by

$$F = 2\beta D_e (1 - e^{-\beta(r-r_0)})e^{-\beta(r-r_0)} \quad (9)$$

where r_0 is the equilibrium distance equals to 0.1421nm and β=2.625x10^{10} m^{-1} and D_e=6.03105x10^{-19} Nm.

4. Parametric studies

Under the assumption of linear behavior of CNTs, the effect of diameter on nanotube's elastic moduli, the effect of wall thickness on nanotube's Young's modulus and the effect of length on nanotube's tensile, bending and torsional rigidities were studied, while under the assumption of non-linear behavior of CNTs, the effect of length on stiffness and tensile strength of defected (vacancy and Stone-Wales) CNTs was studied.

4.1. *The effect of diameter on nanotube's elastic moduli*

Young's modulus and shear modulus of CNTs have been derived by loading the CNTs in axial tension and torsion, respectively. Figs. 3 and 4 plot the elastic moduli against nanotube diameter of armchair, zigzag and chiral CNTs. An evident effect of diameter on Young's modulus of armchair zigzag CNTs, especially for small diameters is obtained. The Young's modulus of chiral CNTs is also affected but not significantly. With increasing tube diameter, the Young's modulus of CNTs increases but not with the same trend for all CNTs. The increase is due to the effect of nanotube curvature. The smaller the nanotube diameter is the higher the curvature, thus resulting in large distortion of the C-C bonds and therefore, in large elongation of the nanotube. As D increases, the effect of curvature diminishes and the Young's modulus of CNTs approaches that of graphene sheet (1.1 TPa).

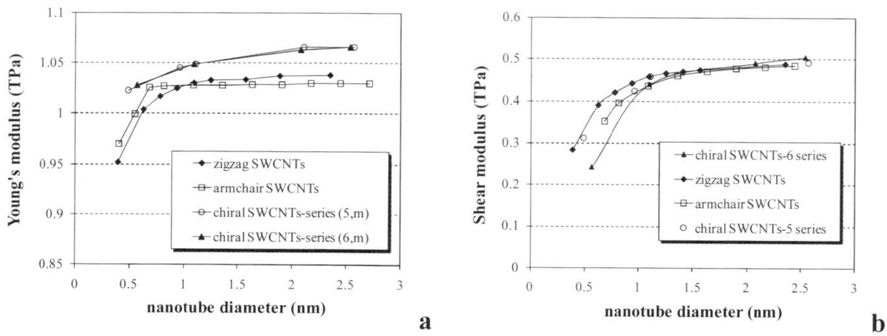

Figure 2. (a) Variation of Young's modulus with regard to nanotube diameter, and (b) Variation of shear modulus with regard to nanotube diameter.

4.2. *Effect of wall thickness*

Several values ranging from 0.064 to 0.69 nm have been adopted for the nanotube's wall thickness in the litetature. In the majority of studies, it has been assumed that the wall thickness of CNTs is equal to the interlayer spacing of graphite (0.34 nm). However, in some studies, values different than the interlayer spacing of graphite have been proposed. In each of the studies, the Young's modulus of CNTs for the specific value of wall thickness has been also derived. The pairs of wall thickness-Young's modulus together with the corresponding predictions of the present FE model are plotted in Fig.3. The figure suggests that the Young's modulus is governed by the wall thickness and that it varies inversely proportional to the wall thickness.

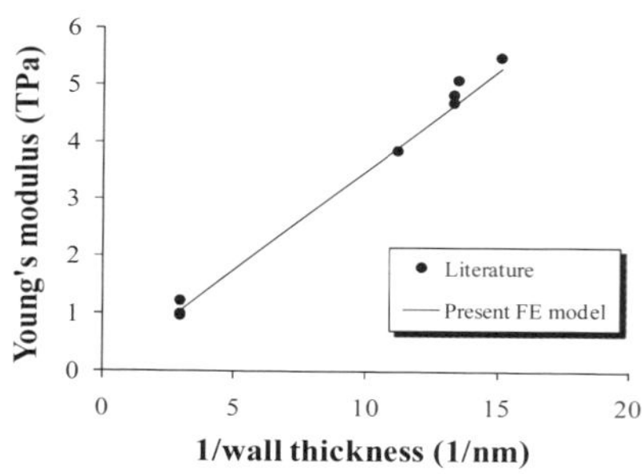

Figure 3. Variation of Young's modulus of the (8,8) nanotube with wall thickness.

4.3. *Effect of length on nanotube's rigidities*

Tensile, bending and torsional rigidities of CNTs have been derived by loading them in axial tension, bending and torsion and applying the following engineering mechanics expressions

$$EA = \frac{FL_n}{u},\ EI = \frac{FL_n^3}{3u},\ GJ = \frac{TL_n}{\phi} \tag{10}$$

where *I*, *J* are the secondary and torsional moments on inertia, respectively, *F* is the applied axial tensile force, *u* the applied normal deflection and φ the rotation angle.

Fig.4 plots the rigidities against tube length *L* for the (20,20) and (20,0) nanotubes, respectively. Variation is observed only at small values of L_n owing to the effect of boundary conditions at the supports. Specifically for bending, the very small initial value is due to the inapplicability of bending theory at small values of nanotube length to diameter ratio. From a certain value of L_n ranging from 40 to 50 nm, all three rigidities stabilize. The same was observed for all other types of nanotubes modeled. Results also show that the rigidities of zigzag nanotubes are smaller than the ones of armchair nanotubes and stabilize at smaller values of L_n. Moreover, it is observed that for both nanotube types bending and torsional rigidities are comparable.

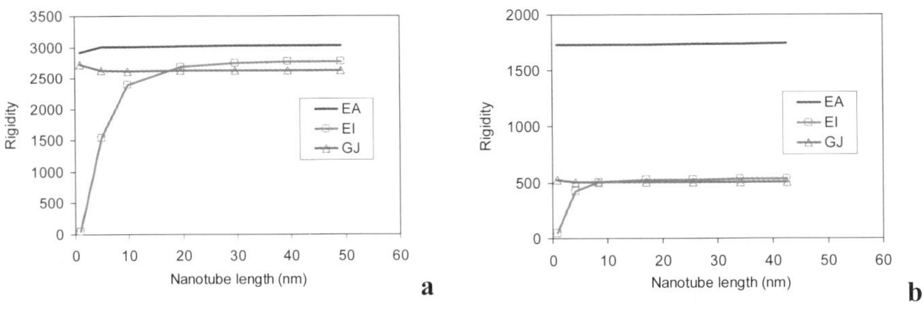

Figure 4. Rigidities against nanotube's length for (a) (20,0) and (b) (20,0) nanotube.

4.4. *Effect of length on the behavior of vacancy defected CNTs*

The vacancy defect considered is the one missing atom which leads to deactivation of three C-C bonds. In the study, aggregated defects are included in order to consider possible interaction between them in large lengths. The predicted stress-strain curves for the vacancy defected (12,12) nanotubes are compared in Fig.5. Comparison shows that there is no effect of length in nanotube's stiffness, while there is minor effect in tensile strength. It is also evident that for the defect patterns considered no defect interaction takes place. It is supposed that vacancy defects may approach in a one-bond distance in order interaction between them to take place.

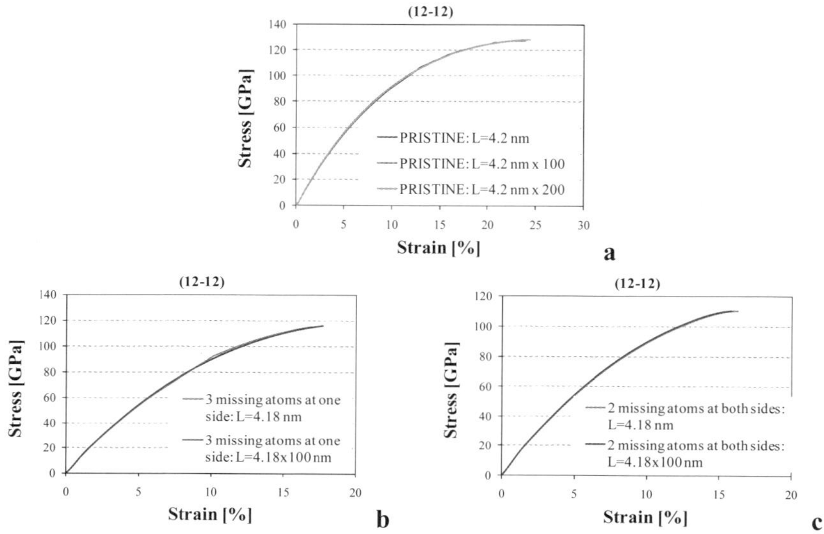

Figure 5. Comparison of predicted stress-strain curves for the vacancy-defected (12,12) tube.

In Fig.6, the predicted stress-strain curves for the vacancy-defected (20,0) nanotubes are compared. There is a decrease in stiffness at large strains for the large lengths and a similar effect in tensile strength. The length effect is more profound in the (20,0) nanotube than in (12,12) because the effect of vacancy defect in this case is more significant since the deactivated bonds are parallel to load and thus, highly-loaded.

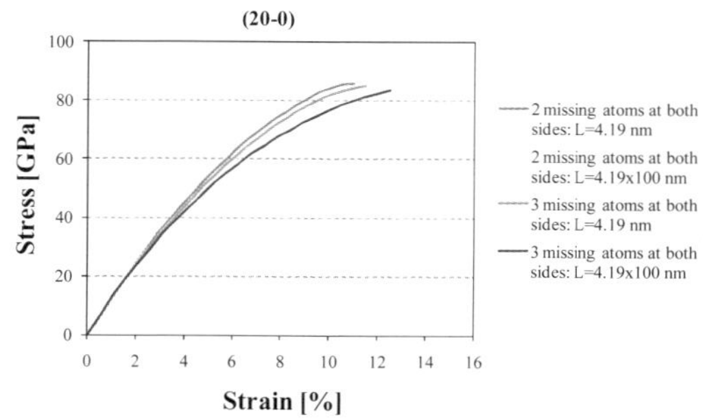

Figure 6. Comparison of predicted stress-strain curves for the vacancy-defected (20,0) nanotube.

Fig.7 compares the predicted stress-strain curves for the SW-defected (20,0) nanotubes. There is a decrease in stiffness at large strains for the large lengths and a similar effect in tensile strength.

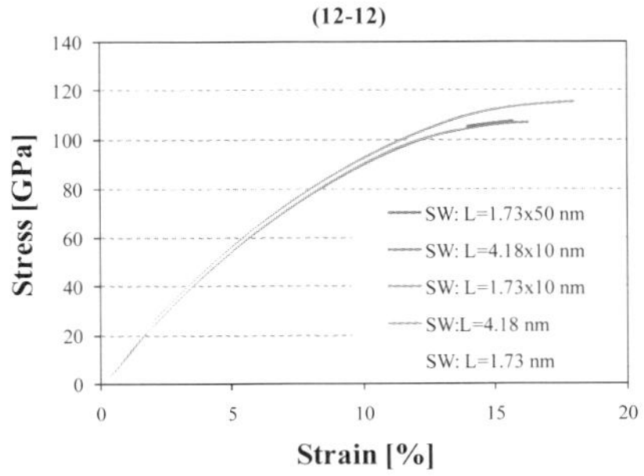

Figure 7. Comparison of predicted stress-strain curves for the SW-defected (12,12) nanotube.

5. Conclusions

An extensive study on the size effect (radius, length) on the elastic properties of pristine and the mechanical behavior of defected CNTs was accomplished. The findings of the study are summarized as follows

- ✓ With increasing the tube diameter, the Young's modulus increases until the effect of tube curvature diminishes.
- ✓ The wall thickness significantly affects the Young's modulus of CNTs. The larger the wall thickness, the smaller the Young's modulus.
- ✓ At small diameters, there is an effect in the nanotube rigidities. At values ranging from 40 to 50 nm all three rigidities stabilize.
- ✓ For the cases of defected CNTs considered, a minor length effect in the mechanical behavior of the nanotubes was found. This is an indication that in the resulted defect-patterns no defect interaction takes place.

References

1. S. Iijima, *Nature (London)* **56-8,** 354 (1991)
2. K.I. Tserpes and P. Papanikos, *Composites Part B.* **36,** 468 (2005)
3. K.I. Tserpes and P. Papanikos, *Composites Part B.* **37,** 662 (2006)

Biomedical Engineering

PHOTON SCATTERING APPLICATIONS IN TISSUE CHARACTERISATION AND IMAGING

Robert Speller,
UCL, London UK

1. Introduction

In general, scattered X-ray photons are regarded as a problem in the medical use of ionising radiation – it creates an unwanted background in images, it leads to poorer definition of radiation beams in therapy and it leads to an increased radiation burden to those involved in delivering X-ray services to the patient. However, scattered photons are an important source of information about the materials they have passed through and can be used to advantage (Speller & Horrocks, 1991). This paper discusses some of these areas of application.
Three topics are covered – coherently scattered photons, incoherently scattered photons and refracted photon beams. Strictly, the latter topic is not usually classed as a scattering process but has been included as the applications it enables can be closely linked to those created by scattering.

2. Coherent scattering

X-ray photons in the energy range of interest in medical imaging undergo one of three interactions. Their dependence upon photon energy is shown in Figure 1. It can be seen that the photoelectric interaction plays a major role at low energy but drops rapidly with increasing energy. Incoherent scattering changes little with energy except at energies below 10 keV which is outside the region of interest in medical applications. Coherent scattering has its greatest contribution in the range of low energy medical examinations and in particular for mammography. Photons undergoing coherent scattering are interacting with the atomic electrons and have wavelengths that ensure that events from different scattering centres can give rise to interference effects. These interference effects generate diffraction patterns that can be observed and used to determine the structure of materials or molecular form factors. Figure 2 shows the diffraction pattern associated with cortical bone expressed as a surface where the functions of energy and scattering angle give rise to varying scattered photon intensity. To make measurements it is normal to fix either scattering angle or energy and determine the diffraction pattern with either an energy resolving (EDXRD – energy dispersive X-ray diffraction) or spatially resolving (ADXRD – angular dispersive X-ray diffraction) detector system respectively. Examples of the output from these two techniques are also shown in figure 2.

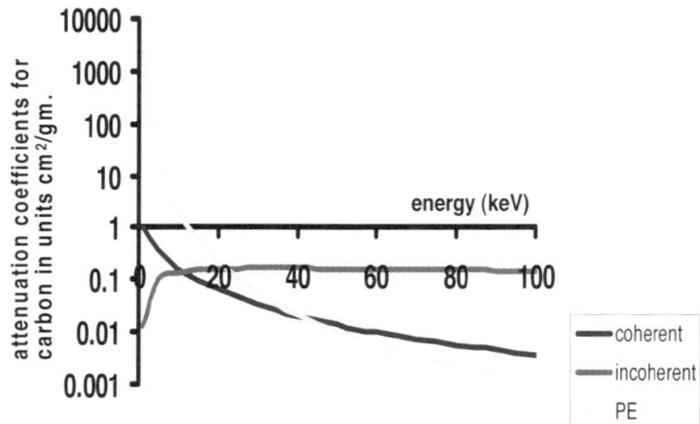

Figure 1:Energy dependence of the important interactions for low energy photons in carbon. Coherent and incoherent are scattering processes, photoelectric is an absorption process.

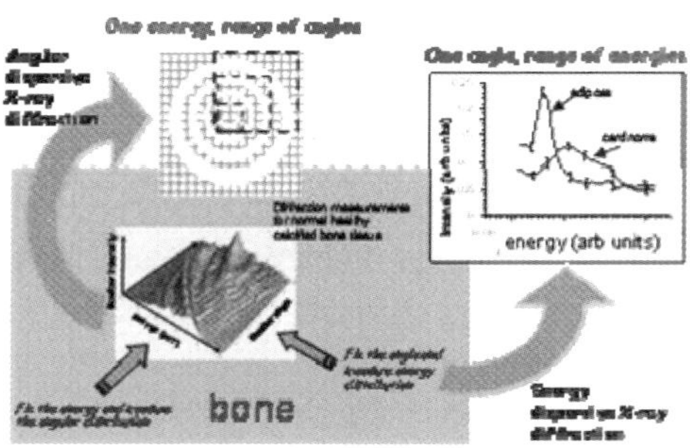

Figure 1: A diffraction surface for cortical bone. The intensity of coherently scattered photons taking into account their phase relationship is plotted against energy and scattering angle. The resultant diffraction information is also shown if energy is fixed (ADRXD) or if angle is fixed (EDXRD).

2.1 *Energy dispersive X-ray diffraction*

EDXRD has been used to characterise tissue in a number of studies (Kidane et al, 1999; Royle et al, 1999; Fernandez et al, 2002; Geraki et al, 2004; Fernandez et al, 2005; Griffiths et al, 2007; King & Johns, 2007). Typical results for excised breast tissue are shown in Figure 3. This plot gives the differential coherent scattering coefficient (γ) as a function of momentum transfer and the differences betweeen adipose, fibroglandular and carcinoma are clearly demonstrated. Suhonen et al (2005) have explained these observations in terms of the disruption to the collagen fibrils that occurs in the early development of cancer at a molecular level. From Figure 3 it can be seen that γ can be used to identify tissues, in particular carcinoma, in a mammographic examination provided the detection system can monitor scattered photons in a way that the intensities are measured at a momentum transfer value of ~1.6 nm^{-1}. One approach is to select energies using a monochromator in a synchrotron beam and to collect the scattered radiation at a given angle. To demonstrate this, a series of excised tissue samples where imaged at the ELLETRA synchrotron facility in Trieste. Using the SYRMEP beamline tuned to energies in the range of 17 to 24 keV and a low light level CCD camera set at 7^0 to the incident beam, images were taken at three different values of momentum transfer. Figure 4 shows an example. The transmission image of the sample demonstrates normal behavour, however, the diffraction images for the region of interest selected shows a contrast reversal typical of carcinoma. This was an undetected tumour in the conventional image.

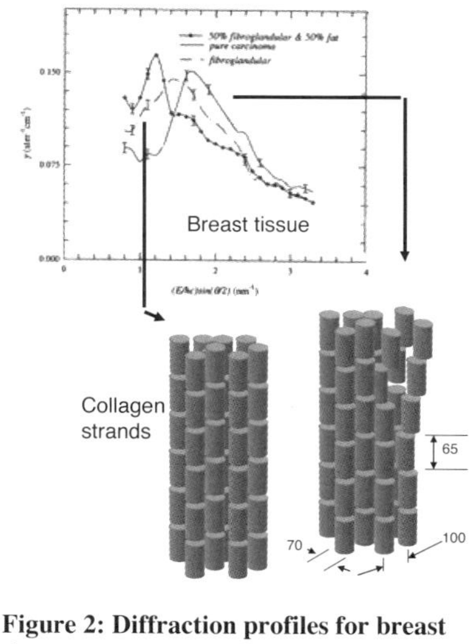

Figure 2: Diffraction profiles for breast tissue - normal and tumour. The disruption of the collagen strands (fibrils), as shown in the figure, explains the change in diffraction profiles. Dimensions are in microns.

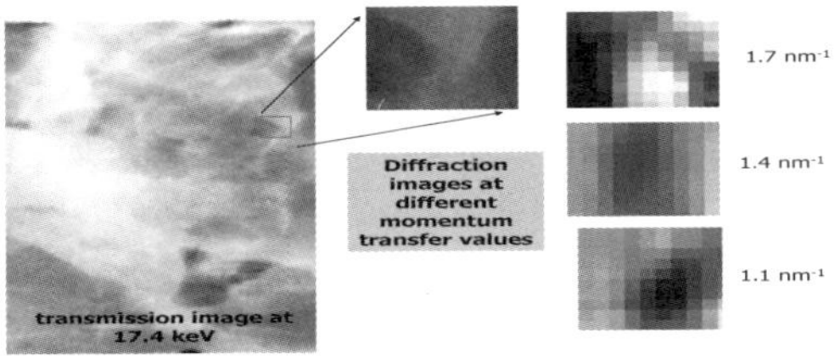

Figure 3: Transmission and diffraction images for a sample of excised breast tissue. The diffraction images indicate a small tumour undetected in the transmission image.

2.2 *Angular dispersive X-ray diffraction*

ADXRD is the technique used in power diffraction systems employed in laboratories for sample analysis. Adapting this technique to tissue characterisation (see Ryan & Farquharson, 2004, King & Johns, 2008) and imaging studies in-vivo requires modifications to the X-ray beam and a new type of spatially resolving detector. Reducing the bandwidth of the incident beam is relatively straightforward by the use of appropriate filters. The detector changes are more significant. Typical diffraction patterns consist of an intense central region and several surrounding regions of much reduced intensity. If detected with a conventional sensor the wide dynamic range cannot be accommodated. This problem has been overcome by using an active pixel sensor with selectable reset times for different regions of interest (Bohndiek et al, 2009). This sensor was designed and built as part of the MI-3 consortium (see http://mi3.shef.ac.uk). Combining the reduced spectral bandwidth with the APS sensor has enabled the recording of diffraction signatures for a range of tissues using clinical X-ray sources. Figure 5 shows the diffraction profiles for the different components of bone tissue and for breast tissue. These data are expressed as scattered intensities versus scattering angle and to be converted to γ versus momentum transfer (cf: Figure 3) requires careful calibration of the of the X-ray beam quality, scattering volume and detector corrections. However, general similarities exist demostrating the ability to characterise tissues using a spatially resolving detector. The appeal of having a detector with these characteristics is that if it can be re-configured in real-time between 'diffraction mode' and 'imaging mode' mammographic screening procedures could include an 'X-ray biopsy' analysis of suspicious regions at the same visit to the clinic. A large area APS is currently under construction by the MI-3 consortium to achieve this goal. Figure 6 shows the general concept that will be followed. Step 1 – a full field mammographic image is taken and searched for suspicious regions. Once located the X-ray beam is collimated, steered and filtered whilst the sensor is re-configured into the 'diffraction mode', i.e., selected regions of interest having different re-set times to allow wide dynamic range diffraction patterns to be collected. Step 2 then acquires the diffraction data and this is done for all suspicious regions. In this way the complete evaluation of the tissue is completed in one examination.

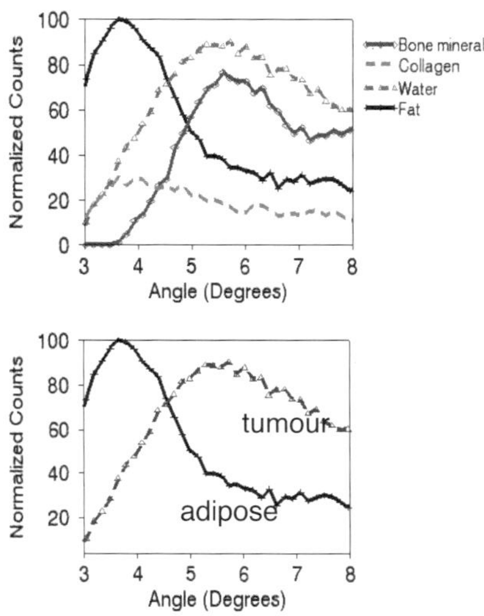

Figure 4: Angular dispersive X-ray diffraction profiles measured with an active pixel sensor for range of materials and tissues. These should be compared with the EDXRD data.

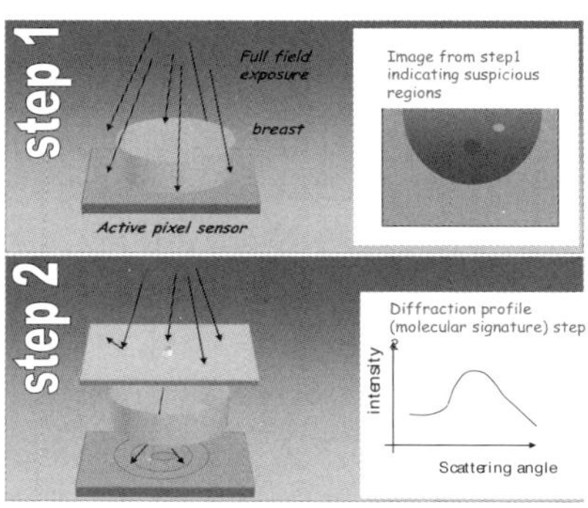

Figure 5: Concept of the X-ray diffraction biopsy system. Image 1 is a transmission image. Reconfiguration of the sensor and X-ray beam allows suspicious regions to be

2.3 *Adaptive intelligent imaging*

In many instances, conventional X-ray examinations are taken using previous experience to control the X-ray exposure conditions. There are some automatic controls applied to mammography, fluoroscopic and CT examinations but these controls are applied globally to the full area of the image. The I-ImaS concept (see, www.i-imas.ucl.ac.uk) was developed to allow the X-ray exposure conditions to be adapted on a local scale in real-time to take into account the tissue under investigation. A simple example might be an increase in photon flux, a change in photon energy or a change in detector resolution if part of the imaged object contains a suspicious region whose characteristics suggest that one of the changes listed would lead to improved detection of the abnormality. An I-ImaS system was built (Fant et al, 2007, Griffiths et al, 2008a) based around a dual-beam scanning system whereby the first beam (known as the scout beam) scanned across the patient followed by an adaptable beam, the so-called I-ImaS beam (see Figure 7). Data recorded from the first beam was used to decide whether or not a suspicious region existed. If it did the second, the I-ImaS beam, was modified by a series of filters. In its first development, these filters adjust the photon flux when used with a Mo-Mo combination as the X-ray source. The important stage in this process is how suspicious regions are recognized. Two approaches have been studied. Firstly, the sensors have been set up such that the scout beam records transmitted photons and statistical analysis of these data have been used to find a measure of texture that can be related to cancer. It was assumed that the disruptive nature of early cancer growth leads to an increase in the variance recorded in photon flux. This first order approach has been modified to take into account density effects and other contributing factors but the computed metric is still a measure of spatial variation in the imaged region (Schulerud et al, 2007, Griffiths et al 2008b) The second approach has been to use the diffraction signal (Esbrand, 2009) – in practice the magnitude of the scattered intensity summed over a narrow window in momentum transfer space. Results from these two approaches are shown in Figures 8, 9 and 10. Figures 8 & 9 show I-ImaS images for dental and mammographic examinations. The I-ImaS image and the dose distribution map are shown for different control functions resulting in reduced overall dose for the I-ImaS images without a loss in diagnostic information. The diffraction driven I-ImaS image is shown in Figure 10 for a mammographic example. The figure contains the I-ImaS image and the diffraction maps used to control dose distribution at two different momentum transfer values. In this case the map derived at 1.6 nm^{-1} was used to control the radiation delivery. Using diffraction has lowered (controlled) the dose more accurately as it allows the edges of the infiltrating tumour to be more closely defined.

Figure 6: The I-ImaS system. The large DAQ board houses the two rows of I-ImaS sensors, power supplies and the FPGA to handle the intelligent control of the system. Between the source (foreground) and the sensors are the filters used to adapt the X-ray beam conditions in real-time.

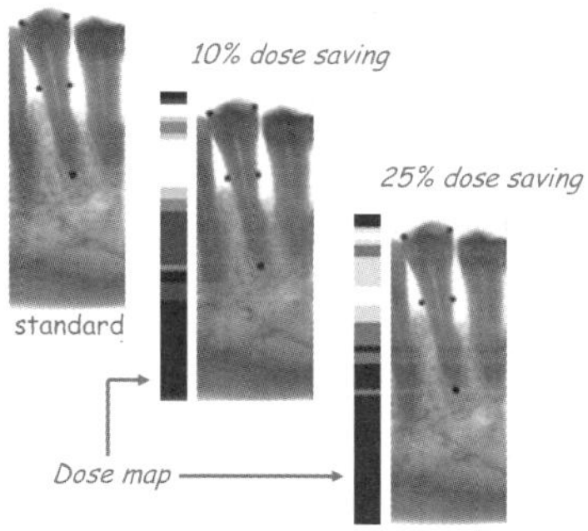

Figure 8: Dental I-ImaS images. Each image is accompanied by the dose distribution map. This distribution has been determined by the algorithms built into the FPGAs and executed by modulating the beam using the filter bank shown in Figure 7.

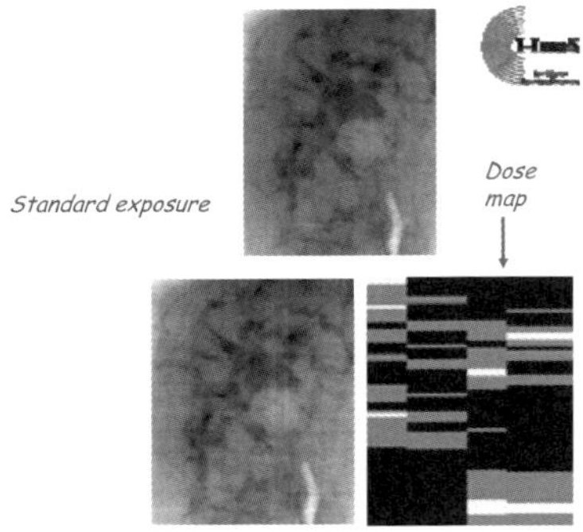

Figure 9: A mammographic I-ImaS image. The dose distribution map shows where the radiation intensity has been adjusted. White represents the highest intensity.

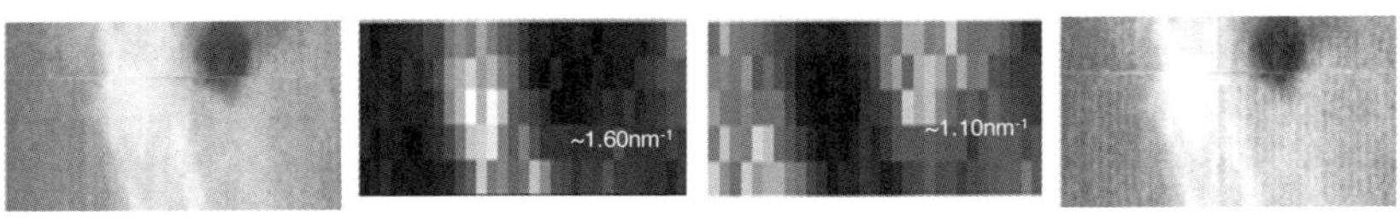

Figure 10: I-ImaS images of breast tissue where the intelligent control has been driven by a diffraction signal rather than a transmission signal. The greater specificity of the diffraction data means a more precise control of dose distribution can be achieved.

3. Incoherent scattering

Incoherent or Compton scattering leads to a change in the photon direction with a loss of energy. The lost energy is given to the electron involved in the interaction and this becomes the recoil electron. The relationship between the scattering angle and loss of photon energy is referred to as the Compton relation and can be used in gamma ray imaging systems to localize a distribution of radioactivity without the use of a mechanical collimator. Such systems are referred to as Compton cameras. These detection systems when compared with conventional gamma cameras (Han et al, 2008) have the advantage that they are light in weight (no lead collimator), have a wide viewing area, allow 3-D determination of activity without the need for moving the system and have high intrinsic efficiency. In their simplest geometry they require a minimum of two detection planes that have both position and energy resolution. They operate in the following way. If an incoming photon (OA in figure 11) undergoes a Compton interaction in one detector plane (A) and then interacts in the second detector plane (B) the two points of interaction lie on the central axis of a cone whose semi-angle is given from the magnitude of the deposited energy in the first interaction. This angle can be computed if the incident photon energy is known and hence the source of that gamma ray must lie on the surface of the cone, Figure11. If a series of such interactions are determined, a series of related cones will be found and gamma ray sources are located where common intersection points occur between the conic surfaces. In view of the Compton kinematics the fraction of energy given to the recoil electron increases with increasing gamma ray energy. Larger depositions can be determined with greater precision and hence Compton cameras have improved performance for high energy isotopes. This makes them ideal as positron imagers (imaging the annihilation gammas of 511 keV) or for looking at Cs-137 / Co-60 for industrial applications.

There are technical challenges to develop detector systems that can be used in this way and there are trade-offs between energy resolution, spatial resolution, efficiency and complexity. Several groups are currently developing systems for a variety of applications (Zhang et al, 2004, Chelikani et al, 2004, Hoover et al, 2006, Dedek et al, 2008). At UCL we have developed cameras using scintillators, silicon, germanium and CZT for applications as diverse as nuclear decommissioning, medical imaging and counteracting the terrorist threat of a 'dirty bomb'.

Figure 12 shows two of these camera systems and sample images are shown in Figure 13. The camera systems shown are a germanium camera consisting of a total of 177 pixels approximately 4 x 4 x 4 mm with 25 in one array and 152 in the other. The 152 pixels are spread over an annulus of width 25 mm and outside diameter of 100 mm. The two planes are 100 mm apart and this geometry has been optimized for medical imaging of either the heart or thyroid. The germanium requires cooling to liquid nitrogen temperatures for correct operation so the two arrays are housed in a cryostat with a liquid nitrogen dewar attached. The other camera uses 98 CZT sensors each of which is 5 x 5 x 5 mm. They are distributed such that approximately half the sensors are in each array. Although the energy resolution of these sensors is approximately 3 times that of the germanium camera they have the advantage that they operate at room temperature. This makes the camera more compact, lighter, and more portable and, as the sensors are not housed in a cryostat, they can be moved to alter the camera geometry to suit the imaging task. This camera has been designed to map the aftermath of a dirty bomb incident.

The images in Figure 13 are examples taken from the germanium camera and all are images of laboratory experiments using Cs-137 to simulate the conditions of imaging with positron emitters. The source distributions were generated using sources mounted on translation stages within scattering materials to simulate the patient. Straightforward backprojection and ITEM (imaginary time expectation maximization, see Pauli et al, 2002) images are shown. Most features are present and depicted with a degree of precision (FWHM of the point spread function = 8.5 ± 1.0 mm) comparable to conventional gamma cameras under similar conditions.

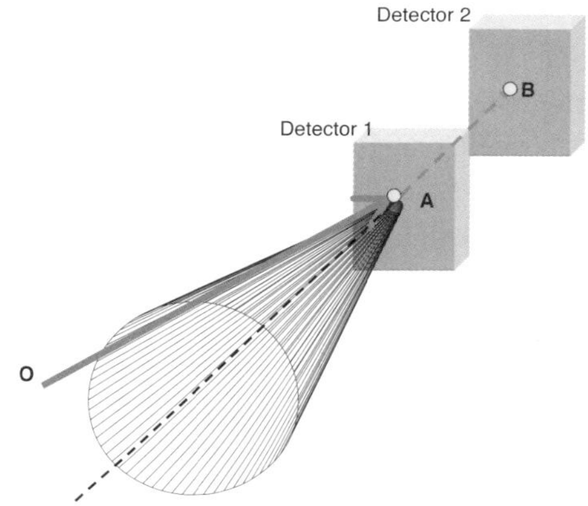

Figure 11: Principle of the Compton camera. Knowing the position of interactions in two detector arrays and the energy deposited, the source of the incident gamma rays can be located to lie on the surface of a cone. Multiple events create multiple cones with local points of interaction. These are the source locations.

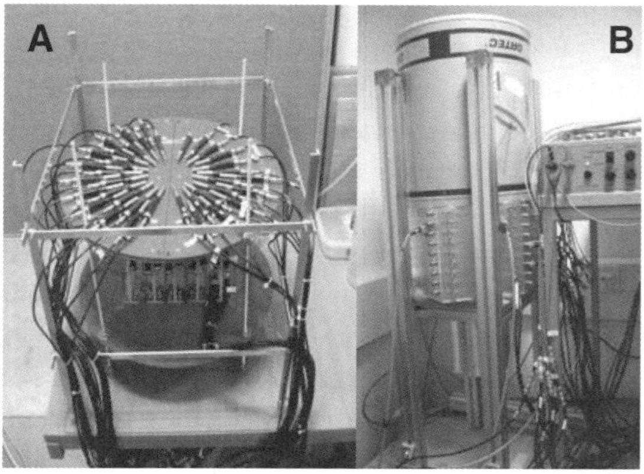

Figure 12: The UCL Compton cameras. 'A' is a CZT based camera built from discret sensors. A total of 98 sensors (5x5x5mm) make up the two arrays. 'B' is a germanium camera consisting of 172 pixels arranged in two planes (25 in the back plane).

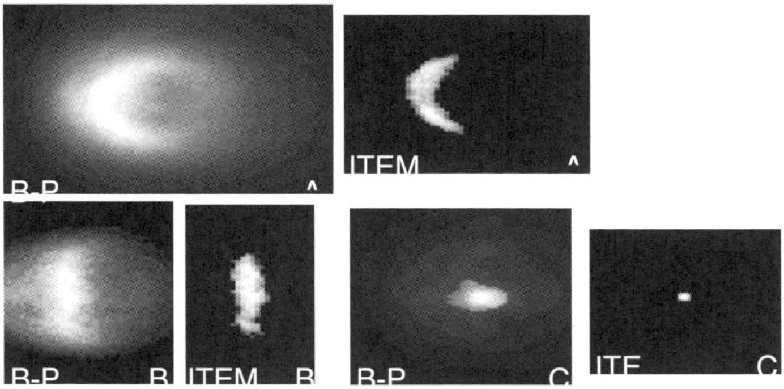

Figure 13: Compton camera images taken from the Ge camera using Cs-137. 'A' is a simulation of a heart with an infarct, 'B' a straight line and 'C' a point source. Both backprojection and ITEM reconstructions are shown.

4. X-ray phase contrast imaging

Treating X-rays as waves and looking at velocity changes as they pass through tissues gives rise to phase contrast imaging. The impact this might have in medicine has been reviewed by Lewis, 2004. The complex refractive index is made up of the imaginary part that accounts for absorption and the real part which is the decrement from unity of the phase velocity. For X-ray photons in tissue the imaginary part is approximately 10^{-3} times the real part and hence the effects created by changes in the real part should be significantly greater than those generated by changes in the absorption component. This is most clearly demonstrated in objects that have very little absorption change. Figure 14 is a demonstration of the potential of X-ray phase contrast to see 'invisible' details. These images of a wasp clearly show features in the phase contrast image that are invisible in the conventional absorption image. The difficulty is finding a suitable way to observe these changes. Figure 15 shows, schematically, how these components create information that can be used in medical imaging. In absorption based (conventional) imaging photons are removed from the incident beam by the sample and generate a change in intensity of the form shown in Figure 15 (left hand side). The magnitude of that change depends upon the attenuation coefficient of the tissues involved but in many of the more challenging cases this difference may be $< 1\%$. In the case when the wave nature of the radiation is considered the wave approaches the object and when the beam enters the object the phase velocity changes (right hand side of Figure 15). Thus at the edge of the object the wavefront will be distorted where those parts of the wave travel faster in the object compared to their near neighbours just outside the object. As the distorted wavefront travels away from the object interference will occur where the distorted and undistorted waves mix. As the change in wavefront direction is of the order of a few micro radians the waves must travel significant distances (~ 1 -2 m) before the interference pattern becomes visible. However, the interference leads to the formation of intense fringes that make detection of the object significantly enhanced. This approach to studying phase contrast effects is called 'free space propagation' or 'in-line holography'. There are other approaches to creating these effects and some are shown in Figure 16. Using an analyzer crystal to select these small deviations has the advantage that the contrast can be controlled by positioning the incident beam at different positions on the rocking curve of the crystal. However, such a technique requires the use of a synchrotron facility. Free space propagation effects can be observed either at synchrotron facilities or by the use of microfocal sources but both have the disadvantage that they are either inaccessible or of low output and hence do not lend themselves to clinical exploitation. The other technique shown in Figure 16 (using coded apertures) can be used with conventional, polyenergetic, divergent X-ray beams provided the focal spot size is 100 μm or less. Such a focal spot provides sufficient spatial coherence that phase effects are seen. The technique relies upon the use of masks placed either side of the object to be imaged. The mask in front of the detector is aligned with the detector pixels so that 'dead spaces' are created between pixels. The one in front of the object divides the X-ray beam into a series of 'micro-shafts' of photons that partially illuminate the pixel edges and partially fall on the dead spaces created by the detector mask. When an object is placed between the masks refraction occurs and causes those micro-shafts at the edges of structures in the object to bend and hence land either well inside the pixel or further into the dead spaces. Either effect will cause a large change in pixel response and hence enhance the detection of edges in the imaged object. This technique has been developed at UCL (Olivo & Speller, 2008, Olivo et al, 2008) and is now being developed into prototype imaging systems for both medical and security applications.

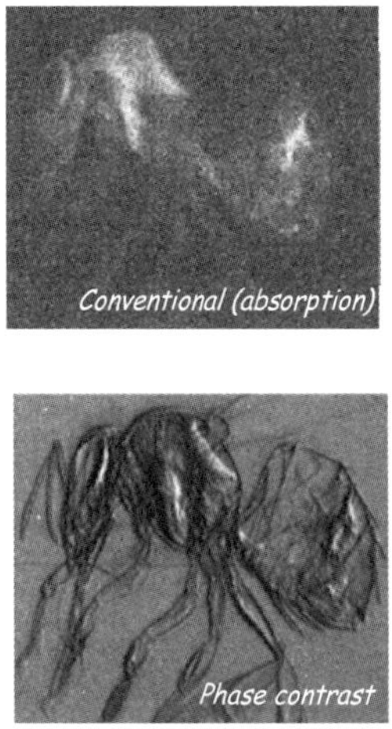

Figure 14: Two images of a common wasp. The conventional absorption based image shows little contrast. The phase contrast image demonstrates all parts

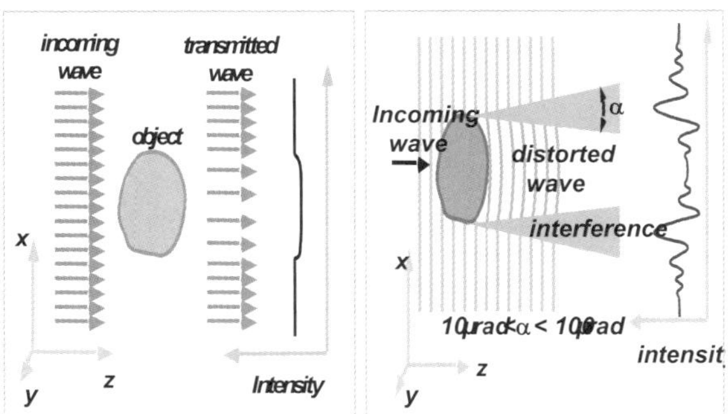

Figure 15: Absorption imaging shown on the left causes a change in intensity due to removal of photons from the beam. Phase contrast imaging, shown on the right, causes the wavefront to be distorted. This distortion creates an interference pattern that is observed.

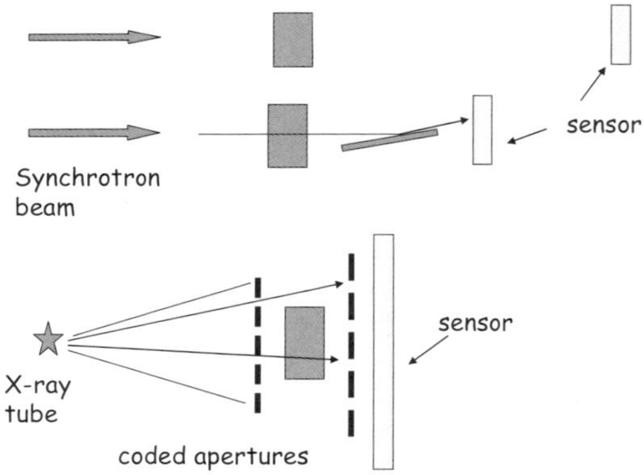

Figure 16: Different techniques for recording phase contrast images. 'A' is free space propagation and optimises the distances to make the interference pattern visible. 'B' uses Bragg scattering from a perfect crystal to identify the directions that the photons are refracted into. 'C' uses coded apertures to analyse the refracted directions.

References

1. Bohndiek, SE, Royle, GJ, Speller, RD.,"An active pixel sensor x-ray diffraction (APXRD) system for breast cancer diagnosis", Phys. Med. Biol., 54, 3513-3527, 2009.
2. Chelikani, S; Gore, J; Zubal, G, "Optimizing Compton camera geometries", Phys. Med. Biol., 49, 1387-1408, 2004.
3. Dedek N, Ghoggali W, Horrocks J.A., Royle G.R. and Speller R.D., 'Reconstruction of UCL Germanium Compton camera data using ITEM', Proc. Symp. Radiation Measurements and Applications, 2008.
4. Esbrand C, PhD thesis, University College London, 2009.
5. Fant A, Gasiorek P, Turchetta R., Avset B., Bergamaschi A., Cavouras D., Evangelou I., French MJ., Galbiati A., Georgiou H., Hall G., Iles G., Jones J., Longo R., Manthos M., Metaxas M., Noy M., Ostby J., Psomadellis F., Royle GJ., Schulerud H., Speller RD., van der Stelt P., Theodoridis S., Triantis F., Venanzi C., "I-IMAS: A 1.5D sensor for high-resolution scanning", Nucl. Instr. Meth. Res. A, 573, 27-29, 2007.
6. Fernandez, M; Keyrilainen, J; Serimaa, R; Torkkeli, M; Karjalainen-Lindsberg, ML; Tenhunen, M; Thomlinson, W; Urban, V; Suortti, P, "Small-angle x-ray scattering studies of human breast tissue samples", Phys. Med. Biol., 47, 577-592, 2002.
7. Fernandez, M; Keyrilainen, J; Serimaa, R; Torkkeli, M; Karjalainen-Lindsberg, ML; Leidenius, M; von Smitten, K; Tenhunen, M; Fiedler, S; Bravin, A; Weiss, TM; Suortti, P, "Human breast cancer in vitro: matching histo-pathology with small-angle x-ray scattering and diffraction enhanced x-ray imaging", Phys. Med. Biol., 50, 2991-3006, 2005.

8. Geraki, K; Farquharson, M; Bradley, D,"X-ray fluorescence and energy dispersive X-ray diffraction for the characterisation of breast tissue", Rad. Phys. Chem, 71, 969-970, 2004.
9. Griffiths, JA; Royle, GJ; Hanby, AM, Horrocks, JA, Bohndiek, SE, Speller, RD. Correlation of energy dispersive diffraction signatures and microCT of small breast tissue samples with pathological analysis", Phys. Med. Biol., 52, 6151-6164, 2007.
10. Griffiths JA., Metaxas MG., Pani S., Schulerud H., Esbrand C, Royle GJ., Price B., Rokvic T., Longo R., Asimidis A., Bletsas E., Cavouras D., Fant A., Gasiorek P., Georgiou H., Hall G., Jones J., Leaver J., Li G., Machin D., Manthos N., Matheson J., Noy M. Ostby JM., Psomadellis F., van der Stelt PF., Theodoridis S., Triantis F., Turchetta R., Venanzi C., Speller RD., 'Preliminary images from an adaptive imaging system', Physica Medica, 24(2), 117-21, 2008a.
11. Griffiths JA., Venanzi C., Speller RD., Royle GJ., Metaxas MG., Esbrand C., van der Stelt P., Li G., Verheij H., Turchetta R., Fant A., Gasiorek P., Theodoridis S., Georgiou H., Cavouras D., Hall G., Noy M., Jones J., Leaver J., Machin D., Greenwood S., Khaleeq M., Østby J., Schulerud H., Triantis F., Asimidis A., Bolanakis D., Manthos N., Longo R, Bergamaschi A., "Design and characterization of the I-ImaS multi-element x-ray detector system", IEEE Transactions in Nuclear Science, 55, 827, 2008b.
12. Han L., Rogers WL., Huh S. and Clinthorne N., "Statistical performance evaluation and comparison of a Compton medical imaging system and a collimated Anger camera for higher energy photon imaging", Phys. Med. Biol., 53, 7029-7045, 2008.
13. Hoover, AS, Sullivan, JP, Baird, B., Brumby, SP., Kippen, RM., McCluskey, CW., Rawool-Sullivan, MW. And Sorensen, EB, "Gamma-ray imaging with a Si/CsI(Tl) compton detector", App Rad. & Isot, 64 , 1648-1654, 2006.
14. Kidane, G; Speller, RD; Royle, GJ; Hanby, AM, "X-ray scatter signatures for normal and neoplastic breast tissues", Phys. Med. Biol., 44,1791-1802, 1999.
15. King, BW, Johns, PC ,"An energy dispersive technique for measuring X-ray coherent scattering form factors", Radiotherapy & Oncology, 84, S75-S76, 264, 2007.
16. King, BW, Johns, PC, "Measurement of coherent scattering form factors using an image plate", Phys. Med. Biol., 53, 5977-5990, 2008.
17. Lewis, RA, "Medical phase contrast x-ray imaging: current status and future prospects", Phys. Med. Biol., 43, 573-3583, 2004.
18. Olivo A and Speller R, "Image formation principles in coded-aperture based x-ray phase contrast imaging", Phys. Med. Biol., 53, 6461-74, 2008.
19. Olivo A, Chana D, Speller R, "A preliminary investigation of the potential of phase contrast x-ray imaging in the field of homeland security", , J. Phys D: App. Phys, 41, 225503, 2008.
20. Pauli J, Pauli EM, Anton G., "ITEM-QM solutions for EM problems in image reconstruction exemplary for the Compton Camera ",Nuc. Inst. & Meth. Phys. A, 488, 323-331, 2002.
21. Royle, GJ; Farquharson, M; Speller, R; Kidane, G, "Applications of X-ray diffraction analysis in crystalline and amorphous body tissues", Rad. Phys. Chem., 56, 247-258, 1999.
22. Ryan, E; Farquharson, M, "dispersive X-ray scattering from breast tissue using synchrotron radiation", Rad. Phys. Chem., 71, 971-972, 2004.

23. Schulerud H., Thielemann J., Kirkhus T., Kaspersen K., Østby JM., Metaxas MG., Royle GJ., Griffiths J., Cook EJ., Esbrand C., Pani S., Venanzi C, van der Stelt PF., Li G., Turchetta R., Theodoridis S., Georgiou H., Hall G., Noy M., Jones J., Triantis F., Asimidis A., Manthos N., Longo R., Bergamaschi A, Speller RD., "Adaptive Image content-based exposure control for scanning applications in radiography", Springer Lecture Notes in Computer Science, vol. 4678/2007, 543-552, 2007.
24. Speller RD and Horrocks JA. 'Photon Scattering: a 'new' source of information in medicine and biology?', Phys. Med. Biol., 36, 1-6, 1991
25. Suhonen, H; Fernandez, M; Serimaa, R; Suortti, P, "Simulation of small-angle x-ray scattering from collagen fibrils and comparison with experimental patterns", Phys. Med. Biol.,50, 5401-5416, 2005.
26. Zhang, LS; Rogers, WL; Clinthorne, NH, "Potential of a Compton camera for high performance scintimammography", Phys. Med. Biol., 49, 617-638, 2004.

LOW FREQUENCY SCATTERING BY A SOFT ACOUSTIC SPHERE EMBEDDED INTO AN ACOUSTICALLY LOSSLESS HALF SPACE

J. Lucas-Lekatsas and V. Kostopoulos

Applied Mechanics Laboratory, Department of Mechanical Engineering and Aeronautics, University of Patras, Patras, GR-26500, Greece

A plane wave is scattered by an acoustical soft sphere embedded into an acoustically lossless half space which disturbs the propagation of the incident wave field. The size of the sphere is much smaller than the wavelength of the incident wave. Low frequency theory reduces this scattering problem to a sequence of potential problems, which can be solved iteratively. Exactly one bispherical coordinate system exists that fits the given geometry of the obstacle and the plane boundary. Exact low frequency coefficients of the zero-th and the first order approximation for the near field as well as the first and second order coefficients for the normalized scattering amplitude are obtained and the scattering cross sections are calculated. Discussion of results and their physical meaning follows.

1. Introduction

The direct and inverse scattering problems by a single object have been extensively investigated in Acoustics [1,2], Electromagnetics [1,2], Elasticity [2] and Thermoelasticity[3,4]. For some single shapes, analytical results have been obtained both for low and high frequency. The present work initiates a study of the effects of a small soft acoustic sphere embedded into an acoustically lossless half space on the propagation of a plane source field. In Section 2 the amplitudes of reflective and diffractive acoustical waves are calculated, the problem of sound diffraction by only a penetrable plane. In Section 3 the geometry of the scattering region is that of a sphere into an acoustically lossless subspace and the diffractive acoustical wave is the incident plane wave. Closed form near field solutions, accurate to $O(k2)$ are obtained if the scatterer behaves as a soft sphere and a bispherical coordinate system is utilized to obtain analytic solution of the scattering problem of a plane wave, incident upon an acoustically soft sphere. In section 4 we provide the low frequency zero-th and first order coefficients of the near field, for a soft scatterer. In section 5, orthogonality arguments lead to the calculation of the surface integrals, are needed to determine the scattering amplitude and cross-section.

2. Sound Diffraction by a penetrable plane

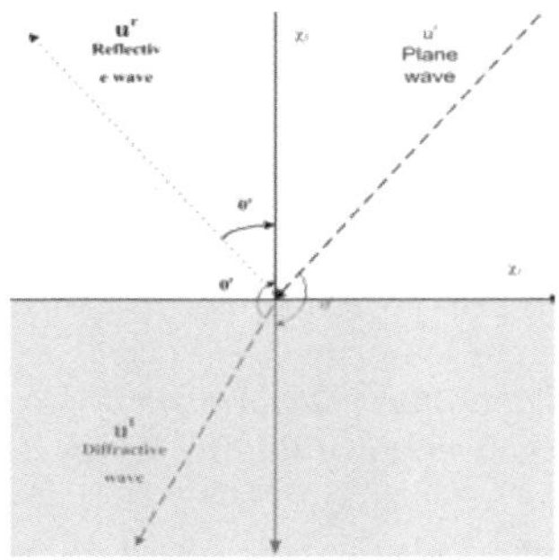

Figure1 Illustration of geometrical configuration of the diffraction problem by a penetrable plane irradiated by a plane wave

Let a penetrable plane (x_3=0) and the spaces V_+, V_- with different densities

$$V_+ = \left\{\underset{\sim}{r} \in R^3 : x_3 \geq 0\right\}, \quad V_- = \left\{\underset{\sim}{r} \in R^3 : x_3 \leq 0\right\} \tag{1}$$

The incident, reflective and diffractive acoustical plane waves in V_+ , V_- are

$$u^i\left(\underset{\sim}{r}\right) = e^{i\underset{\sim}{k}\cdot\underset{\sim}{r}} = e^{i\left(k_1x_1+k_2x_2+k_3x_3\right)}, \quad u^r\left(\underset{\sim}{r}\right) = \mathrm{A}^r e^{i\left(k_1x_1+k_2x_2-k_3x_3\right)}, \quad u_- = A_t \cdot e^{i\cdot \underset{\sim}{k_t}\cdot\underset{\sim}{r}} \tag{2}$$

Were $\hat{\underset{\sim}{k}}$, $\hat{\underset{\sim}{k}}_r$ are the directions of incidence and reflection and the total sound fields in spaces V_+ , V_- are governed by the Helmholtz equations

$$\left(\Delta + k^2\right)u_+ = 0, \quad \left(\Delta + \mu^2 k^2\right)u_- = 0 \tag{3}$$

A_r ,A_t are the unknown amplitudes of the reflective and diffractive waves and the boundary conditions for the penetrable plane x_3=0 are

$$u_+\left(\underset{\sim}{r}\right) = u_-\left(\underset{\sim}{r}\right), \quad \frac{\partial u_+}{\partial x_3} = \beta\frac{\partial u_-}{\partial x_3} \tag{4}$$

By solving the system of equations that transmission conditions (4) lead and from the energy conservation law finally the incident, reflective and diffractive acoustical waves will be [2]

$$u^i\left(\underset{\sim}{r}\right) = e^{i\underset{\sim}{k}\cdot\underset{\sim}{r}} = e^{i\left(k_1x_1+k_2x_2+k_3x_3\right)}, \quad u^r\left(\underset{\sim}{r}\right) = \frac{\beta-\mu^2}{\beta+\mu^2}e^{i\left(k_1x_1+k_2x_2-k_3x_3\right)}, \quad u_-^t\left(\underset{\sim}{r}\right) = \frac{2\beta}{\beta+\mu^2}e^{i\left(k_1x_1+k_2x_2+\frac{\mu^2}{\beta^2}k_3x_3\right)} \tag{5}$$

3. Diffraction of Sound by a soft sphere embedded into an acoustically lossless half space

Consider a sphere of radii R_1 with center located at $(0,0,K_3)$ under a plane. Our initial task is to adopt a bispherical coordinate system in such a way as to be able to identify the given sphere with a specified value of one parameter of the system variables.

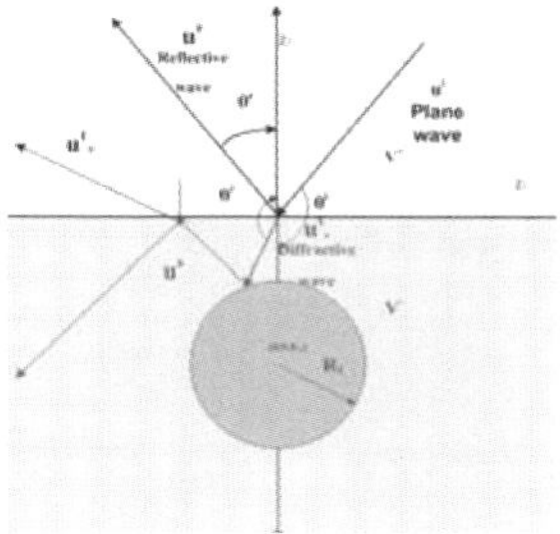

Figure2: Illustration of geometrical configuration of the diffraction of Sound by a soft sphere embedded into an acoustically lossless half space

Taking into account the geometry of the problem, we employ the bispherical system of coordinates [7,8]. For the given geometry of the problem i.e. the radius of a sphereR_1 and the distance of a plane d there is exactly one bispherical system that fits the given sphere scattering obstacle, as shown in Figure 2.

The coordinate surface η =constant is a sphere. Therefore, the region V^+upper the plane is where the incident, reflective and scattered acoustic wave propagates.

$$V^+ = \{(\eta,\theta,\phi) | \eta \in (0,+\infty), \theta \in [0,\pi], \phi \in [0,2\pi)\} \tag{6}$$

As it is shown in Figure 2, the region V^- where the diffractive and scattered acoustic wave propagates is the exterior of the sphere S^- and under the plane P (η=0) and corresponds to the domain.

$$V^- = \{(\eta,\theta,\phi) |\ \eta \in [-\eta_1, 0], \theta \in [0,\pi], \phi \in [0,2\pi)\} \tag{7}$$

Suppressing the harmonic time excitation of the form exp{-iωt}, where ω denotes the angular frequency, the wave equation is reduced to the Helmholtz equation.

Assuming plane incident and reflective waves(5) in V^+ where k stands for the wave number in V^+ *and* $\hat{\underset{\sim}{k}}$ is the unit propagation vector and *the diffractive* plane *wave*(5) as the incident wave scattered by the soft sphere for half space V^-, where k' stands for the wave number in V^- *and* $\hat{\underset{\sim}{k}}_t$ is the unit propagation vector, the scattering problem is then expressed in the following form:

$$u^+(\underset{\sim}{r}) = u^i(\underset{\sim}{r}) + u^r(\underset{\sim}{r}) + u^t_+(\underset{\sim}{r}), \qquad \underset{\sim}{r} \in V^+ \tag{8}$$

$$u^{-}(\underset{\sim}{r}) = u_{-}^{t}(\underset{\sim}{r}) + u^{s}(\underset{\sim}{r}), \qquad \underset{\sim}{r} \in V^{-} \tag{9}$$

which solve the Helmholtz equations

$$(\Delta + k^{2})u^{+}(\underset{\sim}{r}) = 0, \quad \underset{\sim}{r} \in V^{+}, \ (\Delta + \mu^{2}k^{2})u^{-}(\underset{\sim}{r}) = 0, \quad \underset{\sim}{r} \in V^{-} \tag{10}$$

The scattered fields u_{+}^{t}, u^{s} satisfy the Sommerfeld type radiation conditions

$$\lim_{r\to\infty} r\left(\frac{\partial u_{+}^{t}}{\partial r} - iku_{+}^{t}\right) = 0, \qquad \lim_{r\to\infty} r\left(\frac{\partial u^{s}}{\partial r} - ik'u^{s}\right) = 0 \tag{11}$$

Uniformly over the unit sphere S^2 of three dimensional Euclidean space. u^- satisfies the Dirichlet condition on S^-

$$u^{-}(\underset{\sim}{r}) = 0, \qquad \underset{\sim}{r} \in S^{-} \tag{12}$$

and u^+, u^- satisfy the transmission conditions of the penetrable plane P (η=0) and lead to an equivalent set of boundary conditions as below

$$u_{+}^{t}(\underset{\sim}{r}) = u^{s}(\underset{\sim}{r}), \quad \frac{\partial u_{+}^{t}(\underset{\sim}{r})}{\partial n} = \beta\frac{\partial u^{s}(\underset{\sim}{r})}{\partial n}, \quad \eta = 0 \tag{13}$$

Low-frequency scattering

In the case where the wavelength of the incident field is much larger than the radius of the exterior sphere (S^+ $kR_1 << 1$) low frequency techniques can be applied and the scattering problem is reduced to a sequence of potential problems that can be solved iteratively.

The low-frequency approximations satisfy the subsequent problems in potential theory. The leading low-frequency coefficient, the so-called Rayleigh approximation, solves the following boundary value problems:

The Zero-th order approximation

$$\Delta u_{0}^{+}(\underset{\sim}{r}) = 0 \ , \underset{\sim}{r} \in V^{+}, \quad \Delta u_{0}^{-}(\underset{\sim}{r}) = 0 \ , \ \underset{\sim}{r} \in V^{-} \tag{14}$$

$$(u_{0})_{+}^{t}(\underset{\sim}{r}) = u_{0}^{s}(\underset{\sim}{r}) \ \ \eta = 0, \quad \frac{\partial (u_{0})_{+}^{t}(\underset{\sim}{r})}{\partial n} = \beta\frac{\partial u_{0}^{s}(\underset{\sim}{r})}{\partial n}, \ \eta = 0 \tag{15}$$

$$u_{0}^{-}(\underset{\sim}{r}) = 0 \ , \ \eta = -\eta_{1} \tag{16}$$

$$u_{0}^{-}(\underset{\sim}{r}) = \frac{2\beta}{\beta + \mu^{2}} + O(1/r) \ \ r \to \infty \tag{17}$$

The first-order approximate

$$\Delta u_{1}^{+}(\underset{\sim}{r}) = 0 \ , \underset{\sim}{r} \in V^{+}, \quad \Delta u_{1}^{-}(\underset{\sim}{r}) = 0 \ , \underset{\sim}{r} \in V^{-} \tag{18}$$

$$(u_1)_+^t(\underset{\sim}{r}) = u_1^s(\underset{\sim}{r}) \quad \eta = 0, \quad \frac{\partial (u_1)_+^t(\underset{\sim}{r})}{\partial n} = \beta \frac{\partial u_1^s(\underset{\sim}{r})}{\partial n}, \quad \eta = 0 \tag{19}$$

$$u_1^-(\underset{\sim}{r}) = 0, \quad \eta = -\eta_1 \tag{20}$$

$$u_1^-(\underset{\sim}{r}) = \frac{2\beta}{\beta+\mu^2} \hat{\underset{\sim}{k}}_t \cdot \underset{\sim}{r} - \frac{1}{4\pi} \oint_{S^-} \frac{\partial u_0^-(\underset{\sim}{r})}{\partial n} ds(\underset{\sim}{r}) + O(1/r) \quad r \to \infty \tag{21}$$

Where $\frac{\partial}{\partial \eta}$ stands for the normal derivative in the outer direction and

4. The Rayleigh approximation.

In bispherical coordinates, the boundary value problems lead to the axially symmetric low-frequency approximation u_0 (η, θ) which solves the Laplace equation independent of azimuthal angle φ. It accepts R-separation and typical eigenfuction expansion, regular on the axis. In order to solve the above stated boundary value problem it is necessary to expand one in bispherical eigenfunctions[7].

Using the form of the solution of the Laplace equation described earlier and the basic identity (in order to incorporate the R.C. in the expression of the solution in V^+, V^- domain) the following expressions hold true for the fields

$$u_0^-(\eta,\theta) = \sqrt{\cosh\eta - \cos\theta} \sum_{n=0}^{\infty} \left[\sqrt{2}\frac{2\beta}{\beta+\mu^2} e^{-(n+1/2)\cdot|\eta|} + A_n^- e^{(n+1/2)\cdot\eta} + B_n^- e^{-(n+1/2)\cdot\eta} \right] P_n(\cos\theta) \tag{22}$$

$$u_0^s = \sqrt{\cosh\eta - \cos\theta} \sum_{n=0}^{\infty} \left[A_n^- e^{(n+1/2)\cdot\eta} + B_n^- e^{-(n+1/2)\cdot\eta} \right] P_n(\cos\theta) \tag{23}$$

Is the total and scattered field in space V^-

$$(u_0)_+^t = \sqrt{\cosh\eta - \cos\theta} \sum_{n=0}^{\infty} B_n^+ e^{-(n+1/2)\cdot\eta} P_n(\cos\theta) \tag{24}$$

Is the scattered field in space V^+

On surface S^- of the sphere we assume the (16) **Dirichlet** boundary condition and applying the transmission conditions (15)integrating and recalling the orthonormality relations for Legendre polynomials, we find the coefficients A_n^-, B_n^- for n=0, 1, 2, 3…

$$A_n^- = -\frac{\zeta e^{-(2n+1)\cdot\eta_1}}{1+\zeta e^{-(2n+1)\cdot\eta_1}} \frac{2\sqrt{2}\beta}{\beta+\mu^2}, \qquad B_n^- = -\frac{e^{-(2n+1)\cdot\eta_1}}{1+\zeta e^{-(2n+1)\cdot\eta_1}} \frac{2\sqrt{2}\beta}{\beta+\mu^2} \tag{25}$$

For $\eta < 0$ the total field

$$u_0^-(\eta,\theta) = \frac{2\sqrt{2}\beta}{\beta+\mu^2}\sqrt{\cosh\eta-\cos\theta}\sum_{n=0}^{\infty}\left[\frac{1-e^{-(2n+1)\cdot\eta_1}e^{-(2n+1)\cdot\eta}}{1+\zeta e^{-(2n+1)\cdot\eta_1}}\right]e^{(n+1/2)\cdot\eta}\mathrm{P}_n(\cos\theta), \text{ were } \zeta = \frac{\beta-1}{\beta+1}$$

The first–order approximation:

The first-order approximation u_1 is not azimuthally independent anymore and it is R-separable solution of the general form of Laplace equation in bispherical coordinates [2], [6] and [8]

Now, since the asymptotic behaviour of the problem appears [7] we conclude that the solutions of (18)-(21) are expressed in the form

(26)

$$(u_1)_-^s = 2\sqrt{2}a\sqrt{\cosh\eta-\cos\theta}\left\{\sum_{n=0}^{\infty}\left[E_n^- e^{\left(n+\frac{1}{2}\right)\eta} + G_n^- e^{-\left(n+\frac{1}{2}\right)\eta}\right]\mathrm{P}_n(\cos\theta)\right.$$
$$\left. + \sum_{n=1}^{\infty}\left[C_n^- e^{\left(n+\frac{1}{2}\right)\eta} + D_n^- e^{-\left(n+\frac{1}{2}\right)\eta}\right]\mathrm{P}_n^1(\cos\theta)(\kappa_1'\cos\phi+\kappa_2'\sin\phi)\right\}$$

(27)

$$(u_1)_+^t = 2\sqrt{2}a\sqrt{\cosh\eta-\cos\theta}\left\{\sum_{n=0}^{\infty} G_n^+ e^{-\left(n+\frac{1}{2}\right)\eta}\mathrm{P}_n(\cos\theta) + \sum_{n=1}^{\infty} D_n^+ e^{-\left(n+\frac{1}{2}\right)\eta}\mathrm{P}_n^1(\cos\theta)(\kappa_1\cos\phi+\kappa_2\sin\phi)\right\}$$

$$u_1^-(\underset{\sim}{r}) = 2\sqrt{2}a\sqrt{\cosh\eta-\cos\theta}\left\{\sum_{n=0}^{\infty}\left[\frac{2\beta}{\beta+\mu^2}\left(n+\frac{1}{2}\right)\kappa_3'(\operatorname{sgn}\eta)+C^-\right]e^{-\left(n+\frac{1}{2}\right)|\eta|}\mathrm{P}_n(\cos\theta)\right.$$
$$+\sum_{n=0}^{\infty}\left[E_n^- e^{\left(n+\frac{1}{2}\right)\eta} + G_n^- e^{-\left(n+\frac{1}{2}\right)\eta}\right]\mathrm{P}_n(\cos\theta)$$
$$\left.+\sum_{n=1}^{\infty}\left[\frac{2\beta}{\beta+\mu^2}e^{-\left(n+\frac{1}{2}\right)|\eta|} + C_n^- e^{\left(n+\frac{1}{2}\right)\eta} + D_n^- e^{-\left(n+\frac{1}{2}\right)\eta}\right]\mathrm{P}_n^1(\cos\theta)(\kappa_1'\cos\phi+\kappa_2'\sin\phi)\right\}$$

On surface S^- of the sphere we assume the Dirichlet boundary condition (20) and applying the transmission conditions (19) integrating and recalling the orthonormality relations for Legendre polynomials, we find the unknown coefficient of the field's u_1^+ and u_1^-combining the three sets of equations

$$E_n^- = \frac{\zeta\left[\frac{2\beta}{\beta+\mu^2}\left(n+\frac{1}{2}\right)\kappa_3'-C^-\right]}{1+e^{(2n+1)\eta_1}},\quad G_n^- = \frac{\left[\frac{2\beta}{\beta+\mu^2}\left(n+\frac{1}{2}\right)\kappa_3'-C^-\right]}{\zeta+e^{(2n+1)\eta_1}} \tag{28}$$

for n=0,1,2,…

(29)

$$C_n^- = -\frac{2\zeta\beta}{\left(\beta+\mu^2\right)\left[\zeta+e^{(2n+1)\eta_1}\right]},\quad D_n^- = -\frac{2\beta}{\left(\beta+\mu^2\right)\left[\zeta+e^{(2n+1)\eta_1}\right]}$$

for n=1,2,3,

5. The far field

The explicit form of scattering amplitude $g\left(\hat{\underset{\sim}{r}}\right)$ in the Dirichlet problem is given by the expression and using the zero-th and the first order approximation of the total field to calculate, we obtain [6]:

(30)

$$g\left(\hat{\underset{\sim}{r}},\hat{\underset{\sim}{k}}\right) = -\frac{ik}{4\pi}\oint_{S^-}\frac{\partial u_0^-\left(\underset{\sim}{r}'\right)}{\partial n'}ds\left(\underset{\sim}{r}'\right)+\frac{k^2}{4\pi}\oint_{S^-}\frac{\partial u_1^-\left(\underset{\sim}{r}'\right)}{\partial n'}ds\left(\underset{\sim}{r}'\right)+$$
$$-\frac{k^2}{4\pi}\oint_{S^-}\left(\hat{\underset{\sim}{r}}\cdot\underset{\sim}{r}'\right)\frac{\partial u_0^-\left(\underset{\sim}{r}'\right)}{\partial n'}ds\left(\underset{\sim}{r}'\right)\ +O\left(k^3\right)$$

for $\hat{\underset{\sim}{r}} \in S^2$

In order to obtain an analytic expression for the low-frequency approximation of the scattering amplitude, the three integrals in (30) must be evaluated. Calculating these integrals, using orthogonality arguments for the Legendre polynomials as well as the basic formula we arrive at:
Where $\hat{\underset{\sim}{r}} = (O_1, O_2, O_3)$ the unit position vector

(31)

$$g\left(\hat{\underset{\sim}{r}}\right) = ika\sqrt{2}A-(ka)^2\sqrt{2}\left[G+2O_3\tilde{A}\right]\ +O\left(k^3\right),\ ka\longrightarrow\infty$$

Where

(32)

$$A=\sum_{n=0}^{\infty}B_n^-,\quad G=2\sqrt{2}\sum_{n=0}^{\infty}G_n^-$$
$$\tilde{A}=\frac{1}{6}\sum_{n=0}^{\infty}\left\{\left[\coth\eta_1-(2n+1)\right]\left(\sqrt{2}\frac{2\beta}{\beta+\mu^2}+A_n^-\right)e^{-(2n+1)\cdot\eta_1}+\left[\coth\eta_1+2(2n+1)\right]B_n^-\right\}$$

and the relative expression for scattering cross section is provided by the relation

(33)

$$\sigma=\frac{1}{k^2}\oint_{S^2}\left|g\left(\hat{\underset{\sim}{r}}\right)\right|^2 ds\left(\hat{\underset{\sim}{r}}\right)$$

For the explicit form of **total cross section** σ, the integral involved in the expression must be determined and we arrive at

(34)

$$\sigma=8\pi a^2\left[A^2+(ka)^2\left(G^2+\frac{4}{3}\tilde{A}^2\right)\right]+O\left((ka)^4\right)\ ,ka\to 0$$

Note that the leading approximation of the scattering amplitude is proportional to the first power of the wave number and the next approximation is proportional to the second order of the wave number. Nevertheless, because of the exponential behaviour of the terms defining the series, the convergence is so fast that the results are satisfactory for any practical purpose.

6. Conclusions

In the present work Sound Diffraction by a penetrable plane is first solved and the amplitudes of the reflective and diffractive plane waves have been calculated. Bispherical coordinate system provides with appropriate choice of bispherical parameters the sphere and the plane as boundaries of the problem. Assuming the diffractive plane wave as incident wave which is scattered by the soft sphere in subspace we provide low frequency zero-th and first order coefficients of near field, for a soft sphere scatterer According to scattering theory, all results obtained for far fields are expressed in series form. Low-frequency coefficients of the leading two low-frequency approximations as well as the zero-th and first order approximations of the scattering amplitude and scattering cross-section are also obtained.

7. References

[1] D. Colton and R. Kress, Integral Equation Methods in Scattering Theory, Wiley, New York, 1983.

[2] Dassios G, Kleinman R. Low-frequency Scattering., Oxford Mathematical Monographs, OXFORD science publications, Oxford University Press, USA (March 2, 2000)

[3] G. Dassios and V. Kostopoulos, On Rayleigh Expansions in Thermoelastic Scattering, SIAM Journal of Applied Mathematics, Vol. 50,1990, pp.1300-1324

[4] G. Dassios and V. Kostopoulos, The Scattering Amplitudes and Cross Sections in Theory of Thermoelasticity, SIAM Journal of Applied Mathematics, Vol. 48,1988, pp.79-98

[5] M. Hadjinicolaou and V. Kostopoulos, Point Source Excitation for Scattering by Two Spheres, Hermes' 96, 3rd Hellenic-European Conference on Mathematics and Informatics, Ed. E. Lipitakis, September 1996, Athens, Greece.

[6] Charalambopoulos A, Dassios G, Hadjinicolaou M. Analytic solution for low-frequency scattering by two spheres. *SIAM Journal on Applied Mathematics*, Society for Industrial and Applied Mathematics, vol. 58(2), 1998; 370–386.

[7] P. Moon and D. E. Spencer, Field Theory Handbook, Springer-Verlag, Berlin, 1961

[8] P.M. Morse and Feshbach, Methods of Theoretical Physics, Vols. I, II, McGraw-Hill, New York, 1953.

AN AUTOMATED METHOD FOR ATHEROSCLEROTIC PLAQUE CHARACTERIZATION AND QUANTIFICATION USING INTRAVASCULAR ULTRASOUND IMAGES

L.S. ATHANASIOU AND D.I. FOTIADIS
Unit of Medical Technology and Intelligent Information Systems, Dept of Materials Science and Engineering, University of Ioannina, Biomedical Research Institute, Foundation for Research and Technology-Hellas (BRI-FORTH), University of Ioannina, GR 45110 Ioannina, Greece

V.D. TSAKANIKAS
Biomedical Research Institute, Foundation for Research and Technology-Hellas (BRI-FORTH), University of Ioannina, GR 45110 Ioannina, Greece

K.K. NAKA AND L.K. MICHALIS
Michaelideion Cardiac Center, University of Ioannina, Dept of Cardiology, Medical School, University of Ioannina, GR 45110 Ioannina, Greece

In this work an automated method is designed and developed for computer-aided characterization of plaque components using Intravascular Ultrasound Images (IVUS) is presented. We propose a hybrid approach which utilizes both machine learning and image processing techniques. The plaque can be classified as: soft plaque, hard plaque and hard calcified plaque. For the classification between the hard plaques (both calcified and non-calcified) and the soft plaques, several machine learning algorithms are employed. Random Forest algorithm outperformed both Support Vector Machines and Neural Networks. The classification between hard calcified and hard non calcified plaque formations is performed using an image processing algorithm, which automatically detects the acoustic shadow implied by the calcium in the IVUS signal. The classification algorithms are tested using 40 annotated IVUS images, and resulted in to 89% accuracy for hard calcified plaque formations, 84% for soft plaque formations and 88% accuracy for hard non-calcified plaque formations.

1. Introduction

Atherosclerosis is a disease of arteries which refers to the hardening and fattening of the arterial wall blocking the blood flow. The initial stage of the progress of atherosclerotic plaque is the accumulation of lipids in the intima of the artery. An inflammatory response to the oxidation of low density lipoproteins

(LDL) is initialized and monocytes are moved to the region of lesion. A lipid core is developed and a fibrous cap is formed. The smooth muscle cells proliferate and calcium is often accumulated in the plaque. The different components of atherosclerotic plaque are responsible for the different kind of outcomes of the disease. Also the treatment of atherosclerosis is based on the plaque ingredients. This means that it is very important to characterize the atherosclerotic plaque components. Imaging techniques are used to visualize the artery's lumen, wall and plaque borders. By using a specialised catheter with an ultrasound sensor, IVUS (intravascular ultrasound image) images of the artery walls are created and allow various components of the plaque to be identified. Manually plaque characterization is time-consuming and sensitive to observer variability. For this reason several algorithms have been developed to characterize the different types of atherosclerotic plaque using IVUS.

Using IVUS images Zhang et *al.* [1] characterized different plaque formations. Three different plaque types: soft plaque, non-calcified hard plaque and calcified hard plaque were indentified. Gray-level-based texture descriptors, co-occurrence matrices, run – length measures and fractal based measures were the extracted features. The radial profile, the long run emphasis and the fractal dimension were the descriptors that gave the best features for plaque definition. A classifier with linear discrimination functions was used for classification and each region was classified as containing hard or soft plaque. Then hard plaque was divided further to hard and hard calcified plaque based on hard plaque's gray level intensity. Pujol et *al.* [2] extracted features from IVUS images and made a comparison between different features. They studied two different classes of texture descriptors. The first class consisted of co-occurrence matrices descriptors and statistical descriptors and the other class consisted of local binary patterns and cumulative moments. Adaptive boosting (AdaBoost) was applied and classification was achieved characterizing plaque to soft, mixed, hard calcified (calcium) and hard non-calcified (fibrous) plaque. Brunenberg et *al.* [3] developed a new geodesic snake named Stop & Go Snake in an attempt to develop an automatic tissue characterization algorithm for IVUS images. Local binary patterns, gray level co-occurrence matrices and Gabor filters were the extracted features. The feature data were classified using Adaptive boosting technique with decision trees and decision stumps as classifiers. The classification results were post-processed using the Stop & Go Snake. Filho et *al.* [4] presented a method to detect and quantify the calcified regions in IVUS. They applied Adaptive thresholding in IVUS images using Otsu's method and analyzed the acoustic shadow that usually follows the calcified regions to identify the calcium. The quantification process was made by measuring the arc of calcium. In another approach Filho, et *al.* [5] presented a more accurate thresholding method. For each image the threshold value was different because they observed that the brigthnes of the calcification regions in images is changing from image to image. Zhang et *al.* [6] in their attempt to detect the calcified regions in IVUS images used the contourlet transform and active contours. Each image was decomposed with the contourlet transform into low-pass and into band-pass bands. Performing adaptive thresholding on the low-bass bands using 2-D Reny's entropy the regions of interest were detected. Using the acoustic shadow that follows the calcifications they detected the initial calcium in the regions of interest. They suppressed speckle and enhanced calcifications edges with a technique based on the on the anisotropic diffusion of the contourlet band-pass bands. The initial boundaries of calcium were deformed in the boundary vector fields of the image. Vince et *al.* [7] compared different texture analysis techniques for plaque characterization. The law's texture energy, the Haralick's, the neighborhood gray-tone different matrix (NGTDM), the first order statistics and the texture spectrum features were studied using discriminant analysis. Three plaque types were defined in IVUS images: calcified, fibrous and lipidic.

In this work we present a three stage scheme for the classification of plaque formations in IVUS frames in soft, hard non-calcified and hard calcified. Soft plaques consist mainly of soft lipid tissue with loose connectivity and are more prone to rupture. Soft plaques have low echoreflectivity and have good transmission coefficient. In IVUS images soft plaques are texturally homogeneous. Hard plaques have medium echoreflectivity that is similar to the adventitia echoreflectivity and consist of fibrous tissue. In IVUS images hard plaques are texturally heterogeneous. Hard calcified plaques have high echoreflectivity and in IVUS images the composition behind calcium is not depicted. In the first stage the detection of plaque regions is performed. In the second stage the plaque is classified as soft and hard and in the third stage the hard plaque is classified as hard non-calcified and hard calcified. The proposed approach is evaluated using 40 IVUS frames and the achieved accuracy is 86, 14 %.

2. Materials and Methods

A hybrid approach is employed for plaque characterization which combined data-driven and image processing techniques. The stages of the methodology are shown in Fig. 1 and are further analyzed below.

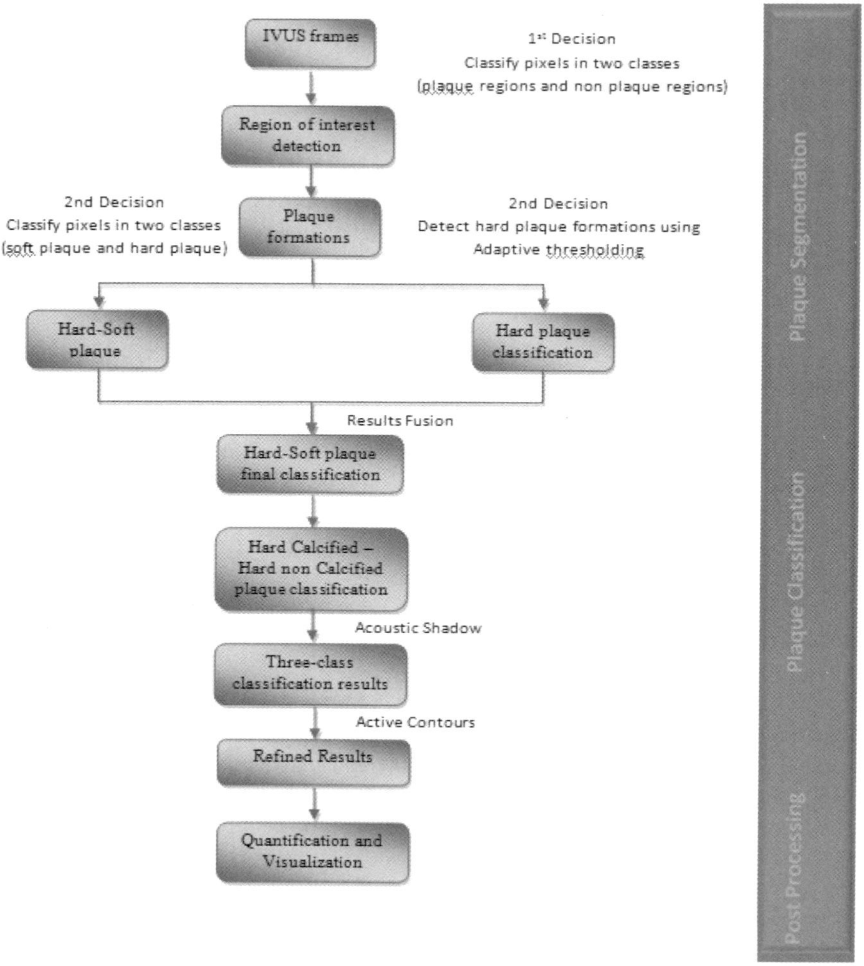

Figure 1. Methodology for Plaque Characterization using IVUS.

2.1. *Plaque regions detection*

In order to initialize the plaque characterization algorithm, the lumen and the media adventitia borders are required, for the isolation of the region of the artery where plaque formations are developed. Bourantas et *al.* [8] method is used to locate the Region of Interest (ROIs). In order to train the machine-learning algorithms, a set of annotated IVUS frames was used. The images came from 8 different patients, and for each IVUS examination, 15 frames from each patient were annotated by experts. This resulted to a set of 120 annotated frames. From this set, 80 frames were used for training the algorithms, while 40 frames were used for testing. From the 80 annotated images, all plaque formations and the normal tissue from the regions of interest were extracted. From each image, 340 pixels where randomly selected, from which 50% of them belong to normal tissue and 50% to plaque formations (regardless of the plaque type). From this procedure, two "pixel banks" where created. Namely, the first "pixel bank" contained 13,600 pixels of normal tissue and the second "pixel bank" contained 13,600 pixels of plaque formations. The combination of these two banks resulted into to the creation of a balanced dataset, which contained 27,200 pixels of the IVUS examinations. This dataset was used for the training of the machine learning algorithms. The next step of the algorithm is to calculate for each pixel in a 21x21 neighbourhood the following features: Co-occurrence Matrices [9], Local Binary Patter [10], the Mean Filter and the Entropy [11] of pixel's neighbourhood. The aforementioned calculations produce a set of 28 features per pixel. We employed three classifiers. Namely, two – class Support Vector Machine [12], Neural Network [13] and Random Forest [14] algorithm were trained. The first stage of the proposed approach is shown in Fig. 2.

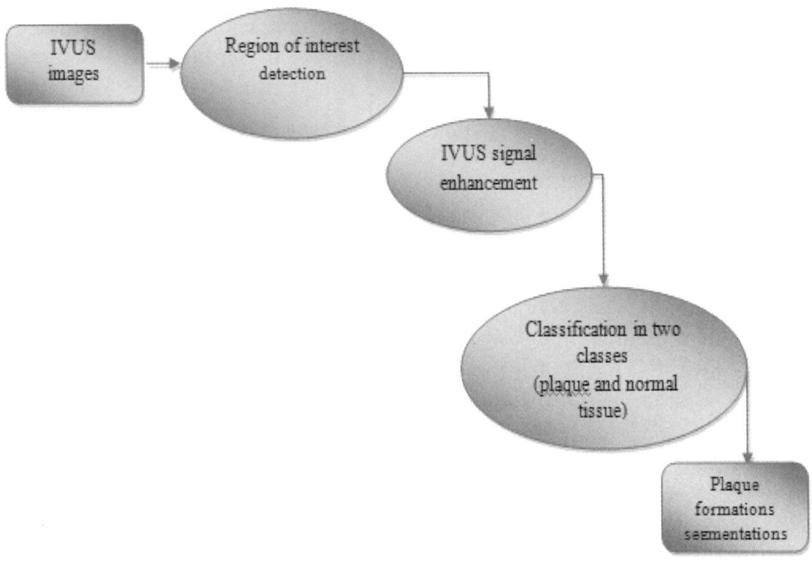

Figure 2. Plaque regions detection stage.

Summarizing the studies for the three different classifiers (Table 1), we can notice that the Random Forest algorithm provides the best results, in terms of accuracy.

Table 1. Classifiers accuracies for plaque formation and normal tissue classification.

Classifier	Accuracy	Classifier's parameters
Support Vector Machine	73.42 %	γ=3.4
Neural Network	75.6296 %	Hidden layers=18 Training time=600
Random Forests	89.8593 %	Number of attributes=11 Number of trees=145

2.2. Hard and soft plaque classification

In the IVUS image, the hard plaque formations appear as bright sections. Several studies [4], [5], [6] have used image processing techniques to detect the hard plaque formations. On the other hand, soft plaques are represented by a pixel luminance. Most of the studies which characterize soft plaques utilize texture features and machine learning techniques. In our method, we combine the aforementioned techniques to the method which consists of the following three steps (Fig. 3):

1. Use a classifier in order to classify hard and soft plaques
   - Calculate texture based features
   - Train the classifier
   - Classify hard and soft plaques.
2. Perform Adaptive Thresholding to the plaque regions, to detect hard plaques.
3. Fuse the results for the hard plaques from step 1 and 2, to exclude possible artefacts and inconsistencies.

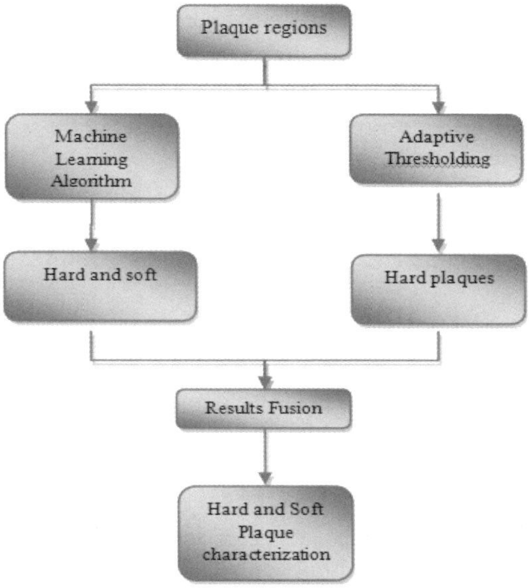

Figure 3. Hard and soft plaque classification stage.

For the classification into between soft and hard plaques three different classifiers are employed: two – class Support Vector Machine, Neural Network and Random Forests. The classification algorithm that provided the highest accuracy was chosen to classify hard and soft plaques. To create the dataset the 80 annotated images that were used for training the algorithms of the first classification stage were obtained. From these 80 annotated IVUS frames, the pixels of 30 soft plaque formations and 30 hard plaque formations (15 hard calcified and 15 hard non calcified) were selected. For each plaque formation 340 random pixels were selected from which 50% of them belong to soft plaque and 50% to hard plaque (both hard non calcified and hard calcified). The result of this procedure was a fully balanced dataset of 20,400 pixels of the IVUS examinations with 10200 pixels representing soft plaque and the other 10200 representing hard plaque. Features for each pixel of the dataset in a 21x21 neighborhood are extracted. The calculated features are: Co-occurrence Matrices [9], Local Binary Patterns [10], the Mean Filter and the Entropy [11] of the pixel's neighborhood. The Random Forests algorithm provides the best results, in terms of accuracy (Table 2).

Table 2. Obtained accuracy for soft and hard plaque classification.

Classifier	Accuracy	Classifier's parameters
Support Vector Machine	89.84 %	γ=5.4
Neural Network	90.7333 %	Hidden layers=18 Training time=800
Random Forests	96.34 %	Number of attributes=8 Number of trees=165

2.2.1. Hard Plaque Classification Refinement - Adaptive Thresholding

The procedure described in the previous classification stage produces a classification into soft and hard plaques. This procedure can produce artefacts, especially to small plaque regions. In order to minimize the possibility of the appearance of these artefacts, we apply an Adaptive Thresholding Technique [5] to Regions of Interest, to produce an estimation of the hard plaque formations. When we calculate these regions, we fuse them with the results coming from the machine learning algorithm, to compute the final hard plaque regions. A threshold th is calculated using the Otsu method. Then we apply to the image a high pass filter using th. We repeat this procedure three times, and the remaining regions represent the hard plaque formations. More precisely, for the IVUS frames that the classification algorithm has detected hard plaque formations, we apply the Adaptive Thresholding Technique. The contours derived from this step are compared, in terms of overlapping area, with the hard plaque formations detected from the classification algorithm. Then, a decision is taken, depending on the percentage of the overlapping area. When the overlapping area is greater than 90% of the contour's area (which was derived from the Adaptive thresholding algorithm), then the methodology verifies the appearance of a hard plaque formation.

2.3. Hard calcified and hard non-calcified plaque classification

In an IVUS examination, the hard plaque formations appear as bright regions. Thus, it is extremely difficult to classify hard calcified and hard non-calcified plaque formations using image processing or machine learning methods. As it is mentioned in previous research studies [4], [5], [6] calcium has the physical property of absorbing larger proportion of the transmitted ultrasound signal than any other material of the arterial wall. Due to this fact, in an IVUS examination, the regions behind a calcified formation appear darker than the rest of the adjacent regions. These dark regions are called acoustic shadows. We utilize the aforementioned property of calcium to classify the calcified hard plaques from the non-calcified hard plaque formations as follows:

1. Transform the IVUS image into polar coordinates.
2. Calculate the median gray level of the pixels in each column of the transformed image.
3. Compare the median gray levels which correspond to the hard plaque regions with the median gray levels of the adjacent regions.
4. If the median gray levels of a hard plaque region are lower than a threshold th, then the hard plaque formation corresponds to a calcified plaque formation. Otherwise, it corresponds to a non-calcified hard plaque formation.

To calculate the threshold value we obtained 15 ROI's annotated as hard calcified plaques and 21 areas annotated as hard non calcified plaques and we transformed the images that corresponded to these ROI's to polar images. In these 36 ROI's we calculate the threshold value for each ROI. The threshold value is the ratio of the sum of the median values of the pixels in a ROI and the sum of the pixels of the ROI. The range of the possible threshold values was from 10-85 and we choose the threshold value that provides the best sensitivity and specificity.

2.4. Post processing

To improve the accuracy of the characterized plaque regions, we perform a post processing step to each plaque region, utilizing active contours in order to detect with higher precision the plaque formations. The procedure includes three steps:

1. Set as initial active contour the result of the classification scheme.
2. Calculate the active contour's deformation.
3. Set as the final plaque formation's region the final deformed active contour.

The described post processing algorithm is applied on the results of the plaque classification scheme previously analyzed. The active contour deformation performed substantially in the results of hard calcified plaques and increased the overlapping areas by 8.9% between the experts' annotations and the algorithms results. This post processing step did not perform well in the algorithm's results for hard non calcified plaques and soft plaques. This was due to the fact that the edge map of these plaque formations showed similar results with the edge map of the tissue surrounding the plaque formations and thus does not permit to the active contour to deform on the plaque formation border. Post processing with active contours is used only for hard calcified plaque regions.

3. Results

To validate our plaque characterization algorithm we used 40 IVUS frames from the initial dataset that were not included in the training step. In these 40 frames there were 21 ROI's annotated as hard non calcified plaque formations, 12 ROI's annotated as hard calcified plaque formations and 15 ROI's annotated as soft plaque formations. To assess the performance of our automated plaque characterization algorithm we determined the overall accuracy as the ratio of the number of the correct classified pixels and the number of the pixels that were classified in the 40 tested images. The overall accuracy reached 86.14 %.

3.1. Plaque formations and normal tissue classification

In the first stage of our plaque characterization algorithm 40 plaque regions were detected. To compare the performance of the algorithm, the areas of both the algorithm-detected plaque regions and the expert-annotated plaque regions are calculated. The correlation between expert's annotations and algorithm's results for plaque detection is shown in Fig. 4, where the plaque area denoted in mm^2 is determined as the number of pixels multiplied by the area of one pixel (in mm^2).

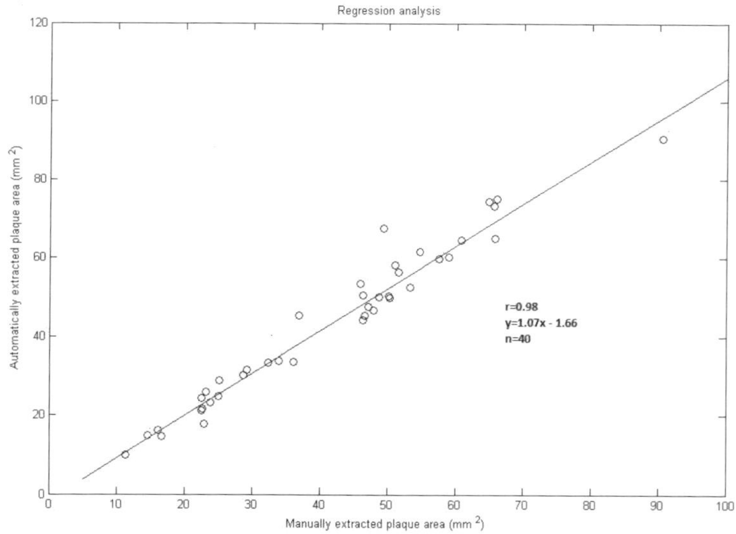

Figure 4. Comparison of expert's extracted plaque area and algorithm's extracted plaque area.

3.2. Soft plaque, hard non-calcified and hard calcified plaque classification

The correlation in the 21 ROI's that were annotated as hard non calcified plaques and were detected as hard non calcified plaques by our classification algorithm is shown in Fig. 5. The results through a frame example for each step of the algorithm are shown in Fig. 8.

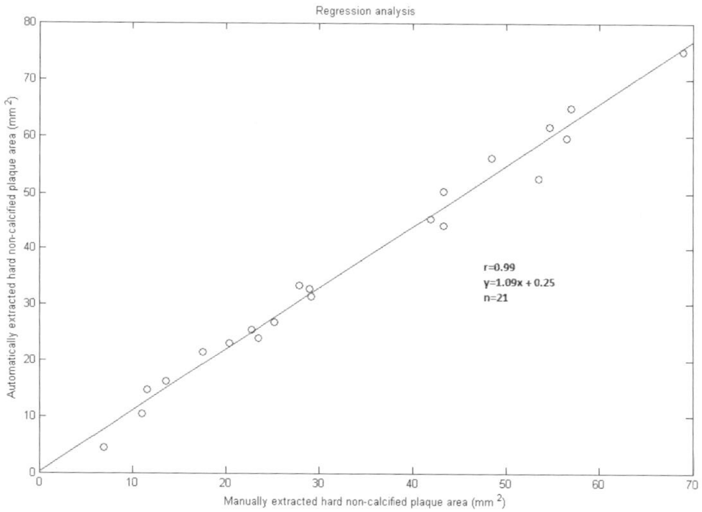

Figure 5. Comparison of expert's extracted hard-non calcified plaque area and algorithm's extracted hard-non calcified plaque area.

In Fig. 6 and in Fig. 7 we can see the correlation of the manually extracted and automatically extracted plaque area for the 12 hard calcified and the 15 soft plaque areas.

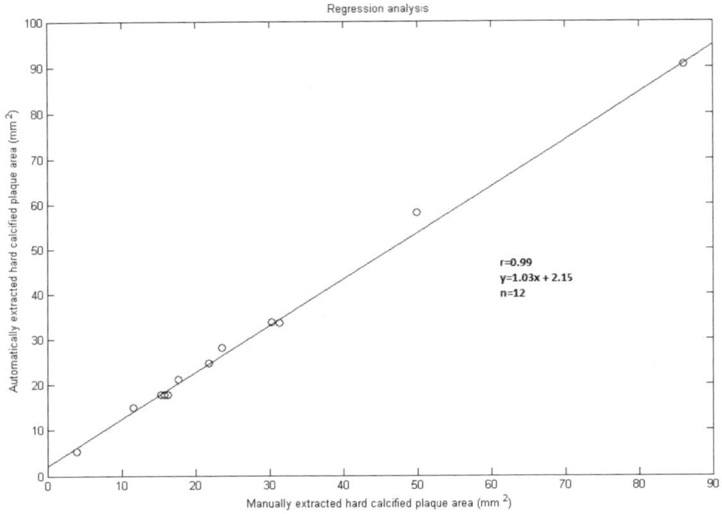

Figure 6. Comparison of expert's extracted hard calcified plaque area and algorithm's extracted hard calcified plaque area.

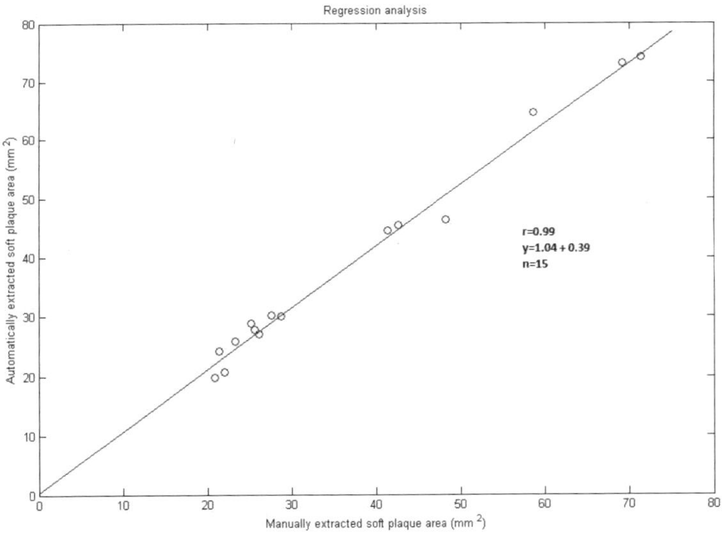

Figure 7. Comparison of expert's extracted soft plaque area and algorithm's extracted soft plaque area.

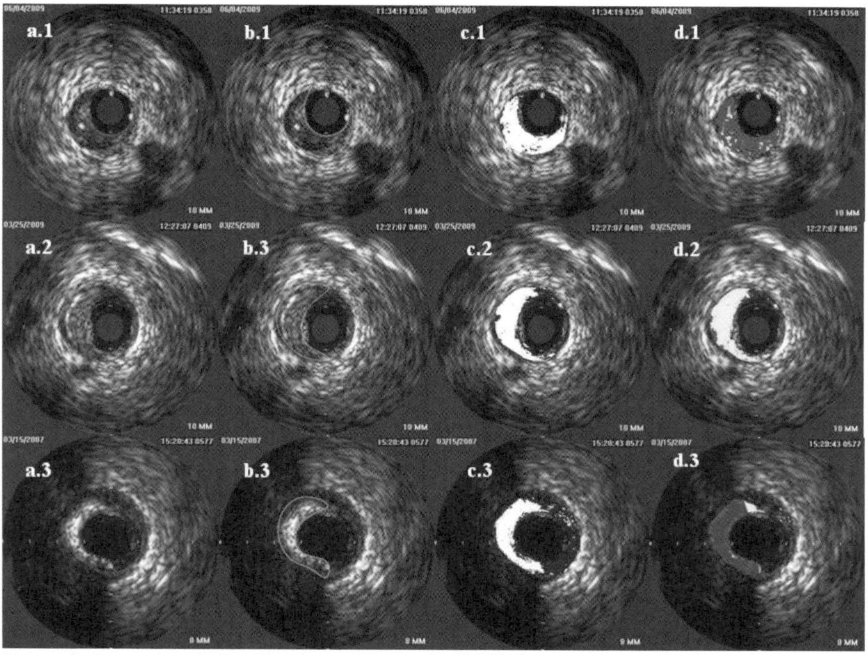

Figure 8. Example of classification evolution in IVUS frames annotated with soft, hard non-calcified and hard calcified plaque. (a.1, a.2, a.3) Original frames. (b.1, b.2, b.3) Annotated frames by an expert as containing soft (b.1), hard non-calcified (b.2) and hard calcified plaque (b.3). (c.1, c.2, c.3) Resulting frames for plaque formations and normal tissue classification stage. The plaque formations pixels are marked with white by the algorithm and the normal tissue pixels are marked with blue. (d.1, d.2, d.3) Resulting frames for soft, hard non-calcified and hard calcified plaque classification stage. The soft plaque pixels are marked with purple, the hard non-calcified pixels are marked with yellow and the hard calcified pixels are marked with red.

3.3. Threshold level of the median values in hard calcified plaques and hard non – calcified plaques classification

As we shift in the range of the possible threshold values (10-85) we chose as threshold value 45 that result in sensitivity 92% and specificity 95 %. Table 3 shows the resulting sensitivity and specificity for 5 different threshold values and the correct classified ROI's for each threshold value and Fig. 9 shows the receiver operating characteristic curve for these values. The area under curve (AUC) of the receiver operating characteristic curve (ROC) is 0.934.

Table 3. Results for different threshold values in hard calcified plaque and hard non calcified plaque classification.

threshold value	sensitivity	specificity	correctly classified ROI's as hard non calcified plaque	correctly classified ROI's as hard calcified plaque
15	0.08	1	21	1
30	0.58	1	21	7
45	0.92	0.95	20	11
60	0.92	0.57	12	11
75	1	0.14	3	12

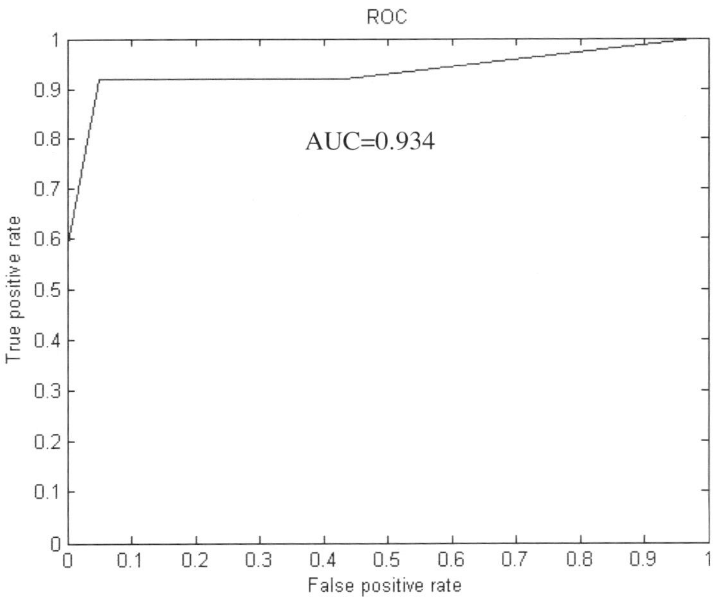

Figure 9. Receiver operating characteristic curve for different threshold values.

4. Discussion

In the present work a methodology for plaque characterization using IVUS images is described. The methodology is based on a hybrid algorithm, which utilizes both machine learning and image processing algorithms. Three plaque types can be indentified: soft plaques, hard plaques and hard calcified plaques.

We have accomplished to provide classification of hard calcified plaque formations with accuracy of 87.81 %, classification of hard plaques with accuracy of 91.33 % and classification of soft plaques with accuracy 84.05 %.

About the limitations of our method, it must be mentioned that, due to the fact that calcium absorbs high amounts of the ultrasound signal, our method cannot detect plaque formations behind calcified formations, due to acoustic shadow.

5. Acknowledgments

This work is part funded by the European Commission (Project ARTREAT: Multi-level patient-specific artery and atherogenesis model for outcome prediction, decision support treatment, and virtual hand-on training, FP7-224297).

References

1. X. Zhang, C. R. McKay, and M. Sonka, "*Tissue Characterization in Intravascular Ultrasound Images*", IEEE Transactions on Medical Imaging vol 17, no 6, (1998).
2. O. Pujol, P. Radeva, J. Vitria, and J. Mauri, "*Adaboost to Classify Plaque Appearance in IVUS Images*", Progress in Pattern Recognition, Image Analysis and Applications, Volume 3287, 0302-9743, (2004).
3. E. Brunenberg, O. Pujol, B. Romeny, and P. Radeva, "*Automatic IVUS Segmentation of atherosclerotic Plaque with Stop & Go Snake*", Medical Image Computing and Computer-Assisted Intervention – MICCAI, (2006).
4. E. Santos Filho, Y. Saijo, A. Tanaka, T. Yambe, S. Li and M. Yoshizawa, "*Automated Calcification Detection and Quantification in Intravascular Ultrasound Images by Adaptive Thresholding*", IFMBE Proceedings Vol.14/3, (2007).
5. E. Santos Filho,Y. Saijo,A. Tanaka, and M. Yoshizawa, "*Detection and Quantification of Calcifications in Intravascular Ultrasound Images by Automatic Thresholding*", Ultrasound in Medicine & Biology, Volume 34, Issue 1, (2008).
6. Q. Zhang, Y. Wang, W. Wang, J. Ma, J. Qian, and J. Ge, "*Automatic Segmentation of Calcifications in Intravascular Ultrasound Images Using Snakes and the Contourlet Transform*", Ultrasound in Medicine & Biology, (2009).
7. D.G. Vince, K.J. Dixon, R.M. Cothren, J.F. Cornhill, "*Comparison of texture analysis methods for the characterization of coronary plaques in intravascular ultrasound images*", Computerized Medical Imaging and Graphics, Volume 24, Issue 4, Pages 221-229, (2000).
8. C. Bourantas, I. Kourtis, M. Plissiti, D. Fotiadis, C. Katsouras, M. Papafaklis, L. Michalis, "*A method for 3D reconstruction of coronary arteries using biplane angiography and intravascular ultrasound images*", Computerized Medical Imaging and Graphics, (2005).
9. T., Mihran, and A. K. Jain, "*Texture Analysis -The Handbook of Pattern Recognition and Computer Vision*", (2nd Edition), Edited by C. H. Chen, L. F. Pau and P. S. Wang. World Scientific Publishing Co. , (1998).

10. T. Mäenpää, and M. Pietikäinen, "*Texture Analysis with Local Binary Patterns - Handbook of Pattern Recognition and Computer Vision*", 3rd. Edited by C. H. Chen and P. S. Wang. World Scientific, (2005).
11. C. G. Raphael, E. W. Richard, "*Digital Image Processing*", Addison-Wesley publishing company, (2002).
12. C. W. Hsu, C. C. Chang, and C. J. Lin. "*A Practical Guide to Support Vector Classification*", Department of Computer Science, National Taiwan University, (2009).
13. C. M. Bishop, "*Pattern Recognition and Machine Learning*", Springer, (2006).
14. B. Leo, "*Random Forests*" Machine Learning 45: 5-32, (2001).

THE IMPACT OF SCATTERING AND PEAK SPECTRUM OF I-123 IN SCINTIGRAPHY BY I-123 MIBG

M.E. LYRA, N. LAGOPATI AND M. SOTIROPOULOS

University of Athens, Dep. of Radiology, Radiation Physics Unit
Athens 11528, Greece
mlyra@med.uoa.gr

CH. CHATZIGIANNIS AND M. GAVRILELLI

Medical Imaging Center, Nuclear medicine section
Athens 11525, Greece
mlyra@medimaging.gr

Metaiodobenzylguanidine (MIBG) scintigraphy is used to image tumors of neuroendocrine origin and study disorders of sympathetic innervation of the myocardium. Scintigraphic imaging methods offer the possibility of qualitative and quantitative assessment of tracer concentration. Accurate correction for the physical degrading factors (attenuation, scatter, partial volume effects) is demanded. An analytical study of the I-123 energy spectrum, scattering and attenuation contribution to the resulting image of an I-123 MIBG scintigraphy, has been undergone, so as to improve the data that could be collected by I-123-MIBG use. In this work, quantitative data were gathered at various source depths, volumes and crystal to phantom distances to determine the effect of these variables on source activity data. The final image is being extracted from the combination of three different images, each and every one being acquired by a different energy window. The middle one has been obtained from the main I-123 photopeak and the others, from the left and right peaks respectively. The problem of Compton scattering as the dominant photon interaction phenomenon and its impact on both the quality of clinical images and the accuracy of quantitative analysis is taken into consideration for a scatter modeling in non-uniform media.

1. Introduction

The main objective of the X-ray imaging is the optimization of the diagnostic procedure for a better image quality/patient dose ratio. The most important factor affecting the image quality is the scattered component of the radiation resulting from the interactions of photons within the object being radiographed. It is therefore of great importance to investigate both theoretical and experimental scatter data such as scatter fractions, spectral and angular distributions of the scattered radiation under diagnostic imaging conditions.

The contribution of the scattered radiation to the image quality has been estimated by many investigators (Barnea and Dick, 1986 [1]; Dick et al., 1986; Chan and Doi, 1983) using the scatter fraction ***F*** defined as:

$$F = \frac{N_T - N_U}{N_T} \qquad (1)$$

where N_T is the total number of (scattered and unscattered) photons at the detector in place and N_U is the unscattered component of N_T. Theoretical investigation of the properties of scattered radiation have been attempted.

As Compton scattering plays a key role, we will recall some of its properties. The Compton Effect discovered in 1923 [2] had served to confirm the particle (photon) nature of radiation, as proposed by A. Einstein. Thus, energetic radiation under the form of X- or gamma-rays behaves like particles and scatter with electrically charged particles in matter [3], [4]. In biomedical domains, X- or gamma-photons scatter electrons in the biological media they traverse. More specifically, in classical (Thomson and Rayleigh) scattering, the scattered radiation has the same wavelength as the incident radiation and no energy is transferred [5]. In a Compton collision or an incoherent scattering event, energy transferred to an electron, which recoils from the collision. Since energy and momentum are conserved, we may analyze the process in detail using the fig.1.

Let *hv* and *p* represent the energy and the momentum of the incident photon, *hv′* and *p′* corresponding parameters for the scattered photon. The momentum, *p*, of a photon of energy *hv* is simply *hv/c*. Let the electron recede from the collision with energy *E*, momentum *q*, and velocity *v*. [6], [7]. Since the electron may have velocity comparable with *c* it is necessary to use the relativistic expressions for its kinetic energy and momentum thus:

$$hv - hv' = E = m_0 c^2 \left\{ \frac{1}{\sqrt{1 - v^2/c^2}} - 1 \right\} \qquad (2)$$

Since momentum is also conserved, the sum of the vectors *p′* and *q* must equal *p*; this means *p*, *q*, and *p′* form the sides of a triangle, so that

$$q^2 = p^2 + (p')^2 - 2pp'\cos\theta \qquad (3)$$

or we can write down expressions for the conservation of momentum in the forward direction

$$\frac{h\nu}{c} = \frac{h\nu'}{c}\cos\theta + \frac{m_0 v}{\sqrt{1 - v^2/c^2}} \times \cos\varphi \tag{4}$$

and in the direction at right angles to give

$$\frac{h\nu'}{c}\sin\theta = \frac{m_0 v}{\sqrt{1 - v^2/c^2}} \times \sin\varphi \tag{5}$$

By eliminating v and φ, the reader may show that:

$$E = h\nu \cdot \frac{a(1 - \cos\theta)}{1 + a(1 - \cos\theta)} \tag{6}$$

$$h\nu' = h\nu \cdot \frac{1}{1 + a(1 - \cos\theta)}, \tag{7}$$

where

$$a = \frac{h\nu}{m_0 c^2} = \frac{h\nu(MeV)}{0,511} \tag{8}$$

The parameter α is the ratio of the energy of the photon to the rest energy of the electron ($m_0 c^2$). I f the energy of the photon is impressed in MeV, then $m_0 c^2 = 0,511$ MeV [7], [8]. By adding the electron energy, E, to the energy of the scattered photon, $h\nu'$ one obtains the incident energy $h\nu$ in agreement with the first equation.

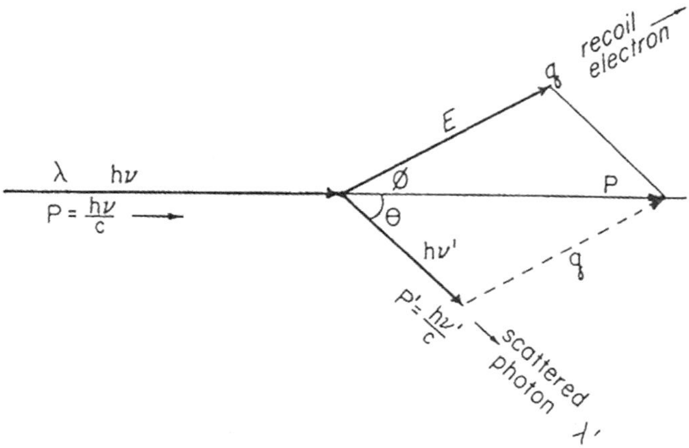

Figure1. Compton collision between a photon of energy *hv* and momentum *p=hv/c*, with a free electron. The recoil electron acquires momentum *q* and energy *E* [6].

In this work, a preliminary study in Metaiodobenzylguanidine (MIBG) scintigraphy which is used to image tumors of neuroendocrine origin [9] and study disorders of sympathetic innervation of the myocardium has undergone. Scintigraphic imaging methods, generally offer the possibility of qualitative and quantitative assessment of tracer concentration. Then, accurate correction for the physical degrading factors (attenuation, scatter, partial volume effects) is demanded. An analytical study of the I-123 energy spectrum, scattering and attenuation contribution to the resulting image of an I-123-MIBG scintigraphy, has been tried to increase and improve the data that could be collected by I-123 -MIBG use.

The endmost plan is to maximize sensitivity for a given crystal geometry, by choosing regions of interest at x -ray gamma and coincidence peaks. Measurements of source activity were found to be independent of the Source Volume [10]. Coincidence counting of measuring I-123 is independent of the counting geometry, the volume and the depth of the source. Corrections for scattering photons improve image quality.
Several sources of inaccuracy and inconsistency as radionuclide purity have been shown to affect the quality of imaging, the estimation of uptake measurements and the absorbed dose of the patient. Depths, position, volume of the organ are other factors that increase the scattering effect.

More than one energy range, in the spectrum, scattering, x-ray, gamma-peak, impurities' peaks and coincidence peak, were selected to be studied. Aim was to increase statistics of the data, improve the image of the organ, without increase of the absorbed dose in the patient.

2. Materials and methods

Phantom studies were performed to determine the feasibility of exploiting the scattering, x-ray and coincidence peaks as well as the low peaks of I-123 impurities. Quantitative data were gathered at various source depths, volumes and crystal to phantom distances to determine the effect of these variables on source activity data.

2.1 *Triple Energy Window method [TEW]*

In this method the final image is being extracted from the combination of three different images, each and every one being acquired by a different energy window. Planar images were acquired by GE tomographic gamma - camera.

The middle one has been obtained from the main I-123 photopeak and the others are from the left and right peaks respectively. Many different window widths have been proposed. The most useful configuration is a 20% central window with a 10 keV upper and lower sub windows, as proposed by ref. [11].

When the counts on lower, central and upper window are defined as *CL*, *CC*, *CU*, and the corresponding widths as *WL*, *WC* and *WU* respectively then the corrected image can be written as:

$$C_{corrected} = CC - (\frac{CL}{WL} + \frac{CU}{WU}) \cdot \frac{WC}{2} \quad (9)$$

As it is shown from the former formula, the counts of a trapezoid inside the photopeak are being estimated. Furthermore the counts of this trapezoid can estimate the Compton contamination of the photopeak (figure 2).

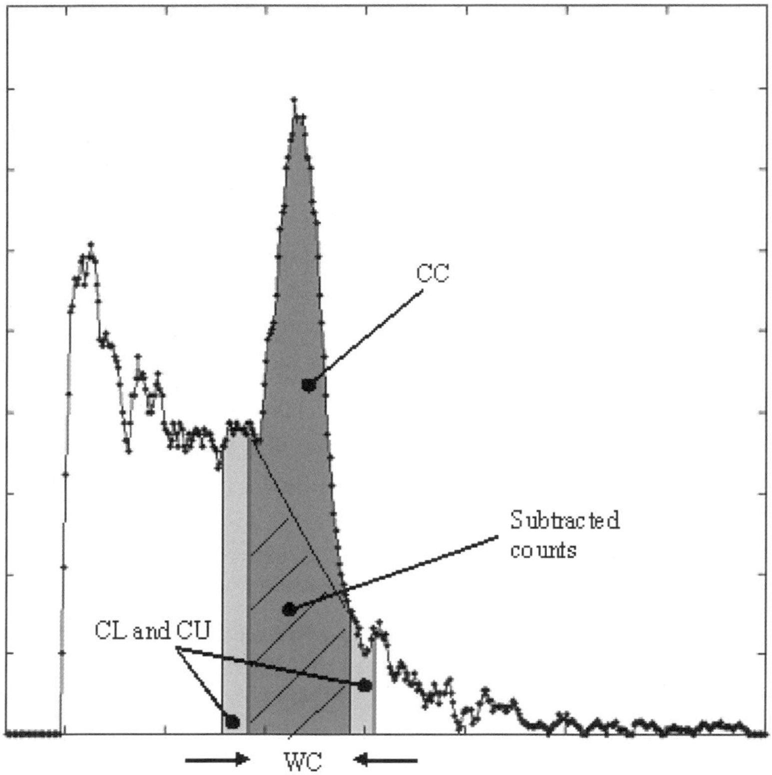

Figure2. This is a schematic representation of the I123 triple energy window [TEW] method.

2.2 *Spectra Simulation*

A simple Monte Carlo code was developed to simulate Compton scattering. The energy spectra can be assumed that consists of the main peak photons and the summation of the sequentially scattered photons [13], [14].

$$F(E_0, E) = W_0 \cdot P(E_0, E) + \sum W_i F_i(E_0, E) \quad (10)$$

where i determines the Compton scattering order and $P(E_0,E)$ is the energy spread function of the gamma camera at energy E_0, which, can approximated be expressed by the Gaussian form of

$$P(E_0, E) = \exp[\frac{-4 \ln 2 (E - E_0)^2}{R^2}]^2 \qquad (11)$$

Where, R is the energy resolution of the γ-camera.

Compton scattering was simulated by Klein-Nishina cross section formula [6], [14] and an analytical form for the incoherent scattering factor [15], [16], which is widely used in many Monte-Carlo codes. No photoelectric absorption was simulated, due to the high probability of the Compton effect.

In this work, the scattering part of the spectra has been simulated by the first six scatter spectra. The spectrum consists of 512 discrete energies, which makes satisfying a simulation of 100000 initial events.

Furthermore the best fit of this ideal spectrum was estimated, by optimizing of the Wi parameters of the equation (10) making possible the estimation of the scattered photons in the photopeak.

3. Results

3.1 Image analysis

Image processing at the scintigraphic images of a 5 years-old girl, were undergone using the TEW method. Contrasts of background and kidney as well as heart to kidney ratio (HKR) were calculated (figure 3 and 4).

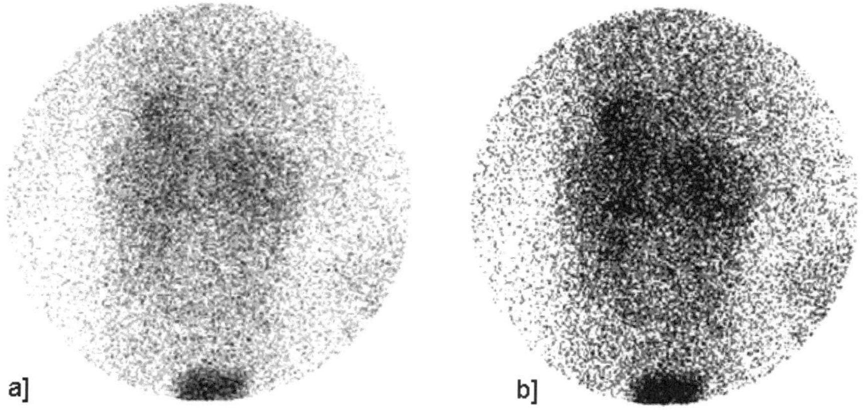

Figure3. Images of I-123-MIBG acquisitions of a patient a] before and b] after TEW method application. Heart to kidney ratio (HKR) calculation.

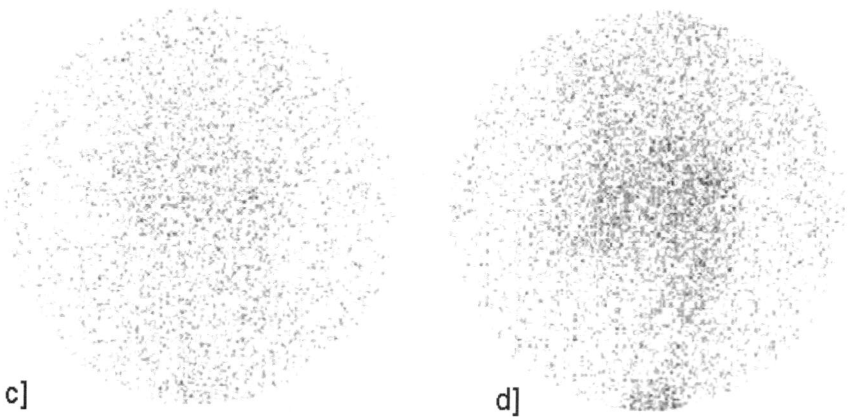

Figure4. Images of low c] and upper d] energy window acquisitions of a patient injected by I-123-MIBG

Contrast, defined as 100 %*(Bg-ROI)/Bg), is improved from 42% to 72% and HKR from 3.56 to 4.48.

3.2 *Quantitative analysis*

The ratio of the total counts of the spectrum to the scattered counts (TS) in the width of the used photopeak window has been calculated in order to quantify the Compton counts in the photopeak and the total effect of it, as well as. Background has been subtracted in both methods. No events from the 689 keV (1.4%) peak have been simulated, as mentioned above. TEW method estimates TS of 2.15 while the simulated data estimate TS of 7.69. Figure5 shows the calculated spectrum. (measured, simulated, photopeak, total scatter and scattering components)

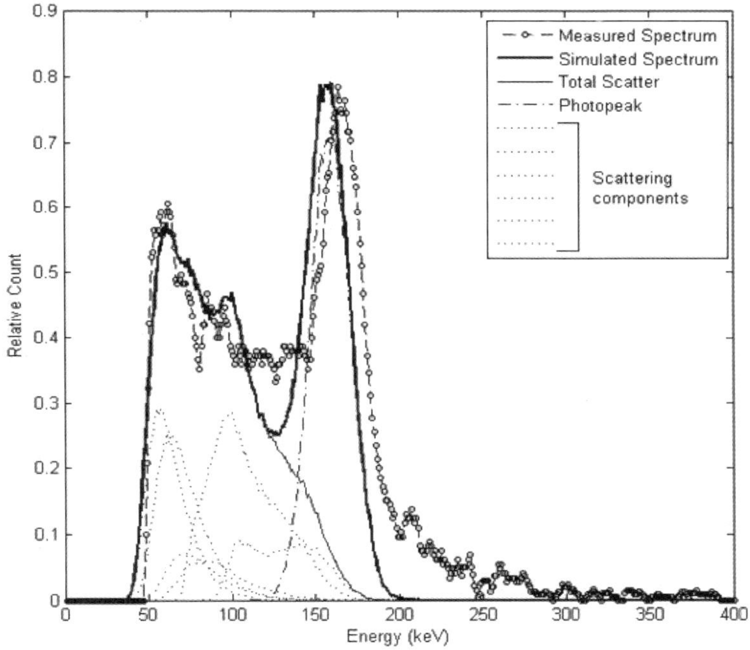

Figure5. The spectrum as has been calculated (measured and simulated)

4. Discussion

As several sources of inaccuracy and inconsistency as radionuclide purity have been shown to affect the quality of imaging, the estimation of uptake measurements and the absorbed dose of the patient, it is of crucial importance, for us to be able to quantify them, in order to eliminate the negative effect at the resulting image.

Accurate correction for the physical degrading factors (attenuation, scatter, partial volume effects) is demanded [17]. Depths, position, volume of the organ are other factors that increase the scattering effect. Thus, corrections for scattering photons improve image quality and are supposed to be obligatory.

By increasing statistics of the data, we could improve the image of the organ, without increase of the absorbed dose in the patient, which is the ideal outcome.

This was an analytical study of the I-123 energy spectrum, scattering and attenuation contribution to the resulting image of an I-123 MIBG scintigraphy, as an attempt to increase and improve the data that could be collected by I-123-MIBG use.

5. Conclusion

We obtained preliminary evidence that the quality of the image is, by far, better and more accurate, when we choose the specific energy windows, with the triple energy window method, which was analytically described in this work. The further analysis, which is going to give the final touch in the final image, is under investigation.

References

1. G. Barnea, and C.E. Dick, Med. *Phys*. **13**, 490, (1986).
2. A. H. Compton, *Phys. Rev*, vol 21, **5**, 483, (1923).
3. G. Beyklin, vol 37, **5**, 579, (1984).
4. S. R. Deans, J. *Math. Phys*, vol. 19, **11**, 2346, (1978).
5. H. H. Barrett, *Progress in Optics*, E. Wolf, Ed., **21**, 219, (1984).
6. H.E. Johns, and J.R.Cunningham, C6, 173
7. H.E. Johns, and J.R.Cunningham C7, 235
8. S. Helgason, *The randon transform,* pubs Birkhauser, Boston 158, (1999).
9. Lyra M, Phinou P, Papanikolos G et al, *WJNM,*vol.4, S1,33,(2005)
10. T. T. Truong, M. K. Nguyen, and H. Zaidi, *International J Biomed Imag*, **10**, 1155, (2007).
11. K. Nakajima, K. Matsubara et al, *J. Nucl. Card. **14***, 843 (2007).
12. E. Chen and C. Lam, *Comput. Biol. Med.* **24**, 229 (1994).
13. J.M. Boone, and J.A. Seibert, *Med. Phys*. **15**, 713, (1988a).
14. J. Baro, M. Roteta, J. Fernandez-Varea and F. Salvat, *Radiat. Phys.Chem.* **44**, 531 (1994).

15. G. Beylkin, *Math. Phys*, *International Press*, 4, **71**, (1994).
16. A. M. Cormack, *Math. Phys*, *International Press*, 4, **105**, (1994).
17. R. Van Holen, S.Staelens and S.Vandenberghe, *Med. Phys.,***36**, 9, (2009)

Predictive Medicine: CFD Techniques in severe Renal Artery Stenosis to facilitate Therapeutic Decision-Making

E. D. Skouras
Institute of Chemical Engineering and High Temperature Chemical Processes, Foundation for Research and Technology, GR-26504, Rion, Greece

G. C. Bourantas
Department of Medical Physics, School of Medicine, University of Patras, GR 26500, Rion, Greece

V. C. Loukopoulos
Department of Physics, School of Natural Sciences, University of Patras, GR 26500, Rion, Greece

E. N. Liatsikos
Department of Urology, School of Medicine, University of Patras, GR 26500, Rion, Greece

G. C. Nikiforidis
Department of Medical Physics, School of Medicine, University of Patras, GR 26500, Rion, Greece

Abstract: We investigated fluid responses to pulsatile Newtonian blood flow through a healthy and a severely stenosed renal artery (RAS). Anatomically realistic models of the aorta and the renal arteries of two New Zealand rabbits were reconstructed from CT scans and used to conduct blood flow simulations with CFD. Earlier studies were concentrated on the effects of the stenosis shape and severity and the effects of transient conditions on the flows. Blood is assumed to be laminar for shear rates greater than $100 s^{-1}$. This situation only occurs for blood flow in large arteries, such as the aorta, while in smaller arteries the blood flow is usually turbulent and often exhibits multiphasic characteristics. For the determination of the most suitable flow regime, laminar and turbulent flow models for single phase and multiphase fluid were tested and the numerical results were validated with experimental data. Blood flow was studied *in vivo* under both in absence (post-treatment) and in presence (pre-treatment) of RAS, and even in simulated cases before the development of RAS. The gold standard method for the assessment of RAS is intra-arterial Digital Subtraction Angiography (DSA), although modern multi-detector CT systems (MDCT) are gradually being used instead, due to their non-invasive nature. The only available functional indicator for hemodynamically significant RAS is the Renal Artery Differential Pressure (RADP), measured with the aid of a catheter. The pressure gradients in the RAS case were at the same levels as in the single-phase experiments, and many orders of magnitude larger than in the healthy case. The presence of RAS increased flow resistance, which led to considerably lower blood flow rates. The wall shear stresses, the differential pressure profiles, and the net forces exerted on the surface of the atherosclerotic plaque at peak pulse were shown to be of relevant high distinctiveness, so as to be considered potential indicators of hemodynamically significant RAS.

Keywords: Multiphase, blood flow, Newtonian, Computational fluid dynamics, Hemodynamics, Renal artery, renal artery stenosis, Computed tomography; Segmentation; Simulated reconstruction; Turbulent flow

1. Introduction

Renal artery stenosis (RAS) afflicts patients with atherosclerotic disease and leads to renal malperfusion, renovascular hypertension, and ischemic nephropathy [1-3]. RAS is typically caused by ostial plaque formation with an eventual occlusion rate up to 18% over 5 years [4] and is currently considered to be an independent predictor of death regardless of the extent and severity of systemic atherosclerosis [5]. The gold standard method for the assessment of RAS is intra-arterial Digital Subtraction Angiography (DSA), although modern multi-detector CT systems (MDCT) are gradually being used instead, due to their non invasive nature. The only available functional indicator for hemodynamically significant RAS (severe stricture or a moderate one with a pressure gradient of at least $5-10\%$) is the renal artery differential pressure (RADP), which is measured with the use of a catheter [2, 3, 6]. However, it is not widely used because it involves catheterization of the target vessel and its correlation with the outcome of the elected therapy is disputed [7, 8].

An important issue regarding to RAS is the percentage of occlusion and its impact on the blood flow conditions. The development of atherosclerosis in humans begins by the adhesion of monocytes to the endothelium [9, 10, 11]. The trigger for the response is not yet known in spite of decades of study [12]. Although, the hemodynamic stresses and forces play important roles in the initial stages of cardiovascular disease [13]. The biomechanical environment in the vascular system almost certainly plays a key role in arterial disease, including atherogenesis [14]. Shear stress can activate platelets and the biomechanical factors are important in monocyte rolling and adhesion. One of the earliest reviews of atheroma [15], claims its association with arterial blood mechanics. There are several hypotheses as to the cause of and progression of atherosclerosis. These include (1) high hemodynamic wall shear stress (WSS), (2) low WSS, (3) high arterial wall stress, (4) high WSS gradients (WSSG), and (5) damage to the endothelium.

In [12] authors concluded that the lack of progress in the last three decades is because most studies, until recently, have been retrospective using idealized or averaged hemodynamic models rather than being prospective using local measured hemodynamic indices and correlated in a noninvasive manner using CFD models. In [14] it was stressed that the specific causative link between biomechanical factors and arterial pathogenesis remains to be identified in part because of the substantial complexity of the highly unsteady, three-dimensional (3-D) biomechanical arterial environment. The confirmation of various hypotheses is lacking and the exact cause eludes investigators. To our opinion, the major reason for this situation is the use of single-phase CFD in the analyses and, the absence of the application of multiphase theory. Additionally, in terms of the blood flow conditions present in RAS cases, several local flow phenomena affect the RADP behavior, as well as other hemodynamic parameters. Turbulence consists of fluctuations in the flow field in time and space. It is a complex process, mainly because it is three dimensional, unsteady and consists of many scales. Turbulence occurs when the inertia forces in the fluid become significant compared to viscous forces, and is characterized by a high Reynolds number. Turbulence may be observed along vessel stenoses, depending on the local geometry and physical conditions, and influences blood flow. It can have a significant effect on the characteristics of the flow; flow resistance, shear stress, and mass transport on the vessel wall, tensile stress on the endothelial cell membrane, hemostasis, and cell proliferation and migration, as well [16, 17]. Some groups have investigated the relationship between wall shear stress (WSS) localization and development of atherosclerotic plaque by applying laminar flow CFD models [18, 19]. Numerical studies on steady and pulsatile flow through stenotic vessels raise several turbulence issues, including fluid-structure interaction and rheology effects [20–25]. In experimental studies of pulsatile flow through constricted tubes [26], vortical and helical structures were observed within mild stenoses. Transition to turbulence was associated with the breakdown of waves, and streamwise vortices occurred at these stenoses. Hemodynamic parameters, such as pressure, velocity, and shear stress are spatially and temporally dependent. Plaque encroachment into the lumen induces shear stress variations along the vessel wall, affecting both disease progression and renal perfusion [27, 28].

This study presents steady simulations of blood flow in the aorta and in the right and left renal arteries for single phase and multiphase blood flow considerations, while transient single phase simulations have been presented elsewhere [29]. The idea behind this paper is, by comparing the numerical data with the experimental *in vivo* measurements at rabbits to nominate the impact of multiphase models of the blood. Additionally, the effects of different blood models on the wall shear stress distributions in the aorta and the renal arteries during the cardiac cycle are also investigated. This comparison will allow us to assess whether or not it is necessary to include single or multi-phase models when investigating blood flow in large vessels, such as aorta and renal arteries. This signal study is a significant first step in the systematic analysis of multiphase cardiovascular hemodynamics.

2. Methods

Tremendous developments in computer hardware and software during the last decade have made it more feasible to create a user-oriented computational system to reveal the individual patterns of hemodynamic flow in patients with vascular diseases. These systems aim to display individual physiologic variations in a practical and time-efficient way and provide a bridge between medical professionals and flow dynamicists. We sought to determine how 3D imaging data collected in DICOM format can be used to create numerical grids for computer simulations of blood flow in a relatively simple way. The steps of a patient-specific computational grid reconstruction and blood flow numerical simulation were presented and, the main tools of the procedure and the respective output formats. The whole process consists of five logical steps (I–V). In step I, primary medical images in DICOM format are obtained. In step II, they are transformed into a format suitable for the software that will be extracting the luminal surface of the blood vessels of interest in step III. In step IV, the surface data are then imported by a grid generator, which creates a volume grid. The resulting grid model is used at step V for the numerical simulation of blood flow.

The reconstruction of the aorta and the renal arteries model was obtained from a stack of slices that were acquired in vivo using CT spiral imaging scanning (Department of Radiology, University Hospital of Rion-Patras, Greece). A patient with severe RAS underwent contrast-enhanced spiral CT (Siemens, Somatom Plus 4A). The examination protocol was set to 120 kVp, 180 mAs, with concurrent administration of 140 ml non-ionic iodinated contrast agent at 3.5 ml/s (time delay was set to 16 s). Scanning coverage was 6 cm with 2 mm/s table feed (total examination time was 30 s). A DICOM (Digital Imaging and Communications in Medicine) format file was created. A total of 370 from the tip of the aortic arch to the iliac arteries were generated. A field of view of 422×422 mm^2 and a 512×512 matrix resulted in a pixel size of 0.8242 mm. Independently from the image acquisition techniques employed, the stack of the 2D images are further processed. Reconstruction was performed with a soft-reconstruction kernel at 1mm intervals. Thresholding and seeded region growing techniques were subsequently applied in order to segment the renal arteries and the aorta. The triangulated surface was then exported in a format that was 'understandable' by the CFD software (STL format). Thus, the anatomical structure examined, possess three-dimensional geometrical and topological characteristics.

3. Computational Fluid Dynamics

3.1. *Multiphase hemodynamic model*

The Navier-Stokes equations (momentum and mass conservation) governing the fluid motion were solved utilizing the CFD commercial code ANSYS CFX. A multiphase non-Newtonian 3-D flow model for describing the hemodynamics in vascular vessels is developed. It uses the principles of mass, momentum, and energy conservation for each phase based on the generalization of the well-known Navier–Stokes equations [30, 31]. The principle difference compared with a single-phase model is the appearance of the volume fraction for each phase, as well as mechanisms for the exchange of mass, momentum, and energy between the phases.

In the two-phase Newtonian hemodynamic model considered, the continuous phase is plasma, which is a Newtonian fluid [32]. The predominant particulate phase suspended in the plasma comprises the red blood cells (RBCs) having a hematocrit, H, or volume fraction, ε_{RBC}, in the normal range of $35-50\%$. Platelets and white blood cells constitute an aggregate volume fraction less than 1%. In this paper, RBCs are modeled more realistically since they control the blood rheology. This is important when modeling monocyte adhesion.

The continuity equation for each phase ($k =$ plasma, RBCs) is given by

$$\frac{\partial(\rho_k \varepsilon_k)}{\partial t} + \nabla \cdot (\rho_k \varepsilon_k \boldsymbol{u}_k) = 0 \tag{1}$$

where ρ is density, ε is volume fraction, t is time, and $\boldsymbol{u}$ is velocity. The sum of the volume fractions for each phase must sum to one. The momentum equation for each phase ($k =$ plasma, RBCs) is given by

$$\frac{\partial(\rho_k \varepsilon_k \boldsymbol{u}_k)}{\partial t} + \nabla \cdot (\rho_k \varepsilon_k \boldsymbol{u}_k \boldsymbol{u}_k) = \rho_k \varepsilon_k \boldsymbol{g} - \varepsilon_k \nabla p + \nabla \cdot \varepsilon_k \overset{=}{\tau}_k + \sum_{l \neq k} \beta_{kl} (\boldsymbol{u}_l - \boldsymbol{u}_k) + \boldsymbol{F}_k \tag{2}$$

The terms of the RHS of Eq.(2) represent gravity $\boldsymbol{g}$, shear stress $\overset{=}{\tau}_k$, pressure p, drag force between the carrier fluid and particulates, and external forces $\boldsymbol{F}$, such as virtual mass, rotational and shear lift, electric, and magnetic. In the drag force, k and l represent plasma or RBCs, and β_{kl} are the interphase momentum exchange coefficients.

Single and two-phase Newtonian models were considered. Blood, was considered as the first phase having viscosity of 4 *cP*. The rest of the fluid volume considered as thin and thick blood phases. The thin phase (plasma) located near the vessels walls as a film-like orientation (of height *ca* 20% relative to the radius of the inlet face of the vessel), having low velocity and approximate unity viscosity. On the other hand, the thick phase (RBCs) was considered as the predominant particulate phase suspended in plasma, having a hematocrit of 45%. The velocity of the RBCs was considered higher than that of the plasma, flowing at the center of the vessel.

4. Numerical implementation

We used the kernel of the commercial CFD code ANSYS CFX, which contains the multiphase 3-D Eulerian–Eulerian model. The numerical solution method uses a finite volume, unstructured mesh, staggered grid arrangement. The scalar variables are located at the cell centers and the vector variables are located at the cell boundaries. The momentum equations are solved using a staggered mesh, while the continuity equations are solved using a donor cell method (Fig. 1).

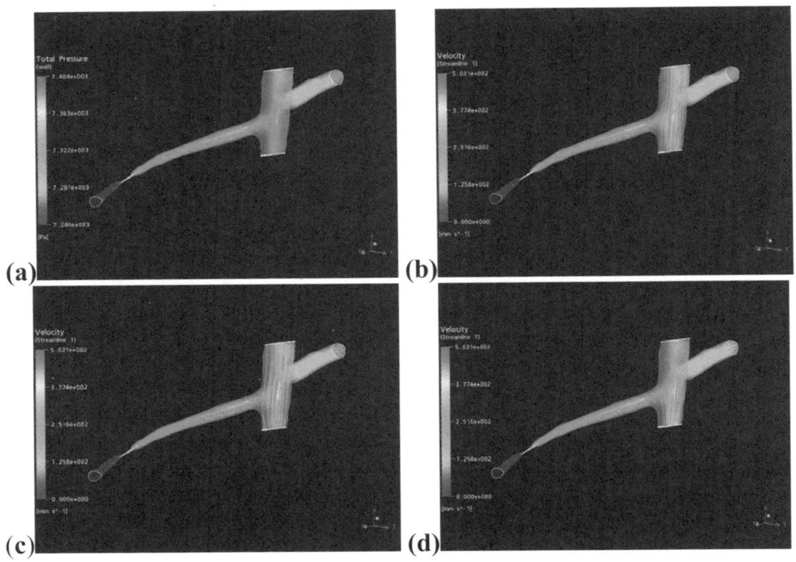

Fig. 1. (**a**) Wall pressure distribution, (**b**) Streamlines starting from inlet, (**c**) Streamlines ending at outlet, (**d**) Streamlines starting from outlet with RAS.

All simulations for the idealized rigid wall renal artery and aorta were carried out using the 3-D computational domain. The problem of accurately reconstructing 3D models of vascular segments from volumetric angiographic images was solved. The discretization of the domain volume into small elements is necessary. Thus, the STL segmented data, obtained from the imaging software, were fully imported into the advanced mesher, where all the segments of irrelevant surface portions were cut-off, leaving only the ones representing the aorta and the renal arterial segments. Additional inlet and outlet surfaces were properly mounted at each opening of the arterial segment, with normal surface along the main flow direction. These inlet and outlet mouths were carefully attached to the respective wall segments and tightly tailored to the adjacent wall STL surfaces, in order to have consistent representation of the blood flow. Meshing was achieved by using the ANSYS ICEM software.

A predefined inlet average pressure was used as inlet and outlet boundary conditions for both the RBCs and plasma, based upon the waveform measured with a thin catheter at the prescribed locations in the aorta and the renal ostium. Zero-slip velocity boundary condition was employed for both RBCs and plasma. No mass transfer was considered between phases.

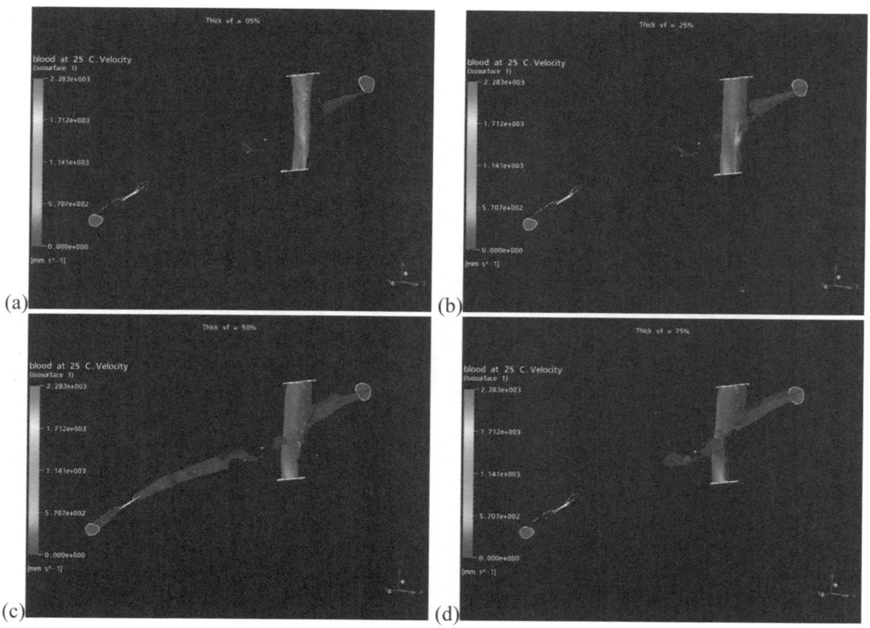

Fig. 2. Isosurfaces of constant volume fraction **(a)** 5% **(b)** 25% **(c)** 50% **(d)** 75%.

5. Results and Discussion

The distribution of the two phases are shown in Fig. 2 . The pressure gradients in the RAS case were at the same levels as in the single-phase experiments [29], and many orders of magnitude larger than in healthy cases. The presence of RAS increased flow resistance, which led to considerably lower blood flow rates. The wall shear stresses, the differential pressure profiles, and the net forces exerted on the surface of the atherosclerotic plaque at peak pulse were shown to be of relevant high distinctiveness, so as to be considered potential indicators of hemodynamically significant RAS.

6. Acknowledgments

Work was supported by the program "K. Karatheodoris 2008" - Funding of research group in University of Patras with P.C. C.596.

7. References

1. Harding M.B., Smith L.R., Himmelstein S.I., et al, *J Am Soc Nephrol* 2, 1608 (1992).
2. Rundback J.H., Sacks D., Kent K.C., et al, *Circulation* 106, 1572 (2002).
3. Rundback J.H., Sacks D., Kent K.C., et al, *J Vasc Interv Radiol* 14, S477 (2003).
4. Zeller T., *J Interv Cardiol* 18, 497 (2005).
5. White C.J., *Circulation* 113, 1464(2006).
6. Yim P.J., Cebral J.R., Weaver A, et al, *Magn Reson Med* 51, 969 (2004)
7. Martin L.G., Rundback J.H., Sacks D., et al, *J Vasc Interv Radiol* 14, S297 (2003).
8. Rees C.R., *J Vasc Interv Radiol* 10, 689 (1999).
9. Fung Y.C., *Biomechanics:Circulation*. Springer, New York (1997).
10. Nichols W.W., O'Rourke M.F., McDonald's Blood Flow in Arteries. Arnold, London (1997).
11. Taylor C.A., Draney M.T., *Annual Rev Fluid Mech* 36, 197 (2004).
12. Steinman D.A., *Annals Biomed Eng* 30, 482 (2002).
13. Malek A.M., Alper S.L., Izumo S., *J American Med Assoc* 282, 2035 (1999).
14. Steinman D.A., Vorp D.A., Ethier C.B., *J Vasc Surgery* 37, 1118 (2003).
15. Caro C.G., Fitz-Gerald J.M., Schroter R.C., *Nature* 223, 1159 (1969).
16. Varghese S.S., Frankel S.H., *J Biomech Eng* 125, 445 (2003).
17. Younis B.A., Berger S.A., *J Biomech Eng-Trans ASME* 126, 578 (2004).
18. Wentzel J.J., Gijsen F.J., Stergiopulos N., et al, *J Biomech* 36, 681 (2003).
19. Wentzel J.J., Janssen E., Vos J., et al, *Circulation* 108, 17–23 (2003).
20. Bathe M., Kamm R.D., *J Biomech Eng* 121, 361–9 (1999).
21. Buchanan J.R., Kleinstreuer C., Comer J.K. *Comput Fluids* 29, 695 (2000).
22. Qiu Y., Tarbell JM. *J Vasc Res* 37, 147 (2000).
23. Tang D., Yang C., Huang Y., Ku D.N. *Comput Struct* 72, 341 (1999).
24. Tang D., Yang C., Ku D.N. *Comput Struct* 72, 357 (1999).
25. Tang D., Yang J., Yang C., Ku D.N. *J Biomech Eng* 121, 494 (1999).
26. Ojha M., Cobbold R.S.C., Johnston K.W., Hummel R.L., *J Fluid Mech* 203, 173 (1989).
27. Wentzel J.J., Kloet J, Andhyiswara I., et al, *Circulation* 104, 91 (2001).
28. Wentzel J.J., Gijsen F.J., Schuurbiers J.C., et al, *J Biomech* 38, 1551 (2005).
29. Kagadis G.C., Skouras E.D., Bourantas G.C., et al, Med Eng Phys 30, 647 (2008).
30. Anderson, T.B., Jackson, R., *I & EC Fundamentals* 6, 524 (1967).
31. Gidaspow D., *Multiphase Flow and Fluidization: Continuum and Kinetic Theory Descriptions*, Academic, NY (1994).
32. Lightfoot, E.N., *Transport Phenomena and Living Systems: Biomedical Aspects of Momentum and Mass Transport*, Wiley, New York (1974).

POINT-OF-CARE MONITORING AND DIAGNOSTICS FOR RHEUMATOID ARTHRITIS AND MULTIPLE SCLEROSIS

F.G. KALATZIS, T.P. EXARCHOS, N. GIANNAKEAS,
P. RIZOS, D.I. FOTIADIS
Unit of Medical Technology and Intelligent Information Systems, Dept. of Material Sciences & Engineering,
University of Ioannina, GR 45110, Ioannina, Greece

S. MARKOULA, E. HATZI, I. GEORGIOU
Genetics Unit, Dept. of Obstetrics and Gynaecology, Medical School,
University of Ioannina, GR 45110, Ioannina, Greece

In this work the POCEMON diagnostic platform is presented. The platform aims to providing early prognosis and diagnosis of rheumatoid arthritis (RA) and multiple sclerosis (MS) autoimmune diseases at the point of care. The objective of the POCEMON platform is the development of a diagnostic lab-on-chip device based on genomic microarrays of HLA-typing. The POCEMON will advance and promote the primary health care across Europe by supporting: a) point-of-care diagnostics, b) monitoring of immune system status and c) management of the chronic MS and RA autoimmune diseases. The platform combines high-end computational tools based on microfluidics, microelectronics, microarrays and intelligent diagnosis algorithms. Clinical and genomic data are collected from patients with the RA and MS diseases. The clinical data and the selected associated SNPs are modelled using data mining techniques to allow the knowledge modeling framework to provide the diagnosis for new patients performing the point-of-care test. The microfluidic LOC device supplies the diagnostic component of the platform with a set of SNPs associated with the diseases. The knowledge-based decision support system combines this genomic information with the clinical data of the patient to obtain the diagnostic outcome.

1. Introduction

Personalized diagnosis and treatment is driven by the rapid development of information and communication technologies and the advances on both hardware and software development. Those efforts promote new diagnostic procedures based on Lab-on-a-chip (LOC) technologies [1]. The LOC technologies are capable to perform a wide range of proteomic and genomic tests by using a sample of blood or other body fluid such as saliva. These tests aim to facilitating healthcare at preferred environments and point of care disease diagnosis at primary healthcare level. The

LOC technologies expose a tremendous potential to improve the health of people worldwide. The use of LOC and microfluidic technologies in remote settings has been perceived as potentially one of the most powerful applications of the technology by taking advantage of its small size, low volume requirement for samples, and rapid analysis. Indeed, portable LOC devices are now beginning to be used in remote settings as a result of developments in integrating fluid actuation, sample pre-treatment, sample separation, signal amplification and signal detection into a single device. These advances place the field of LOC research in a prime position to tackle the profound issue of global health where the challenges in device design are arguably the most demanding and the need for new health technologies the greatest.

Autoimmune disorders [2] develop when the immune system destroys normal body tissues. This is caused by a hypersensitivity reaction similar to allergies, where the immune system reacts to a substance that normally would ignore. The disorder may affect only one organ or tissue type or may affect multiple organs and tissues. Organs and tissues commonly affected by autoimmune disorders include blood components such as red blood cells, blood vessels, connective tissues, endocrine glands such as the thyroid or pancreas, muscles, joints, and skin.

There are many diseases which are believed to be autoimmune. Those include Multiple Sclerosis (MS) [3], Rheumatoid Arthritis (RA) [4], Addison's Disease, Ankylosing Spondylitis, Autoimmune Haemolytic Anaemia, etc.

The HLA (human leukocyte antigens) are proteins found in the membranes (outer coating) of nearly every cell in the body (all cells that have a nucleus). HLA antigens are the major determinants used by the body's immune system for the recognition and differentiation of self from non-self (foreign substances). There are many different major histocompatibility (HLA) proteins and individuals possess only a small, relatively unique set that is inherited by their parents. It is unlikely that 2 unrelated people will have the same HLA make-up. Many HLA molecules exist but some are of special interest because they are more common in certain autoimmune diseases. For example, HLA-B27 antigen is found in 80-90% of people with ankylosing spondylitis and Reiter's syndrome and can aid in the diagnosis of these diseases. HLA-B27 is also present in 5-7% of people without autoimmune disease. Thus, the mere presence of this HLA molecule is not indicative of a disease. HLAs detectable in blood provide clues to immune conditions.

There have been some reports on HLA genotyping by microarray [5,6]. Most of these studies focused on tissue transplantation and association between HLA genotypes and disease or pathological process [7,8]. A genotyping microarray approach can analyze polymorphisms at multiple-sites in a single gene or in multiple genes rapidly and in a parallel way [9]. The genomic HLA microarray scanning data can also be computerized and shared with a Personal Digital Assistant (PDA) device which will be hosted in the diagnostic POCEMON platform. Although, the feasibility of microcantilever based DNA biosensors has been demonstrated [10,11], as well as, the integration of a electronic interface realized using CMOS technology [12,13], a complete system including the detection module and the microfluidic structures for the sample handling is still not available in the market for specific applications in the genomic field.

In this paper, we describe the methodology which is applied to the huge amount of data produced by the gene discovery phase (the Whole Genome Association Study) of the POCEMON platform to select the most informative SNPs which are directly related with the RA & MS autoimmune disorders. Finally, the lab-on-chip technology which is used to develop the innovative diagnostic LOC platform based on HLA-typing microarrays is demonstrated.

2. The POCEMON diagnostic system

2.1. *Methodology*

A major part of the POCEMON platform is the development of specific HLA-typing microarrays capable to diagnose the MS & RA autoimmune disorders. To complete this task the following steps are included:

a) Genome-wide association study: It includes genotyping of 3000 controls and cases DNA samples both from MS & RA patients and production of a huge amount of raw data. This process is based on the iLLumina technology [14].
b) Selection of the most informative SNPs using an approach based on data mining methods and genetic algorithms.
c) Development of a diagnostic lab-on-chip device based on genomic microarrays of HLA-typing to allow the early diagnosis of diseases such as multiple sclerosis (MS) and rheumatoid arthritis (RA).

Genome Wide Association Study: It is estimated that more than 10 million SNPs exist in the human genome. SNPs are known to contribute to population diversity and phenotypic differences between individuals, and cause predispositions to diseases. Whole-genome association studies hold the promise of identifying SNPs associated with a certain phenotype of interest as well as those that can serve as diagnostic markers. Genes in the vicinity of SNPs that appear to cause the phenotype could be qualified as new drug targets for pharmaceutical companies. To successfully identify candidate SNPs using whole genome association analysis, the

researchers must consider sample size, multiple testing correction, SNP selection and genotyping quality.
A genome-wide association study (GWA study) - also known as whole genome association study (WGA study) - is an examination of genetic variation across the human genome, designed to identify genetic associations with observable traits, such as blood pressure or weight, or why some people get a disease or condition.
These studies require two groups of participants: people with the disease (cases) and sex-/age-matched unaffected individuals (controls). After obtaining samples from an individual, the set of markers such as SNPs are scanned nd searched for markers of genetic variation.
If genetic variations are more frequent in people with the disease, the variations are said to be "associated" with the disease. The associated genetic variations are then considered pointers to the region of the human genome where the disease-causing problem resides. Since the entire genome is analyzed for the genetic associations of a particular disease, this technique allows the genetics of a disease to be investigated in a non-hypothesis-driven manner.
A genome-wide association study is an approach that involves rapidly scanning markers across the complete sets of DNA, or genomes, of many people to find genetic variations associated with a particular disease. Once new genetic associations are identified, researchers can use the information to develop better strategies to detect, treat and prevent the disease. Such studies are particularly useful in finding genetic variations that contribute to common, complex diseases, such as asthma, cancer, diabetes, heart disease and mental illnesses.
The Illumina Infinium II technology allows the fast and reliable genotyping of up to 1.000.000 SNPs per individual at once. Whole Genome Genotyping (WGG) is based on the Sentrix BeadChip platform [15] on which 13 million DNA-immobilized Beads are randomly dispersed and assembled into wells created on a single slide. The location and identification of each bead is obtained through a decoding process. On average a redundancy of around 20 beads per bead type is obtained. The first step of Infinium WGG is a Whole Genome Amplification of genomic DNA (WGA). WGA is based on Multiple Displacement Amplification (MDA) which uses the φ29 DNA polymerase and random primers to amplify the entire genome generating hundreds of micrograms of amplified DNA starting from an initial input of several hundred nanograms (currently, 750 ng). WGA gives a uniform locus representation that enables the access to almost all the SNPs in the genome. The amplified DNA is then fragmented to an average size of around 300 bp using an enzymatic fragmentation protocol and locus-specifically hybridized to each individual target on the beads. The specificity of the hybridization is assured by the use of long full-length 50-mer oligonucleotides. After hybridization, each SNP is characterized

through a single-base enzymatic extension assay using two-color labeled nucleotides. After the extension, the labels are visualized by staining with a sandwich-based immunohistochemistry assay that increases the overall sensitivity of the assay. BeadChips are imaged using a two-color confocal laser system with 0.8μm resolution. The Bead intensities are extracted and genotypes are calculated with the BeadStudio 3.2.29 software using a "cluster" file, based on a set of reference samples supplied by Illumina.

Selection of the most informative SNPs: Genetic profiles of the MS & RA individuals are constructed using the data produced by the gene discovery phase using the iLLumina technology. The analysis of the produced huge amount of data, leads to genes/SNP patterns which are responsible for the RA & MS diseases as well as genetic risks. In order to handle such a dataset there is a need to select the most informative genes/SNPs for further analysis.
While the target of POCEMON aims to the early prognosis and diagnosis of the RA & MS autoimmune diseases the selection of the most informative SNPs is necessary. The approach which is used in the POCEMON platform is based on data mining and genetic algorithms. Artificial intelligent methods such as global search and weighted decision trees are employed for the selection of the most significant SNPs. The SNPs selection approach is applied to datasets to provide the early prognosis and diagnosis of the RA & MS autoimmune disorders. It is expected that the number of the features is significantly reduced while the quality of knowledge is enhanced. The selection approach provides the most significant SNPs. Data mining methods support all the processes of discovering interesting and previously unknown patterns in the genotyped MS & RA datasets. The main advantage of the data mining is related to the study of an individual rather than the population, providing routes for personalization in early prognosis and diagnosis. The goal of the selection is to identify the minimum set of non-redundant features (e.g., SNPs, genes) which are useful in the classification phase.

Lab-On-Chip implementation and integration: The integration of LOC technology with PDA - using microelectronics - will provide new diagnostic tools at the primary care level. The implementation and the development of the portable POCEMON diagnostic LOC is achieved by following a modular approach of the various sub-parts of the integrated platform. The different components are developed according to the following steps:

- Identification of the major relationship between HLA alleles and the specific multiple sclerosis and rheumatoid arthritis autoimmune diseases.

- Design of the HLA microarray probes based on the technology of oligonucleotide microarrays for autoimmune diseases.
- Implementation of the HLA genotyping microarrays on appropriate materials.
- Design of the microelectronics LOC with mechanisms to handle blood and/or saliva samples.
- Incorporation of LOC techniques to automatically analyze oligonucleotide microarrays through hybridization inside the diagnostic platform using microelectronics.
- Employment of portable and mobile technology in the diagnostic LOC using PDAs.
- Implementation of appropriate desktop PDA software for signal intensity analysis of the microarray.
- Extraction of HLA probe's results in the PDA screen. In this phase the LOC platform provides a preliminary diagnosis for autoimmune diseases.
- Integration of all of the above in the final diagnostic product, an autonomous portable diagnostic LOC platform.

Development of the Diagnostic Lab-On-Chip: Starting from the state-of-the-art a suitable configuration for the LOC is studied in order to provide the handling of DNA samples, PCR amplification and HLA typing with a labelfree approach based on a microcantilever array of appropriate size functionalized with oligonucleotide probes. The microfluidics include sample ports, channels for sample delivering and reservoirs with temperature control for PCR amplification. The selected technology for the realisation of the LOC fluidics involves bulk micromachining techniques (wet and dry etching steps) and wafer-to-wafer bonding to create channels and reservoirs, deposition of thin film metallization to implement heaters and thermometers for the thermal control of the PCR reactor. The label-free detector is realised with a microcantilever array working in the bending mode (i.e. in the stress-detection mode) which is more suitable for operating in the liquid phase. The beam surface is provided with a gold electrode to address and bond the specific DNA probes based on thiol chemistry. The a) signal conditioning electronics (including a switching matrix), b) signal amplification and c) ADC are all integrated on-chip with a CMOS technology. The technologies involved deal with both ad hoc Silicon-On-Insulator (SOI) approaches for the realisation of thin suspended Silicon beams with tight thickness control and the implementation of thin film cantilever compatible with CMOS processes. In the framework of this system implementation, the design of single system modules (microfluidic module for sample handling and PCR amplification, detector module and signal conditioning electronic module) is studied as well as a technological approach suitable for the realisation of a single module and system integration. First, each module is developed as stand-alone device for validation of the selected technological processes and preliminary efficiency testing. Secomd, a definitive design is realized to implement the final modules in a monolithic chip. The signal conditioning interface and device specifications is addressed in the development on a PDA platform of the software tools needed to control the LOC operation and data acquisition and analysis. Summarizing, the approach aimed to develop a diagnostic HLA Lab-on-Chip is based on:

- The development of the microfluidics components for DNA sample handling, PCR amplification and reservoirs.
- The development of detector modules based on cantilevers array for preliminary studies of the sensitivity to specific DNA chains.
- The development of optimized detector modules based on a high density cantilevers array on CMOS technology. Related to the ultra-high expected sensitivity of the sensors, the PCR amplification module could be substituted with a less sophisticated microfluidics pre-concentrator.
- The development of an integrated control unit (electronic circuit) for sensor read out and signals conditioning.

3. Conclusions

The integration of LOC technology with genomic microarrays of HLA-typing provides new diagnostic tools at the primary care level. Also, the most informative selected SNPs are used to design and produce new microarray chips containing the appropriate probes to diagnose the RA & MS autoimmune diseases. The application of computational algorithms to datasets produced by other similar genome wide association studies provides the scientific and research community with innovative tools for the early diagnosis of some autoimmune disorders.

References

1. P. Yager, T. Edwards, E. Fu, K. Helton, K. Nelson, M.R. Tam and B.H. Weigl, Nature, **442**, 412, (2006).
2. P. Parham, New York: Garland Science, **344**, (2005).
3. G. Rosati, Neurol. Sci., **22**,, 117, (2001).
4. V. Majithia and S.A. Geraci, Am. J. Med. **120**, 936, (2007).
5. F. Zhang, S.W. Hu, J. Huang, H. Wang, W. Zhang, Y.Y. Geng, et al., Tissue Antigens, **65**, 467, (2005).
6. C. Consolandi, E. Busti, C. Pera, L. Delfino, G.B. Ferrara, R. Bordoni, B. Castiglioni, L. Rossi Bernardi, C. Battaglia and G. De Bellis, Hum Immunol. **64**, 168, (2003).
7. N. Dhiman, R.G. Bonilla, R.M. Jacobson, D. O'Kane and A. Poland, Scandinavian Journal of Infectious Diseases, **35**, 332, (2003).

8. L.M. Rimsza, R.A. Roberts, T.P. Miller, J.M. Unger, M.LeBlanc, R.M. Braziel, D.D. Weisenberger, W.C. Chan, H.K. Muller-Hermelink, E.S. Jaffe, R.D. Gascoyne, E. Campo, D.A. Fuchs, C.M. Spier, R.I. Fisher, J. Delabie, A. Rosenwald, L.M. Staudt and T.M. Grogan, Blood, **103**, 4251, (2004).
9. S. Wang, Progress in Biotechnology, **19**, 45, (1999).
10. J. Fritz, M.K. Baller, H.P. Lang, H. Rothuizen, P. Vettiger, E. Meyer, H.J. Güntherodt, Ch. Gerber, and J.K. Gimzewski, Science, **288**, 316, (2000).
11. K.M. Hansen, H.F. Ji, G.H. Wu, R. Datar, R. Cote, A. Majumdar, and T. Thundat, Anal. Chem., **73**,, 1567, (2001).
12. J. Verd, G. Abadal, J. Teva, M.V. Gaudó, A. Uranga, X. Borrisé, F. Campabadal, J. Esteve, E.F. Costa, F. Pérez-Murano, Z. J. Davis, E. Forsén, A. Boisen, and N. Barniol, Journal of Microelectromechanical Systems, **14**, 508, (2005).
13. X. Yu, Y. Tang, H. Zhang, T. Li, and W. Wang, IEEE Sensors Journal, **7**, 489, (2007).
14. F.J. Steemers, K.L. Gunderson, Pharmacogenomics, **6(7)**, 777, (2005).
15. K.L. Gunderson, F.J. Steemers, G. Lee, L.G. Mendoza, and M.S. Chee, Nat.Genet. **37**, 549, (2005).
16. M.A. Burns, B.N. Johnson, S.N. Brahmasandra, K. Handique, J.R. Webster, M. Krishnan, T.S. Sammarco, P.M. Man, D. Jones, D. Heldsinger, C.H. Mastrangelo, D.T. Burke, Science, **282**, 484 , (1998).

DNA: THE AXIOMS OF CELL GEOMETRY
THE DIFFERENTIAL GEOMETRY OF ITS FUNCTION
THE AFFINE GEOMETRY OF ITS STRUCTURE

Ioannis L. Arahovitis
iarahov@math.uoa.gr
Association of Emeritii Professors
National and Kapodistrian University of Athens
www.somotimon.uoa.gr

Abstract

The DNA (Deoxyribonucleic acid) is commonly known as the Double Helix because of its shape. Therefore, one wonders why "helix" and not another space curve?. In what follows, using arguments from Differential Geometry, we proceed to show that Nature was obliged to choose this particular space curve, as being the unique one fulfilling the necessary and sufficient conditions enabling DNA to accomplish the reason of its existence.

On the other hand we will prove that DNA has the structure of an affine plane completing this way the "matematization" of DNA.

The appearance of Affine Geometry in this important area of Biology and farther its appearance in Physics (Quantum Mechanics, Newtonian Mechanics, Astrophysics Relativity, Space Technology, Black Holes, LHC, etc) encourage our hope that Affine Geometry is what we are looking for, i.e. the Unification Geometry of the Universe and gives a new orientation to our research.

*

To be more analytic, the DNA molecule contains the genetic code of all living organisms i.e. all the instructions needed to construct indispensable components of the cells. Therefore, mathematically speaking, it constitutes **the system of axioms** on which Molecular Biology of the Cell – and hence human existence especially – is founded. Its importance is obvious, since any intervention on it will cause the inevitable collapse of the cell's geometry.

After describing the DNA's shape and function and giving the necessary mathematical information, we will proceed to describe the **differential geometry** of its function, as well as, the **affine geometry** of its structure.

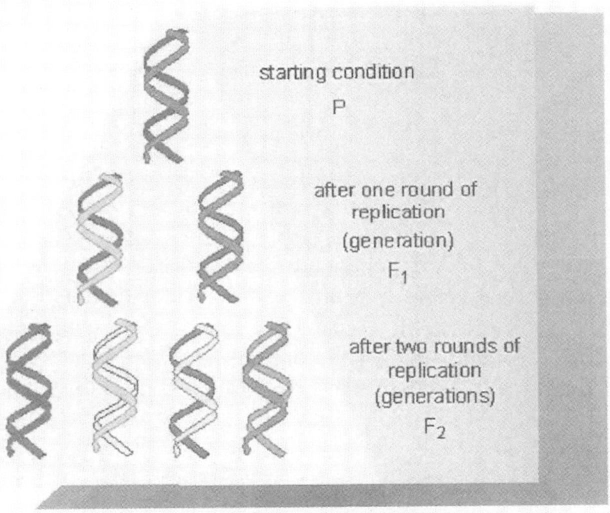

The preceding figures describe the fact that the 1,8m long molecule of the DNA – of the Double Helix – replicates and every new molecule must be identical to its parent one. The DNA's helices are antiparallel (one runs 5' to 3' while the other runs 3' to 5') and after every replication each helix is an exact copy of the other parent helix, i.e. it is parallel to his parent helix.

Now the question arises : Why Helix?

Before, we need some information from Differential Geometry concerning the theory of parallel curves.

Let z(t)be a regular curve in parametric form and k be a real number. By the parallel curve zk(t) at distance k we mean the curve defined by

$$zk(t) = z(t) + kN(t)$$

where N(t) is the unit normal vector at t. That is zk(t) is the locus of points at distance k from z(t) measured along the normal to the tangent line of the given curve C. If the equation of C is in Cartesian coordinates f(x , y) = 0 and (x_0 , y_0)

is any point M of the parallel curve, then the vector form of the equation of the parallel is

$$r - rM = -kN$$

as in the following figure

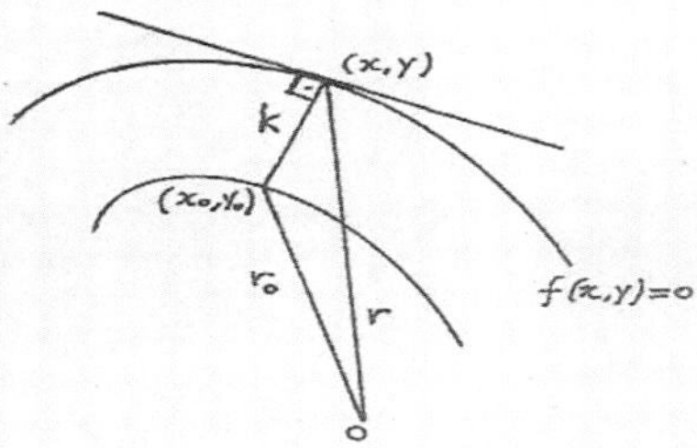

The tangent vector T = (dx , dy) and a normal vector $N = (f_x , f_y) = \nabla f$, where f_x , f_y are the partial derivatives of f. A unit normal vector (denoted again by N) is $N = \nabla f / || \nabla f|| = (f_x / R , f_y / R)$ where $R = \text{sqrt}(f_x^2 + f_y^2)$. In fact the slope of the tangent is dy / dx and that of the normal –dx / dy, since T.N = 0, which means $(dx , dy).(f_x , f_y) = 0$, or that $f_x dx + f_y dy = 0$. It follows that

$$x = x_0 - kf_x / R$$
$$(*)$$
$$y = y_0 - kf_y / R$$

As an application of (*) to the circle $(x\text{-}a)^2 = (y - b)^2 = c^2$, we obtain for the equation of its parallel

$$(x_0 - a)^2 + (y_0 - b)^2 = (c + k)^2,$$

which is again the equation of a circle homocentric to the given one, i.e. the parallel curve is a curve that belongs to the same family with the given curve.
But the parallel curve of a given curve is not always in the same family with the given one.

The following figure (cf. [6], p. 168) shows a parabola (and its evolute) and some parallel curves which are not parabolas because they have irregular points. The **curvature** of a curve C, y = f(x) is the rate of change (the derivative) of θ along the arc s of the curve , where θ is the angle of the tangent of C with the axe ox. Therefore, the curvature is a measure of how much “pointed” is the curve.
It can be proved that

$$\rho(\theta) = d\theta / ds = f''(x) / [1+((f'(x))^2]^{3/2} = \rho(x).$$

Note that the radius of curvature of the circumference of a circle is constant, while that of a parabola is not . Therefore, if the distance k is equal to the **radius of curvature** $1 / |\rho(t)|$, then (and only then) the parallel curve has a singular point, given that the equation of the curve parallel to z(t) is by definition $z_k(t) = z(t) + kN(t)$. It follows that

$$z_k'(t) = z'(t) + kN'(t) = z'(t) - k\rho(t)z'(t) = z'(t)(1 - k\rho(t)).$$

If $k = 1/\rho(t)$ and since $z'(t) \neq 0$ we obtain $z'_k(t) = 0$, i.e. $z_k'(t)$ has an irregularity.

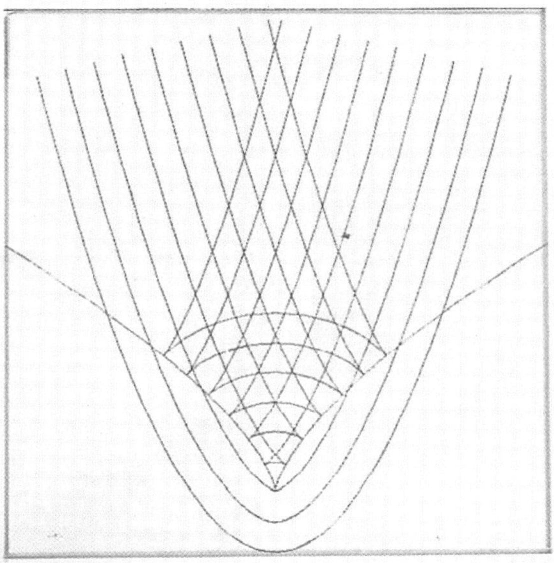

The parametric equations of a circular helix are as follows:

$$\textbf{(1)} \qquad \begin{aligned} x &= a\cos t \\ y &= a\sin t \\ z &= bt \end{aligned}$$

with **a** and **b** coefficients. In the molecule of the DNA there are two such antiparallel helicies.

Taking into account the existing bases in the middle of the two strands, we can thought of as forming the rungs of a helicoidal ladder. Each rung is composed of two base pairs. Either an adenine – thymine pair that form a 2-hydrogen bond together, or a cytosine – guanine pair that form a 3-hydrogen bond. Thus, we can consider the molecule as being a part of the surface of a right helicoid, the parametric equations of which are

$$\mathbf{x = u\cos(v)\ ,\ y = u\sin(v)\ ,\ z = v}$$

The following figure is the mathematical model of the DNA **:**

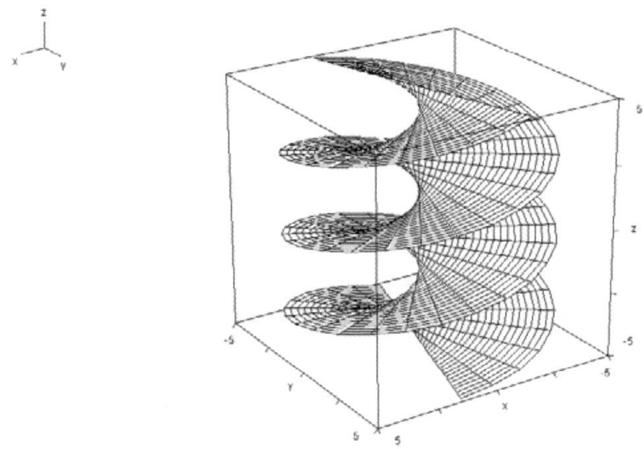

Consider now the following figure:

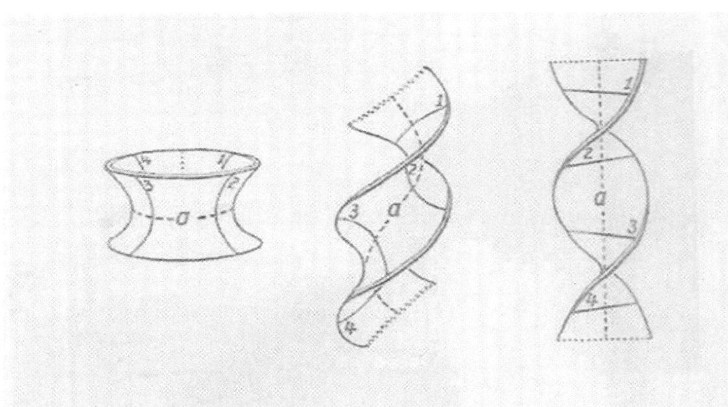

From Differential Geometry of Curves we remind the following "Two curves which, at any of their points, have a common principal normal are called **Bertrand** curves" (cf. e.g. [3], 23, p. 67).

Two Bertrand curves are **parallel,** in the sense that the distance measured on their common principal normal between corresponding points of the two Bertrand curves, is constant(cf. [1], p. 462), in analogy to Leibniz definition for plane curves(cf. [3], p. 67).

One can prove (cf. [3], Th. 23.1, p. 68) the following

<u>Theorem:</u> For a given curve C there is a curve C* so that C and C* are Bertrand curves if and only if the curvature $\kappa(t)$ and the torsion $\tau(t)$ of C satisfy a linear relation with constant coefficients

$$c_1\kappa(t) + c_2\tau(t) = 1.$$

Using the above theorem we also can prove (cf. [3], Th. 23.2, p. 70, [1], p. 462)

<u>Theorem:</u> A circular helix C is the only nonplanar biregular curve having more than one curve C* such that C* is a Bertrand curve for C and therefore parallel to the helix and which has constant curvature and torsion (cf. [4a], σ. 229,ασκ. 144)

Furthermore, we can see (cf. [3], p. 70) that there are infinitely many curves which must also be circular helices, so that C and any of these curves are Bertrand curves and hence parallel to C.

Conclusion**: <u>The helix is the only space curve that can have unlimited replications.</u>**

Turning to the helicoid, we remind from Differential Geometry of Surfaces the following:

"A surface is called **minimal** if it has an identically vanishing mean curvature H, i.e. if $H = 0$".

The problem of finding a minimal surface is known as the **Plateau Problem** (cf. [3], p. 245). Minimal surfaces are sometimes defined as surfaces with the property that any portion of them bounded by a closed curve C has the smallest area compared with any portion of a surface bounded by C. But this definition by means of a minimum condition of the Calculus of Variations is not always satisfactory(cf. [3], p. 244 and [5], p. 182). So we adopt the definition above. However minimal surfaces have the property of minimal area (cf. [2], p. 220).

Now, we have the following

<u>Theorem:</u> Catenoids are the only surfaces of revolution which are minimal surfaces (cf. [3], 81.2, p. 247). Also we have the

<u>Theorem</u>: The catenoid $f(v,u) = (u\cos(v) , u\sin(v) , h(u))$ can isometrically mapped onto a portion of the right helicoid(cf. [3], 81.3, p. 249).

The isometric mapping which is equiareal (cf. [3], p. 245) can be realized as in the

last figure, which therefore is an informal proof.

Portions of the right helicoid (: $z = \mathrm{arctg}(y/x)$, cf. below) are isometric to catenoids which are the only minimal surfaces among all surfaces of revolution, according to the above theorem.

From all the above we conclude that **in DNA we discover again the Nature's wisdom since the helicoid** (as well as the catenoid) **is a surface of minimal area** (cf. [4], σ. 136).

To deal with the geometric structure of the DNA molecule we need Klein's e.g. definition of an affine space of n-dimension and of some relevant definitions:

" Let E be a set $\neq \varnothing$, V a (real) vector space of n-dimension and δ an action of the additive group V_+ of V on E. If δ is properly transitive, then (E, V, δ) is called an **affine space of n-dimension**".

"We say that **a group G (left) acts on a set X**, if there exists a map δ, called the **(left) action of G on X**,

$$\delta : G \times X : (g , x) \rightarrow \delta(g , x) \equiv gx$$

such that

$$\Delta_1 : \delta(e , x) = x , \forall x \in X , \text{ where e is the neutral element of G}$$
$$\Delta_2 : \delta(g_1 , \delta(g_2 , x)) = \delta(g_1 . g_2 , x) , \forall g_1 , g_2 \in G , x \in X$$

and that the action δ is **properly transitive** if $\forall x_1 , x_2 \in X , \exists ! g \in G : \delta(g , x_1) = x_2$".

Now we proceed to prove that the DNA's molecule is (part of) an affine plane (: an affine space of dimension 2), i.e. a plane not of the Euclidean Geometry, but of another- one, non-Euclidean Geometry : the **Affine Geometry,** probably the Geometry of the Universe.

Let $V = \mathbf{R}^2$ and $E = \{(x,y,z) \in \mathbf{R}^3 | z = \text{barctg}(y/x)\}$ a (right for b=1) helicoid : from the equations of the helix $x = a\cos t$, $y = a\sin t$, $z = bt$ we obtain $y/x = \text{tg}t$ and therefore $z = \text{barctg}(y/x)$, since $t = \text{arctg}(y/x)$. Define

$$\delta : \mathbf{R}^2_+ \times E \rightarrow E : ((k,l) , (x,y,z)) \rightarrow \delta((k,l) , (x,y,z)) := (x+k , y+l , \text{barctg}(y+l/x+k)).$$

Then δ is an action . In fact

$$\Delta_1 : \delta((0,0) , (x,y,z)) = (x+0 , y+0 , \text{barctg}(y+0/x+0)) = (x,y,z).$$

$$\Delta_2 : \delta((k_1,l_1) , \delta((k_2,l_2) , (x,y,z))) = \delta((k_1,l_1) , (x+k_2 , y+l_2 , \text{barctg}(y+l_2/x+k_2))) = (x+k_2+k_1 , y+l_2+l_1 , \text{barctg} (y+l_2+l_1/x+k_2+k_1)) = \delta((k_1+k_2 , l_1+l_2) , (x,y,z)) = (x+k_1+k_2 , y+l_1+l_2 , \text{barctg}(y+l_1+l_2/x+k_1+k_2)),$$ since $\mathbf{R}^2_+$ is abelian.

For $(x_1,y_1,z_1) , (x_2,y_2,z_2) \in E$, consider the unique $k = x_2 - x_1$, and $l = y_2 - y_1$. Then

$$\delta((x_2 - x_1 , y_2 - y_1 , (x_1, y_1,z_1)) = (x_1+x_2 - x_1, y_1+y_2 - y_1, \text{barctg}(y_1+y_2 - y_1/x_1+x_2 - x_1)) =$$
$$(x_2, y_2, \text{barctg}(y_2 /x_2) = (x_2, y_2 , z_2),$$ i.e the proper transitivity of δ, **QED.**

References

[1] Gallier, J. *Geometric Methods and Applications For Computer Science and Engineering* Springer, 2001.

[2] Guggenheimer, H.W. *Differential Geometry* ,Dover, 1977.

[3] Kreyszig, E. *Differential Geometry,* Dover, 1991.

[4] Μπρίκας, Μ. *Μαθήματα Θεωρίας Επιφανειών ΙΙ*, Αθήνα, 1963.

[4$^{\alpha}$] Μπρίκας, Μ. *Μαθήματα Θεωρίας Καμπύλων,* Αθήνα, 1965.

[5] Struik, D.J. *Lectures on Classical Differential Geometry,* Sec. Ed. Dover, 1988.

[6] Zimmermann, W.- Cunningham, S. (edit.) *Visualization in Teaching and Learning Mathematics*, AMS, 1991.

http://nobelprize.org, DNA-The Double Helix, The Official Web Site of the Nobel Foundation, Nobel Web AB, 2008.

http://www.biologie.uni-hamburg.de

http://en.wikipedia.org/wiki/Double_helix

http://207.207.4.198/pub/flash/24/mnu.swf, DNA Replication

http://users.rcn.com/jkimball.ma.ultranet/BiologyPages/D/DoubleHelix.html

PREDICTION OF ATHEROSCLEROTIC PLAQUE FORMATION BASED ON LDL TRANSPORT FOR A 3D PATIENT SPECIFIC CORONARY ARTERY WITH DEFORMABLE WALLS

A. I. SAKELLARIOS
University of Ioannina
Dept of Materials Science and Engineering
University Campus of Ioannina, 45110, Ioannina, Greece

M. I. PAPAFAKLIS
Michaelideion Cardiac Centre
University Campus of Ioannina, 45110, Ioannina, Greece

C. V. BOURANTAS
Michaelideion Cardiac Centre, University of Ioannina, Ioannina, Greece
Dept. of Cardiology, University of Hull, Castle Hill Hospital, East Yorkshire, UK

L. K. MICHALIS
University of Ioannina
Medical School, Dept of Cardiology
University Campus of Ioannina, 45110, Ioannina, Greece

D. I. FOTIADIS
University of Ioannina
Dept of Materials Science and Engineering
University Campus of Ioannina, 45110, Ioannina, Greece

Several studies investigate the effect of blood flow on the transport of Low Density Lipoproteins (LDL) in the arteries. These studies performed simulations on 2D arterial models with rigid walls. In this study, the LDL transport is studied using a realistic 3D coronary artery model. This work is based on Fluid – Structure Interaction (FSI) which provides the possibility to both study the effect of blood flow on the arterial wall and to investigate how the arterial deformation affects the LDL accumulation next to the arterial wall. The geometrical model used for the simulation is obtaining from IVUS coupled with biplane angiography. The arterial wall is assumed to be linear elastic, homogenous and impermeable. The blood flow is assumed to be Newtonian, laminar and incompressible. The Navier-Stokes equations are used to calculate the blood velocity and pressure. In order to study the LDL transport the Convection-Diffusion equation are used and we assume that the blood is a homogenous dilution with LDL and the LDL movement does not affect the blood flow. Finite elements are used for the solution of differential equations of blood flow and the arterial structural mechanics. The results indicate that the LDL accumulation is related to the wall shear stress. More specifically, regions with low wall shear stress have increased levels of LDL concentration. The results of LDL concentration are used to make a prediction about the region that is prone to develop the atherosclerotic plaque. It is also shown that regions with large abnormalities on the arterial wall surfaces indicate elevated concentration of LDL.

1. Introduction

In western societies, atherosclerosis is the most common cause of death. Atherosclerosis is the accumulation of "atheroma" in the arterial wall, while many are the responsible factors for this progress. High LDL concentration, hypertension, genetic predisposition, diabetes mellitus and smoking are some of the leading factors [1]. The progression of the atherosclerotic plaque is associated with mechanical, biochemical and biological reactions. Blood flows in artery's lumen and forces to arterial wall with effects to endothelial permeability, gene expression and arterial wall mechanics. It is assessed that the atherosclerotic plaque is developed at complex regions such as the arterial bifurcations or curves [2, 3]. Also, atherosclerosis is detected at regions with low wall shear stress while the initial stage of the atherosclerotic plaque formation is the accumulation of Low Density Lipoprotein (LDL) into the arterial wall. LDL's permeability at endothelium is related to its size, concentration, pressure and lesions on the endothelial membrane. As reported in experimental results the LDL accumulation occurs at regions where low WSS exists [4]. In the other hand blood interacts with the arterial wall, which is not rigid but deformable. Thus, the blood flow dynamics and the solute flux dynamics are related to the deformation of the arterial wall.

Until now many studies have been developed to model the mass transport in arteries. These models could be classified into three categories. First the wall free models where the arterial wall is not permeable and the concentration of the solute is calculated only in the lumen [5]. Second the lumen-wall models where the arterial wall consists of one homogenous porous layer and the concentration of the solute is calculated both to the arterial lumen and wall [5-8]. Third the multilayer models where the arterial wall consists of the media, the intima and the adventitia and interaction between these layers exists [9].

The arterial wall dynamics and the interaction with the blood flow were studied widely. Holzapfel *et al.* presented a work on the arterial wall mechanics considering non-linear models [10]. Yang [11] studied the fluid structure interaction (FSI) in pulmonary artery bifurcations while Li [12] studied the deformation of the carotid arterial wall. Leach and Scotti considered non-linear behavior of the arterial wall and developed FSI models in atherosclerotic arteries [13, 14].

Although many models have been proposed for the mass transfer in arteries, none of these studies consider the interaction of the blood flow with the deformable arterial wall. In this study we develop the LDL transport in realistic coronary arteries with deformable walls. The arterial wall is considered to be homogenous and linear elastic. Also, the endothelium is not permeable to the LDL. The coronary arteries are commonly disease from atherosclerotic plaque.

2. Methods

2.1. *3-D reconstruction of coronary arteries*

The simulations are performed in realistic coronary arteries which are reconstructed using the fusion of Bi-plane Angiographic and Intravascular Ultrasound (IVUS) data. The methodology was first proposed by Bourantas [15]. Concisely, IVUS frames are pre-processed to extract the lumen and media-adventitia borders. In the next step a 3-D catheter path is extracted from the angiographic data and the previously detected borders are placed on it. In fig. 1 is shown a reconstructed coronary artery.

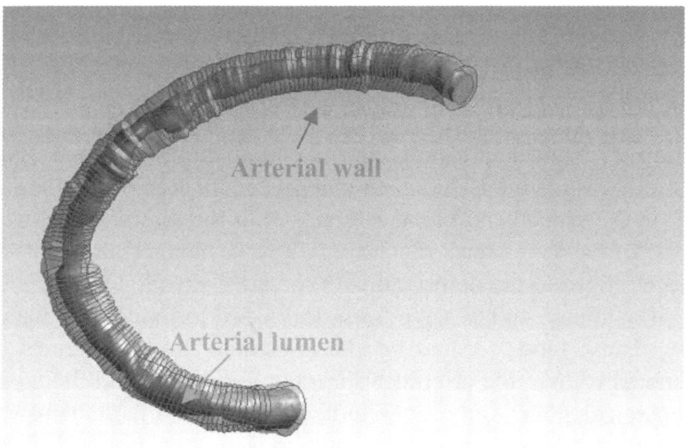

Fig. 1: Reconstructed realistic coronary artery.

2.2. *Modeling fluid dynamics*

The Navier – Stokes equation and the continuity equation are used to model the blood flow in the arterial lumen:

$$-\mu\nabla^2 u + \rho(u \cdot \nabla)u + \nabla p = 0 \tag{1}$$

$$\nabla u = 0, \tag{2}$$

where u is the blood velocity, p is the pressure, μ is the viscosity of the blood and ρ is the blood density.

At the inlet of the coronary artery we consider a fully developed parabolic velocity profile with flow rate of 60 ml/min. At the outlet we define constant pressure condition equal to zero.

2.3 *Modeling LDL transport*

In this study we assume that blood is a homogenous dilution with LDL which does not affect the flow. Thus, we use the advection – diffusion equation to describe the LDL transport in the arterial lumen:

$$\nabla \cdot (-D\nabla c + cu) = 0, \tag{3}$$

where c is the concentration of the LDL and D is the LDL diffusivity in the arterial lumen. The boundary conditions we use to model the LDL transport is a constant LDL concentration at the inlet and convective flux at the outlet. We remind that the arterial endothelium is not-permeable. At the inlet we use normal rates of LDL concentration equal to 10^{-3} mol/m^3 and also pathogenic high LDL concentration equal to 10^{-2} mol/m^3.

2.4. *Static structural analysis of the arterial wall*

The arterial wall is considered to be linear elastic and it consists of one homogenous layer. The deformation of the arterial wall is modeled using the following equations:

$$\nabla \cdot \boldsymbol{\sigma} + \mathbf{F} = \rho\ddot{\boldsymbol{\upsilon}} \tag{4}$$

$$\varepsilon = \frac{1}{2}\left[\nabla\boldsymbol{\upsilon} + (\nabla\boldsymbol{\upsilon})^T\right] \tag{5}$$

$$\boldsymbol{\sigma} = \mathbf{C} : \boldsymbol{\varepsilon}, \tag{6}$$

where eqn. (4) is the equation of motion, eqns. (5) are the strain – displacement equations and eqn. (6) is the constitutive equation. Also, $\boldsymbol{\sigma}$ is the Cauchy stress tensor, ε is the infinitesimal strain tensor, $\boldsymbol{\upsilon}$ is the displacement vector, $\mathbf{C}$ is the fourth-order tensor of material stiffness, $\mathbf{F}$ is the body force per unit volume and ρ is the mass density of the arterial wall.

At the part of the coronary artery which is connected to the aorta we define fixed support as boundary condition. An FSI interface is considered at the interface of the arterial wall with the blood flow.

2.5 *Modeling the fluid structure interaction*

In the first step the blood flow velocities and pressure are calculated. The loads which are arrived from blood to the arterial wall are calculated and then the deformation of the arterial wall is determined based on the current loads. In this step if the convergence of the overall system is reached the blood velocity, pressure and fluid domain are updated for a new calculation. Otherwise the calculation starts from the beginning. In the fig. 2 is shown the flowchart of the proposed methodology.

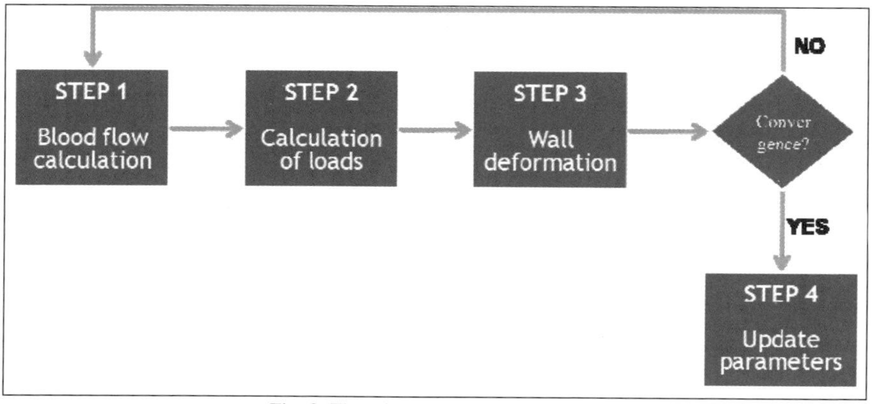

Fig. 2: Flowchart of the FSI modeling.

2.6 *Meshing method and calculation*

The simulations are performed using a finite element package. The fluid domain is discretized in hexahedral elements due to the non-linear Navier-Stokes equations while the arterial wall domain is discretized in tetrahedral elements. Minimum element size is defined for the domains but the element size at the FSI interface is lower. Also, the splines of the geometries are divided to finite points as shown in the fig. 3.

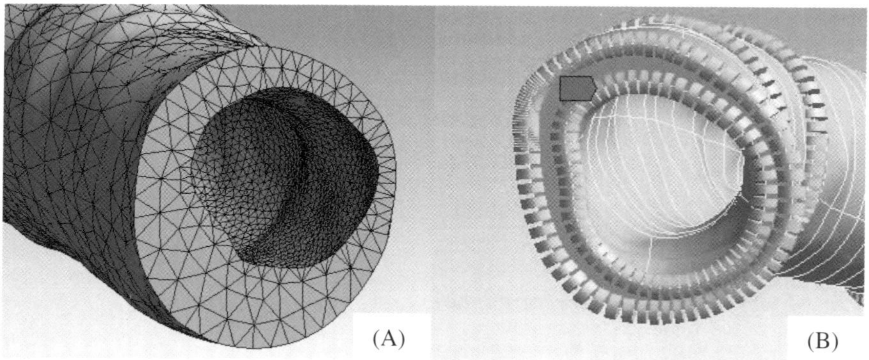

Fig. 3: (A) The arterial wall discretized in tetrahedral elements. (B) The splines are divided for better quality mesh.

3 Results

According to experimental studies we expect that the LDL accumulation would be higher at regions with low wall shear stress. Our simulation results agree entirely with these data. In the fig. 4 is shown the wall shear stress distribution in a coronary artery with many atherosclerotic regions. Also, it is presented that the wall shear stress is lower mainly at the regions after atherosclerotic plaque, where the flow has abnormalities such the recirculation.

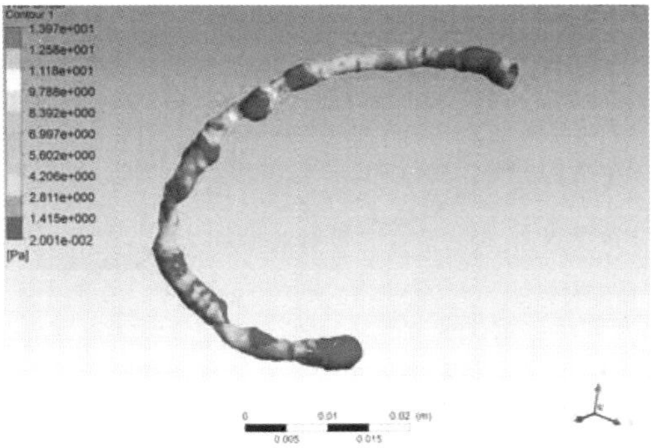

Fig. 4: Wall shear stress distribution in atherosclerotic coronary artery.

In fig. 5 is shown the LDL concentration at the regions of low wall shear stress of the same coronary artery. It is indicated that the higher value of LDL concentration is found at the regions of low wall shear stress. So, the wall shear stress relates to LDL accumulation and also this result is much more acute at the regions after atherosclerotic plaque. These results justify to some degree the configuration of the plaques (thicker at the end of the region).

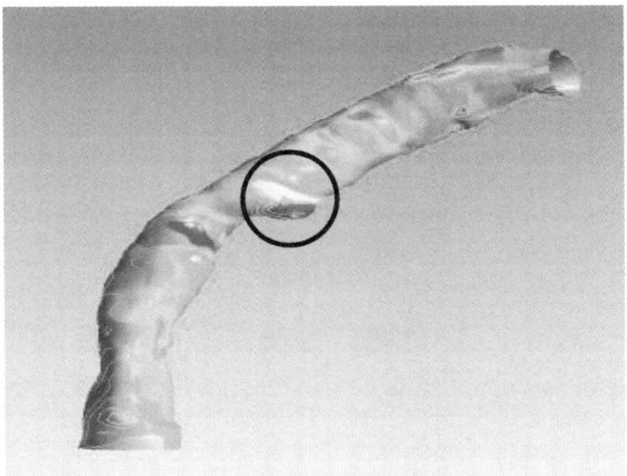

Fig. 5: Region with high LDL accumulation after the atherosclerotic plaque.

In fig. 6 is shown the relation of the wall shear stress to the LDL concentration. It is demonstrated that as the wall shear stress is getting lower the LDL accumulation is increased and vice versa.

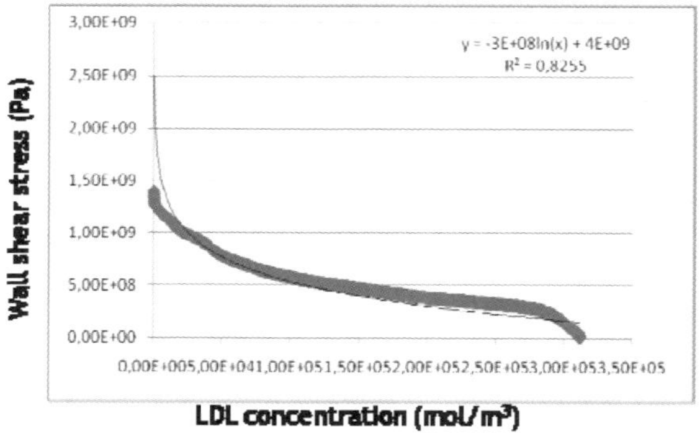

Fig. 6: Wall shear stress relation to the LDL accumulation

As regards to the structural analysis of the arterial wall the greater deformation is found at the free part of the artery as shown in the fig. 7. The deformation of the arterial wall contributes to the accumulation of the LDL at the specific regions after the atherosclerotic plaque and makes the effect brighter.

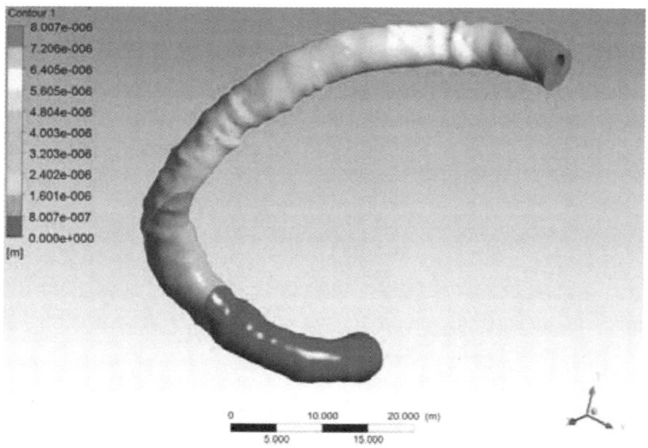

Fig. 7: Total mesh deformation of the arterial wall. The highest value is indicated at the free end.

4 Conclusions

The LDL transport in realistic coronary arteries is studied and for the first time the arterial wall is deformable. The Navier – Stokes equations are used to model the blood flow and the advection – diffusion equation to model the LDL transport. The arterial wall considered to be linear elastic and impermeable. The results indicate the increased LDL accumulation at the regions after atherosclerotic plaque, where the wall shear stress is lower. Our model predicts the region of high LDL accumulation, which is the initial stage of the atherosclerotic plaque formation.

There should be no doubt that several challenges remain. The non – linear behaviour of the arterial wall should be considered. Also, the LDL penetrates the endothelium and accumulates in the intima. This stage would be the next step of our work to model totally the accumulation of the LDL and the initial stage of plaque progress.

References

1. Braunwald's Heart Disease. A Textbook of Cardiovascular Medicine. 7th Edition.
2. Chatzizisis Y et al. (2007) Role of endothelial shear stress in the natural history of coronary atherosclerosis and vascular remodeling. Molecular, cellular and vascular behavior. Journal of the American College of Cardiology. 49:2379-2393
3. White C. Frangos J. (2007) The shear stress of it all: the cell membrane and mechanochemical transduction. Philosophical Transactions of Royal Society. 362:1459-1467
4. Prosi M. Zunino P. Perktold K. Quarteroni A. (2005) Mathematical and Numerical Models for Transfer of Low – Density Lipoproteins through the Arterial Walls: A new Methodology for the Model Set up with applications to the Study of Disturbed Luminal Flow. Journal of Biomechanics. 38:903-917
5. Sakellarios A. Fotiadis D. Michalis L. 2008, Finite Element Modeling of LDL Transport in Carotid Artery Bifurcations. EMBEC Conference, Antwerp
6. Ross E. (2002) Computational Modeling of Mass Transfer and Links to Atherosclerosis. Annals of Biomedical Enginnering 30:461-471
7. Hong J. Wei L. Fu C. Tan W. (2007) Blood Flow and Macromolecular Transport in Complex Blood Vessels. Clinical Biomechanics. Article in press.
8. Sun N. Wood N. Hughes A. Thom S. Yun Xu X. (2006) Fluid – Wall Modeling of Mass transfer in an Axisymmetric Stenosis: Effects of Shear – Depended Transport Properties. Annals of Biom. Engineering. 34:1119-1128
9. Yang N. Vafai K. (2006) Modeling of Low Density Lipoprotein (LDL) Transport in the Artery – Effects of hypertension. Inter. J. Heat and Mass Transfer. 49:850-867
10. Holzapfel G. Gasser T. Ogden R. 2000, A New Constitutive Framework for Arterial Wall Mechanics and a Comparative Study of Material Models. Journal of Elasticity. 61:1-48
11. Li M. Beech – Brandt J. John L. Hoskins P. Easson W. (2007). Numerical analysis of pulsatile blood flow and wall vessel mechanics in degree of stenosis. Journal of Biomechanics. 40:3715-3724
12. Yang X. Liu Y. Yang J. (2007). Fluid-structure interaction in a pulmonary arterial bifurcation. Journal of Biomechanics.40:2694-2699
13. Scotti C. Finol E. (2007). Compliant biomechanics of abdominal aortic aneurysms: A fluid–structure interaction study. Computers and Structures. 85:1097–1113
14. Leach J. Rayz V. Mofrad M. Saloner D. (2009). An efficient two-stage approach for image-based FSI analysis of atherosclerotic arteries. Biomechanics and Modeling in Mechanobiology.
15. Bourantas C. Kourtis I. Plissiti M. Fotiadis D. et al. 2005, A Method for 3D reconstruction of coronary arteries using biplane angiography and intravascular ultrasound images. Computerized Medical Imaging and Graphics, 29:597-606

SETTING A RATIONAL FRAMEWORK FOR EXPERIMENTAL DESIGN AND ANALYSIS OF HIGH-THROUGHPUT DNA MICROARRAY EXPERIMENTS AND DATA[†]

GEORGE I. LAMBROU
Hematology/Oncology Unit, 1st Department of Pediatrics
University of Athens
Goudi, Athens, 11527, Greece[1]
E-mail: glamprou@med.uoa.gr

ARISTOTELIS CHATZIIOANNOU
Institute of Biological Research & Biotechnology
National Hellenic Research Foundation
48 Vassileos Constantinou Ave.
E-mail: achatzi@eie.gr

EMMANOUIL SIFAKIS
Biomedical Engineering Laboratory
National Technical University of Athens
Zografou, Athens, 15773, Greece
E-mail: sifakise@biomed.ntua.gr

ANDRIANA PRENTZA
Department of Digital Systems
University of Piraeus
Piraeus, 18534, Greece
E-mail: aprentza@unipi.gr

DIMITRIS KOUTSOURIS
Biomedical Engineering Laboratory
National Technical University of Athens
Zografou, Athens, 15773, Greece
Email: dkoutsou@biomed.ntua.gr

ELEFTHERIA KOULTOUKI
Hematology/Oncology Unit, 1st Department of Pediatrics
University of Athens
Goudi, Athens, 11527, Greece
E-mail: glamprou@med.uoa.gr

FOTINI TZORTZATOU-STATHOPOULOU
(a) University Research Institute for the Study and Treatment of Childhood Genetic and Malignant Diseases
(b) Hematology/Oncology Unit, 1st Department of Pediatrics
University of Athens
Goudi, Athens, 11527, Greece
E-mail: ftzortza@med.uoa.gr

[†] This work was co-funded by the European Social Fund and National Resources – (General Secretariat for Research and Technology) PENED 2003.

Microarray experimentation has been widely used for screening and high throughput discovery of mechanisms underlying biological systems. However, there are many factors that need to be taken into account for an effective microarray experimental design. Such factors are the abundance of starting material (RNA), number of replicates or experimental cost. The present work proposes rational strategies for microarray experimental design, with the scope to minimize signal variance and improve precision, with the minimum amount of technical replicates needed, thus alleviating the burden of biological material collection and the cost of the experiment. In this sense, alternative microarray design strategies are tested (direct comparisons, all paired comparisons, indirect comparisons through the implementation of various loop strategies), targeting to minimize the wet lab experimental effort needed, for the more precise interpretation of the biological question investigated. In this work we propose a methodological approach for the computational implementation of comparisons between different microarray experimental samples which are not co-hybridized together. Moreover, this approach is tested *in-silico*, on a showcase of a simple *"loop"* experimental design, including 3 biological triplicates. The proposed methodological approach introduces a robust method for calculating and comparing gene expression profiles between sets of microarray experiments by using the minimum number of experiments on one hand and getting robust results on the other. Further extension of this work could form a rational operational computational framework for the design of microarray or more general high-throughput biological experiments.

1. INTRODUCTION

DNA microarrays are a high-throughput technology which enables the simultaneous measurement of thousands of genes in one experiment, thereby representing an excelling methodology for the massive screening of biological time-points or "snap-shots". Since its advent, many advances have taken place both on the level of experimentation as well as on the level of analysis. Microarray data present some disadvantages though, which nevertheless, compared to the amount of information they provide, can be considered tractable. Specifically the problems have mainly to do with the impact of various sources of noise and the bias they incur in the analysis of the experimental results. For example, microarrays incorporate the use of fluorescent dyes based on certain chemical and analytical methods and protocols, which introduce their intricacies regarding binding specificity or light emission upon laser excitation. Yet, one must always keep in mind that the vast amount of information as regards instantaneous genome expression provided by DNA microarray analyses represents an invaluable asset. Therefore, it is imperative that rational strategies are formulated as regards the design and interpretation of microarray experiments, in order to ensure maximized reusability of the given experimental material, minimize experimental costs and improve analytical efficacy. As microarray technology goes more mature though, many design, analysis and normalization approaches have been developed to successfully address more complicated questions [1, 2]. The present work attempts to contribute in the formulation of a rational framework for microarray design and analysis.

An alternative to the classical *"reference" design*, in microarray experiments is the *"loop" design*, where more than two fluorescent dyes can be used [1, 3-6]. Another work focuses in the comparison of the two methodologies i.e. *"loop" designs vs. "reference" designs* [3]. There are pros and cons for every design. The *"reference" design*, albeit simple to the point of being self-evident, has the disadvantage that the reference sample is used multiple times thus minimizing noise variation for it while the other samples are measured only once, exposed thus in heavier impact of noise. In this sense, it has been suggested by several authors the superiority of the *"loop"* compared to the *"reference" design* in terms of reliability and robustness [1, 7]. However, the drawback in the use of *"loop" design* is its complexity and the lack of methods to render channel ratios among experiments comparable [1, 8]. The present work focuses in setting a rational framework for the analysis of *"loop"* designed microarray experiments, exploiting the vectorial nature of microarray sample instances. It also provides the necessary background for the new approach of analyzing microarray data.

2. DATA TRANSFORMATION AND *"LOOP" DESIGN*

2.1 The Logarithmic Transformation

As mentioned above, several approaches have been proposed for the transformation, normalization and analysis of microarray data and in particular for *"loop"* experimental designs. By far the most popular and known transformation used is the logarithmic. In particular, the logarithm of the ratio of two intensities *R* and *G,* where *R* stands for the *Cy3* and *G* for the *Cy5* intensities respectively, or $\log_2\left(\frac{G}{R}\right)$ (**1**), is a very popular transformation, due to the fact that it renders the space of the population of values symmetric. Also it is suitable for numerical computing as due to the properties of the logarithms the operations of multiplications and division are instead becoming additions and subtractions which are computed much faster than the former. Logarithmic transformation is applicable to any type of experimental design, either *"reference"* or *"loop"* design. Since designing a *"loop"* experimental setup encompasses the purpose of minimizing the number of experiments performed, a plausible rationale is to compare the calculation of ratios derived either from directly linked between them experiments or distant ones. As shown in Figure 1, experiments ***A*** through ***F*** serve to the derivation of values from direct comparisons (solid lines) whereas dashed line connections indicate a computational comparison. It has been proposed that the ideal way of performing such comparisons is the "*loop*" design experimental setup with interwoven comparisons [9]. In essence, it is possible to derive experiment ***C*** computationally by exploiting estimated ratios from the previous two experiments (***A*** and ***B***).

Let *R* and *G* be the intensities of *Cy3* and *Cy5* respectively for any three of the consecutive experiments of Figure 1 and *i* be the same gene measured on each array then:

$$\log_2\left(\frac{G_{j+1,i}}{R_{j,i}}\right)+\log_2\left(\frac{G_{j+2,i}}{R_{j+1,i}}\right)=\log_2\left(\frac{G_{j,i}}{R_{j+2,i}}\right) \quad \textbf{(2)}$$

where it is implied that $Rj=Gj$ for $j=A,..,F$ if no systematic technical errors are taken into account. This statement can be generalized to *n* number of iterations, with the limitation, however, as reported by *Kerr* and *Churchill* that the technical limit for obtaining reliable results is at the scale of eight nodes for a loop. In particular from Eq. (2) we get:

$$\log_2\left(\frac{G_{2,i}}{R_{1,i}}\right)+\ldots+\log_2\left(\frac{G_{n,i}}{R_{n-1,i}}\right)=\log_2\left(\frac{G_{1,i}}{R_{n,i}}\right) \quad \textbf{(3)}$$

This hypothesis was tested on a three-loop design experimentally[10]. Subsequently, this hypothesis was used as a filtering criterion for gene selection. In particular, genes in experiment ***A*** were compared

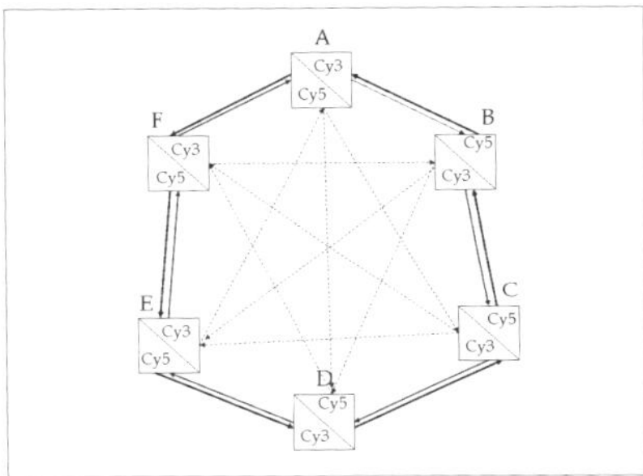

Figure1."*Loop*"microarray design adopted from *Kerr and Churchill [6]*

to ***B*** and ***B*** were compared to ***C***. In addition, genes in experiment ***A*** were compared to ***C*** and for further analyses we used those genes that conformed to the rule as defined by Eq. (2). We were able to identify genes that really conformed to this hypothesis indicating that a computational approach is possible without direct comparison.

2.2 The Vectorial/Geometric Transformation Approach for a Three-fold "Loop" Design

Besides applying logarithmic transformations raw intensities should be filtered and normalized. Methods and algorithms of normalization vary and it is not within the scope of this paper to describe them.

However, we mention here those methods for data filtering and normalization that we use in the following steps of this work. A common filtering methodology that we have also applied to our microarray data is the *Signal-to-Noise Ratio (SNR)* given by:

$$SNR=\frac{\mu_{R,G}-\mu_B}{\sigma_B} \quad \textbf{(4)}$$

Besides that, another factor that can provide substantial information is the possibility for multiple experiments and the calculation of the average values from replications of one experiment. Typically, the mean ratio r_μ of a number of triplicates *j* is given by:

$$r_\mu = \frac{1}{j}\sum_{i=1}^{j}\log_2\left(\frac{G_i}{R_i}\right) \quad \textbf{(5)}$$

which represents the *geometric mean* of the j replicates. This can be equally stated as:

$$r_\mu = \log_2\left(\sqrt[j]{\prod_{i=1}^{j}\frac{G_i}{R_i}}\right) \quad \textbf{(6)}$$

These are some indicative ways of filtering microarray data and finding the mean values of replicates. After, studying the works of *Churchill* and *Altman* [5-7] we have asked the question whether there would be another way which could be used as a method in order to analyze raw or normalized intensities by taking into account the individual channel signal intensities. The *log* ratio gives an estimate of the ratio itself but it does not yield any information as regards the individual channel signal intensities comprising this ratio. Therefore, we extended our analysis to include the properties of signal vectors. Each microarray spot consists of two channel signals. These, can be represented on a *Cartesian* system of coordinates and the resulting representation is a vector $\boldsymbol{v}$ with magnitude $|v|$ and direction θ (more precisely $tan\theta$) as presented in Figure 2 for three independent experiments concerning the same gene.

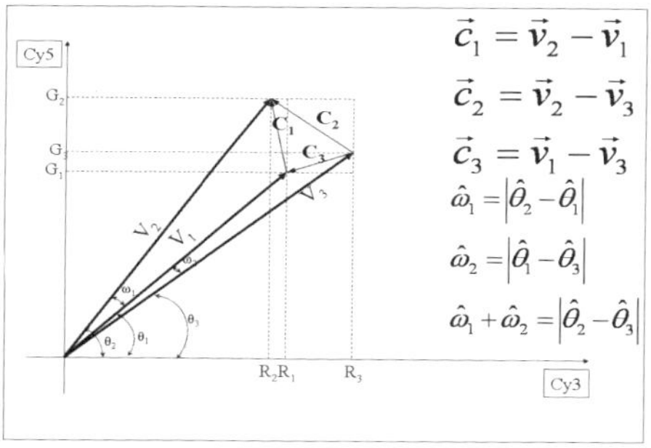

Figure 2. Each vector v_1, v_2, v_3 with directions θ_1, θ_2, θ_3 respectively represent the expression of the same gene in three independent experiments. We define three new vectors c_1, c_2, c_3 as defined above which are the difference between each pair of experiments.

To proceed with our analysis we posed the following question: given two known vectors $\boldsymbol{v}_1$ and $\boldsymbol{v}_2$ is it possible to calculate a third vector $\boldsymbol{v}_3$ as a function of the previous two if it is known that the three are interconnected? We reduced our analysis into a case study of three experiments in a "loop" design and we tried to express the third experiment as a function of the other two. In order, to achieve this we define three new vectors $\boldsymbol{c}_1$, $\boldsymbol{c}_2$, $\boldsymbol{c}_3$ which are defined as presented in Figure 2. In order to solve this problem we combined algebraic and geometric methods. In particular, each of the measures of vectors $\boldsymbol{c}_1$, $\boldsymbol{c}_2$, and $\boldsymbol{c}_3$ can be written as:

$$\begin{aligned} \| c_1^{\ 2} \| &= (G_2 - G_1)^2 + (R_1 - R_2)^2 \\ \| c_2^{\ 2} \| &= (R_3 - R_2)^2 + (G_3 - G_2)^2 \\ \| c_3^{\ 2} \| &= (G_3 - G_1)^2 + (R_3 - R_1)^2 \end{aligned} \tag{7}$$

which is easily derived from the Pythagorean triangles formed between the three vectors $\boldsymbol{c_1}$, $\boldsymbol{c_2}$, $\boldsymbol{c_3}$ and the coordinates of the vectors $\boldsymbol{v_1}$, $\boldsymbol{v_2}$, and $\boldsymbol{v_3}$. In order to derive more equations and render the system solvable, we exploited the geometrical properties of the triangles formed by $\triangle v_3 v_2 c_2$, $\triangle v_1 v_3 c_3$ and $\triangle v_2 v_1 c_1$ which gives:

$$\begin{aligned} c_1^2 &= v_1^2 + v_2^2 - 2\| v_1 \| \| v_2 \| \cos \omega_1 \\ c_2^2 &= v_2^2 + v_3^2 - 2\| v_2 \| \| v_3 \| \cos(\omega_2 + \omega_1) \\ c_3^2 &= v_1^2 + v_3^2 - 2\| v_1 \| \| v_3 \| \cos \omega_2 \end{aligned} \tag{8}$$

Also:

$$\begin{aligned} \omega_1 &= |\theta_2 - \theta_1| \\ \omega_2 &= |\theta_1 - \theta_3| \\ \omega_1 + \omega_2 &= |\theta_2 - \theta_3| \end{aligned} \tag{9}$$

of which the first two are readily derived from the values of the respective coordinates R_1, G_1, R_2, G_2. Here we should also make use of the relations that:

$$\begin{aligned} R_2 &= G_1 \\ R_3 &= G_2 \\ R_1 &= G_3 \end{aligned} \tag{10}$$

which represent the interconnection relations. From Eq. (7) to Eq. (10) not only we derive estimations of the third experiment ($\boldsymbol{v_3}$), but we also can minimize the inference of experimental measurement error.

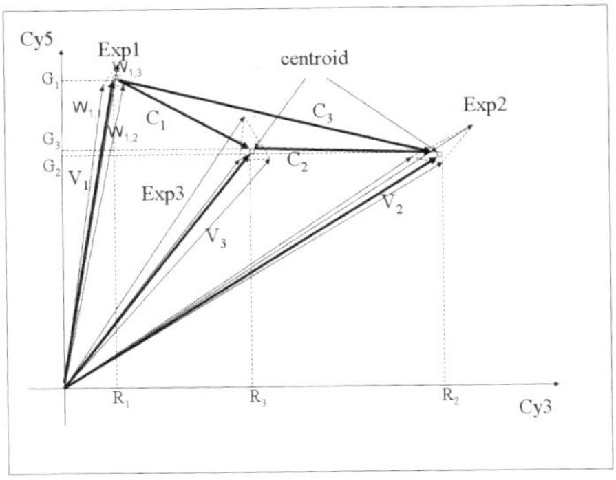

Figure 3. Interpretation of three experiments, each in triplicate, where the vectors $\boldsymbol{V_1}$, $\boldsymbol{V_2}$, $\boldsymbol{V_3}$ are the mean vectors of $\boldsymbol{w_{1,i,j}}$, $\boldsymbol{w_{2,i,j}}$ and $\boldsymbol{w_{3,i,j}}$ of the ith gene in the jth experiment with coordinates $x_1(0.0)$ to $x_2(x_n, y_n)$. Where x_n, y_n are the coordinates of the centroid of each replicate.

2.3 Incorporating the role of Variance for The Vectorial/Geometric Interpretation

In the aforementioned relations we have not taken into account the possibility of replicates among similar experiments and their respective variance. The above hypotheses are based on the fact that the signal intensities of two different dyes of the same sample are equal (or at least approach equality). This of course is not the case in real experiments, therefore it is necessary to define a way for measuring variance within the approach we describe in the present work. In Figure 3 we present an interpretation of variance in a vector and geometric level. Let $\mathbf{w}_{1,i,j}$, $\mathbf{w}_{2,i,j}$, $\mathbf{w}_{3,i,j}$ be the vectors describing the expression intensities of the ith gene in the jth replicate experiment with $\mathbf{w}_{1,i,j}\ (x_{1,i,j}, y_{1,i,j})$, $\mathbf{w}_{2,i,j}\ (x_{2,i,j}, y_{2,i,j})$, $\mathbf{w}_{3,i,j}\ (x_{3,i,j}, y_{3,i,j})$. We can then define three new vectors as $V_1\ (R_1, G_1)$, $V_2\ (R_2, G_2)$, $V_3\ (R_3, G_3)$ where R and G are the coordinates of each centroid formed by the three vectors $\mathbf{w}_{1,i,j}$, $\mathbf{w}_{2,i,j}$, $\mathbf{w}_{3,i,j}$. In particular, the three new vectors V_1, V_2, V_3 would be defined by:

$$V_1\left(\frac{x_{1,i,1}+x_{2,i,1}+x_{3,i,1}}{3}, \frac{y_{1,i,1}+y_{2,i,1}+y_{3,i,1}}{3}\right) = V_1(R_1, G_1)$$

$$V_2\left(\frac{x_{1,i,2}+x_{2,i,2}+x_{3,i,2}}{3}, \frac{y_{1,i,2}+y_{2,i,2}+y_{3,i,2}}{3}\right) = V_2(R_2, G_2)$$

$$V_3\left(\frac{x_{1,i,3}+x_{2,i,3}+x_{3,i,3}}{3}, \frac{y_{1,i,3}+y_{2,i,3}+y_{3,i,3}}{3}\right) = V_3(R_3, G_3) \quad \textbf{(11)}$$

In other words, as the surface area S of the triangle formed by $\mathbf{w}_{1,i,j}\ (x_{1,i,j}, y_{1,i,j})$, $\mathbf{w}_{2,i,j}\ (x_{2,i,j}, y_{2,i,j})$, $\mathbf{w}_{3,i,j}\ (x_{3,i,j}, y_{3,i,j})$ approaches the respective centroids as variance σ approaches zero. This is another method of looking at the variance between the microarray experiments. Again, this method takes into account both channel signal intensities probably allowing comparisons and testing for the individual channel signals.

2.4 The Vectorial/Geometric Transformation Approach for a Multiple-fold "Loop" Design

We have encountered and solved the problem of the vector transformation for a three-fold *"loop" design.* But, how can the problem be addressed for larger (meaning more than three) *"loop" design*? In Figure 4 we present a geometric solution to that question.

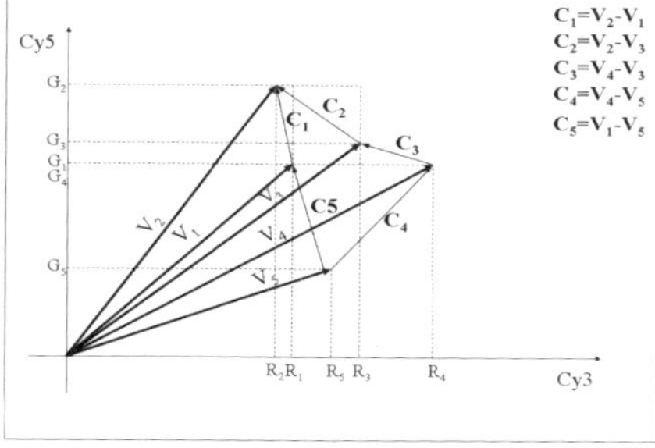

Figure 4. Example of a *"loop" design* with five experimental setups.

From the analysis we described in section 2.2 we can derive a generalized formulation of the problem for more than three nodes in a *"loop"* experiment. In particular the general formulation would be:

$$\| c_n^2 \| = (G_{n+1} - G_n)^2 + (R_n - R_{n+1})^2 \quad \textbf{(12)}$$

$$\begin{aligned} c_n^2 &= v_n^2 + v_{n+1}^2 - 2\| v_n \| \| v_{n+1} \| \cos(\omega_n) \\ c_{n+1}^2 &= v_{n+1}^2 + v_{n+2}^2 - 2\| v_{n+1} \| \| v_{n+2} \| \cos(\omega_n + \omega_{n+1}) \\ c_{n+2}^2 &= v_n^2 + v_{n+2}^2 - 2\| v_n \| \| v_{n+2} \| \cos(\omega_{n+1}) \end{aligned} \quad \textbf{(13)}$$

and:

$$R_i = G_{i-1} \text{ for i=1,...,n and } R_1 = G_n \quad \textbf{(14)}$$

$$\begin{aligned} c_n^2 &= v_n^2 + v_{n+1}^2 - 2v_n v_{n+1} \cos(\omega_n) \\ c_{n+1}^2 &= v_{n+1}^2 + v_{n+2}^2 - 2v_{n+1} v_{n+2} \cos(\omega_n + \omega_{n+1}) \\ c_{n+2}^2 &= v_n^2 + v_{n+2}^2 - 2v_n v_{n+2} \cos(\omega_{n+1}) \end{aligned} \quad \textbf{(15)}$$

giving the generalized form of calculating the nth vector as a function of the previous two.

3. CONCLUSIONS

The present work was concerned with the development of a new mathematical framework for the analysis of *"loop"* microarray experimental setups. We propose a method utilizing the vectorial properties of microarray dual-channel intensities. This method takes into account not only channel ratios but also channel signal intensities It is moreover representing each gene measurement as a vector, making use of its geometric properties. The present work sets a framework for the development of a more robust and reliable microarray data analysis methodology in the future.

Keywords: Microarrays, vectors, gene expression profiles, methodology, dual-channel microarrays

REFERENCES

[1] J. P. Townsend, *BMC Genomics,* vol. 4, p. 41, Oct 2 (2003).
[2] A. Butte, *Nat Rev Drug Discov,* vol. 1, pp. 951-60, Dec (2002).
[3] V. Vinciotti, R. Khanin, D. D'Alimonte, X. Liu, N. Cattini, G. Hotchkiss, G. Bucca, O. de Jesus, J. Rasaiyaah, C. P. Smith, P. Kellam, and E. Wit, *Bioinformatics,* vol. 21, pp. 492-501, Feb 15 (2005).
[4] Y. Woo, W. Krueger, A. Kaur, and G. Churchill, *Bioinformatics,* vol. 21 Suppl 1, pp. i459-67, Jun (2005).
[5] G. A. Churchill, *Nat Genet,* vol. 32 Suppl, pp. 490-5, Dec (2002).
[6] M. K. Kerr and G. A. Churchill, *Biostatistics,* vol. 2, pp. 183-201, Jun (2001).
[7] N. S. Altman and J. Hua, *Genet Res,* vol. 88, pp. 153-63, Dec (2006).
[8] K. Dobbin and R. Simon, *Bioinformatics,* vol. 18, pp. 1438-45, Nov (2002).
[9] M. Pirooznia, P. Gong, J. Y. Yang, M. Q. Yang, E. J. Perkins, and Y. Deng, *BMC Genomics,* vol. 9 Suppl 2, p. S11, (2008).
[10] G. I. Lambrou, S. Vlahopoulos, C. Papathanasiou, M. Papanikolaou, M. Karpusas, E. Zoumakis, and F. Tzortzatou-Stathopoulou, *Leuk Res,* vol. 33, pp. 1684-95, Dec (2009).

A CLUSTERING – BASED METHOD FOR GENE EXPRESSION COMPARISON IN EMBRYONIC STEM CELL DERIVED HEAMNGIOBLASTS AND ADULT BONE MARROW STROMAL CELLS

LEANDROS MAGLARAS AND DIMITRIOS FOTIADIS
Unit of Medical Technology and Intelligent Information Systems, Dept. of Engineering and Material Science University of Ioannina, GR 45110 Ioannina, Greece

IOANNIS GEORGIOU
Genetics Unit, Department of Obstetrics and Gynecology, Medical School, University of Ioannina, GR 45110 Ioannina, Greece

It is of great significance to discover commonly expressed genes in embryonic stem cells and into the tissues that those cells are developed. Microarray technology is widely used in stem cell research to identify new cell lines, to study stem cell function (functional genomics) and to define their role in cancer development. The proposed method focuses on the discovery of similarities between progenitor stem cells and the corresponding adult tissue. Initially, the most significantly expressed genes are identified using the t and the p-value test. Next, four different mathematical validation techniques are used to calculate the optimal number of gene clusters. Finally, the K – means algorithm is employed to find the common genes within the clusters. For the evaluation of the proposed approach microarray data from the Gene Expression Omnibus and the Stembase databases are employed. These datasets contains embryonic stem cell-derived hemangioblasts (Blast – Colony – Forming Cells) and adult bone marrow stromal cells from mus musculus. The application of K - means relates the expressed genes in the progenitor embryonic stem cells with the genes which express in the tissue of the full-grown organism which is developed. Ten distinct clusters are extracted for the data which are analyzed and the similarities of the most significantly expressed genes are discussed.

1. Introduction

DNA microarray technology is used to simultaneously measure the gene expression of a large number of genes. This technology relies on the hybridization properties of nucleic acids to monitor DNA or RNA abundance on a genomic scale in different types of cells. A miniaturized array has up to tens of thousands of single stranded DNA attached to it (probes) that are in truth a short section of a gene. DNA microarrays can be used to detect changes in gene expression level, Single Nucleotide Polymorphisms (SNPs), gene expression patterns in healthy or diseased biological tissues. In a typical microarray chip when the hybridization is finished the chip is scanned at two different wavelengths since two different paints are used to tag the cDNA used. Then the chip is scanned at two different wavelengths and the intensities of each probe are measured. The final chip can inform us about the expression of a gene in a particular experimental condition. The data extracted from a typical microarray experiment can be represented by a matrix where the rows are the genes and the columns are the experiments as it is shown in Fig 1.

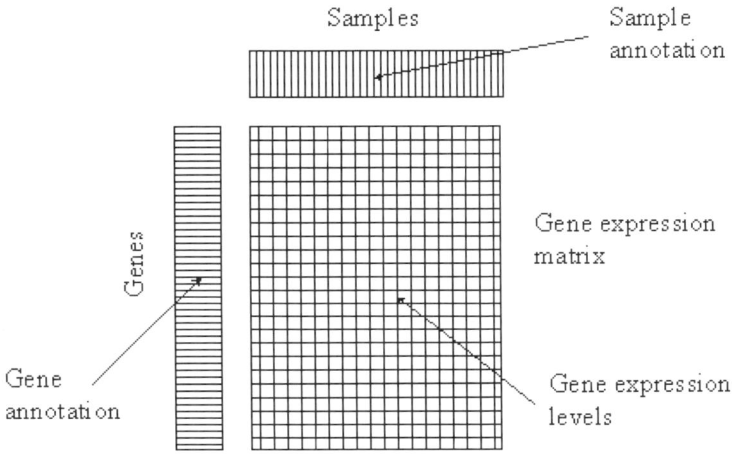

Figure 1. Data matrix of a typical microarray experiment.

Stem cells are characterized by some unique properties. More specifically, in order to characterize a cell as stem cell it must have the self – renewal and the potency properties. The self – renewal is the ability to go through numerous cycles of cell division while maintaining the cell's undifferentiated state. The potency is the capacity of a cell to differentiate into specialized cell types. The stem cells used in this work are of the hematopoietic type and come from the embryonic mesoderm where primitive embryonic stem cells are evolved in the adult hematopoietic stem cells. These two stem cell types (embryonic mesodermal and hematopoietic stem cells) are also present in a process known as angiogenesis. Angiogenesis is the process of the forming of blood vessels from primal angiogenetic cells (stem cells in mesoderm)[1,2].

A microarray data analysis for gene expression consists of several steps the first of which is the normalization[3]. Normalization adjusts for differences in labeling and detection and for differences in the quantity of initial RNA from the samples examined. The expected mean intensity ratio between the two channels should be one. These problems can cause a shift in the average ratio of the scanned intensities and they must be rescaled before an experiment can be properly analyzed. Most of the proposed methods are based on the Robust Multi Array Normalization (RMA), GCRMA or MAS 5.0 algorithms and target more specific types of data[4, 5,6].

The commonly used clustering methods include K – means clustering, a method which uses a predefined number of clusters. This algorithm is commonly used when there is biological knowledge concerning the data to be examined[7, 8, 9]. Some algorithms propose a method to implement biological knowledge in the K –means algorithm in order to use this information to calculate the number of clusters[8, 10]. Hierarchical clustering creates a hierarchy of clusters which may be represented in a tree structure. The root of the tree consists of a single cluster containing all observations, and the leaves correspond to individual observations[11, 12, 13]. Hybrid methods which use both hierarchical and k – means clustering trying to obtain the benefits of both algorithms have also been proposed [14, 15, 16].

The proposed methodology tries to find biologically meaningful relationship between gene expression in embryonic stem – cell derived hemangioblasts and adult hemangioblasts. We use the raw cell intensities from publicly available microarray data and then we normalize our data with the RMA algorithm. Then, we filter the genes to narrow our dataset and calculate the number of clusters using four mathematical indexes. Finally, we cluster the genes using K - means clustering and find the common biological role in the genes within a cluster. In our work we try for the first time to compare gene expression between progenitor hematopoietic and adult stem cells. The data used are from microarray experiments and are available through known databases. We propose a methodology whicht clusters the data in a predefined number of clusters without a priori biological knowledge but with the help of mathematical indexes. Clustering allows us to understand common biological functions of genes within the same cluster.

2. Method

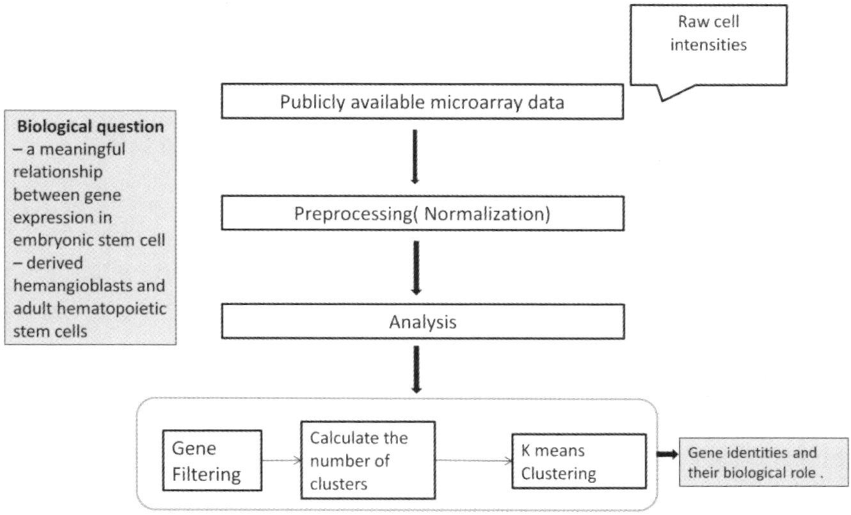

Figure 2. Flowchart of the proposed methodology.

The flowchart of our methodology is shown in Fig. 2. The proposed methodology tries to find biologically meaningful relationship between gene expression in embryonic stem – cell derived hemangioblasts and adult hemangioblasts. We use the raw cell intensities from publicly available microarray data and then we normalize our data with the RMA algorithm. Then, we filter the genes to narrow our dataset and calculate the number of clusters using 5 mathematical indexes. Finally, we cluster the genes using K - means clustering and find the common biological role in the genes within a cluster.

2.1 Preprocessing

In the preprocessing step we firstly remove the non – expressed genes from our data matrix. These are the genes that have zero intensity values in the experiment. After that, the normalization procedure takes place. We use a technique to reduce biological and laboratory variation between experiments. This technique takes place in three steps. In the first step, we define the perfect match (PM) and mismatch (MM) variables in a microarray experiment that is repeated j times where i are the different genes, j are the different samples and k are the different probes. The algorithm then proceeds by assuming a background correction model, where s_{ij} is the hybridization signal that is distributed exponentially and bg_{ijk} is the background noise modeled by a normal distribution.

$$PM_{ijk} = bg_{ijk} + s_{ijk} \quad . \qquad (1)$$

The second step is the quantile normalization between different microarrays that is applied to the background - corrected hybridization signals. The final step is the additive model that is applied to the background – corrected, quantile – normalized and log – t transformed values of Pm_{ijk}, where Y_{ijk} is the background – corrected, quantile – normalized and log – transformed values of PM, y_{ij} is the final normalized log transformed expression leve,l a_{ik} is the probe affinity and e_{ijk} is random Gaussian noise.

$$Y_{ijk} = y_{ij} + a_{ik} + e_{ijk} \, . \qquad (2)$$

The above model is fitted to the data and the final, normalized values of the logarithms of the expression signals are obtained. A histogram showing the distribution of Perfect Match probe intensities values (black) and the convoluted probability distribution function (grey) is shown in Fig 3.

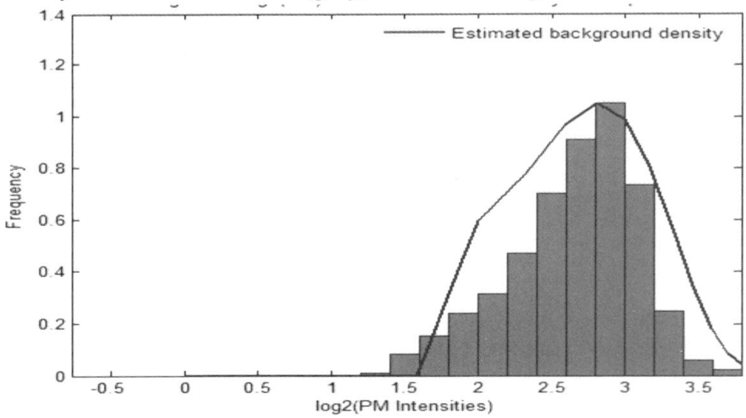

Figure 3. Distribution of Perfect Match probe intensities values (black) and the convoluted probability distribution function (qrey).

2.2 Gene filtering

After the normalization, the gene filtering process takes place to narrow our dataset. During this procedure, we aim to find and remove a number of genes that are statistically insignificant. This is done by performing an unpaired t –

test for differential gene expression with a standard two – tailed and two – sample t – test for every gene in our two different cell types. With a p – value cutoff of 0.05 we find the genes that have statistical significance. Finally, we remove the housekeeping genes that are typically needed for the maintenance of the cell because they are transcribed at a relatively constant level and their expression are unaffected by experimental conditions. The normal quantile plot of the t – score is shown in figure 4a. A volcano plot where we can see with the deeper blue color the genes that are outside of our cutoff value of 0.05 is shown in fig 4b.

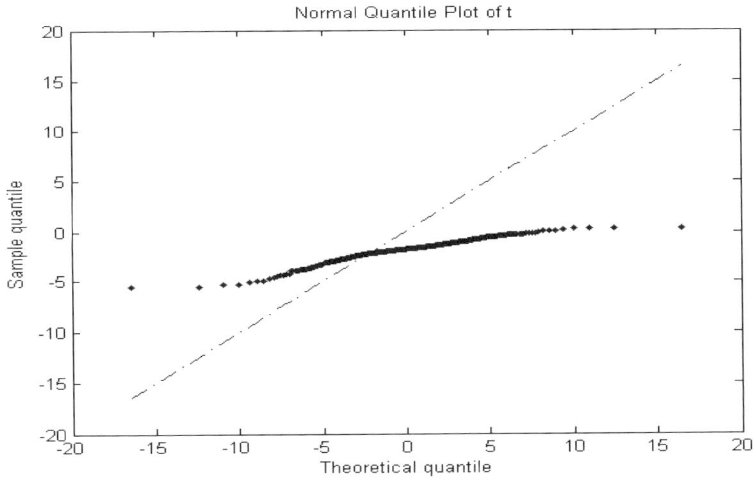

Figure 4a. Normal quantile plot of the t – score.

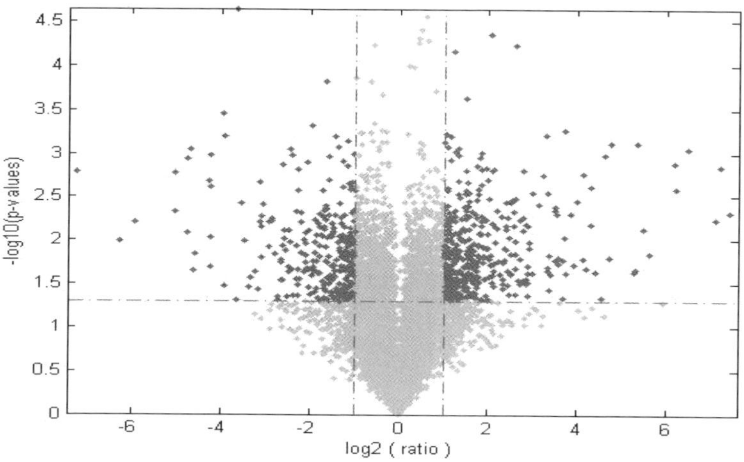

Figure 4b. Volcano plot

2.3 Calculating the number of clusters

K - means, is a divisive clustering method that needs a priori knowledge to define the number of clusters. In microarray data analysis this a priori knowledge is usually biological but in our method this is the biological relationship we try to prove. For this reason we use mathematical validation indices to calculate the number of clusters and four indexes are used for higher accuracy.

2.3.1 Silhouette Index[17]

The Silhouette Index describes the average $s(i)$ of a cluster which is a measure of how tightly grouped all the data in the cluster are. In the equation that follows $a(i)$ is the average dissimilarity of i with all the other data within the same cluster, $b(i)$ is the cluster with the lowest average dissimilarity to i. Thus the average $s(i)$ in the entire dataset is a measure of how appropriately the data has been clustered. If there are too many or too few clusters, such as when a poor choice of k in the K - means algorithm has been made, some of the clusters will display much narrower silhouettes than the rest:

$$s(i) = \frac{b(i) - a(i)}{\max\{a(i), b(i)\}} \quad . \tag{3}$$

2.3.2 Rand Index[18]

The second index used is the Rand index where given a set of n elements S and two partitions of S to compare $X = \{x_1, \ldots, x_n\}$ and $Y = \{y_1, \ldots, y_n\}$ the index is given by the following equation (4) where a is the number of pairs of elements in S.

$$R = \frac{a+b}{a+b+c+d} = \frac{a+b}{\binom{n}{2}} \,. \tag{4}$$

Where a, the number of pairs of elements in S that are in the same set in X and in the same set in Y, b, the number of pairs of elements in S that are in different sets in X and in different sets in Y, c, the number of pairs of elements in S that are in the same set in X and in different sets in Y, d, the number of pairs of elements in S that are in different sets in X and in the same set in Y. $a + b$ is the number of agreements between X and Y and $c + d$ as the number of disagreements between X and Y. The Rand index has a value between 0 and 1, with 0 indicating that the two data clusters do not agree on any pair of points and 1 indicating that the data clusters are exactly the same.

2.3.3 Davis and Bouldin and Calinski and Harabash Indexes[17,19]

The Davis and Bouldin Index is given by the following equation

$$R = (1/n)\sum_{i=1}^{n}(R_1) \,. \tag{5}$$

where R_1 stands for the maximum value of R_{ij},

$$R_{ij} = (SSW_i + SSW_j) / DC_{ij} \,. \tag{6}$$

and SSW is the sum of squares within clusters and DC_{ij} is the distance between the centers of two clusters i,j and k *is* the number of clusters.

The Calinski and Harabash index is given by Eq. (7) where k and n are the data points, SSB, SSW the sum of squares between and within clusters respectively;

$$\frac{SSB/(k-1)}{SSW/(n-k)} \tag{7}$$

2.4 K – means clustering

We then cluster our data of 726 genes within 5 clusters as suggested by the validation indices and using the euclidean distance metric k means algorithm.

Given a set of observations $(x_1, x_2, \ldots, x_n)$, where each observation is a d-dimensional real vector, then K-means aims to partition the n observations into k sets $(k < n)$ $S = \{S_1, S_2, \ldots, S_k\}$ so as to minimize the within-cluster sum of squares that is given by this equation.

In Fig. 5 we can see our data grouped in 5 different clusters. The first cluster contains 128 genes, the second cluster 167 genes, the third cluster 75 genes, the fourth 190 genes and the fifth cluster 184 genes.

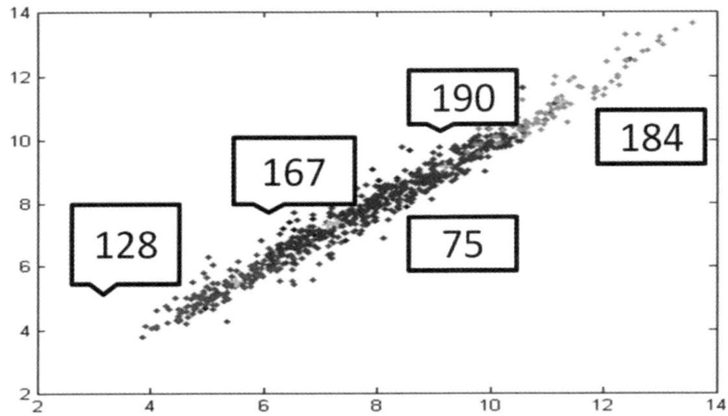

Figure 5. Results of k –means clustering.

3. Datasets

The dataset used for this method is composed of 8 microarray experiments that use the Affymetric MG – U74A chip. The 4 experiments are embryonic mesodermal stem cells (hemangioblasts), they are multipotent stem cells, progenitors of the hematopoietic cells and are isolated from the mus musculus[20]. The other 4 are Hematopoietic stem cells of the CD34-cell type and are usually isolated from the umbilical cord or from the bone marrow of mus musculus[21]. Both of these experiments are available through the Stembase database.

4. Results

We present in Fig. 6 an overview of the gene expression profiles, the genes that have been clustered and the five distinct groups we have divided the data into.

Each probe has been centered on its median expression values across the samples analyzed. Measurements that are above the median are colored black and those below the median are colored grey. The statistically and biologically significant genes are 726 out of 12488 in 8 microarray hybridizations, 4 samples with cd34 hematopoietic stem cells and 4 samples of mesodermal stem cells (hemangioblasts). We also describe the genes with a distinct biological function in each cluster that are discovered with the help of the Gene Annotation database (GO).

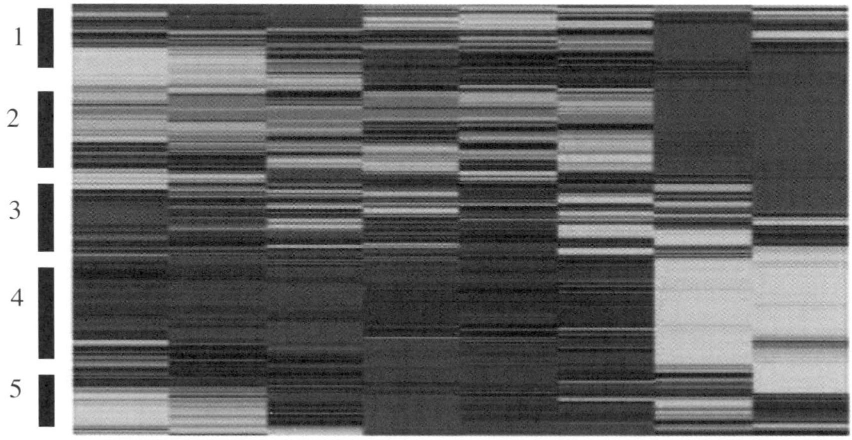

Figure 6: Overview of the gene expression profiles. Each probe has been centered on its median expression values across the samples analyzed. Measurements that are above the median are colored black and those below the median are colored grey. The intensity is proportional to the fold change.

In the first cluster we find the following cycle – related genes Cyclin D1, Cyclin D3, Cycclin E1, Cyclin B1, CDK 2, CDK 7 CDK 8 P21 cip1/waf and P27 kip1. In the second cluster we discern apoptosis – related genes Bcl 2, Bfl 1, Mcl1, Bac, Bad, Bmf, Bid, Bid3, Bok that are members of the bcl - 2 family subgroup and the FADD, TRAIL and Fas genes that are members of the death pathway subgroup. In the third cluster we have the genes that are hematopoietic transcription factors Ranx, Lmo2, E2A, Myb, Bml -1, Gata, Fog, Bax5 and the myeloid subgroup that contains the Pu1, Fli 1 and Elf genes. In the fourth cluster we discern hematopoietic – receptor type molecules Kit – R, TransferinR, Fet, M – CS – F – R, Osm, Cntf – R genes. In the fifth and final cluster we find the genes that are related to Wnt signaling: Fzd, Lrp, Kremen. Ror, Vangl, Sfrp, Dvl, CK, Gpb, Frat, Axin, Dapper, Dab2.

The accuracy in each is shown in Table 1. As accuracy we define the gene percentage of the whole number of genes in each cluster, related to the distinct biological function associated with that particular cluster.

Table 1: Accuracy in each cluster.

Clusters	Total number of genes	Number of genes with distinct biological function	Accuracy %
1	128	97	78
2	167	115	67
3	75	49	65
4	190	172	90
5	184	155	84

5. Discussion

We proposed a clustering – based methodology to compare gene expression between embryonic stem cell derived hemangioblasts and adult bone marrow stromal cells. Our method classifies accurately 726 genes within 5 clusters with distinct biological functions each using K – means clustering. Each of these biological functions plays a role in cell development and evolution from cell to adult stem cells. To define the number of clusters that we divided our data into we used 4 mathematical indexes for greater accuracy in our prediction, because there was no prior biological knowledge to define the number of clusters.

We obtained genes that have a common biological function within the same cluster. The accuracy percentages of the clusters were higher in the first, fourth and fifth cluster (78%, 90% and 84%, respectively) and lower in the second and third cluster (67% and 65%, respectively).

Our method needs to be tested in a larger dataset and more particularly, we need to obtain more experimental microarray data from embryonic hematopoietic stem cells. There is no comparison using experimental data from microarrays between adult and embryonic stem cells in literature thus, we cannot compare our method in terms of accuracy and gene classification. In the future, it is possible to compare other stem cell types from embryonic and adult tissues. It is also possible to try a comparison between different stem cell types of the same origin e.g. adult hematopoietic cells with adult neural stem cells. Finally, there is interest in a comparison between stem cells and cancer cells since it has been suggested that they share some properties.

6. References

1. Fujiwara H., Hayasi Y., Sanzen N., Weber C.N., Emto T.,Futaki S., Niwa H., Murray P., Edgar D., Sekiguchi K., Regulation of mesoderm differentiation of mouse embPacini S., Trombi L., Fazzi R., Montali M., Ikehara S., Abraham NG. Identification and purification of mesodermal progenitor cells (MPCs) from human adult bone marrow. 18(6): 857-866. doi:10.1089/scd.2008.0291 (2009).
2. Pacini S., Trombi L., Fazzi R., Montali M., Ikehara S., Abraham NG. Identification and purification of mesodermal progenitor cells (MPCs) from human adult bone marrow. 18(6): 857-866. doi:10.1089/scd.2008.0291 (2009).
3. Bolshakova N., F. Azuaje. Estimating the number of clusters in DNA microarray data. Methods of Information in Medicine 45(2):153-7 (2006).
4. Barbacioru C., Wang Y., Canales R., Sun Y., Keys D., Chan F., Poulter K.,Samaha R. Effects of various normalization methods on Applied Biosystems expression array system data. BMC Bioinformatics 7:533 (2006).
5. Dabney A., Dtorey D., Normalization of two - chammel microarrays accounting for experimental design and intensity - dependent relationships. Genome Biology, doi:10.1186/gb-2007-8-3-r44. Normalization of microarray data: single labelled and dual-labelled arrays, (2006).
6. Lim W., Wang K., Lefenbre C. and Califano A. Comparative analysis of microarray normalization procedures: effects on reverse engineering gene networks.Bioinformatics 23(13):i282-i288 (2007).
7. Dougherty, E.R. The fundamental role of pattern recognition for gene-expression/microarray data in bioinformatics. PR(38), No. 12 pp. 2226-2228 (2005).
8. Sung LY, Gao S, Shen H, Yu H et al. Differentiated cells are more efficient than adult stem cells for cloning by somatic cell nuclear transfer. Nat Genet; 38(11):1323-8(2006).
9. Eisen MB, Chiang DY, Brown PO Discovery of regulatory elements from microarray data using genome mean expression profiles. 9th International Conf on Intelligent Systems for Molecular Biology, Copenhagen (2001).
10. Tibshirani R, Hastie T, Eisen M, Ross D, Botstein D, Brown P. Clustering methods for the analysis of DNA microarray data. Technical Report, Department of Statistics, Stanford U (1999).
11. Bandyopadhyay S., Mukhopadhyay A., and Maulik U. An improved algorithm for clustering gene expression data, Bioinformatics, 23(21):2859-2865 (2007).

12. Boratyn G., Datta S., Datta S. Incorporation of biological knowledge into distance for clustering genes, Bioinformation, 1(10):396-405 (2007).
13. Burkart F., Wren D., Herschkowitz D., Perou C., Garner H. Clustering microarray-derived gene lists through implicit literature relationships, Bioinformatics, 23:1995-2003 (2007).
14. Bréhélin L., Gascuel O., and Martin O., Using repeated measurements to validate hierarchical gene clusters, Bioinformatics, 24(5):682-688 (2008).
15. Chen X. Curve-based clustering of time course gene expression data using self-organizing maps. Journal of Bioinformatics and Computational Biology, 7(4):645-661 (2009).
16. Du Z., Wang Y., Ji Z. PK-means: A new algorithm for gene clustering, Computational Biology and Chemistry, 32(4):243-247 (2008).
17. Bolshakova N., F. Azuaje. Estimating the number of clusters in DNA microarray data. Methods of Information in Medicine, February; 45(2):153-7 (2006).
18. Dudoit, J. Fridlyand. A prediction-based resampling method for estimating the number of clusters in a dataset. Genome Biology. 3(7): 0036.1-21(2002).
19. Strehl. Relationship-based Clustering and Cluster Ensembles for High-dimensional Data Mining. Ph.D thesis, The University of Texas at Austin (2002).
20. Lugus JJ, Chung YS, Mills JC, Kim SI et al. GATA2 functions at multiple steps in hemangioblast development and differentiation. Development; 134(2):393-405 (2007).

MORPHOLOGICAL IMAGE DENOISING AND RANDOM-WALKS SPOT DETECTION IN MICROARRY IMAGES

A. MASTROGIANNI, E. DERMATAS

Dept. of Electrical Engineering and Computer Technology, University of Patras, HELLAS

A. BEZERIANOS

Dept. of Medical Physics, School of Medicine, University of Patras, HELLAS

In this paper, a novel method for noise reduction and spot detection in microarray images, based on mathematical morphology and Random-Walks model is presented and evaluated. Initially, an automatic method detects the subarrays and a morphological denoising method is used to reduce artifacts at each individual subarray. In the enhanced images, a novel segmentation algorithm based on Random-Walker algorithm (RWA) derives the spot positions. The proposed method is validated in two different databases of real distorted microarray images, giving quite satisfying results both in noise elimination and spot detection accuracy.

1. Introduction

DNA microarray technology is a high-throughput technique that determines how a cell can control the expression of large numbers of genes simultaneously. In a typical microarray setting, up to 10,000 different cDNA segments are spotted onto a glass slide. When a solution of fluorescently labeled DNA fragments is added to the glass slide, fluorescence occurs in the spots where gene activity is present. The output of a single microarray experiment is a digital image. The whole microarray image, known as a meta-array, is composed by a matrix of almost equally spaced blocks, called subarrays or subgrids, each one consisting of a certain number of rows and columns. In an ideal microarray image, the spots in each subarray are arranged at equal space. However, in real microarray images the subarray spots are not spaced uniformly, because of variations occurring during printing and fixation processes. Moreover, due to the nature of the acquisition process and the reusability capabilities of the glass slides, microarray experiments involve a large number of error-prone procedures that lead to a high level of noise in the resulting data. A very noisy microarray image and its lower-left subarray are shown in Fig.1.

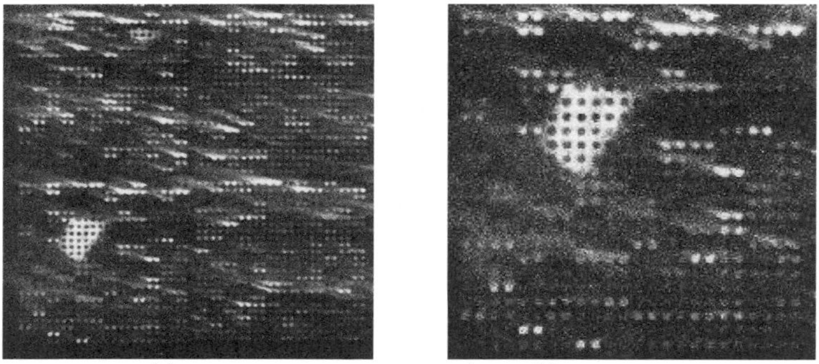

(a) (b)

Figure 1. A real noisy microarray image: (a) The image contains four subarrays distorted by a noise known as "comet tails", (b) The lower left subarray with high background noise and weak spots.

The problem of estimating the spot intensity values is decomposed into three major tasks [1]:

1. *Addressing or Gridding*: the process of assigning coordinates to each spot.
2. *Segmentation*: each spot pixel is classified to either as foreground or background.
3. *Intensity extraction*:
   i. extraction of foreground spots intensity,
   ii. extraction of background intensity.

Noise elimination and spot segmentation, both constitute challenging problems in microarray image analysis as they lead to the recovery of the gene expression measurements. In the last decade, microarray image denoising and spot detection problems have been, thoroughly, studied and a lot of robust methods have been proposed, giving various solutions for the enhancement of the microarray image [2-5].

Mathematical morphology (MM) is a powerful non-linear image analysis technique with applications in several fields, including medical diagnostics, histology, industrial inspection, computer vision and character recognition [6-7]. The Random-Walks model is used to derive pixel-based annotation probabilities by estimating a random walker's probability to reach a known identity pixel, starting from the pixel of interest. The results of Random-Walks model have been applied to physics, economics, computer science, image processing and a number of other fields as a fundamental model for random processes in time [8-9]. Recently, the Random-Walker algorithm (RWA) has been introduced as a close-form, high-quality segmentation algorithm requiring only an initial annotation for a small number of pixels, the seeds [10].

This paper is proposing a novel denoising and spot detection method for microarray images, based on morphological operations and RWA. During the experimental process, where images from two databases are used to evaluate the proposed method, noise level is eliminated and accurate spot detection is obtained, even in case of relatively-high distorted images.

In the next section a detailed presentation of the noise reduction and spot segmentation method is presented. In section 3, the experimental database and a number of representative examples are shown. Finally, a short discussion on the results is given.

2. Microarray image denoising and spot detection method

Based on the popular MM tools and the RWA, the proposed method is decomposed into three main processes:

A. Subarray detection and rotation.

Due to the fact that each subarray may contain a different type of noise and have different rotation angles, comparing to the other subarrays in a microarray, it is absolutely necessary to detect their individual position and correct the rotation angle separately (Fig.1). In this work, the subarrays position is automatically estimated using a recently proposed subgrid detection method [11]. Initially, in the detected microarray area, an adaptive profiling projection-based method derives the subarray grid. During grid detection, spot spacing is estimated and used for exact subarray location. The estimated subarrays are rotated, if needed, using an alignment process proposed in [12].

B. Image enhancement and noise elimination.

Each two-channel (red-green) subarray is converted into a gray-scale image and a median filter is applied reducing "salt&pepper"-like noise. The conversion into the grayscale space enhances the real spot areas and the noise density is decreased in cases where the noise distorts individually the red or the green channel. The subarray image is contrast enhanced and converted into a binary image using Otsu's method [13].

In each binary subarray, all components i.e. spots and noisy artifacts, are labeled using a MM function. A binary image is annotated by a grouping process, associating the same label in neighbor pixels that share similar pixel intensity values.

After labeling, a clustering function, classifies the produced labeled areas A, using a threshold value, S_l, based on the number of pixels contained in each component of the subarray. If the expected spot radius R is known, the spot area must contain $S_2=\pi R^2$ pixels. All components, whose area A satisfies the criterion $A <= S_1$, are classified as noisy artifacts, small spots or portions of spots. The remaining isolated regions of interest (ROI) contain the spots which cover the large areas and a small number of artifacts. The new segmented subarray image contains only the highly fluorescent spots and is almost noise-free. The removal of the noisy artifacts is of major importance to the RWA applied to the subarray image for accurate location of the real spots.

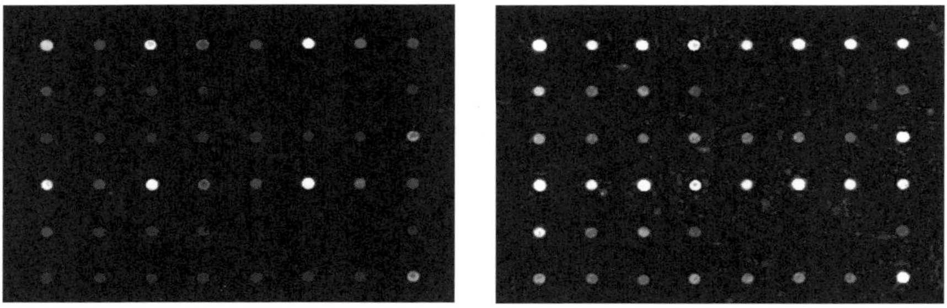

(a) (b)

Figure 2. (a) Gray-scale subarray image from the"Human Sarcomas", (b) Contrast enhanced version.

C. Location of spot positions.

The RWA spot segmentation is obtained by annotating each unseeded pixel to the seed-label that a random walker starting from that pixel would be most likely to reach first. The RWA probabilities are estimated by solving the Dirichlet problem, which leads to a system of linear equations for each subarray under evaluation. In order to represent the subarray image by random walker biases, a function F_{rwa} that maps a change in image intensities to edge weights must be defined.

At each ROI, the pixel closest to the centroid is estimated and is annotated as spot seed. The term spot-seed is denoting neighbor pixels or a unique pixel belonging to a spot. In the proposed method, the RWA is used to segment the areas near centroids. Therefore only one spot pixel for each ROI is assumed to be a spot-seed. Taking into account that the efficiency of the RWA is greatly affected by the accuracy and the number of spot and background seeds, the following methodology is used:

- The spot seeds are defined by the centroid pixels,
- The background seeds are the low-brightness pixels defined by the pixels with intensity lower than a threshold value T_{int} ,
- In the proposed method, the function F_{rwa} is represented by the typical Gaussian weighting function:

$$w_{ij} = exp\,(-\beta\,(g_i\text{-}g_j)^2),$$

where g_i is the pixel i intensity. The value of β represents the only user - adjustable parameter of the RWA.

3. Experimental results and dataset

The proposed method is evaluated in two databases, the "Human Sarcomas" containing 34 microarray images of 32 subarrays [14], and 36 cDNA microarray images from the "Lymphoma/Leukemia Molecular Profiling Project" ("LLMPP"). In this database each image contains 16 subarrays [15].

The spot radius R is set to five pixels and satisfies all spot radiuses on both databases [16]. All spot areas containing less or equal than S_l=25 pixels are removed from the ROI set. The threshold value T_{int}, used for the assignment of the background seeds, is set equal to $T_{int}=0.001$. All pixels whose image intensity (limited in [0, 1]) is less than T_{int} are classified as background seeds.

The presented image denoising and spot detection method has been used to decrease the level of "salt&pepper"-like noise and small artifacts. Furthermore, the proposed method has the capability of detecting small spots in the available subarray images. Two representative examples are shown in Fig.3 and Fig.4. The segmentation accuracy is significantly reduced in strongly distorted images, i.e. extremely low brightness spots are not detectable and high-rate of false positive events are encounter, especially in cases where high level of non-"salt&pepper" noise is met (Fig.5).

4. Conclusions

The proposed denoising and subarray segmentation method is based on the popular MM and the recently proposed RWA. In evaluation experiments, carried out in the subarrays of two different microarray databases, it has been shown that both noise reduction and an accurate detection of spot size and position can be obtained. In case where the subarray images are strongly distorted and a large number of spots are low-fluorescenced, the number of ROI is reduced and consequently the RWA accuracy is decreased.

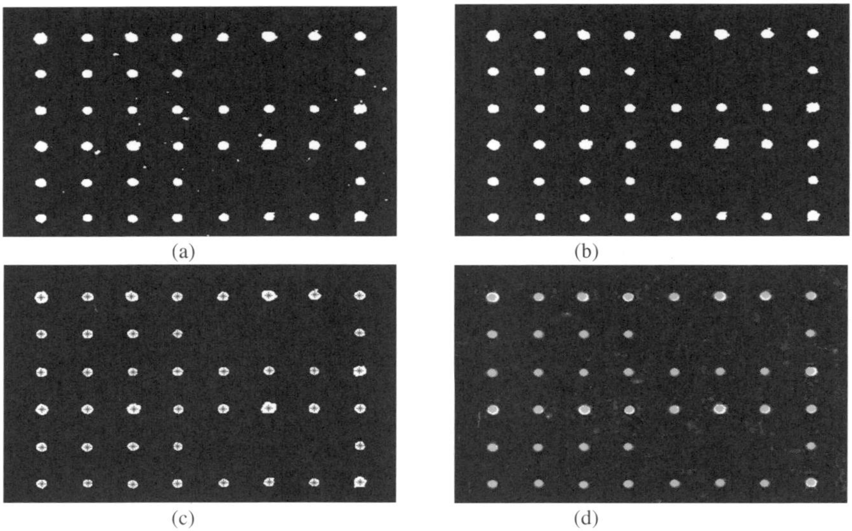

Figure 3. (a) Binary subarray image from the"Human Sarcomas", (b) MM denoised subarray image, (c) Spot-seeds, (d) Spot segmentation on the original gray-scale subarray image.

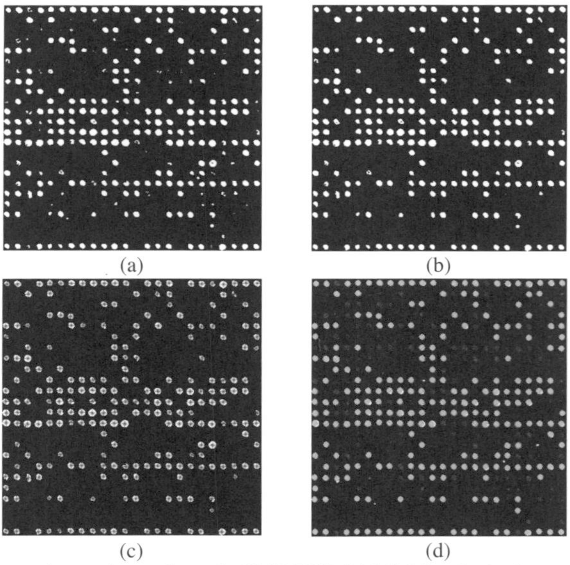

Figure 4. (a) Binary subarray image from the "LLMPP", (b) MM denoised subarray image, (c) Spot-seeds, (d) Spot segmentation on the original gray-scale subarray image.

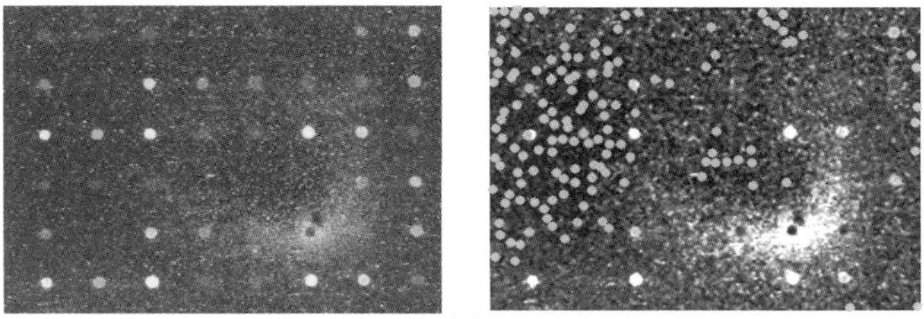

(a)

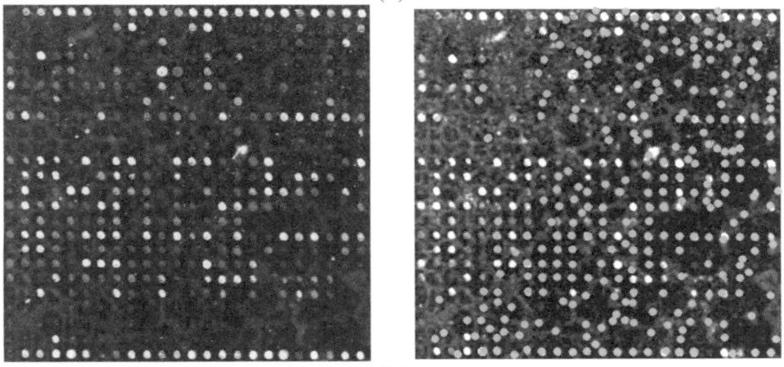

(b)

Figure 5. (a) Highly distorted subarray example from the“Human Sarcomas” and the segmented image, (b) Highly distorted subarray example from the“LLMPP” and the segmented image.

Acknowledgments

This paper is part of the 03ED013 research project, implemented within the framework of the “Reinforcement Programme of Human Research Manpower” (PENED) and co-financed by National and Community Funds (20% from the Greek Ministry of Development-General Secretariat of Research and Technology and 80% from E.U.-European Social Fund).

References

1. Robert S. H. Istepanian, “Microarray Image Processing: Current Status and Future Directions”, IEEE, Transactions on Nanobioscience, Vol.2, No.4, (2003).
2. Th. Margaritis, K. Marias, D. Kafetzopoulos, “Improved Microarray Spot Segmentation by combining two Information Channels”, Proceedings of the 28th IEEE EMBS Annual International Conference, New York, USA, (2006).
3. Jinn Ho, Wen-Liang Hwang, “Segmenting Microarray Image Spots using an Active Contour Approach”, IEEE International Conference on Image Processing, ICIP, (2007).

4. X. H. Wang, Robert S. H. Istepanian, Yong Hua Song, "Microarray Image Enhancement by Denoising Using Stationary Wavelet Transform", IEEE Transactions on Nanobioscience, Vol.2, No.4, (2003).
5. A. Mastrogianni, E. Dermatas, A. Bezerianos, "Microarray Image Denoising using Spatial Filtering and Wavelet Transformation", Proceedings of the 13th International Conference on Biomedical Engineering, ICBME, Singapore, (2008).
6. Felix Meyenhofer, Olivier Schaad, Patrick Descombes, Michel Kocher, "Automatic analysis of microRNA Microarray images using Mathematical Morphology", Proceedings of the 29th Annual International Conference of the IEEE EMBS Cité Internationale, Lyon, France, (2007).
7. Chiao-Ling Shih, Hung-Wen Chiu, "Automatic spot detection of cDNA microarray images using mathematical morphology methods", IEEE EMBS Asian-Pacific Conference on Biomedical Engineering, (2003).
8. I.F. Akyildiz, Yi-Bing Lin, Wei-Ru Lai, Rong-Jaye Chen, "A new random walk model for PCS networks", IEEE Journals on Selected Areas in Communications, Vol. 18, No.7, (2000).
9. Tae Hoon Kim, Kyoung Mu Lee, Sang Uk Lee, "Generative Image Segmentation Using Random Walks with Restart", Proceedings of the 10th European Conference on Computer Vision: Part III, Marseille, France, (2008).
10. Grady, L., 'Random Walks for Image Segmentation', IEEE Trans. on Pattern Analysis and Machine Intelligence, Vol.28, pp. 1768-1783, Nov., (2006).
11. A. Mastrogianni, G. Doukas, E. Dermatas, A. Bezerianos, 'Automatic subarray detection for microarray images', Proceedings of the 4th IAPR International Conference on Pattern Recognition in Bioinformatics, Sheffield, United Kingdom, (2009).
12. A. Mastrogianni, E. Dermatas, A. Bezerianos, "Robust pre-processing and noise reduction in microarray images", Proceedings of the 5th IASTED International Conference on Biomedical Engineering, Innsbruck, Austria, (2007).
13. N. Otsu, 'A threshold selection method from gray-level histograms', IEEE Trans. Sys. Man., Cyber, vol.9, pp. 62-66, (1979).
14. Human Sarcomas database. Available at http://smd.stanford.edu/
15. The Lymphoma/Leukemia Molecular Profiling Project database. Available at http://llmpp.nih.gov/.
16. Peter Bajcsy, http://algdocs.ncsa.uiuc.edu/PR-20050204-1.pdf.

Electromagnetic Scattering and Biomedical Applications

REMOVING NON UNIFORM ILLUMINATION EFFECTS IN INFRARED IMAGES

M. VLACHOS AND E. DERMATAS

Department of Electrical Engineering & Computer Technology

University of Patras, Patras, Greece

{mvlachos,dermatas}@george.wcl2.ee.upatras.gr

In this paper a novel brightness normalization method is used to eliminate the non-uniform illumination effects in infrared images. The algorithm is based on a fuzzy c-means algorithm (FCM), where the objective function is modified to take into account local information of each pixel related both on multiplicative and additive shading effects. At each iteration, the modified FCM algorithm re-estimates the cluster centroids and the membership weights by processing the adopted shading features. Brightness correction is performed by the inverse of the image formation model after FCM convergence.

Experiments were conducted in a database of 10 artificial and 5 real infrared images. The real data consist of images of a human finger acquired by a low-cost CCD camera when the finger is illuminated with an array of five leds in the near-infrared spectrum. The experimental results show that the proposed method decreases significantly the non-uniform illumination effects and does not introduce brightness variations if the background illumination is uniform.

1. Introduction

The correction of non-uniform illumination and the elimination of shading artifacts is an important preprocessing task, employed in a great number of image processing applications such as segmentation, registration, object recognition or quantitative analysis. Although, a careful and accurate set up of the image acquisition system may degrade the importance of a brightness normalization algorithm, non-uniform illumination appears due to light reflection in the scene. The problem of shading and the presence of intensity inhomogeneities are common in many image modalities. Although, these problems have minor impact on human vision, due to the fact that human perception has the ability to deduce the real image content from a distorted image, their impact in automatic image processing or analysis cannot be deteriorated. Thus, the correction of shading and the compensation of non uniform illumination is an important preprocessing step in many image processing tasks. In this paper, a totally different shading effect is also studied: the problem of near-infrared light equalization of images capturing the light propagated through human fingers.

In general, shading can be either object-independent or object-dependent. In the first case, an imperfect set up of the image acquisition system introduces the shading and may be corrected by calibration methods, while the latter shading effect is introduced from the interaction between the objects in the scene. The object-dependent shading can be corrected only retrospectively by using information from the acquired images. An efficient retrospective correction method must correct the shading only when it is present and keep unchanged the shading-free images.

Several methods have already been proposed [1-5] for retrospective shading correction using a linear image formation model, which consists of an additive and a multiplicative shading component. The formation of an image Y is an interaction between objects and illumination in the scene. At each pixel, the relation between the acquired image and the shading-free image X can be described by the following linear model:

$$Y = A \cdot X + B, \qquad (1)$$

where A and B are the pixel-dependent multiplicative and additive shading parameter respectively. In shading correction, the image X is derived from the acquired image Y by estimating the multiplicative A and additive B components.

Many approaches [1, 2, 4] simplify the problem by assuming that only one component, either the multiplicative [2, 4] or the additive [1, 2], has distorted the shading-free image X. In object-independent shading correction, the multiplicative or the additive component is estimated by acquiring one calibration image. In object-dependent shading correction, the solution requires a retrospective method, i.e. the pixel dependent shading components are estimated from the acquired image Y.

A numerous retrospective shading correction methods have already presented in the literature [1-5]. Among them, linear filtering [1, 2] assumes that only the additive shading component has distorted the image and the correction parameter is estimated using low pass filtering. In a similar approach, homomorphic filtering [2, 4] assumes that only the multiplicative shading component is present in the acquired image and the correction parameter is estimated using low pass filtering in the log-intensity of the acquired image. In both approaches the success in the estimation of the shading components depends on the adopted distortion model and the use of an appropriate low pass filter. In [6], a *YUV* shading correction technique applicable to colour images is implemented using a linear filtering approach only to the intensity values of channel *Y*. Experiments conducted on images with synthesized shading show feasibility of the approach.

More advance solutions use morphological filtering [5], assuming that the background has darker or brighter intensity values than the other objects of the scene and the objects size is smaller than the brightness variation in the background area. Thus, the selection of an appropriate morphological operator can be used for the estimation of the background pixels, which represents either the multiplicative or the additive shading component.

Due to the specific image acquisition process, very often retinal images are non-uniformly illuminated and exhibit local luminosity and contrast variability and thus the automatic analysis for diagnostic purposes becomes very difficult. In these images, the luminosity and the contrast variability is estimated from the background part of the image and the subsequent compensation of this variability is applied in the whole image [7]. A method to correct non-uniform illumination in transmission electron microscopy (*TEM*) is proposed in [8]. *TEM* images suffer from brightness variations due to electron imaging defects or non-uniform support films and specimen staining. These variations render image processing operations such as segmentation more difficult. The derived close-form correction is computational efficient, requiring the estimation of the global illumination field which is achieved using only image intensity gradients.

A fuzzy c-means (FCM) based algorithm that simultaneously estimates the shading distortion while segmenting the image is presented in [9]. A multiplier field term, used to approximate the spatial intensity variation, is incorporated into the FCM objective function which is minimized iteratively. The estimation of the bias field is based on the current tissue class centroids and the membership values of the voxels and an iterative low-pass filtering is then used for fine tuning. The algorithm is evaluated on clinical breast *MRI*.

A novel algorithm is used for fuzzy segmentation of magnetic resonance imaging data and estimation of intensity inhomogeneites using fuzzy logic [10]. This algorithm is formulated by modifying the objective function of the standard FCM algorithm to compensate for such inhomogeneites and to allow the pixels annotation to be influenced by the labels in its immediate neighbors. Experimental results are conducted on both synthetic images and *MRI* data and demonstrate the effectiveness and efficiency of the proposed algorithm.

In this paper, a novel FCM brightness normalization method is used to estimate both multiplicative and additive shading components of the non-uniform illumination effects. The FCM objective function is modified to take into account local information of each pixel and to include the linear model parameters of the shading distortions. The modified FCM algorithm is iterative and at each iteration the model parameters are re-estimated based on the cluster centroids and the pixel memberships. Brightness correction is performed by the inverse of the image formation model after FCM convergence.

2. Linear shading correction based on the FCM algorithm

If the relation between the acquired image Y and the shading-free image X is described by the linear model of (1), where A, B are the unknown pixel-dependent multiplicative and additive shading components, the objective function of the FCM algorithm for separating the data x into c clusters is given by:

$$J = \sum_{i=1}^{c}\sum_{k=1}^{N} u_{ik}^{p} \cdot \left\| x_k - v_i \right\|^2 , \tag{2}$$

where $\{v_i\}_{i=1}^{c}$ are the cluster centroids. The memberships $[u_{ik}] = U$ satisfy the following restrictions:

$$\mathcal{U}\left\{ u_{ik} \in [0,1] \mid \sum_{i=1}^{c} u_{ik} = 1 \ \forall\ k \text{ and } 0 < \sum_{k=1}^{N} u_{ik} < N \ \forall\ i \right\}. \tag{3}$$

The fuzzifier p is a weighting exponent on each fuzzy membership and determines the amount of fuzziness of the resulting classification. According to restrictions described in (3), the objective function is minimized when high membership values are assigned to pixels. In the FCM segmentation, each pixel label is estimated using the greater membership criterion.

Exploiting the fact that neighbor pixels are spatially correlated and frequently belonging to the same class [10], the objective function of the FCM algorithm is modified by introducing a factor which allows correlation between labels of neighbour pixels [11]. This factor biases the FCM convergence and consequently forces the solution toward piecewise-homogenous labelling. The modified objective function is given by:

$$J_{ab} = \sum_{i=1}^{c}\sum_{k=1}^{N} u_{ik}^{p} \cdot \left\| x_k - v_i \right\|^2 + \frac{n_R}{N_R} \cdot \sum_{i=1}^{c}\sum_{k=1}^{N} u_{ik}^{p} \cdot \left(\sum_{x_r \in N_k} \left\| x_r - v_i \right\|^2 \right), \tag{4}$$

where N_k represents the set of pixels belonging to the intermediate neighbourhood of pixel x_k and N_R is the cardinality of N_k. The influence of the second additive term of (4) on the pixel annotation is weighted by the parameter n_R. When the image suffers from strong noise presence, a high value for the parameter n_R must be used and a low value for the parameter n_R is used for robust image segmentation in case of large *SNRs*.

Substituting (1) to (4) yields:

$$J_{ab} = \sum_{i=1}^{c}\sum_{k=1}^{N} u_{ik}^{p} \cdot \left\| \frac{y_k - b_k}{a_k} - v_i \right\|^2 + \frac{n_R}{N_R} \cdot \sum_{i=1}^{c}\sum_{k=1}^{N} u_{ik}^{p} \cdot \left(\sum_{x_r \in N_k} \left\| \frac{y_r - b_r}{a_r} - v_i \right\|^2 \right). \tag{5}$$

The new minimization problem of J_{ab} must be solved taking into account the restrictions of (3). Thus, the introduction of one Lagrange multiplier forms an equivalent minimization problem:

$$F_{ab}=\sum_{i=1}^{c}\sum_{k=1}^{N}\left(u_{ik}^{p}\cdot D_{ik}+\frac{n_R}{N_R}\cdot u_{ik}^{p}\cdot\gamma_i\right)+\lambda\cdot\left(1-\sum_{i=1}^{c}u_{ik}\right), \tag{6}$$

where $D_{ik}=\left\|\frac{y_k-b_k}{a_k}-v_i\right\|^2$ and $\gamma_i=\sum_{y_r\in N_k}\left\|\frac{y_r-b_r}{a_r}-v_i\right\|^2$ (7)

The derivative of F_{ab} with respect to u_{ik}, at the minimum of J_{ab} gives an estimator for parameter u_{ik}*:

$$u_{ik}^{*}=\left(\frac{\lambda}{p\cdot\left(D_{ik}+\frac{n_R}{N_R}\cdot\gamma_i\right)}\right)^{1/(p-1)}. \tag{8}$$

However,

$$\sum_{j=1}^{c}u_{jk}=1\ \forall\ k\Rightarrow\sum_{j=1}^{c}\left(\frac{\lambda}{p\cdot\left(D_{jk}+\frac{n_R}{N_R}\cdot\gamma_j\right)}\right)^{1/(p-1)}=1 \tag{9}$$

$$\Rightarrow\lambda=\frac{p}{\left(\sum_{j=1}^{c}\left(\frac{1}{p\cdot\left(D_{jk}+\frac{n_R}{N_R}\cdot\gamma_j\right)}\right)^{1/(p-1)}\right)^{p-1}}. \tag{10}$$

Substituting (10) in (8):

$$u_{ik}^{*}=\frac{1}{\sum_{j=1}^{c}\left(\frac{D_{ik}+\frac{n_R}{N_R}\cdot\gamma_i}{D_{jk}+\frac{n_R}{N_R}\cdot\gamma_j}\right)^{1/(p-1)}}. \tag{11}$$

If the derivative of F_{ab} with respect to v_i is set to zero, the following equation is derived:

$$v_i^* = \frac{\sum_{k=1}^{N} u_{ik}^p \cdot \left(\left(\frac{y_k - b_k}{a_k} \right) + \frac{n_R}{N_R} \cdot \sum_{y_r \in N_k} \left(\frac{y_r - b_r}{a_r} \right) \right)}{(1+n_R) \cdot \sum_{k=1}^{N} u_{ik}^p}. \tag{12}$$

Similarly, if the derivative of F_{ab} with respect to a_k is set to zero the following equation is derived:

$$a_k^* = \left(y_k - b_k \right) \cdot \frac{\sum_{i=1}^{c} u_{ik}^p}{\sum_{i=1}^{c} u_{ik}^p \cdot v_i}. \tag{13}$$

and the derivative of F_{ab} with respect to b_k produces:

$$b_k^* = y_k - a_k \cdot \frac{\sum_{i=1}^{c} u_{ik}^p \cdot v_i}{\sum_{i=1}^{c} u_{ik}^p}. \tag{14}$$

The algorithm for shading correction can be summarized as follows:

1. *Initialization*: Two randomly selected pixels constitute the centroids of the two clusters (c=2, background and scene objects). The a_k are set in near to one random values and b_k are set in near to zero random values.
2. *Update Membership:* The partition matrix U is estimated using (11).
3. *Centroids estimation*: At each cluster the corresponding prototypes are estimated using (12) and the partition matrix U estimated in the previous step.
4. *Estimation of multiplicative shading component*: The parameters a_k are derived from (13).
5. *Low pass filtering*: The multiplicative shading component is filtered using an iterative low pass filter.
6. *Estimation of additive shading component*: The parameters b_k are estimated from (14).

If $\left\| v_{new} - v_{old} \right\| < \varepsilon$, where ε is a small real number, the algorithm is terminated and the last shading parameters are used to estimate the shading-free image using (1), otherwise the FCM algorithm is repeated from step 2.

In the proposed algorithm the low pass filtering of step 5 is applied to prevent the algorithm from converging on a bad local minimum.

3. Experimental Results

The proposed algorithm is evaluated on both real infrared images of a human finger and artificial created images. The real images are acquired using infrared leds at 840 nm and a low cost CCD camera. The finger is placed between the camera and a row of five infrared leds with adjustable illumination. Although, a careful image acquisition setup is achieved by selecting wide-angle leds, the acquired infrared images suffer from shading and non-uniform illumination mainly due to tissue structure of the finger. The artificial images are created using the weighed sum of two artificial images. The first image is constructed using an artificial vessel-like network containing connected lines of different widths with junctions and bifurcations, and multiple low pass filtering is applied to simulate the blurriness of the edges which is apparent to the real images due to the blood flow and scattering. The second artificial image is created by applying an iterative spatial low pass Gaussian filter with a large window size to the original infrared image and used to simulate the non uniform image background. In Fig. 1, an artificial created image and an estimation of the corresponding shading-free image are shown, derived after multiplicative and additive shading component elimination using the proposed algorithm.

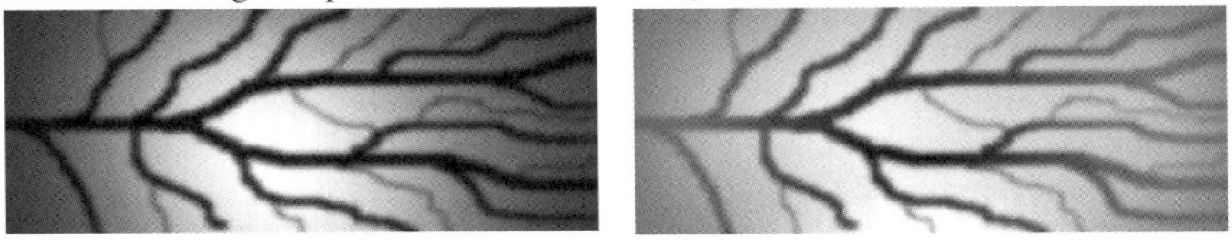

Figure 1. Artificial infrared image of a finger and the corresponding estimation of the shading-free image using the proposed algorithm.

In Fig. 2 two typical examples of the proposed shading elimination process are shown when real infrared images are processing.

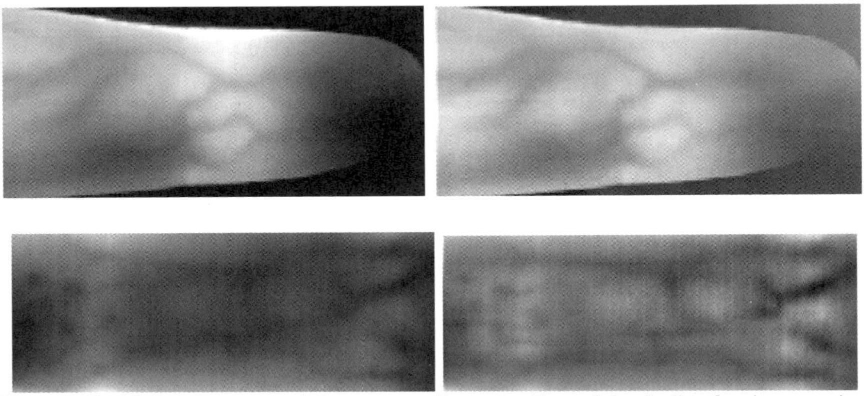

Figure 2. Real infrared images and the corresponding estimations of the shading-free images using the proposed algorithm.

Acknowledgments

This work is supported by grant KARATHEODORIS of the University of Patras.

References

1. J.C. Russ, The image processing handbook, Second Edition, IEEE press, Boca Raton, 1995.
2. I.T. Young, J.J. Gerbrands, and L.J. Vliet, Image Processing Fundamentals, CRC press, Boca Raton, 1998, pp 51.1-51.81.
3. B. Likar, J.B.A. Maintz, M.A. Viergever, and F. Pernus, Retrospective shading correction by entropy minimization, Journal of Microscopy, 2000, 197:3, pp. 285-295.
4. S.E. Umbaugh, Computer vision and image processing, Prentice-Hall, New York, 1998.
5. R.M. Haralick, S.R. Sternberg and X. Zhuang, Image Analysis Using Mathematical Morphology, IEEE Trans. Pattern Anal. Machine Intell., 1987, 9: 532-550.
6. C. Munzenmayer, F. Naujokat, S. Muhldorfer, and T. Wittenberg, Enhancing Texture Analysis by Color Shading Correction, in 9. Workshop Farbbildverarbeitung, K.-H. Franke, ed., Zentrum für Bild- und Signalverarbeitunge.V., (Ilmenau), 2003.
7. M. Foracchia, E. Grisan and A. Ruggeri, Luminosity and contrast normalization in retinal images, Medical Image Analysis, 2005, pp. 179-190.
8. T. Tasdizen, E. Jurrus and R. Whitaker, Non-uniform illumination correction in transmission electron microscopy.
9. W. Chen, M.L. Giger, A fuzzy c-means (FCM) based algorithm for intensity inhomogeneity correction and segmentation of MR images, IEEE International Symposium on Biomedical Imaging: Nano to Macro, 2004, vol.2, pp. 1307-1310.
10. M.N. Ahmed, S.M. Yamany, N. Mohamed, A.A. Farag, T. Moriarty, A modified fuzzy c-means algorithm for bias field estimation and segmentation of MRI data, IEEE Transactions on Medical Imaging, 2002, vol. 21, pp. 193-199.
11. M.N. Ahmed, S.M. Yamany, N. Mohamed, A.A. Farag, T. Moriarty, Bias field estimation and adaptive segmentation of MRI data using modified fuzzy C-means algorithm, in Proc. IEEE Int. Conf. Computer Vision and Pattern Recogn., 1999, vol. 1, pp. 250-255.

ESTIMATION OF HUMAN RBC SIZE USING LIGHT SCATTERING IMAGES AT MULTIPLE WAVELENGTHS

G. APOSTOLOPOULOS

Department of Electrical and Computer Technology, University of Patras, KatoKastritsi, Patras, 26500, Greece

S. TSINOPOULOS

Mechanical Department, Technological Educational Institute of Patras, Patras, Greece

E. DERMATAS

Department of Electrical and Computer Technology, University of Patras, KatoKastritsi, Patras, 26500, Greece

In this paper, a novel method for the estimation of the human Red Blood Cell (RBC) size using light scattering images at multiple wavelengths is presented. The information retrieval process includes, image normalization, a two-dimensional discrete Wavelet transformation (DWT2), and a Radial Basis Neural Network (RBF-NN) is used to estimate the RBC geometrical properties. The proposed method was evaluated in both regression and identification tasks by processing images of simulated device used to acquire scattering phenomena of moving RBCs. The database includes 1575 images of 50x50 pixles, is created with the boundary element method (BEM) in 9 different wavelengths. The experimental setup consists of nine, equally spaced, light beams from 632.8 to 1032.8 nm moving RBCs in a thin glass. When artificial white Gaussian noise is added to the simulated images at 10 dB SNR, the mean absolute error between the actual geometrical properties and the RBF based estimations is less than 1% of the actual RBC size. In the identification experiments correct classification rate better than 99.8% is obtained.

1. Introduction

The Red blood cell (RBC) is the most common type of blood cell, filled with hemoglobin, a bio-molecule that can bind to oxygen. In humans, RBCs take the form of flexible biconcave disks, lack a cell nucleus, sub-cellular organelles and the ability to synthesize protein, and live for about 120 days. Several blood diseases change the typical size and distribution of red blood cells, including: Anemia, malaria parasites, microangiopathic diseases etc, these pathologies generate fibrin strands that sever RBCs as they try to move past a thrombus.

The scattering of electromagnetic waves by dielectric particles is a problem of great significance for a variety of applications, ranging from particle sizing and remote sensing to radar meteorology and biological sciences [1]. In the area of medical diagnosis, understanding how a laser beam interacts with blood suspensions or a whole-blood medium is of paramount importance in quantifying the RBCs inspection process in many commercial devices. The correlation between the obtained patterns and the physical characteristics of blood, such as the concentration of hemoglobin and the degree of oxygen saturation, is determined with the aid of simple multiple-scattering theories when dilute suspensions of blood are used [2-4].

Light scattering has been used for efficient and accurate measurement of the geometrical properties of micro-particles [6-8]. Most of those methods based on light scattering spectroscopy are being investigated in a back-scattering geometry, since it has been shown that the spectrum over wave-number of backscattered light has a periodic component with an oscillation frequency proportional to the particle size [9].

A great variety of analytical, semi-analytical, and numerical methodologies are available in the literature for the implementation of such a scattering problem. Analytical methods based on theories such as those of Mie, Rayleigh, Fraunnhoffer, and Rayleigh-Gans and on anomalous-diffraction, ray-optics [12]. Numerical methods, such as the method of moments, the boundary element method (BEM), the surface and volume integral-equation methods, the finite-element method can solve with no significant restrictions very complicated scattering problems.

Recently, a device for acquisition of scattering images of RBC-flow is presented [5,15]. The block diagram of the experimental device is shown in Fig.1, consisting of a monochromatic light source, two iris diaphragms, a Fourier convex lens, the flow chamber, a semi-transparent screen, a beam stopper, and a Charge Couple Device (CCD) camera. The light source is a 5-mW vertically polarized laser, and the beam diameter is 0.8 mm. The suspension is circulated continuously into the flow chamber at a constant flow rate by a syringe pump. Particles in the flow chamber are illuminated by the laser beam, and the forward light-scattering pattern appears on the semitransparent screen. The beam stopper is used to prevent the light of the central beam from blinding the CCD camera. In this paper, a novel method for the estimation of the human RBC size using light scattering images of the simulated device [15] at multiple wavelengths is presented and evaluated. The goal of this paper is twofold, to identify human RBCs and to estimate the diameter, the maximum and minimum thickness of the RBC from multiple digital images, derived when narrow-band light beams illuminate the RBCs-flow, using image processing and non-linear regression techniques.

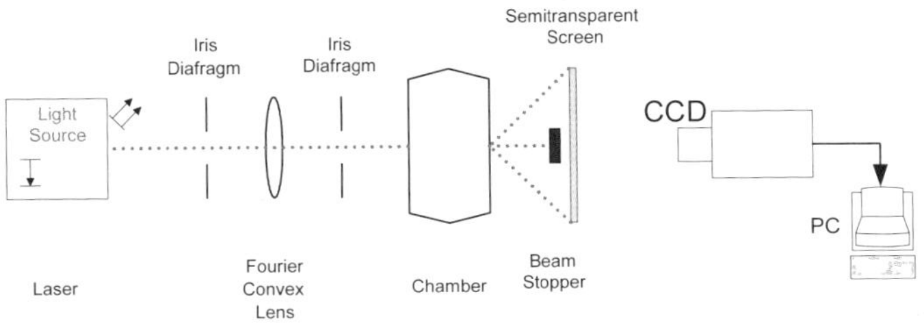

Figure 1. The schematic presentation of the simulated device used to acquire scattering images in a RBC-flow [5,15].

The structure of this paper is as follows: In section 2, the proposed estimation and identification method of human RBC uses DWT2 based-features and the RBF neural network is presented. The image database, the experiments, the regression and identification results are presented and discussed in Section 3. Finally, a conclusion is given.

2. Estimation of RBC size

In each image of the simulated device, scattering phenomena of a single RBC type at a unique wavelength are recorded. Taking into account that the gain information is not related to the RBC size, a linear normalization of the pixels' intensity according to the brightest value is applied to remove this irrelevant information. In the feature extraction module, the discrete two-dimensional Wavelet Transform (DWT2) has been used to compress and extract the scattering information from the image. Thus, in each image at each wavelength a DWT2 is applied, and the corresponding wavelet coefficients $\mathbf{W}_{\mathbf{I}_n}^{\lambda_k}$ are derived. The feature vector $\mathbf{V}$, composed by the most important wavelet coefficients, is used by the radial-basis functions neural network (RBF-NN) to derive the actual RBC size.

2.1. *Features Extractions*

The basic DWT2 is a filter bank of row and column directions, decomposing an image into four sub-images. The approximated image LL is obtained by low-pass filtering in both row and column directions. The LH, HL, and HH images contain high frequency components. By decomposing the approximated image of each level into four sub-images iteratively, a pyramidal image tree is acquired. During decomposition [19], there is neither loss nor redundancy of information between the levels. The energy (L2-norm) of each sub-image can be a favorable feature of texture because it's close-related to the dominant spatial frequencies of the original image. An efficient method to extract the salient information is to detect and retain only the significant coefficients [16]. In the proposed feature extraction method, the energy of the high frequency sub-images decomposed up to M decomposition level is estimated, and is assigned to the component of the feature vector including also the energy of the LL sub-image $P_1^{(LL)}$ in the first level only. Therefore, the feature vector includes 3M+1 components:

$$\left(P_1^{(LL)}, P_1^{(LH)}, P_1^{(HL)}, P_1^{(HH)}, P_2^{(LH)}, P_2^{(HL)}, P_2^{(HH)}, \ldots, P_M^{(LH)}, P_M^{(HL)}, P_M^{(HH)}\right)$$

where, $P_i^{(XX)}$ denotes the energy of the XX sub-image in the i level.

2.2. *Features Normalization*

The generalization capabilities and the training efficiency of the RBF network can be improved using a feature normalization process [20]. Therefore, each DWT2 coefficient of the feature vector is divided by the corresponding standard deviation estimated in the complete set of training data.

2.3. *Estimation of RBC size using Neural Network*

Artificial neural networks have been used widely in many regression and classification problems [13,14] using computing elements simulating the lowest level processing unit of the human brain, the neuron.

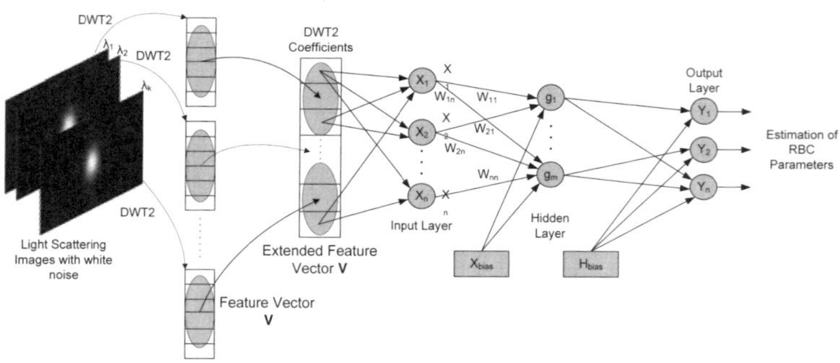

Figure 2. Structure of the non-linear regression model.

Among the most important advantages of the RBF, a very small number of parameters need to defined, i.e the number of hidden neurons, the centre and the radius of the adopted radial-basis functions, denoted as spread. In Fig. 2, the structure of the proposed RBF-NN-based non-linear regression method is shown.

According to [17,18], healthy human RBCs have an axisymmetric geometry with z indicating the axis of the symmetry, as shown in Fig. 3. Typical values of R1, R2 and R3/R2 varies from 4.5 to 10.5 μm, 1.5 to 3 μm and 0.4 to 0.8, respectively. The image database has been designed to cover the suggested ranges in equal spaced steps of 0.250 μm for R1 and R2 and 0.05 for the ratio R3/R2, producing 1575 images for each wavelength λ_k, of 50 x 50 pixels each. Taking into account both axisymmetric geometry of the scatterer and the non-axisymmetric boundary conditions of the problem, the simulated images are estimated by solving the forward scattering problem of EM plane wave at specific wavelength and the described human RBCs [15].

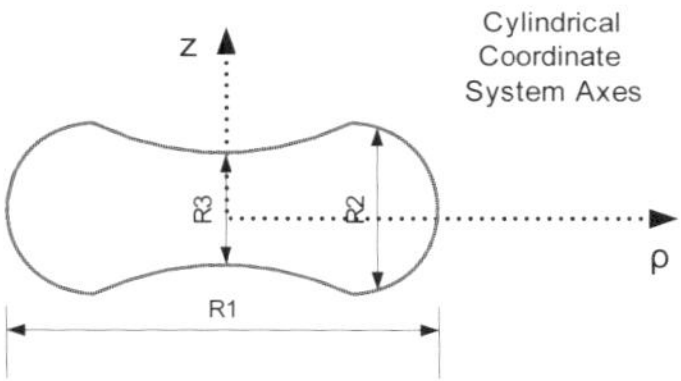

Figure 3. A representation of a normal RBC according to [17,18].

The original images are distorted using additive white Gaussian noise at 10 dB SNR. This type of noise approximates several phenomena in real image acquisition systems, including thermal-noise in CCD cameras and distortion from optical elements. The structure of the RBF-NN is used to estimate the geometrical properties of the human RBC as illustrated in Fig 2. In the training of the RBF-NN 788 images for each wavelength λ_k, are used, and for the regression and identification experiments the remaining 787 images are selected. Both training and testing sets are uniformly distributed in the RBC sizes. The identification process is completed by the nearest classification rule of the Euclidean distance between the regression values of the RBF-NN and the RBC sizes used to build the image database.

2.4. *Estimation of RBC size*

In the regression experiment the DWT2 features are evaluated using the orthogonal basis Daubechies-3 wavelet family. Due to the fact that the FFT of the scattering images contain few dominant low-pass frequencies, the DWT2 coefficients of the first two decomposition layers are used. Therefore, the Feature Vector **V** has a total number of 7 DWT2 coefficients which derived from each image. The Extended Feature Vector **V** (EF -V) is composed by each image's feature vector including a total number of *7 x k* (63) DWT2 coefficients for each wavelength. In the same set of features, the regression error is estimated in the evaluation images set. The regression error versus the number of hidden neurons and different wavelengths of the three geometrical properties of RBC is shown in figures 4, 5 and 6. Better regression rate is obtained when the number of neurons is increased, while the best accuracy is achieved when the Extended Feature Vector **V** is used against Feature Vector **V** of scattering images at each wavelength λ_k. The discrimination power of the RBF-NN output is used to build an RBC identification system. In figure 7, the mean correct identification rate of the RBC size versus wavelength is shown. The best mean identification rate of 99.8 % is achieved when manipulate the entire beneficial information of the scattering images from each wavelength.

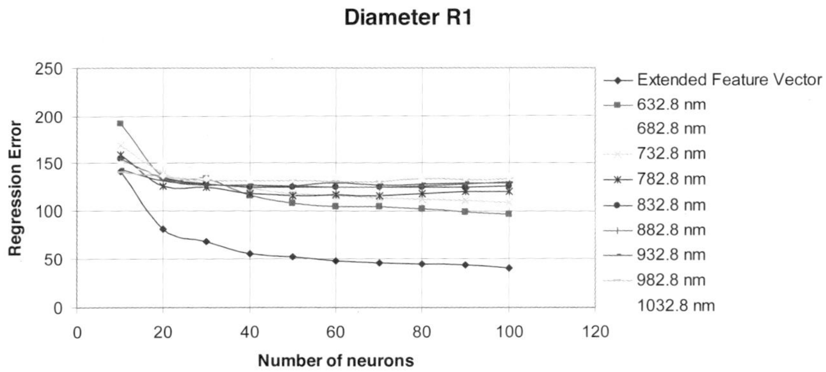

Figure 4. Regression error of R1 in various wavelengths.

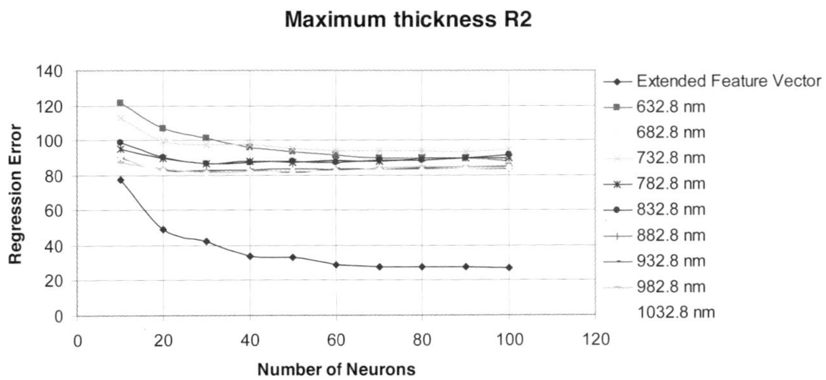

Figure 5. Regression error of R2 in various wavelengths.

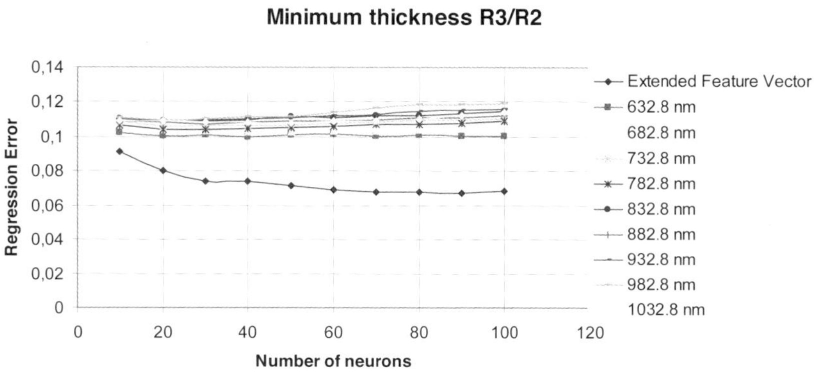

Figure 6. Regression error of R3/R2 in various wavelengths.

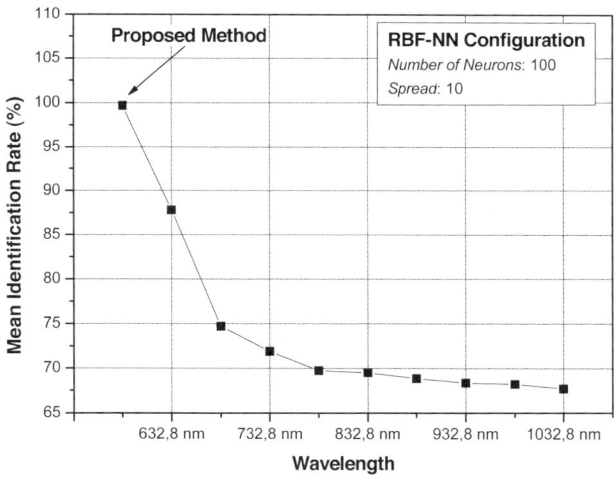

Figure 7. Mean identification rate versus wavelength.

3. Conclusions

The main goal of this research was to implement two fully automated methods for the estimation and classification of human RBC from scattering images using image processing and supervised neural network techniques. The evaluation of the proposed methods show that the excellent detection and identification rates for three important geometrical properties of human RBC can be achieved by processing the scattering phenomena of laser light, even in cases where the simulated images are distorted by white Gaussian noise at 10 dB SNR. The excellent performance of the proposed method in all experiments is an important guideline for medical implementation in real-life applications.

References

1. C. F. Bohren and D. R. Huffman, Absorption and scattering of light by small partcles, Wiley, New York, 1983.
2. J. Plasek and T. Marik, Determination of undeformable erythrocytes in blood samples using light scattering, Appl. Opt. 21, 4335-4338, 1982.
3. J. M. Steinke and A. P. Shepherd, Comparison of Mie theory and the light scattering of red blood cells, Appl. Opt. 27, 4027-4033, 1988.
4. M. Hammer, D. Schweitzer, B. Michel, E. Thamm, and A. Kolb, Single scattering by red blood cells, Appl. Opt. 37, 7410-7418, 1998.
5. G. N. Constantinides, D. Gintides, S.E. Kattis, K. Kiriaki, C.A. Paraskeya, A.C. Payatakes, D. Polyzos, S.V Tsinopoulos and S. N. Yannopoulos, Computation of light scattering by axisymmetric nonspherical particles and comparison with experimental results, Appl. Opt. 37, 7310-7319, 1998.
6. V. Backman, M. B. Wallace, L. T. Perelman, J. T. Arendt, R. Gurjar, M. G. Muller, Q. Zhang, G. Zonios, E. Kline, J. A. McGilligan, S. Shapshay, T. Valdez, K. Badizadegan, J. M. Crawford, M. Fitzmaurice, S. Kabani, H. S. Levin, M. Seiler, R. R. Dasari, I. Itzkan, J. Van Dam, and M. S. Feld, Detection of pre-invasive cancer cells, Nature 406, 35, 2000.
7. J. R. Mourant, T. M. Johnson, S. Carpenter, A. Guerra, T. Aida, and J. P. Freyer, Angular Dependent light scattering from multi-cellular spheroids, J. Biomed. Opt. 7, 378, 2002.
8. A. Katz, A. Alimova, M. Xu, E. Rudolph, M. Shah, H. E.Savage, R. Rosen, S. A. McCormick, and R. R. Alfano, Bacteria size determination by elastic light scattering, IEEE J. Sel. Top. Quantum Electron. 9, 277, 2003.
9. L. T. Perelman, V. Backman, M. Wallace, G. Zonios, R. Manoharan, A. Nusrat, S. Shields, M. Seiler, C. Lima, T. Hamano, I. Itzkan, J. Van Dam, J. M. Crawford, and M. S. Feld, Observation of periodic fine structure in reflectance from biological tissue: A new technique for measuring nuclear size distribution, Phys. Rev. Lett. 80, 627, 1998.
12. H.C. Van Hulst, Light scattering by small particles, Dover, New York, 1981.
13. L. Song and R. M. Donovan, Segmentation of cell image using an expert system, 10th Annual Inter. Conf. , IEEE EMBS 1383-1384, 1988.
14. F. Arman and J. A. Pearce, Unsupervised classification of cell images using pyramid node linking, IEEE Trans. BME (37)6: 647-652, 1990.
15. S. V. Tsinopoulos and D. Polyzos, Scattering of He-Ne laser light by an average-sized red blood cell, Appl. Opt. 25, 5499, 1999.
16. C.S. Lu, P. C. Chung, and C. F. Chen, Unsupervised texture segmentation via wavelet transform, Pattern Recognit., vol.30, no.5,pp.729-742,1997.
17. P.W. Kuchel and E.D. Fackerell, Parametric equation representation of biconcave erythrocytes, Bull. Math. Biol. 61:209-220, 1992.
18. S. Munoz San Martin, J.L Sebastian, M. Sanchol and G. Alavarez, Modeling Human Erythrocytes shape and size abnormalities, q-bio. QM/0507024, 2005.
19. Seisuke Fukuda and Haruto Hirosawa, A Wavelet-Based Texture Feature Set Applied to Classification of Multifrequency Polarimetric SAR Images, IEEE Trans. On Geoscience and Remote Sensing, vol,37,no.5,pp.2282-2286,1999.
20. S. Molau, Normalization in the Acoustic Feature space for improved speech recognition, in Faculty of Mathematics, Information and Natural Sciences. Aachen: RWTH, 2003.

ELECTROMAGNETIC WAVE SCATTERING BY QUASI-HOMOGENEOUS OBSTACLES

N. L. TSITSAS
Department of Mathematics,
School of Applied Mathematical and Physical Sciences,
National Technical University of Athens, GR-15780, Athens, Greece,
E-mail: ntsitsas@esd.ntua.gr

C. A. VALAGIANNOPOULOS
School of Electrical and Computer Engineering,
National Technical University of Athens, GR-15780, Athens, Greece,
E-mail: valagiannopoulos@gmail.com

In this work we investigate the electromagnetic wave scattering phenomena by quasi-homogeneous obstacles, namely obstacles with wavenumber functions not exhibiting large variations from an average value $\overline{k}$. First, we express the field coefficients by means of a T-matrix method for the corresponding piecewise-homogeneous scatterer and then perform the best linear approximation by differentials to express these coefficients as linear combinations of the distances of the wavenumber samples from $\overline{k}$. Moreover, the total far-field pattern of the quasi-homogeneous scatterer is decomposed into that of the respective homogeneous scatterer with wavenumber $\overline{k}$ plus the perturbation far-field pattern, depending exclusively on the wavenumber's deviations from $\overline{k}$. Numerical results are presented concerning (i) the far-field patterns, computed by the proposed technique and the T-matrix method, (ii) the variations of the perturbation far-field pattern, and (iii) the prediction of each layer's contribution to the far-field.

1. Introduction

Radially inhomogeneous obstacles arise in various applications, including optics, combustion, chemical engineering, environmental science, and biology[1,2,3]. The investigation of wave scattering phenomena from such obstacles initiated in the 1960's with the work of Kerker[4] and has thenceforth developed intensively. The main established methodologies concern the approximation of the radially inhomogeneous refractive index by step functions. More precisely, in 5 and 6 a finely stratified spherical and cylindrical model has been investigated respectively. The radially inhomogeneous refractive index distribution has been approximated by a large number of concentric spherical/cylindrical layers with constant material parameters in each layer. Besides, in 7 the Debye potential method has been utilized to represent the scattering coefficients of a radially inhomogeneous sphere. Moreover, concerning light scattering by multilayered spheres recursive algorithms to calculate the scattered field have been developed in 8.

In this paper we consider a plane electromagnetic wave impinging on a quasi-homogeneous obstacle, namely an obstacle with wavenumber function not exhibiting large variations from an average value $\overline{k}$. The solution of the scattering problem involves the following two steps: (i) the unknown field coefficients are expressed analytically by means of a T-matrix method[9] for the corresponding piecewise homogeneous (layered) scatterer (ii) in the derived expressions of the field coefficients the *best linear approximation by differentials* (first-order Taylor expansion) is performed with respect to the samples of the wavenumber function around the fixed point $\overline{k}$. Thus, we express the field coefficients as finite linear combinations of the distances of the wavenumber samples from $\overline{k}$. The weights of these linear combinations are named *layer factors* and each one describes the contribution of the specific layer in the total scattered field. The developed method actually constitutes a multi-dimensional linearization of the T-matrix solution, reducing the original finite product of matrices, to a finite sum. Moreover, the total far-field pattern of the quasi-homogeneous scatterer is decomposed into the far-field pattern of the respective homogeneous scatterer with wavenumber $\overline{k}$ plus the *perturbation far-field pattern*, depending only on the deviations of the wavenumber function from the average value $\overline{k}$.

Numerical comparisons of the far-field as computed by the proposed technique, and the exact T-matrix method are given which exhibit the small relative differences between the results of the two methods. Furthermore, we investigate the perturbation far-field pattern as well as the layer-factors and examine their dependence on the scatterer's parameters. The numerical results concerning the layer-factor variations are expected to be utilized in potential applications in the sense of predicting the far-field by altering suitably the parameters of the quasi-homogeneous obstacle.

2. Solution of the scattering problem

2.1. *The scatterer and the incident field*

Consider an inhomogeneous infinite cylinder of external radius a, composed of a dielectric material with ρ-dependent wavenumber $k(\rho)$ not exhibiting large variations from an average value $\bar{k}$. The entire structure has constant magnetic permeability μ_0. The cylinder is illuminated by the z-polarized normally incident plane wave of unit amplitude[10]

$$\mathbf{E}^{inc}(\rho,\phi) = \exp(-ik_0\rho\sin\phi)\hat{\mathbf{z}} = \sum_{n=-\infty}^{+\infty} (-1)^n \exp(in\phi) J_n(k_0\rho)\ \hat{\mathbf{z}} \quad (\rho > a)$$

The scattering problem is analyzed for the above described 2-D cylinder. Methodological modifications and extensions in order to handle the cases of a slab (1-D) and a spherical (3-D) scatterer are discussed in 11.

2.2. *Implementation of a T-matrix method*

First, we impose a T-matrix method for the computation of the scattered field. To this direction the cylinder is discretized by the cylindrical surfaces $\rho = a_j = (U - j + 1)a/U$ $(j{=}1,\ldots,U)$ into U homogeneous layers V_j of equal thickness a/U and wavenumber $k_j = k(\frac{a_j + a_{j+1}}{2})$. The total electric field inside layer V_j $(a_{j+1} < \rho < a_j,\ j = 1,\ldots,U-1)$ is expressed as

$$\mathbf{E}^j(\rho,\phi) = \sum_{n=-\infty}^{+\infty} (-1)^n \exp(in\phi) \left[\alpha_n^j H_n(k_j\rho) + \beta_n^j J_n(k_j\rho)\right] \hat{\mathbf{z}} \tag{1}$$

where J_n and H_n are the cylindrical Bessel and Hankel of first kind functions of order n and α_n^j and β_n^j are under determination coefficients. The total field in the exterior V_0 and the core V_U of the cylinder is given by Eq. (1) with $\beta_n^0 = 1$ and $\alpha_n^U = 0$, valid for $\rho > a$ and $0 < \rho < a_U$, respectively.

By imposing the boundary conditions on the cylindrical surfaces $\rho = a_j$, the field's coefficients in layer V_{j-1} are connected to those in V_j as

$$\begin{bmatrix} \alpha_n^j \\ \beta_n^j \end{bmatrix} = \mathbf{A}_n^j \begin{bmatrix} \alpha_n^{j-1} \\ \beta_n^{j-1} \end{bmatrix} \tag{2}$$

where

$$\mathbf{A}_n^j = \mathbf{A}_n^j(k_{j-1},k_j) = -\frac{\pi}{2i}\cdot$$
$$\begin{bmatrix} -y_j J_n(x_j)H_n'(y_j) + x_j J_n'(x_j)H_n(y_j) & -y_j J_n(x_j)J_n'(y_j) + x_j J_n'(x_j)J_n(y_j) \\ y_j H_n(x_j)H_n'(y_j) - x_j H_n'(x_j)H_n(y_j) & y_j H_n(x_j)J_n'(y_j) - x_j J_n(x_j)H_n'(y_j) \end{bmatrix}$$

with $x_j = k_j a_j$ and $y_j = k_{j-1}a_j$.

Now, by applying successively Eq. (2) for $j{=}1,\ldots,U$ we find

$$\alpha_n^0 = -\frac{(\mathbf{A}_n^U\mathbf{A}_n^{U-1}\cdots\mathbf{A}_n^2\mathbf{A}_n^1)_{12}}{(\mathbf{A}_n^U\mathbf{A}_n^{U-1}\cdots\mathbf{A}_n^2\mathbf{A}_n^1)_{11}} \tag{3}$$

where $(\mathbf{X})_{pq}$ denotes hereafter the (p,q) element of a matrix $\mathbf{X}$.

2.3. *Best linear approximation of the field coefficients*

The described T-matrix analysis holds so far for any piecewise-homogeneous cylinder. Now, we focus on the case of a quasi-homogeneous cylinder with wavenumber "close" to $\bar{k}$. To this end, we approximate the products $\mathbf{A}_n^U \cdots \mathbf{A}_n^2$, appearing in Eq. (3), by applying the best linear approximation method. More precisely, we consider $\mathbf{A}_n^U \cdots \mathbf{A}_n^2$ as a matrix function of the $U-2$ variables k_j $(j = 2, \ldots, U-1)$, and approximate it by its $(U-2)$-dimensional first-order Taylor expansion around the fixed vector $(k_2, \ldots, k_{U-1}) = (\bar{k}, \ldots, \bar{k}) \equiv \bar{\mathbf{k}}$. For this reason, we need the explicit forms of the derivatives $\partial(\mathbf{A}_n^U \cdots \mathbf{A}_n^2)/\partial k_j$ evaluated at $\bar{\mathbf{k}}$. Note that for a fixed j the variable k_j appears exclusively in the matrices $\mathbf{A}_n^j$ and $\mathbf{A}_n^{j+1}$, while for every index j and wavenumber κ holds $\mathbf{A}_n^j(\kappa,\kappa) = \mathbf{I}$, where $\mathbf{I}$ the 2×2 identity matrix. Thus, we conclude to the following simplified form

$$\left.\frac{\partial(\mathbf{A}_n^U \cdots \mathbf{A}_n^2)}{\partial k_j}\right|_{\bar{\mathbf{k}}} = \left.\frac{\partial \mathbf{A}_n^{j+1}}{\partial k_j}\right|_{(\bar{k},\bar{k})} + \left.\frac{\partial \mathbf{A}_n^{j}}{\partial k_j}\right|_{(\bar{k},\bar{k})}$$

by means of which we obtain the best linear approximation

$$\mathbf{A}_n^U \cdots \mathbf{A}_n^2 \simeq \mathbf{I} + \sum_{j=2}^{U-1} (k_j - \bar{k}) \left[\left.\frac{\partial \mathbf{A}_n^{j+1}}{\partial k_j}\right|_{(\bar{k},\bar{k})} + \left.\frac{\partial \mathbf{A}_n^{j}}{\partial k_j}\right|_{(\bar{k},\bar{k})} \right] \tag{4}$$

Moreover, by combining Eqs. (3) and (4) we obtain

$$\alpha_n^0 \simeq -\frac{(\mathbf{A}_n^1)_{12} + \sum_{j=2}^{U-1} (k_j - \bar{k})(\mathbf{C}_n^j \mathbf{A}_n^1)_{12}}{(\mathbf{A}_n^1)_{11} + \sum_{j=2}^{U-1} (k_j - \bar{k})(\mathbf{C}_n^j \mathbf{A}_n^1)_{11}} \tag{5}$$

where

$$\mathbf{C}_n^j = \left.\frac{\partial \mathbf{A}_n^{j+1}}{\partial k_j}\right|_{(\bar{k},\bar{k})} + \left.\frac{\partial \mathbf{A}_n^{j}}{\partial k_j}\right|_{(\bar{k},\bar{k})}$$

Now, by imposing a new first-order Taylor expansion in Eq. (5), we get

$$\alpha_n^0 \simeq -\frac{(\mathbf{A}_n^1)_{12}}{(\mathbf{A}_n^1)_{11}} - \frac{1}{[(\mathbf{A}_n^1)_{11}]^2} \sum_{j=2}^{U-1} (k_j - \bar{k})(\mathbf{C}_n^j)_{12} \tag{6}$$

3. Far-field decomposition and the layer-factors

The scattered field in the radiation zone ($\rho \to \infty$) is expressed as

$$\lim_{\rho\to\infty} E^{sc}(\rho,\phi) = \sqrt{\frac{2}{\pi k_0 \rho}} \exp\left[i\left(k_0\rho - \frac{\pi}{4}\right)\right] g_{2\mathrm{D}}(\phi)$$

where $g_{2\mathrm{D}}$ is the far-field pattern given by

$$g_{2\mathrm{D}}(\phi) = \sum_{n=-\infty}^{+\infty} i^n \exp(in\phi)\alpha_n^0 \tag{7}$$

and α_n^0 is computed with the aid of Eq. (6).

By combining Eqs. (6) and (7) we get the decomposition

$$g_{2\mathrm{D}} = g_{2\mathrm{D}}^{\mathrm{hom}} + g_{2\mathrm{D}}^{\mathrm{per}} \tag{8}$$

where

$$\begin{aligned} g_{2\mathrm{D}}^{\mathrm{hom}} &= -\sum_{n=-\infty}^{+\infty} i^n \exp(in\phi)\frac{(\mathbf{A}_n^1)_{12}}{(\mathbf{A}_n^1)_{11}}, \\ g_{2\mathrm{D}}^{\mathrm{per}} &= \sum_{j=2}^{U-1} \mathcal{LF}_j\ (k_j - \bar{k}), \quad \mathcal{LF}_j = -\sum_{n=-\infty}^{+\infty} \frac{i^n \exp(in\phi)}{[(\mathbf{A}_n^1)_{11}]^2}(\mathbf{C}_n^j)_{12} \end{aligned} \tag{9}$$

By Eq. (8) the total far-field pattern $g_{2\mathrm{D}}$ is decomposed into the far-field pattern $g_{2\mathrm{D}}^{\mathrm{hom}}$ of a homogeneous cylinder with wavenumber $\bar{k}$ and the *perturbation far-field pattern* $g_{2\mathrm{D}}^{\mathrm{per}}$ depending only on the deviations of the wavenumber function $k(\rho)$ from $\bar{k}$. In particular, for an unperturbed cylinder (that is $k(\rho) = \bar{k}$) holds $g_{2\mathrm{D}}^{\mathrm{per}} = 0$. Hence, the influence of the wavenumber profile perturbations from $\bar{k}$ is incorporated exclusively in $g_{2\mathrm{D}}^{\mathrm{per}}$.

Importantly, the far-field decomposition (8) is feasible due to the consideration of first-order Taylor expansions. On the other hand, the utilization of higher-order Taylor expansions would increase the approximation accuracy, but would not make a decomposition of the form of Eq. (9) feasible.

Moreover, $\mathcal{LF}_j$ $(j = 2, \ldots, U-1)$ in (9) is the *layer-factor* of layer V_j and determines the contribution of each specific layer V_j to $g_{2\mathrm{D}}^{\mathrm{per}}$. The layer-factor depends on $\phi, \bar{k}$, a, U, but not on $k(\rho)$.

4. Numerical results and discussion

First, we compare the far-field pattern by using the proposed approximate technique, and the exact T-matrix method; the respective approximate and exact computed quantities are denoted by $g_{2\mathrm{D}}^{\mathrm{ap}}$ and $g_{2\mathrm{D}}^{\mathrm{ex}}$.

Figs. 1a and b depict the far-field pattern relative error, that is the relative difference between $g_{2\mathrm{D}}^{\mathrm{ex}}$ and $g_{2\mathrm{D}}^{\mathrm{ap}}$, with respect to $\bar{k}/\Delta k_{\max}$ for $k_1(\rho) = \bar{k} + \Delta k_{\max}\sin(4\pi\rho/a)$ with $k_0 = 2\pi$, $\bar{k} = 3k_0$ and (a) $k_0 a = \pi, 2\pi, 4\pi$, with $\phi = 60^{\circ}$ and (b) $\phi = 0^{\circ}, 90^{\circ}, 270^{\circ}$ with $k_0 a = 2\pi$. Note that $\Delta k_{\max}$ denotes the maximum deviation of $k_1(\rho)$ from $\bar{k}$.

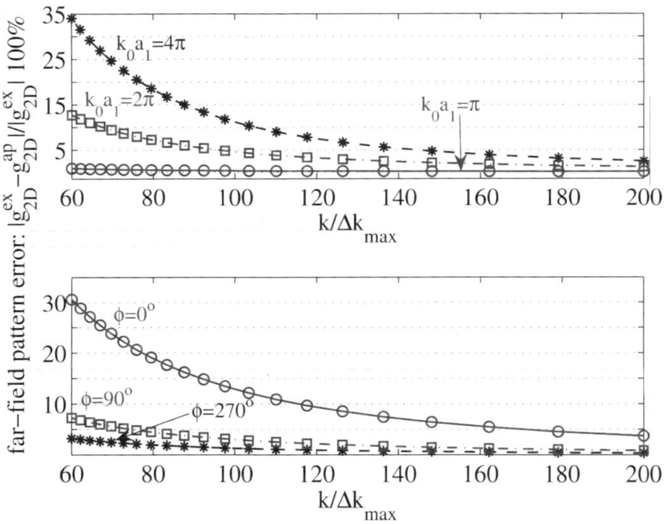

Figure 1. Relative difference between g_{2D}^{ex} and g_{2D}^{ap} as a function of $\overline{k}/\Delta k_{\max}$ for $k_0 = 2\pi$, $\overline{k} = 3k_0$, $U = 20$ and (a) $k_0 a = \pi$ (solid), 2π (dash-dot), 4π (dash), with $\phi = 60^o$, and (b) $\phi = 0^o$ (solid), 90° (dash-dot), 270° (dash) with $k_0 a = 2\pi$.

The far-field error decreases with increasing $\overline{k}/\Delta k_{\max}$, namely the approximation of the exact solution by the proposed method becomes more accurate as the wavenumber profile approaches $\overline{k}$. Fig. 1b shows that the relative far-field pattern error is smaller in "strong" scattering directions like $\phi = 90^o$ and $\phi = 270^o$, corresponding to back- and forward-scattering compared to "weak" scattering directions, like $\phi = 180^o$.

As a conclusion, the respective relative differences between the far-field pattern of the exact T-matrix and the proposed approximate method are small for a wide range of the problem parameters involved.

Second, we investigate the effect of perturbation in the wavenumber profile on the scattered field by the developed method. We consider sinusoidal and Lunenburg-type wavenumber profiles, corresponding respectively to $k_2(\rho) = \overline{k} - \Delta k_{\max} + 2\Delta k_{\max} \sin(\pi\rho/a)$ and $k_3(\rho) = \overline{k} + \Delta k_{\max} - 2\Delta k_{\max}(\rho/a)^2$. Figs. 2a and b depict the far-field patterns $|g_{2D}^{per}|$ as functions of ϕ for (a) sinusoidal and (b) Lunenburg-type profile, both with $\Delta k_{\max}/\overline{k} = 0.02, 0.06, 0.1$, $\overline{k} = 3\pi$, and $k_0 a = 2\pi$.

The far-field patterns for $k_2(\rho)$ attain two local maxima at $\phi = 90^o$ and 270° corresponding to the back- and forward-scattering directions. The values of $|g_{2D}^{per}|$ are larger for the Lunenburg-type profile $k_3(\rho)$ compared

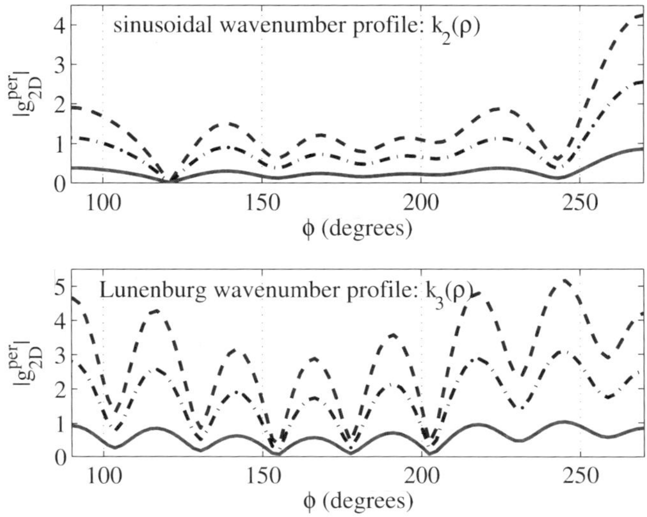

Figure 2. Far-field patterns $|g_{2\mathrm{D}}^{\mathrm{per}}|$ due to the wavenumber profile perturbations as functions of ϕ for (a) sinusoidal $k_2(\rho)$ and (b) Lunenburg $k_3(\rho)$ profile, with $\Delta k_{\max}/\overline{k} = 0.02$ (solid), 0.06 (dash-dot), 0.1 (dash), and $\overline{k} = 3\pi$, $k_0 a = 2\pi$, $U = 25$.

to the sinusoidal $k_2(\rho)$. This is because in the Lunenburg-type profile the outer layers (lying near $\rho = a$ and thus having greater influence on the far-field) have larger wavenumber values than those of the linear profile. In addition, for both wavenumber functions $|g_{2\mathrm{D}}^{\mathrm{per}}|$ exhibits oscillations with amplitude increasing with $\Delta k_{\max}$ and frequency independent of $\Delta k_{\max}$.

Finally, we investigate the layer factor variations. To this end, Figs. 3a and b depict the layer factor $\mathcal{LF}_j$ as a function of the layer index j for a quasi-homogeneous cylinder with $k_0 = 2\pi$, $U = 120$, $\phi = 90°$ and (a) $k_0 a = 0.4\pi, \pi, 1.6\pi$ with $\overline{k} = 3\pi$, (b) $\overline{k}/k_0 = 1.1, 2.5, 5$ with $k_0 a = 2\pi$.

The layer factor $\mathcal{LF}_j$ decreases with the layer index j. In other words, the far-field contribution of the layers closer to the cylinder's core V_U is significantly weaker from that of the layers closer to the outer layer V_1. This is mainly due to the fact that the area of the layers V_j (being equal to $\pi a^2(2U - 2j + 1)/U^2$) decreases with j. Notice also that $\mathcal{LF}_j$ is nearly zero for j tending to U, that is for layers very near to the core V_U.

Moreover, from Fig. 3a and for fixed layer index j the layer factor increases with $k_0 a$. For $k_0 a$ large enough the decreasing tendency of the layer factor $\mathcal{LF}_j$ with the layer index j is not monotonic but oscillatory. In fact, the amplitude and frequency of these oscillations increases with $k_0 a$.

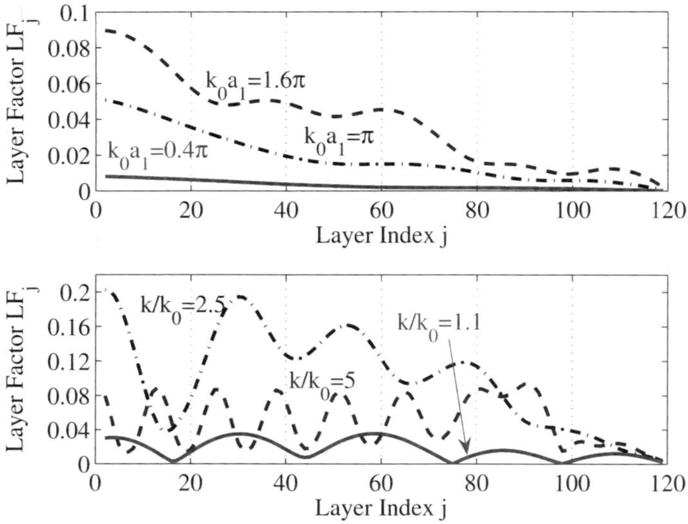

Figure 3. Layer factor $\mathcal{LF}_j$ as a function of the layer index j with $(j = 2, \ldots, U-1)$ for $k_0 = 2\pi$, $U = 120$, $\phi = 90^o$ and (a) $k_0 a = 0.4\pi$ (solid), π (dash-dot), 1.6π (dash) with $\overline{k} = 3\pi$, (b) $\overline{k}/k_0 = 1.1$ (solid), 2.5 (dash-dot), 5 (dash) with $k_0 a = 2\pi$.

Besides, from Fig. 3b we see that for fixed $\overline{k}/k_0$, the layer factor $\mathcal{LF}_j$ is oscillatory with j and the oscillations frequency increases with $\overline{k}/k_0$.

Acknowledgment

The work of N. L. Tsitsas was supported by the State Scholarships Foundation, while he was a post-doctoral research scholar.

References

1. L. Shafai, *Can. J. Phys.* **50**, 749 (1972).
2. Z.-F. Sang, Z.-Y. Li, *Phys. Letters A* **332**, 376 (2004).
3. S. Saengkaew, et. al, *Exp. Fluids* **43**, 595 (2007).
4. M. Kerker, *The Scattering of Light and Other Electromagnetic Radiation* (Academic, 1969).
5. L. Kai, P. Massoli, *Applied Optics* **33**, 501 (1994).
6. I. Gurwich, et. al, *J. Quant. Spectr. Radiat. Transfer* **63**, 217 (1999).
7. A. Y. Perelman, *Applied Optics* **35**, 5452 (1996).
8. W. Yang, *Applied Optics* **42**, 1710 (2003).
9. N. L. Tsitsas, C. Athanasiadis, *Quart. J. Mech. Appl. Math.* **59**, 55 (2006).
10. C. A. Balanis, *Advanced Engineering Electromagnetics* (Wiley, 1989).
11. C. A. Valagiannopoulos, N. L. Tsitsas, *J. Opt. Soc. Am. A* **26**, 870 (2009).

AN ELECTRO-HYDRODYNAMIC MODEL FOR IMPEDANCE CARDIOGRAPHY

P. A. VOLTAIRAS* AND D. I. FOTIADIS AND A. CHARALAMBOPOULOS

Department of Materials Science and Engineering,
University of Ioannina, GR 451 10 Ioannina, Greece
E-Mail: pvolter@cs.uoi.gr

L. K. MICHALIS

Department of Cardiology, Medical School,
University of Ioannina, GR 451 10 Ioannina, Greece

We propose an an-harmonic, electro-hydrodynamic (EHD) model to study pulsatile impedance cardiography measurements. The model explains the similarities between the impedance cardiography pulse and the arterial pressure pulse.

1. Introduction

Bio-electromagnetic measurements lack a firm theoretical basis that associates them with structural, geometrical and physiological factors[1]. This obstructs possible clinical applications. Electrical impedance instrumentation is not relatively costly. Hence, among the first and most popular developed techniques was impedance plethysmography (IPG), a method of determining changing tissue volumes in the body, based on the measurement of the electric impedance of the body surface[2]. In the case of the measurement of the impedance of the thorax, the technique is called impedance cardiography (ICG). In previous works[3,4], we proposed an analytical, an-harmonic, three dimensional, viscoelastic blood flow model to account for the peripheral distortion of the arterial pulse. The purpose of this study is to couple this blood flow model with an an-harmonic electromagnetic (EM) wave propagation solution of Maxwell's equations to interpret ICG pulses.

*Corresponding author.

2. Impedance Cardiography

A typical recording apparatus is given in Fig. 1. The impedance measure-

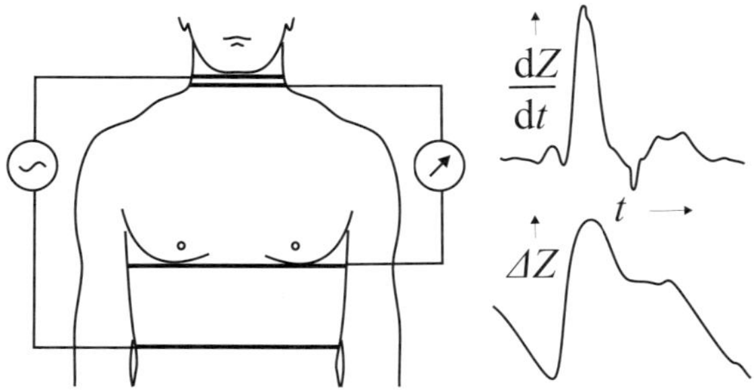

Figure 1. Impedance cardiography measurements.

ment is made by introducing an electric current of 0.5 to 4 mA to the outer pair of electrodes, in the frequency range of 50 to 100 kHz. Then, the inner pair of electrodes records the pulsatile decrease in impedance of 0.1 to 0.2 ohms (decreasing impedance, ΔZ, and negative first impedance derivative, dZ/dt, are shown in an upward direction). To relate the ΔZ obtained on the thorax to the pulsatile blood volume change, the parallel column model is frequently used[2,5]. The model consists of a conducting volume with impedance Z_0 (it contains a heterogenous mixture of the non-time-varying tissues such as bone, muscle, fat, etc.) in parallel with a time varying fluid column (it represents blood) with resistivity ρ_r, length L (the distance between the inner electrodes) and a time varying cross-sectional area which oscillates up to a finite value. Then, an empirical formula for the stroke volume (the volume of blood pumped from one ventricle of the heart with each beat), based on the parallel column model, is[5]:

$$SV = \rho_r \left(\frac{L}{Z_0}\right)^2 t_{ej} \, \frac{dZ_{min}}{dt} \, . \tag{1}$$

t_{ej} is the systolic ejection time and dZ_{min}/dt is the magnitude of the largest negative derivative of the impedance change occurring during systole. The various notches in dZ/dt represent different events in the cardiac circle such as closure and opening of aortic, pulmonic and mitral valves, etc. ICG measurements are interpreted in terms of Geselowitz reciprocity relation[6]:

$$\Delta Z = -\Delta\sigma \int_V \mathbf{L}_1^E \cdot \mathbf{L}_2^E \, dV, \quad \mathbf{L}_i^E = \mathbf{E}_i/\mathcal{I}_i \, , \tag{2}$$

where $\Delta\sigma$ is the change in conductivity due to the imposed current $\mathcal{I}_1$ in the outer electrode pair, $\mathbf{L}_i^E$ are the lead fields, $\mathbf{E}_i$ are the electric fields, $\mathcal{I}_i$ are the currents and V is the volume between the inner electrodes in Fig. 1. The subscripts 1 and 2 denote imposed and induced electric fields and currents, due to the outer and inner electrode pairs, respectively. ΔZ is influenced by many different factors including geometry, tissue conductivity, blood flow, etc. Because of this complexity, it is difficult to reliably measure an isolated physiological parameter, which has been the principle factor limiting its use[5]. Therefore, a more reliable physiological modeling is needed.

3. The EHD model

The problem geometry is depicted in Fig. 2. The inner infinite cylinder of

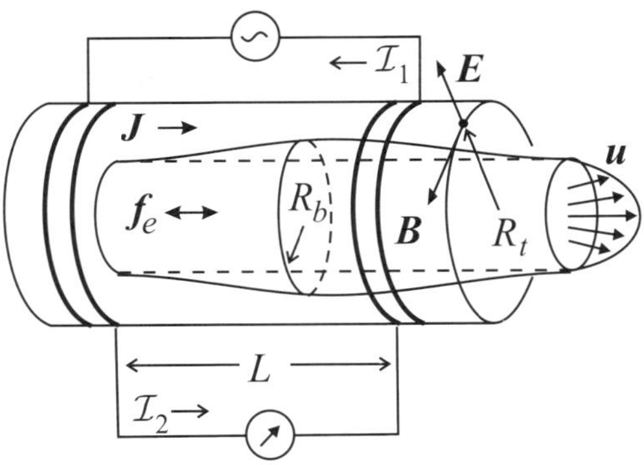

Figure 2. Problem Geometry.

radius R_b represents the total blood flow in the thorax. It is quasi-elastic, since it undergoes small but non-vanishing radial deformations ($R(t) \simeq R_b$). The blood flow problem is the one we developed in order to interpret the peripheral distortion of the arterial pulse[3,4] (blood is considered a non-Newtonian linear Maxwell fluid, see Eqs. (6) and (7) below). The outer infinite cylinder, of constant[a] radius R_t, corresponds to the thorax surface, where the four band electrodes are placed. The blood flow is rotational:

$$\mathbf{u} = \nabla \times \mathbf{A}_b, \quad \mathbf{A}_b = S(r)\,\Psi_b(t,z)\,\mathbf{e}_\theta\,, \tag{3}$$

[a] R_t depends on respiratory dynamics that are not included directly in our model.

which satisfies the continuity equation $\nabla \cdot \mathbf{u} = 0$. We use cylindrical coordinates (r, θ, z). The no-slip boundary condition is satisfied approximately, due to the quasi-elastic character of the arterial wall, namely $S(R_b)/L = u_r^0 \ll u_0$, $S_z(R_b) \equiv S'(R_b) + S(R_b)/R_b = 0$, where prime denotes differentiation with respect to the argument, u_r^0 is a constant radial velocity and $u_0 = L/t_0$ is the characteristic velocity with t_0 a characteristic time constant and S_z/L is the radial part of the axial velocity component. The electric, $\mathbf{E}$, and magnetic, $\mathbf{B}$, fields are expressed in terms of the EM vector potential $\mathbf{A}_e$ as:

$$\mathbf{E} = -\frac{\partial \mathbf{A}_e}{\partial t}, \quad \mathbf{B} = \nabla \times \mathbf{A}_e, \quad \mathbf{A}_e = E_0\, t_0 \Psi_e(t, z)\, \mathbf{e}_r\,, \tag{4}$$

where E_0 is a characteristic electric field constant. The dimensionless spatiotemporal functions Ψ_b, and Ψ_e, are considered equal:

$$\Psi_b(t,\, z) = \Psi_e(t,\, z), \tag{5}$$

although they are solutions of the blood flow:

$$\rho\,(\mathbf{u}_{,t} + \mathbf{u} \cdot \nabla \mathbf{u}) = \nabla \cdot \mathbf{T} - \nabla P + \mathbf{f}_e\,, \tag{6}$$

$$\mathcal{L}\,\mathbf{T} \equiv \mathbf{T} + t_w\, \frac{\partial \mathbf{T}}{\partial t} = \mu_v\, \left[\nabla \otimes \mathbf{u} + (\nabla \otimes \mathbf{u})^T\right], \tag{7}$$

and the EM wave propagation problems:

$$\nabla \times \mathbf{E} = -\frac{\partial \mathbf{B}}{\partial t}, \quad \nabla \cdot \mathbf{D} = 0, \quad \nabla \times \mathbf{H} = \frac{\partial \mathbf{D}}{\partial t} + \mathbf{J}, \quad \nabla \cdot \mathbf{B} = 0, \tag{8}$$

respectively. $\otimes$ denotes tensor product, $\mathbf{T}$ is the stress tensor, $\mathcal{L}$ is the operator of linear viscoelasticity and t_w is the relaxation time. P, μ_v and ρ are the blood's pressure, shear viscosity and density, respectively. The thorax is considered a homogeneous, isotropic conductive medium:

$$\mathbf{B} = \mu \mathbf{H}, \quad \mathbf{D} = \varepsilon \mathbf{E}, \quad \mathbf{J} = \sigma \mathbf{E}, \quad \mu \simeq \mu_0, \quad \varepsilon = \varepsilon_r\, \varepsilon_0, \tag{9}$$

where σ is the conductivity constant and ε_r is the relative electric permittivity. The impulsive EM body force term in Eq. (6) is of the form:

$$\mathbf{f}_e = \mathbf{J} \times \mathbf{B} \simeq \nabla \times \mathbf{a}_e, \quad \mathbf{a}_e \equiv \mathbf{J} \times \mathbf{A}_e, \quad \mathbf{J} = J_c\, G(r)\, \mathbf{e}_z\,, \tag{10}$$

where J_c is a characteristic current density constant. We approximate the convective acceleration as $\mathbf{u} \cdot \nabla \mathbf{u} \simeq \frac{1}{2} \nabla u^2 - \mathbf{u}_c \times \boldsymbol{\omega}$, with $\mathbf{u}_c = u_c\, \mathbf{e}_z$ a constant axial velocity vector. $\boldsymbol{\omega} = \nabla \times \mathbf{u}$ is the vorticity vector. Also we assume that $\mathcal{L}\mathbf{u}_c \cdot \nabla \mathbf{u} \simeq \mathbf{u}_c \cdot \nabla \mathbf{u}$. Then, following a similar methodology to that developed previously[3,4] we obtain the inhomogeneous Klein-Gordon (apart from the fourth convection term) vorticity equation:

$$\mu_v \nabla^2 \mathbf{A}_b - \rho\, t_w \frac{\partial^2 \mathbf{A}_b}{\partial t^2} - \rho \frac{\partial \mathbf{A}_b}{\partial t} - \rho\, \mathbf{u}_c \cdot \nabla \mathbf{A}_b + \mathbf{a}_e = 0\,, \tag{11}$$

and the resembling (Klein-Gordon or telegraph) equation for the propagation of EM waves:

$$\nabla^2 \mathbf{A}_e - \mu\varepsilon \frac{\partial^2 \mathbf{A}_e}{\partial t^2} - \mu\sigma \frac{\partial \mathbf{A}_e}{\partial t} = 0\,. \tag{12}$$

P is given in terms of the blood flow velocity, $\mathbf{u}$, as:

$$P(\mathbf{x},\, t) = P_s - P_b + P_c, \quad P_b = \tfrac{1}{2}\rho\, u^2, \quad P_c = \rho\, \mathbf{u}_c \cdot \mathbf{u}\,, \tag{13}$$

where $\mathbf{x}$ is the position vector and $P_s = P_s(t)$ is an arbitrary temporal function. The generalized Bernoulli equation (13) is the cornerstone of our further analysis. P_b is the well known Bernoulli's pressure. P_c is the pressure, due to the convection of vorticity. Both P_c and P_b emanate from the quasi-linearization of $\mathbf{u}\cdot\nabla\mathbf{u}$, introduced above. The form of the cardiac in origin solitary pulse $P_s(t)$, the value of u_c, as well as the amplitude, the width and the shift of the flow pulse $\mathbf{u}$, with respect to P_s, controls all the changes in the shape of the pressure pulse. Due to the presence of both electric and magnetic fields in our model, the changes in the thoracic impedance are calculated by the generalized Geselowitz reciprocity relation:

$$\Delta Z = -\int_V \left(\Delta\sigma\, \mathbf{L}_1^E \cdot \mathbf{L}_2^E + K^B - K^E\right) dV, \tag{14}$$

where $K^j = a_1^j \mathbf{L}_2^j \cdot \frac{\partial \mathbf{L}_1^j}{\partial t} - a_2^j \mathbf{L}_1^j \cdot \frac{\partial \mathbf{L}_2^j}{\partial t}$, $a_i^E = \epsilon_i$, $a_i^B = \mu_i$, $\mathbf{L}_i^E = \mathbf{E}_i/\mathcal{I}_i$, $\mathbf{L}_i^B = \mathbf{B}_i/(\mu_i \mathcal{I}_i)$, and $i = 1, 2$, $j = E, B$, that includes the classical Geselowitz reciprocity relation (2) as a special case, when $K^E \simeq K^B$.

4. The an-harmonic solution

Let us introduce the dimensionless parameters:

$$\begin{aligned}
&r = r/R_b, \quad \beta = L/R_b, \quad \bar{z} = z/L, \quad \bar{t} = t/t_0\,,\\
&p = P/P_0,\ U_c = u_c/u_0,\ U_p = 1/\sqrt{\mathrm{De\,Re}} = u_p/u_0,\ g = G/P_0\,,\\
&\mathrm{De} = t_w/t_0, \quad \mathrm{Re} = L\,u_0/\nu, \quad T = 2\,\mathrm{De}, \quad s = S/(L\,u_0)\,,\\
&\epsilon = \sqrt{U_p^2\, T^2\, \delta^2 + k^2 - 1},\ k = U_c/U_p,\ \xi = \sqrt{t_m/t_e},\ U_h = u_h/u_0\,,\\
&\tau_b = \epsilon\, \bar{t}/T, \quad \eta_b = \epsilon\, \bar{z}/(U_p T), \quad \tau_e = \xi t/(2t_{em}), \quad \eta_e = \xi z/(2L)\,,\\
&p_h = (\mu_v u_h)/(P_0 R_b), \quad p_r^0 = (\mu_v u_r^0)/(P_0 R_b), \quad U_r^0 = u_r^0/u_0\,,
\end{aligned} \tag{15}$$

where

$$u_p = \sqrt{\nu/t_w}, \quad u_{em} = 1/\sqrt{\mu\varepsilon}, \quad u_h = J_c\, E_0\, t_0\, R_b/\mu_v\,, \tag{16}$$
$$\nu = \mu_v/\rho, \quad P_0 = \rho\, u_0^2, \quad \Delta Z_0 = \Delta\sigma\, E_0^2\, \pi R_t^2\, L/\mathcal{I}_0^2\,, \tag{17}$$
$$t_{em} = L/u_{em} = L\sqrt{\mu\varepsilon} = \sqrt{t_e t_m}, \quad t_e = \varepsilon/\sigma, \quad t_m = \mu\sigma L^2\,. \tag{18}$$

u_p is the pulse wave velocity, u_{em} is the EM wave velocity, u_h is the impulsive heart flow velocity, which depends on the free currents J_c in the heart, ΔZ_0 is an impedance constant, t_{em} is the transit time of EM waves, t_e is the charge relaxation time, t_m is the magnetic diffusion time, ν is the kinematic viscosity of the blood, P_0 is a characteristic pressure, E_0 is a characteristic electric field, $\mathcal{I}_0$ is a characteristic current, De is the Deborah number, Re is the Reynolds number, and β is the characteristic length-to-radius ratio. Then, following steps similar to our previous works [3,4] the spatiotemporal part of the solution, either for the vorticity Eq. (11) or for the EM wave Eq. (12), is given in compact form by:

$$\Psi_i(\eta_i, \tau_i) = \sum_q C_q^{(i)}\, \Psi_q^{(i)}(\eta_i,\, \tau_i), \tag{19}$$

$$\Psi_q^{(i)}(\eta_i, \tau_i) = \frac{\partial \psi_q^{(i)}}{\partial \tau_i} = \frac{1}{2}\left(\psi_{q-1}^{(i)} - \kappa_i\, \psi_{q+1}^{(i)} - \frac{2\kappa_i}{\epsilon_i}\, \psi_q^{(i)}\right), \tag{20}$$

$$\frac{\partial \psi_q^{(i)}}{\partial \eta_i} = -\frac{1}{2}\left(\psi_{q-1}^{(i)} + \kappa_i\, \psi_{q+1}^{(i)} - \frac{2k_i}{\epsilon_i}\, \psi_q^{(i)}\right), \tag{21}$$

$$\psi_q^{(i)}(\eta_i, \tau_i) = e^{(k_i\,\eta_i - \tau_i)/\epsilon_i}\left(\frac{\tau_i - \eta_i}{\tau_i + \eta_i}\right)^{q/2} h_q^{(i)}\left(\sqrt{\tau_i^2 - \eta_i^2}\right). \tag{22}$$

For the blood flow problem the spatiotemporal function in Eq. (19) corresponds to $i = b$ with $\kappa_b = 1$, $k_b = k$, $\epsilon_b = \epsilon > 0$, $h_q^{(b)} = J_q$, while for for the EM wave problem to $i = e$ and $\kappa_e = -1$, $k_e = 0$, $\epsilon_e = -1$, $h_q^{(e)} = I_q$. The radial part of the solution of the vorticity Eq. (11) is given by:

$$s(\tilde{r}) = \frac{2U_h}{\beta}\left[\frac{\tilde{r}}{2} - \frac{1}{\lambda}\frac{J_1(\lambda\tilde{r})}{J_0(\lambda)}\right] - U_r^0\left(\tilde{r}^3 - 2\,\tilde{r}\right), \tag{23}$$
$$g(\tilde{r}) = -p_h\beta\lambda^2\,\tilde{r} + p_r\beta^2\left(\lambda^2\tilde{r}^2 + 8 - 2\lambda^2\right)\tilde{r}\,, \tag{24}$$

where $\lambda = \delta/\beta$ the roots of $\lambda_n\, J_0(\lambda_n)/2 - J_1(\lambda_n) = 0$, $n = 1, 2, \ldots$ We considered only the first physiologically appropriate mode, $\lambda_2 = 5.1356$. J_q and I_q are the Bessel and modified Bessel functions of order q, respectively, $C_q^{(i)}$ are undetermined constants and $\tau_i \geqslant \eta_i$, $(i = b, e)$. The blood flow $(i = b)$ and EM wave $(i = e)$ pulses described by the eigenfunctions in

Eq. (22) are relatively undistorted waveforms, since they maintain their shape although they attenuate.

5. Results and Discussion

For $\mathbf{E}_1 = E_0\,\mathbf{e}_r$, $\mathcal{I}_1 = \mathcal{I}_2 = \mathcal{I}_0$, Eq. (2) becomes:

$$\Delta Z = -\frac{\Delta Z_0}{u_0}\int_0^L \frac{\partial \Psi_e}{\partial t} dz\,. \tag{25}$$

Since thoracic changes in the volume of blood, which is an electrically conductive fluid, act on the measured electrical impedance at the thorax, we may assume that ΔZ is proportional to the arterial pressure P:

$$\Delta Z = -\gamma\,\Delta Z_0\,P\,, \tag{26}$$

where γ is a proportionality constant. Then, Eqs. (25) and (26) give:

$$\frac{1}{u_0}\int_0^L \frac{\partial \Psi_e}{\partial t} dz = \gamma\,P\,. \tag{27}$$

Eqs. (27) and (5) constitute a nonlinear optimization problem, which couples apart from the constants in the eigenfunction expansions (19) and the flow and EM material parameters, which are usually missing from simple models like the parallel column model[5]. We will not proceed to its general solution here. Instead, according to Eq. (26) and for the pressure data for the radial artery, given in our previous works[4,7], we present a typical ICG pulse in Fig. 3(a), for $\Delta\sigma = 6.67$, $\text{ohm}^{-1}\text{cm}^{-1}$, $E_0 = 1\,\text{mV/cm}$, $R_t = 15\,\text{cm}$,

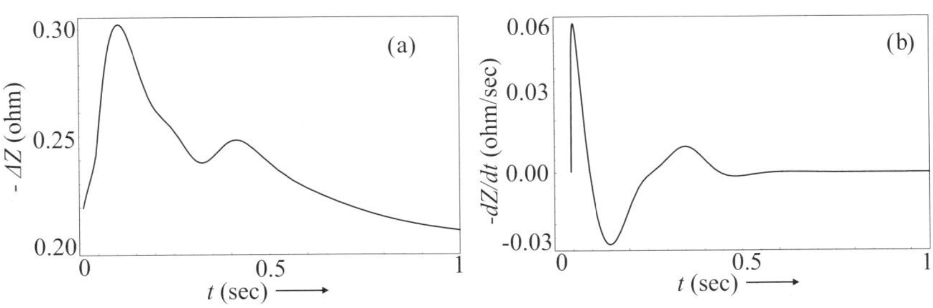

Figure 3. Calculated ΔZ (a) and dZ/dt (b) pulses.

$L = 30\,\text{cm}$, $\mathcal{I}_0 = 4\,\text{mA}$ and $\gamma = 18.2\,\text{cm}^2/\text{N}$. In Fig. 3(b) we present the first derivative of the ICG pulse, according to:

$$\frac{dZ}{dt} = -\gamma\,\Delta Z_0\,\frac{\partial(P - P_s)}{\partial t}\,. \tag{28}$$

Aorta is one of the main contributors to the dZ/dt signal. However, we had no available flow pulse data for the aorta. Therefore, for the fitting of the aortic pressure pulse we used[4] just one term in the eigenfunction expansions (19), which was inadequate for the dZ/dt pulse. Nevertheless, the agreement between the theoretical estimations in Fig. 3 and the ICG measurements in Fig. 1 is quite satisfactory. Due to Eq. (28) we can replace the stroke volume empirical Eq. (1) by:

$$SV = \rho_r \left(\frac{L}{Z_0}\right)^2 \gamma\, \Delta Z_0\, t_{ejec}\, \rho \left| (\mathbf{u}_c - \langle \mathbf{u} \rangle) \cdot \frac{\partial \langle \mathbf{u} \rangle}{\partial t} \right|_{min} . \tag{29}$$

since what is measured in related experiments is the volumetric flow rate[4] through the tube $Q = 2\pi \int_0^{R_b} r\, u_z\, dr$, with $u_z = (S'(r) + S(r)/r)\, \Psi_b(z,t)$, and $\langle u \rangle = Q/(2\,\pi\, R_b^2) = u_A\, \Psi_b(z,t),\ u_A \equiv \beta\, u_r^0$.

6. Conclusions

We developed an EHD framework for interpreting ICG measurements in terms of related pressure and flow data, since thoracic changes in the volume of blood, which is an electrically conductive fluid, act on the measured electrical impedance at the thorax. By solving the proposed nonlinear optimization problem, our model will contribute to the determination of the interrelation between complex (electrical, rheological, etc.) physiological parameters, and their effect on the ICG pulse. Progress in these research directions is important for advancements in biomedical instrumentation.

References

1. C. Polk, E. Postow, *Handbook of Biological Fffects of Electromagnetic Fields*, CRC Press (1996).
2. J. Malmivuo, R. Plonsey, *Bioelectromagnetism*, Oxford University Press (1995).
3. P.A. Voltairas D. I. Fotiadis, L. K. Michalis, in *Proceedings of the Eighth International Workshop on Mathematical Methods in Scattering Theory and Biomedical Engineering*, World Scientific, 380 (2008).
4. P.A. Voltairas, D. I. Fotiadis, A. Charalambopoulos, L. K. Michalis, in the e-book *Single and Two-Phase Flows on Chemical and Biomedical Engineering*, Bentham Science Publishers (2010).
5. R. Patterson, in *The Biomedical Engineering Handbook*, CRC Press, 1470 (2000).
6. D. A. Geselowitz, *IEEE Trans. Biomed. Eng.* **18**(1), 38 (1971).
7. P.A. Voltairas, D. I. Fotiadis, C.V. Massalas, L. K. Michalis, *J. Biomech.* **38**(7), 1423 (2005).

Size Effects and Enhanced Elastic Theories

ACOUSTIC SCATTERING BY OBSTACLES INVISIBLE IN THE BACKSCATTERED DIRECTION

B. D. SLEEMAN
SCHOOL OF MATHEMATICS,
UNIVERSITY OF LEEDS,
LEEDS, LS2 9JT,
UNITED KINGDOM

Motivated by the work of Aleksenko and Plakhov [1] concerning the problem of a parallel flow of particles falling on a body with piecewise smooth boundary and leaving no trace the analogous problem of acoustic scattering of a plane wave illuminating the same body is considered. It is shown that at high frequencies, with use of the Kirchoff approximation and the geometrical theory of diffraction, the back scattered far field is zero for a discrete set of wave numbers. That is the body is acoustically "invisible" in the back scattered direction

1. Introduction

In quantum billiard dynamics one studies the Helmholtz equation in a bounded domain D subject to Dirichlet boundary conditions on the boundary ∂D. Billiard dynamics can be considered from seemingly two distinct points of view. The interior problem is that in which the object is to calculate the spectrum, i.e., the eigenvalues k^2 of the equation $\triangle\phi + k^2\phi = 0$ in D and where ϕ vanishes on ∂D. In the exterior problem one studies acoustic scattering of an incident field from the obstacle defined by the billiard boundary with the same boundary conditions.

It has been conjectured [5] that there is a strong underlying duality between these problems. Informally it states that k^2 is an eigenvalue of the interior Dirichlet problem if and only if the scattering matrix $S(k)$ of the exterior problem has an eigenvalue 1. This statement is true for certain canonical billiards such as circles and ellipses but without further refinement is not true in general. An important consequence of the conjecture is that at an interior eigenvalue the billiard is " invisible" for a certain wavefunction. In the interior of the billiard the wave function corresponds to the associated eigenfunction while in the exterior of the billiard it is the wave function corresponding to the eigenvalue 1 of the scattering matrix.

That is the eigenfunction of the interior problem can be continued to a single valued wave function in the plane. Billiards can be constructed for which continuation is not possible. Of particular relevance to this paper is that for smooth convex billiards the classical Poincaré scattering map [2] is identical to the map which describes the dynamics in the interior. This is simply a consequence of the fact that the normal and the tangent at the point of reflection are common to the reflections from either the interior or the exterior. For an up-to-date review of this "inside- outside" duality see [9] and the references given there.

In contrast to the above problems is the "problem of minimal resistance". In mathematical terms the problem can be stated as follows: Suppose a body D at rest occupies a bounded connected set and has a piecewise smooth boundary ∂D. A parallel flow of particles falls on the body and perfectly elastically reflect when colliding with the boundary ∂D and move freely between successive reflections. The resistance to be minimised is determined by calculating the sum over all momenta imparted to the body D by particles striking the body during an interval of unit time. This problem was originally considered by Isaac Newton [7] in 1687. Very recently Aleksenko and Plakhov [1] have taken up the problem and introduced the notion of a body with zero resistance, a body that leaves no trace and an invisible body. They prove that such bodies exist and construct examples.

Motivated by this work we consider the problem of acoustic scattering of a plane wave by the body D and show, using the Kirchoff approximation of high frequency scattering and the geometrical theory of diffraction, that the body appears "invisible" in the back scattered direction provided the wave number k takes certain discrete values. Although this problem is distinct from the "interior-exterior duality " phenomena of quantum billiard dynamics there appears to be an analogous "duality" present which is deserving of deeper study.

The paper is organised as follows: In section 2 we describe the results of Aleksenko and Plakhov [1] and, in particular, recall their examples of bodies which have zero resistance and are invisible. Section 3 is devoted to the formulation of our acoustic scattering problem, the Kirchoff approximation and the geometrical theory of diffraction. In section 4 we discuss several examples and conclude the paper in section 5 with some open problems and conjectures.

2. Bodies of zero resistance and invisible bodies

Let $D \subset \mathbb{R}^3$ be a bounded connected set with piecewise smooth boundary ∂D and let $v_0 \in S^2$. Let $(x,v) \mapsto (x_D^+(x,v), v_D^+(x,v))$ from a full measure subset of $\mathbb{R}^3 \times S^2$ into itself be the scattering map defined as follows:

Let the motion of a billiard $x(t)$, $v(t)$ satisfy the relations

$$x(t) = \begin{cases} x + vt, & \text{if } t < t_1, \\ x^+ + v^+t, & \text{if } t > t_2. \end{cases}$$

$$v(t) = \begin{cases} v, & \text{if } t < t_1, \\ v^+, & \text{if } t > t_2. \end{cases}$$

(where t_1, t_2 are a pair of real numbers depending on the particular motion.) Then $x^+ \equiv x_D^+(x,v)$, $v^+ \equiv v_D^+(x,v)$.

Denote $\tilde{x}_D^+(x,v) = x^+ - < x^+, v^+ > v^+$ and $t^* = t_D^*(x,v) = - < x^+, v^+ >$. Then $x^+ + v^+t = \tilde{x}^+ + v^+(t - t^*)$, and $\tilde{x}^+$ is orthogonal to v^+. Finally $\{v\}^\perp$ denotes the orthogonal complement to the one-dimensional subspace $\{v\}$, that is the plane containing the origin and is orthogonal to v.

Definition

(1) D has zero resistance in the direction v_0 if $v_D^+(x,v_0) = v_0$ for almost every x.

(2) D leaves no trace in the direction v_0 if, in addition to 1, the mapping $x \mapsto x_D^+(x,v_0)$ from a subset of $\{v_0\}^\perp$ into $\{v_0\}^\perp$ is defined almost everywhere in $\{v_0\}^\perp$ and preserves the two-dimensional Lebesgue measure.

(3) D is invisible in the direction v_0 if, in addition to 2, $\tilde{x}_D^+(x,v_0) = x$.

This definition has the following interpretation. Suppose there is a parallel flow of particles falling on the body D. Initially the velocity of a particle is $-v_0$; it makes several reflections form D and finally moves freely with velocity $v_D^+(x,v_0)$ where x is the initial position of the particle. The resistance of the medium to the bodies motion is proportional to

$$\mathbf{R}_{v_0}(D) = \int_{\{v_0\}^\perp} (v_0 - v_D^+(x,v_0))dx,$$

where dx is the Lebesgue measure in $\{v_0\}^\perp$.

In case (1) of the definition we have $\mathbf{R}_{v_0}(D) = 0$. If the body D has a mirrored surface then in case (3) it is invisible in the direction v_0. In case

(2), if the body is moving, the medium appears to be unchanged after the body has passed.

The main result of Aleksenko and Plakhov [1] is the following theorem. Fix $v_0 \in S^2$.

Theorem 1.

(a) There exists a body with zero resistance in the direction v_0 but leaves a trace.

(b) There exists a body that leaves no trace in the direction v_0 but is not invisible.

(c) There exists a body that is invisible in the direction v_0.

The proof of this result is based on a number of carefully constructed examples. These examples are central to the acoustic scattering problems considered in the paper. A key example in Aleksenko and Plakhov's work is constructed from a pair of equilateral triangles. Specifically consider the two identical coplanar equilateral triangles $A'AO'$ and $C'CO$ shown in Figure 1, where the line OO' is parallel to the initial particle velocity v_0. The body generated by rotation of the triangle ABC about the axis OO' has zero resistance in the v_0 direction. To see this consider Figure 1 representing a vertical cross section of the rotationally symmetric body.

Let a particle with initial velocity v_0 first hit the segment $C'B'$ at the point E'. After reflection, using Snell's Law, the direction of motion forms an angle $\pi/3$ with the vertical. The particle then hits the segment BA at E and after a second reflection moves vertically downwards with a final velocity v_0. This body leaves a trace and so is not invisible. In fact the particles that initially belong to the cylindrical layer of width x (see figure 1) finally occupy the smaller cylindrical layer of the same width x. The density of this smaller layer is greater than that of the initial larger layer.

Suppose we now glue a copy of Figure 1 to itself along $A'A$ to obtain the cavity shown in Figure 2 and carry a corresponding rotation about the line of symmetry. Again consider the particle reflections as before. It is clear that the body so generated is invisible in the v_0 direction.

Aleksenko and Plakhov [1] now develop their arguments further by considering a 1-parameter family of bodies having all the desired properties claimed in theorem 1. To this end consider a pair of isosceles triangles with angles α, α and $\pi - 2\alpha$ where $0 < \alpha < \pi/4$. The triangles are placed symmetrically to eachother with respect to a given point, see Figure 3. The point lies along the symmetry axis of each triangle, at a distance $(\tan 2\alpha - \tan \alpha)/2$ from its obtuse angle and at a distance $(\tan 2\alpha + \tan \alpha)/2$

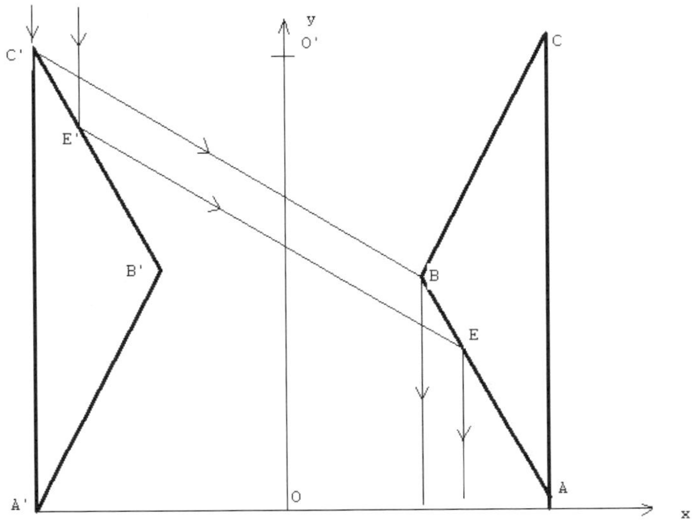

Figure 1. Cavity formed by equilateral triangles.

from its base. The length of the base of each triangle is 2. From the figure we see that the body has zero resistance in the v_0 direction parallel to the bases of the triangle. A zero resistance body, a trackless body and an invisible body are constructed by the above procedures of rotation, translation with gluing and doubling applied to the pair of triangles.

The triangular shapes above and the properties of invisibility will now be studied from the point of view of acoustic scattering.

3. Acoustic scattering, the Kirchoff approximation and the Geometrical theory of Diffraction

Let B be a bounded domain in $\mathbb{R}^2$ with piecewise continuous boundary ∂B. Let the incident field u^i be an entire function, (e.g a plane wave) which satisfies Helmholtz equation and let u^s be the corresponding scattered field. Then the direct sound- soft obstacle scattering problem is to find a total field

$$u = u^i + u^s,$$

where $u \in C^2(\mathbb{R}^2 \backslash \bar{B}) \cap C(\mathbb{R}^2 \backslash B)$ and satisfies Helmholtz equation in $\mathbb{R}^2 \backslash \bar{B}$, i.e. u solves

$$\begin{aligned} \triangle u + k^2 u = 0, \text{ in } \mathbb{R}^2 \backslash B, \\ u = 0, \text{ on } \partial B \end{aligned} \tag{1}$$

and where u^s satisfies the Sommerfeld radiation condition

$$\lim_{r \to \infty} r^{1/2} \left(\frac{\partial u^s}{\partial r} - iku^s \right) = 0. \tag{2}$$

It is well known that the scattered field u^s has the asymptotic behaviour

$$u^s = e^{-ikr} r^{-1/2} u^{\infty} + u', \tag{3}$$

where $u' \in L^2(\mathbb{R}^2 \backslash B)$.

The solution of the above problem when the obstacle B is composed of the triangular pairs D in Figures 1 and 2 does not admit to exact analytical treatment in general and one must use numerical methods or an approximate method. In the high frequency, or short wavelength, regime considerable progress is to be made by use of the Kirchoff approximation.

Consider a sound soft obstacle B with boundary ∂B and suppose the incident field u^i is a plane wave given by

$$u^i = e^{ikx \cdot \omega},$$

where $\omega = (\cos \phi, \sin \phi)$ is the direction of incidence.

The total field is

$$u = e^{ikx \cdot \omega} + u^s, \tag{4}$$

which satisfies the boundary condition $u = 0$ on ∂B, i.e

$$u^s = -e^{ikx \cdot \omega}, \ x \in \partial B.$$

Using Green's theorem the scattered field u^s is expressible in the form

$$u^s(\boldsymbol{x}) = -\frac{i}{4} \int_{\partial B} \left\{ \Phi(\boldsymbol{x}, \boldsymbol{y}) \frac{\partial u^s}{\partial \nu(\boldsymbol{y})}(\boldsymbol{y}) + e^{ik\boldsymbol{y} \cdot \boldsymbol{\omega}} \frac{\partial \Phi(\boldsymbol{x}, \boldsymbol{y})}{\partial \nu(\boldsymbol{y})}(\boldsymbol{y}) \right\} ds(\boldsymbol{y}), \tag{5}$$

where $\boldsymbol{x} \in \mathbb{R}^2 \backslash \bar{B}$,

$$\Phi(\boldsymbol{x}, \boldsymbol{y}) = H_0^{(2)}(k \mid \boldsymbol{x} - \boldsymbol{y} \mid), \ \boldsymbol{x} \neq \boldsymbol{y}$$

and for $\mid \boldsymbol{x} \mid \gg \mid \boldsymbol{y} \mid$

$$H_0^{(2)}(k \mid \boldsymbol{x} - \boldsymbol{y} \mid) \sim \left(\frac{2i}{\pi k r} \right)^{1/2} e^{-ikr} e^{ik(\boldsymbol{x}/|\boldsymbol{x}|) \cdot \boldsymbol{y}}$$

. Now every incident wave of direction $\boldsymbol{\omega}$ divides the boundary ∂B uniquely into two parts-the illuminated part ∂B_I and the shadow part ∂B_S as shown in Figure 4. Under the Kirchoff approximation in the shadow area just outside ∂B_S the total wave is negligibly small and that just outside ∂B_I

$$\partial_\nu u^s \approx ik\boldsymbol{d}\cdot\nu e^{ik\boldsymbol{x}\cdot\boldsymbol{\omega}}.$$

On substituting these approximations into (5) we have

$$u^s(x) \sim \frac{e^{ikr}}{(2\pi ikr)^{1/2}}\left[f_I(\boldsymbol{k}_i,\boldsymbol{k}_s) + f_S(\boldsymbol{k}_i,\boldsymbol{k}_s)\right], \tag{6}$$

where $k\boldsymbol{\omega} \equiv \boldsymbol{k}_i = k(\cos\phi, \sin\phi)$ and $\boldsymbol{k}_s = k(\cos\theta, \sin\theta)$ are the incident and scattered directions respectively and

$$\begin{aligned} f_I(\boldsymbol{k}_i,\boldsymbol{k}_s) &= \frac{1}{2}\int_{\partial D_I} e^{i(\boldsymbol{k}_i-\boldsymbol{k}_s)\cdot\boldsymbol{y}}(\boldsymbol{k}_i-\boldsymbol{k}_s)\cdot\nu ds(\boldsymbol{y}) \\ f_S(\boldsymbol{k}_i,\boldsymbol{k}_s) &= -\frac{1}{2}\int_{\partial D_S} e^{i(\boldsymbol{k}_i-\boldsymbol{k}_s)\cdot\boldsymbol{y}}(\boldsymbol{k}_i+\boldsymbol{k}_s)\cdot\nu ds(\boldsymbol{y}). \end{aligned} \tag{7}$$

In general these integrals are evaluated using the method of stationary phase as indicated in [4] and [9]. For the scattering problems discussed in the next section the integrals can be evaluated directly.

The theory outlined above does not account for the effects of diffraction. Diffracted rays, which may enter shadow regions, are produced by incident rays which strike edges, or vertices of boundary surfaces, or which graze such surfaces. The geometrical theory of diffraction developed by Keller [6] is designed to account diffraction effects. The power of the theory and the insight into wave propagation has been exploited in many fields. For example it has been applied to both homogeneous and inhomogeneous media and used to describe wave propagation in fluid mechanics, elasticity, quantum mechanics, surface waves and elsewhere.

For our purposes we consider fields diffracted by straight edges. Specifically we consider the two dimensional case in which the edge is a straight line and the incident rays all lie in planes normal to the edge. The diffracted rays are also normal to the edge and eminate from it in all directions. It is sufficient to consider only those rays in one plane normal to the edge. If u_i is the incident field meeting an edge then Keller [6] shows that the field diffracted by the edge is given by

$$u_d = Du_i r^{-\frac{1}{2}} e^{ikr}, \tag{8}$$

where r is the distance from the edge and D is the diffraction coefficient.

In the case when the edge is formed by a wedge of angle $(2-n)\pi$, as shown in Figure 5, the diffraction coefficient is;

$$D = \frac{e^{i\pi/4}\sin\pi/n}{n(2\pi k)^{1/2}}\left\{\left(\cos\pi/n - \cos(\frac{\theta-\alpha}{n})\right)^{-1} - \left(\cos\pi/n - \cos(\frac{\theta+\alpha+\pi}{n})\right)^{-1}\right\}, \tag{9}$$

where α is the angle of the incident ray and θ is the angle of a diffracted ray.

4. Obstacles invisible in the back scattered direction

Consider the cavity bounded by the triangle edges ABC and $A'B'C'$ in Figure 1. With the coordinate system indicated in the figure we illuminate the cavity with a plane wave incident field in the direction $(0,1)$, i.e $\boldsymbol{k}_i = k(0,1)$ and observe the scattered field from the back scattered direction $(0,-1)$, i.e $\boldsymbol{k}_s = k(0,-1)$. For this configuration the shadow boundary of the cavity is specified by the set $x = \pm 1/2,\ \sqrt{3}/2 \le y < \infty$. That is the regions $(-1 \le x < -1/2) \times (\sqrt{3}/2 \le y < \infty) \cup (1/2 \le x < 1) \times (\sqrt{3}/2 \le y < \infty)$ comprise the shadow.

By formally applying the Kirchoff Approximation to the cavity we see ,from (7), that $f_S(\boldsymbol{k}_i, \boldsymbol{k}_s) = 0$ and

$$f_I(\boldsymbol{k}_i, \boldsymbol{k}_s) = 4\sqrt{3}k\int_0^{\pi/3} e^{2\sqrt{3}ik\frac{\tan\theta}{(\sqrt{3}+\tan\theta)}}\frac{\sec^2\theta}{(\sqrt{3}+\tan\theta)^2}d\theta, \tag{10}$$

which after the substitution

$$z = \frac{1}{(\sqrt{3}+\tan\theta)}$$

is readily integrated to give

$$f_I(\boldsymbol{k}_i, \boldsymbol{k}_s) = \frac{i}{\sqrt{3}}(1 - e^{i\sqrt{3}k}). \tag{11}$$

From(11) we make the key observation that if

$$k = \frac{2}{\sqrt{3}}n\pi,\ n = 1, 2, \cdots, \tag{12}$$

then, under the Kirchoff approximation, the total radiation pattern and consequently the scattering cross section is zero. In other words at these wave numbers the cavity appears "invisible" and is undetected in the back scattered far-field. This we believe is an approximate acoustic equivalent to the problem of billiards leaving no trace. We may also think of the cavity as being hidden from an observer under the illumination by a plane wave with wave number chosen from (12) for some large value of n.

The wave numbers (12) have a further intriquing significance: If we scale the equilateral triangle $A'AO'$ of side 2 in Figure 1 to one of side 1, then this is equivalent to rescaling the wave number k by 2 to get the wave numbers

$$K_n = \frac{4}{\sqrt{3}} n\pi.$$

that is

$$K_n^2 = \frac{16}{27} 9n^2\pi^2. \tag{13}$$

These numbers are precisely the symmetric eigenvalues of the Laplacian defined in the unit equilateral triangle due to Pinsky [8]. This observation appears to have some resonance with the interior-exterior duality discussed in [9] but whether it has any deeper significance is unkown.

The above analysis is not complete without consideration of the diffracted fields emanating from B' and C, in Figure 1, which result in a propagating ray in the back scattered direction. There will be corresponding rays emanating from B and C'. From Figure 1 we see that there will be a unique diffracted ray from B' which becames an incident field striking C which is again diffracted to give rise to a unique ray propagating in the backscattered direction.

To begin with, the incident plane wave u^i takes the value $e^{ik\frac{\sqrt{3}}{2}}$ at B'. For the wedge defined at B', using the notation of equation (9), $n = 4/3$, $\alpha = 0$ and $\theta = 2\pi/3$. The diffraction coefficient at B' is then

$$D_{B'} = \frac{4e^{i\pi/4}}{3(2\pi k)^{1/2}} \tag{14}$$

and the diffracted field emanating from B' is

$$u_{B'} = \frac{4e^{i\pi/4}}{3(2\pi k)^{1/2}} e^{ik\frac{\sqrt{3}}{2}} r^{-1/2} e^{ikr}. \tag{15}$$

This diffracted ray now enters the shadow region and becomes an incident ray striking the wedge at C. That is the incident ray at C has the value

$$u_i(C) = u_{B'} = \frac{4e^{i\pi/4}}{3(2\pi k)^{1/2}} \frac{e^{ik\frac{3\sqrt{3}}{2}}}{3^{1/4}}. \tag{16}$$

With respect to the wedge at C, $n = 11/6$, $\alpha = 5\pi/3$ and $\theta = \pi/2$ and so the corresponding diffraction coefficient is

$$D_C = \frac{6e^{i\pi/4}\sin 6\pi/11}{11(2\pi k)^{1/2}}\left\{\left(\cos 6\pi/11 - \cos(\frac{9\pi}{11})\right)^{-1} - \left(\cos 6\pi/11 - \cos(\frac{3\pi}{11})\right)^{-1}\right\}, \quad (17)$$

so the diffracted field at C in the back scattered direction is

$$u_C = D_C u_i(C)\frac{e^{iky}}{y^{1/2}}. \quad (18)$$

The radiation pattern for this diffracted wave is easily seen to be of the form

$$f_C = A\frac{e^{ik\frac{\sqrt{3}}{2}}}{(2\pi k)^{1/2}}. \quad (19)$$

Note that there will be a similar radiation pattern arising from diffraction at C'.

The important consequence for the back scattered radiation pattern for the cavity is that (10) is modified to

$$f_I(\boldsymbol{k}_i, \boldsymbol{k}_s) = \frac{i}{\sqrt{3}}(1 - e^{i\sqrt{3}k}) + \mathcal{O}(k^{-1/2}),\ k \gg 1. \quad (20)$$

The upshot of this is that for $k \approx \frac{2}{\sqrt{3}}n\pi$, $n = 1, 2, \cdots$, the cavity is acoustically invisible in the back scattered direction.

We now turn to the cavity configuration depicted in Figure 3. As before the cavity is bounded by the triangle edges ABC and $A'B'C'$. With the coordinate system indicated in Figure 3 the cavity is illuminated with a plane wave incident field in the direction $(0, 1)$, i.e $\boldsymbol{k}_i = k(0, 1)$ and observations of the scattered field are made in the back scattered direction $(0, -1)$, i.e $\boldsymbol{k}_s = k(0, -1)$. For this configuration the shadow boundary of the cavity is specified by the set $x = \pm 1/2(\tan 2\alpha - \tan\alpha,\ 1 \leq y < \infty$. That is the shadow region is $(-1/2(\tan 2\alpha + \tan\alpha) \leq x < -1/2(\tan 2\alpha - \tan\alpha)) \times (1 \leq y < \infty) \cup (1/2(\tan 2\alpha - \tan\alpha) \leq x < 1/2(\tan 2\alpha + \tan\alpha)) \times (1 \leq y < \infty)$.

By again applying the Kirchoff Approximation to the cavity we see that $f_S(\boldsymbol{k}_i, \boldsymbol{k}_s) = 0$ and

$$f_I(\boldsymbol{k}_i, \boldsymbol{k}_s) = k\Omega\cot\alpha e^{ik\Omega\cot\alpha} \cdot \int_0^{\tan^{-1}\left(\frac{2}{\Omega - 2\tan\alpha}\right)} e^{-ik\Omega\frac{\cot^2\alpha}{(\cot\alpha + \tan\theta)}}\frac{\sec^2\theta}{(\cot\alpha + \tan\theta)^2}d\theta, \quad (21)$$

where $\Omega \equiv (\tan 2\alpha + \tan\alpha)$.

By making the substitution

$$z = \frac{1}{(\cot\alpha + \tan\theta)},$$

equation (21) can be integrated to give

$$f_I(\boldsymbol{k}_i, \boldsymbol{k}_s) = i\tan\alpha(1 - e^{2ik}). \tag{22}$$

From this result we conclude that if

$$k = n\pi,\ n = 1, 2, \cdots, \tag{23}$$

then under the Kirchoff approximation the total far field is zero and that the cavity is acoustically "invisible" in the back scattered direction. It is worth noting that if we rescale the problem so that the base of the cavity AA' is scaled to unity; this is equivalent to scaling the wave number k by Ω to obtain the wave number

$$K = \Omega n\pi$$

and in the case $\alpha = \pi/6$ we recover the wave number (13).

The connection with the eigenvalues of interior problems is now tenuous. Indeed the eigenvalues $n^2\pi^2$ deduced from (23) are simply the eigenvalues of the vibrating unit string and seem to have nothing to do with the eigenvalues of vibrating triangles as the result (13) might suggest.

Once again, as in the above, we can apply the geometric theory of diffraction to show that the radiation pattern (24) has to be modified to

$$f_I(\boldsymbol{k}_i, \boldsymbol{k}_s) = i\tan\alpha(1 - e^{2ik}) + \mathcal{O}(k^{-1/2}), \tag{24}$$

leading to the conclusion that this class of cavities are acoustically invisible for $k \approx n\pi,\ n \gg 1$.

5. Conclusions

Motivated by the work of Aleksenko and Plakhov [1] concerning the problem of a parallel flow of particles falling on a body with piecewise smooth boundary and leaving no trace we have considered the analogous acoustic scattering problem of a plane wave illuminating the same body. We have shown that at high frequencies with the use of the Kirchoff approximation together with the geometrical theory of diffraction, the far field (equivalently the total scattering cross section) is zero for a discrete set of wave numbers. That is, the scattering body appears to be acoustically invisible in the back scattered direction. If the body is constructed from equilateral triangles then the wave numbers are approximated by the eigenvalues of the equilateral triangle. However if the constructions are based on certain isosceles triangles the corresponding wave numbers are approximated by those of the vibrating string. This observation is not understood and suggests that the problem be solved numerically for a range of frequencies in order to obtain a deeper understanding of " acoustic invisibility". This will be the subject of further work. In addition, we will examine similar problems in electromagnetic scattering.

The subject is of considerable interest at the present time. Indeed Chen. et.al [3] have proposed the use of transformation optics to design gateways that can block electromagnetic waves but allow the passage of other entities. In other words the device can be thought of as an implementation of a "hidden portal" which allows passage of physical objects but appears to be blocked to electromagnetic waves.

Acknowledgement The author is indebted to Evgeny Lakshtanov of the University of Aveiro, Portugal for bringing the problem of bodies of zero resistance to his attention. Informative discussions with Alexander Plakhov of Aberystwyth University are much appreciated.

References

1. A. Aleksenko and A. Plakhov, Bodies of zero resistance and bodies invisible in one direction. *Nonlinearity,* **22**, 1247-1258, 2009.
2. R. Blümel, B. Dietz, C. Jung and U. Smilansky, *J. Phys. A: Math. Gen.* **25**, 1483-1502, 1992.
3. H. Chen, C. T. Chan, S. Liu and Z. Lin, A simple route to a tunable electromagnetic gateway. *New J. Physics*, **11**, 083012, 2009.
4. B. Dietz and U. Smilansky, A scattering approach to the quantization of billiards-The inside-outside duality. *Chaos*, **3**, 581-589, 1993.
5. E. Doron and U. Smilansky, Quantization of chaotic billards: a scattering theory approach. *Nonlinearity,* **5**, 1055-1084, 1992.
6. J. B. Keller, Geometrical Theory of Diffraction. *J. Opt. Soc. Amer.,* **52**, 116-130, 1962.
7. I. Newton, *Philosophiae Naturalis Principia Mathematica* , 1687.
8. M. A. Pinsky, The eigenvalues of an equilateral triangle. *SIAM. J. Math. Anal.* **11**, 819-827, 1980.
9. B. D. Sleeman , The inverse acoustic obstacle scattering problem and its interior dual. *Inverse Problems and Imaging*, **3**, 211-229, 2009.

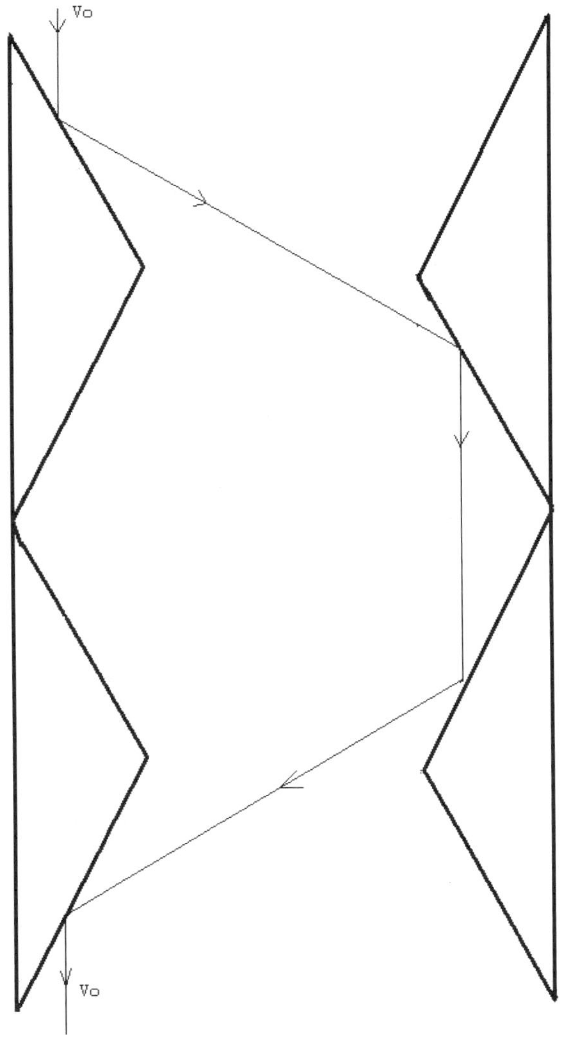

Figure 2. Cavity invisible in the v_0 direction.

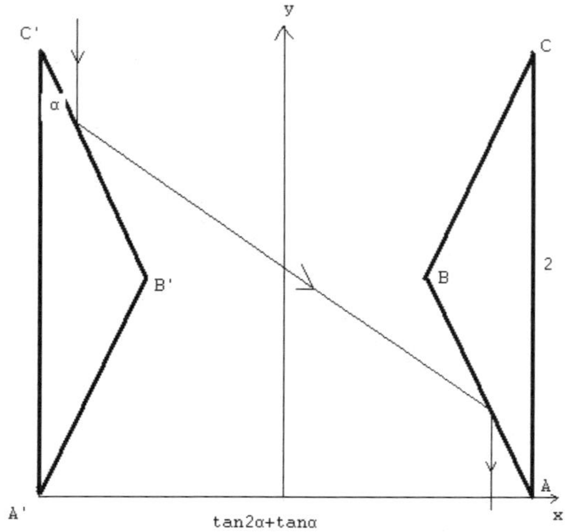

Figure 3. 1-parameter family of cavities.

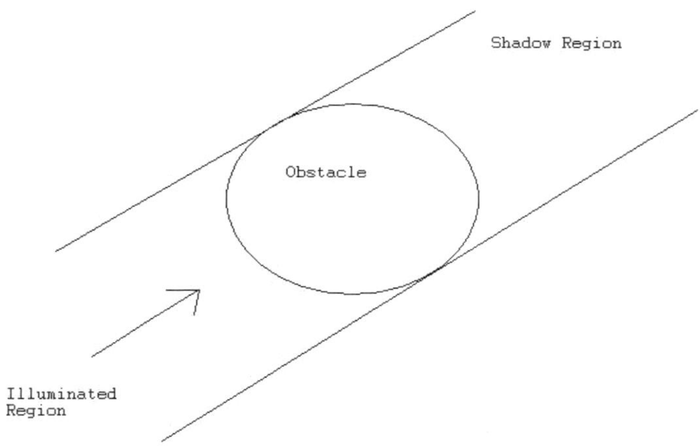

Figure 4. The incident plane wave divides the boundary of the scatterer into an illuminated part and a shadow part.

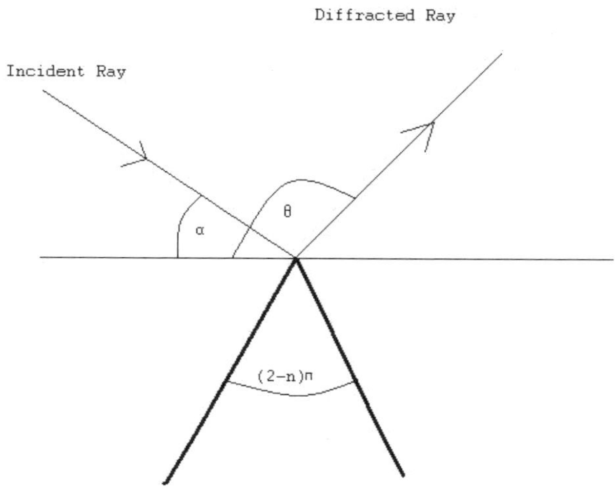

Figure 5. Diffraction by a wedge of angle $(2 - n)\pi$.

THE USE OF STRAIN GRADIENT ELASTICITY IN MODELLING TISSUES: THE CASE OF THE HUMAN HEART

A. E. GIANNAKOPOULOS
Laboratory for Strength of Materials and Micromechanics
Department of Civil Engineering
University of Thessaly, Volos 38334, Greece

V. I. ZAFIROPOULOU
Laboratory for Strength of Materials and Micromechanics
Department of Civil Engineering
University of Thessaly, Volos 38334, Greece

In recent years, strain gradient elasticity has been used to capture size effects in stiffness and toughness that appear in structures of different sizes. An important aspect of such theories, often taken of little notice, is the non-classical boundary conditions that appear as couple-stress tractions. Such tractions can provide deformation of the body they are acting on, without the need for classic type of supports that are usually needed when classic tractions are applied. Moreover such tractions seem to be "invisible", since no external pressure is acting and they can be related to non-mechanical sources, like electricity. The human heart seems to stand without support (bones etc), floating in the human body and electric pulses applied to its surface make it vibrate, inflating periodically the various chambers that circulate the blood in the body. We propose a model for the muscle tissues of the heart, in the context of gradient elasticity, relating the power due to the couple stresses with the electric power provided by chemistry. The effect is to provide a model that describes at the surface of the heart couple stress tractions that can be direct functions of the electrocardiogram (ECG) signals. The heart is modeled as a thick spherical shell that deforms under the fore-mentioned boundary conditions and the problem is solved in closed form, providing useful relations that connect the deformation of the walls of the heart with the ECG. Many applications can benefit from the present approach, like pacers and other medical devices.

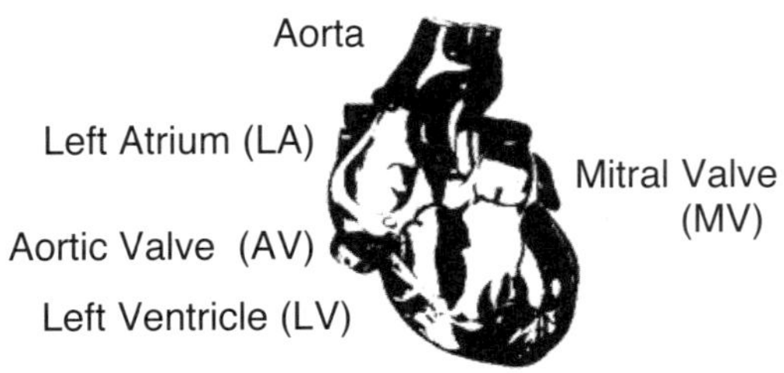

Figure 1. The cross-section of the human heart (From http://www.turburbosquid.co

1. Introduction

The heart is the pumping station of the human body. It has the size of a clenched first, is conically shaped and is a hollow muscular organ [1]. The left-side of the heart, is a self-contained pump which drives the oxygen-rich blood into the aortic semilunar outer valve. It does most of its work at high pressure to drive blood through the entire systemic circulation to every location of the body. The heart is divided into the aortic and the ventricle, Fig. 1. In this work we will focus on the mechanics of the left ventricle.

The heart rate of a normal adult person is 80 beats/min (aortic cycle period of 754ms) and the heart beats about 3×10^9 times in the lifetime of an average person. During the periodic motion of the heart, the entire ventricular wall moves inward, approximately toward the geometric center of the ventricle. The specific gravity of the cardiac muscle is 1.05. The periodic pressure-volume and pressure –time diagrams of the heart response are given schematically in Fig. 2 [2]. The region 1-2 represents the isometric contraction the region 1-4-5 represents the systolic ejection and the region 5-6 represents the isometric relaxation. In this work, we will focus on the region 2-3 (isovolume contraction) which takes place within $T_{isv} = 0.06s$ and is fast enough to ignore the viscoelastic effects. The following values are typical for a healthy adult. The maximum systolic pressure is $P_{sy} \approx 16\text{kPa}(120\text{mm Hg})$. The stroke volume is $V_s = 72cc$ and the ejection fraction is $V_s / V_{ed} = 0.6$. The end-diastolic pressure is $P_{ed}/P_{sy} = 0.2$. The time period for systolic action is $T_s / T_{isr} = 6$. As the heart fills during diastole, work is done by the blood, by distending the ventricle. This work is stored as elastic tension to be given back during systole. After the period of isometric construction, the blood leaves the heart. The mechanical work performed by the heart during each contraction is called cardiac work and is about W=0.8J.

The ventricle wall has a complex three-dimensional muscle-fiber architecture. The myocytes are relatively short and are connected to from a predominant fiber axis that is approximately tangent to the wall. The muscle fiber angle makes a smooth transmural transition from epicardium to endocardium from -60° to 70°. The maximum static muscle stress is 0.35 MPa, which can be held for short periods of time, because the muscles fatigue. The maximum static sustainable stress generated by muscle is about 0.1 MPa. Various constitutive models have been proposed for the elasticity of passive cardiac tissues. The simplest law has been proposed by Schneck [3] and connects the uniaxial tensile stress σ with the uniaxial tensile strain ε

$$\sigma = K\varepsilon^{n} \tag{1}$$

For cardiac muscle, $K = 1.8826 \times 10^{5}\ N/m^{2}$ and $n = 1.6$. The source of the passive tension is the titin protein, which connects the thick myosin filaments. To the passive force-length properties of the muscle, we have to add its active force-length properties [4]. Since we will focus on the analysis of the active stresses, we will assume a tangent elastic modulus

$$E = \frac{d\sigma}{d\varepsilon} = Kn\,\varepsilon^{n-1} \tag{2}$$

which will be evaluated at a passive stress level of 0.1MPa, giving E= 5 kPa.

Muscle fibers are chemomechanical actuators. They range in thickness from 10 to 100 μm and in length from 1 to 30 cm. They are surrounded by a connective tissue sheath (epimysium). Muscle fibers are composed of parallel bundles of myofibrils. The myofibrils are invested by a plasma membrane (sarcolema) and are from repeated units called sarcomeres that act as contractile units. A sarcomere is between the Z-lines and contains centrally a thin filament (actin), which is surrounded by a thick filament (myosin). The heads on the myosin body attach to the thin filament to from cross-bridges. The mechanics of cross-bridge movement is believed to cause relative movements at the myofilaments. Thin filaments have diameters of about 10-20 nm and length of 2.5μm. Thick filaments have diameters of about 15nm and length 1.6μm.

Muscle shortening is caused by sliding of a thin filament past the thick myofilaments, causing the fiber to shorten as a whole and the muscle to contract and the appearance of <u>double forces</u> [5], Fig 3. The energy for the linear motion is provided by the hydrolysis of adenosine triphosphate (ATP). As the ATP is hydrolyzed, the head swings back about 5nm to the locked position where adenosine diphosphate (ADP) and phosphate ion (PI) remain bounded. When PI leaves the myosin, the head binds the actin. The actin bindings are uncovered by the release of Calcium (Ca^{+2}). The head rotates, causing the sliding of the actin filament against the myosin filament and ADP is released to continue the cycle (the ATP binds at the back of the head of the cross-bridge on the myosin filament, causing a conformation which cannot bind actin). The hydrolysis of 1mole ATP releases 10kcal of chemical energy.

Ionic pumps within membranes operate to produce different concentrations of ions in the intracellular and extracellular volumes around cells. In cardiac cells, the sodium-potassium pump produces different ionic concentrations across the membrane. As a result a transmembrane voltage ϕ (polarization) develops across the membrane [6], according to Coldman's equation. A base line voltage of $\varphi = -80mN$ develops at rest (polarized state) and an overshoot voltage $\varphi = 20\text{mN}$ develops at excitation (depolarized state). A plateau voltage of $\varphi = -20\text{mN}$ holds for about 0.3s (systolic ejection), Fig. 4. The cardiac tissue is an electric synctium (all cells are coupled through intercellular junctions). The electrocardiogram (ECG) is the recording on the body surface at the electrical activity generated by the heart.

In the following section, we will present a model that incorporates the cardiac action potential φ into the mechanical response of the heart. To the best of our knowledge, this is a new approach for the active stressing of the heart muscles.

2. A strain gradient elasticity model of living tissues

We will attempt to model the active stresses of living tissues. The physiology of the problem directed us to formulate a coupled electro-mechanical problem (static). We would like to cast a constitutive equation between the stresses (σ_{ij}) and the strains (ε_{ij}), as well as between the electric fields (E_i) and the electric displacements (D_i). We assume that it is possible to homogenize the material response through a continuum with the appropriate boundary conditions. Taking only linear terms, the Helmoltz free energy per unit volume (Ψ) is a function of the strains, the material polarization (P_i) and the spatial gradients of the polarization $\left(P_{i,j} \equiv \frac{\partial P_i}{\partial x_j} \right)$. According to the Landau-Ginsburg-Devonshire theory, we have

$$\Psi = \frac{1}{2} \mathcal{L}_{ijpq}\, \varepsilon_{ij}\, \varepsilon_{pq} + \frac{1}{2\kappa_{ij}} D_i\, D_j + \frac{1}{2} a_{ijpq}\, P_{i,j}\, P_{p,q} - T\,\eta \tag{3}$$

where $\mathcal{L}_{ijpq} = \mathcal{L}_{pqij} = \mathcal{L}_{jipq} = \mathcal{L}_{ijqp}$ is the elastic coefficient tensor, $\kappa_{ij} = \kappa_{ji}$ is the dielectric permeability tensor, $a_{ijpq} = a_{pqij}$ is the exchange energy tensor that determines the magnitude of the regions that become electrically polarized, T is the absolute temperature and η is the entropy per unit volume. The first term in Eq. (3) denotes the elastic energy, the second term the electric energy and the third term the exchange energy.

Denote by $\chi_{ij} = \chi_{ji}$ the electric susceptibility tensor:

$$P_i = \kappa_0 \, \chi_{ij} \, E_j \tag{4}$$

where $\kappa_0 = 8.854 \times 10^{-12} \, \mathrm{C}^2 / \left(\mathrm{N} \cdot \mathrm{m}^2\right)$ is the electric susceptibility of vacuum. The electric constitutive equation is

$$D_i = \kappa_0 \, E_i + P_i = \kappa_0 \left(\delta_{ij} + \chi_{ij} \right) E_j \equiv \kappa_{ij} \, E_j \, . \tag{5}$$

Denote by d_{ijk} the piezoelectric coefficient tensor. Then the piezoelectric constitutive equation is

$$P_i = d_{ipq} \, \sigma_{pq} \qquad \text{(direct piezoelectric response)} \tag{6}$$

$$\varepsilon_{ij} = d_{ijk} \, E_k \qquad \text{(reverse piezoelectric response).} \tag{7}$$

For a homogeneous body:

$$P_{i,j} = d_{ipq} \, \sigma_{pq,j} \, . \tag{8}$$

Therefore, the exchange energy in Eq. (1) can be written as

$$\frac{1}{2} a_{ijpq} \, P_{i,j} \, P_{p,q} = \frac{1}{2} a_{ijpq} \, d_{imn} \, \sigma_{mn,j} \, d_{prs} \, \sigma_{rs,q} \equiv \frac{1}{2} \hat{\mu}_{ijk} \, \hat{\kappa}_{ijk} \, , \tag{9}$$

For isotropic response

$$\hat{\mu}_{ijk} = \ell^2 \, \sigma_{ij,k} \qquad \text{and} \qquad \hat{\kappa}_{ijk} = \varepsilon_{jk,i} \tag{10}$$

with $\sigma_{ij} = \mathscr{L}_{ijpq} \, \varepsilon_{pq}$ the stress-strain relation and ℓ an internal length which characterizes the material and is the outcome of the combination of various electro-mechanical coefficients. Thus Eq. (3) can take the final form.

$$\boxed{\Psi = \frac{1}{2} \mathscr{L}_{ijpq} \, \varepsilon_{ij} \, \varepsilon_{pq} + \frac{1}{2 \kappa_{ij}} D_i \, D_j + \frac{1}{2} \hat{\mu}_{ijk} \, \hat{\kappa}_{ijk} - T \, \eta} \, . \tag{11}$$

In the above equation, the term $\frac{1}{2}\hat{\mu}_{ijk}\hat{\kappa}_{ijk}$ proves that the constitutive model follows the general approach of a strain gradient response that was detailed by Mindlin [7]. The electric problem can be solved independently from the mechanical problem. The internal length for stiff piezoelectric ceramic materials is small (micrometer). For soft materials, the internal length can be large (millimetre), as can be readily detected from Eqs. (9) and (10).

The strains are related to the displacement vector through the geometric relations

$$\varepsilon_{ij} = (u_{i,j} + u_{j,i})/2 \tag{12}$$

and the elastic field vector is related to the electric potential φ

$$\mathrm{E}_i = -\phi_{,i} \tag{13}$$

Note that comma denotes spatial differentiation with respect to the corresponding coordinate x_i and repeated indices imply summation from 1 to 3.

The <u>electrostatic problem</u> has an equilibrium equation

$$D_{i,i} = q \tag{14}$$

where q is the volume charge density and boundary condition $D_i n_i = -\omega$ (surface charge density), with n_i the unit normal vector at the surface.

The <u>mechanical problem</u> (static, without body forces) has an equilibrium equation

$$(\sigma_{ji} - \widehat{\mu}_{kji,k})_{,i} = 0 \tag{15}$$

and "traction" and "double force" boundary conditions.

$$\widehat{P}_i = n_j(\sigma_{ji} - \widehat{\mu}_{kji,k}) - [d_j - (d_p n_p)n_j](n_k \widehat{\mu}_{kji}) \tag{16}$$

$$R_i = n_k n_j \widehat{\mu}_{jki} \tag{17}$$

where $d_j = \partial / \partial x_j - n_k n_j \partial / \partial x_k$ is the surface gradient operator.

Figure 5 shows an approximate model of the left ventricle in the form of a spherical shell. In this case the mechanical problem simplifies using spherical coordinates (r, ∂, φ)

$$\varepsilon_{rr} = \frac{du}{dr} \qquad \varepsilon_{\theta\theta} = \varepsilon_{\varphi\varphi} = \frac{u}{r} \tag{18}$$

$$\sigma_{rr} = \mathrm{E}\varepsilon_{rr} \qquad \sigma_{\theta\theta} = \sigma_{\varphi\varphi} = \mathrm{E}\varepsilon_{\theta\theta} \tag{19}$$

where u(r) is the radial displacement. Equilibrium takes the form

$$\nabla^2 u - \ell^2 \nabla_u^4 = 0 \tag{20}$$

with a general solution:

$$u(r) = c_1 r + c_2 / r^2 + (c_3 / r)\cosh(r/\ell) + (c_4 / r)\sinh(r/\ell) \tag{21}$$

and boundary conditions

$P(r_e) = \sigma_{rr} - \ell^2 \partial^2 \sigma_{rr} / \partial r^2 = 0$

$R(r_i) = \ell^2 \partial \sigma_{rr} / \partial r = 0$

$R(r_e) = R_o$ (electric interaction)

$u(r_i) = 0$ (systolic fluid solid interaction).

where $r_e = 42mm$ (external radius), $r_i = 27mm$ (internal radius) and $\ell = 10mm$ (internal length). Taking the external energy as

$$e = 2\pi r_e^2 R_o \varepsilon_{rr}(r_e) \tag{22}$$

we compute $P(r_i) = -11.15\text{kPa}$ which is in good accord with the measured change of systolic pressure.

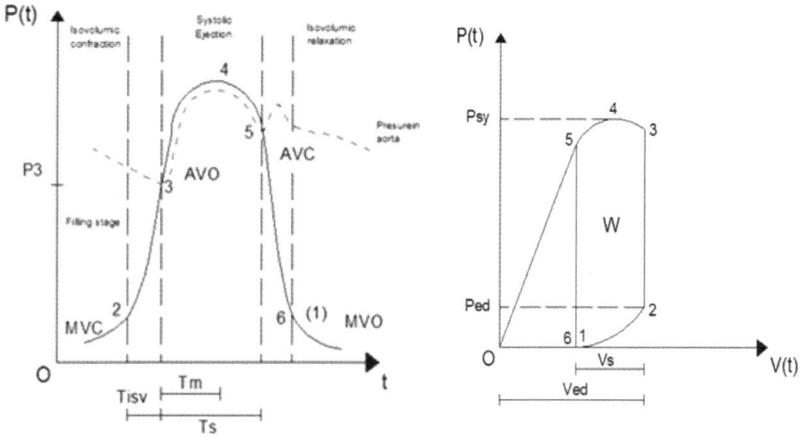

Figure 2. Pressure-volume (P-V) and pressure-time (P-t) periodic response of the left ventricular part of the heart. MVC: mitral valve close. MVO: mitral valve open. AVO: aortic valve open. AVC: aortic valve close. W: cardiac work.

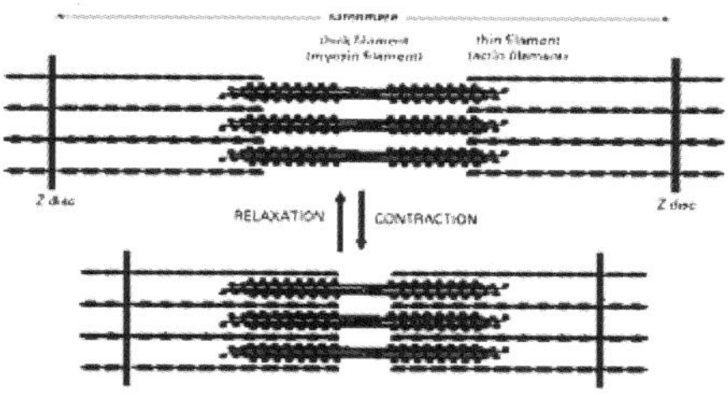

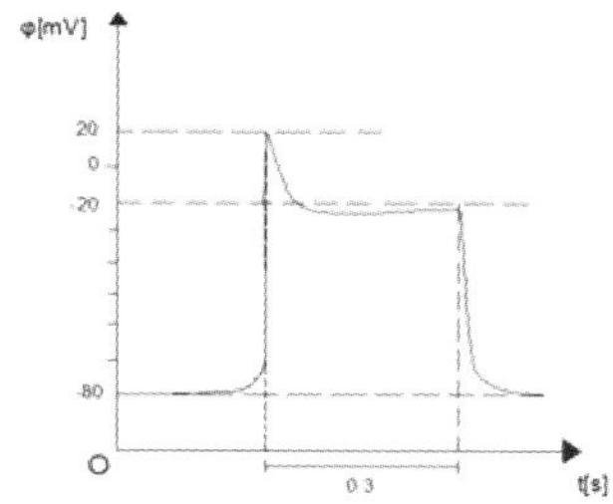

Figure 3. The contraction mechanism of a sarcomere. A dipole-force mechanism is developed due to the action of the myosin head movements and the hydrolysis of ATP.

Figure 4. The transmembrane voltage that develops across the membrane of the heart during the heart cycle.

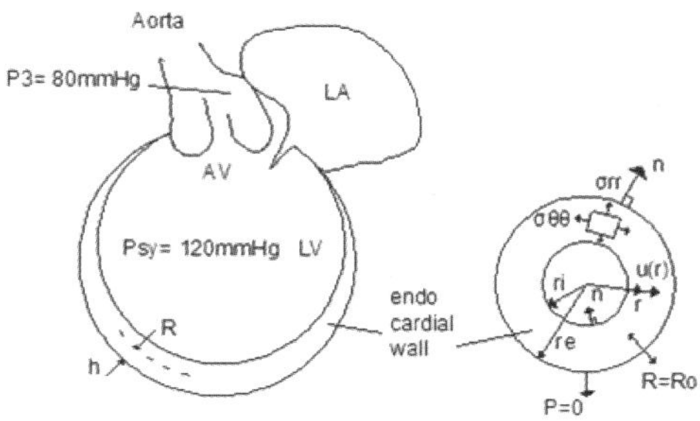

Figure 5. The approximate spherical shell model of the left ventricle.

References

1. D.E. Standness and D.S. Sumber, Hemodynamics for Surgeons, Grune and Stratton Inc, N.Y. (1975).
2. D.N. Chista, Applied Biomedical Engeneering Mechanics, CRC Press, Boca Raton FL (2009).
3. D.J. Schneck, Mechanics of Muscle, 2nd ed., N.Y. University Press (1992).
4. J.P. Humphrey and F.C.P.Yin, *J. Biomechanics Engineering*, **109**, 298 (1987).
5. A.F. Huxley, *Progress and Biophysical Chemistry*, **7**, 255 (1957).
6. R. Plonsey and R.C. Barr, Bioelectricity: A Quantitative Approach, N.Y. Plenum Press (1988).
7. R.D. Mindlin and N.N. Eshel, *Int. J. Solids Struct.* **1**, 109 (1968).

BOUNDARY ELEMENT SOLUTIONS FOR FREQUENCY DOMAIN PROBLEMS IN MINDLIN's STRAIN GRADIENT THEORY OF ELASTICITY

A. PAPACHARALAMPOPOULOS, D. POLYZOS
Department of Mechanical Engineering and Aeronautics
University of Patras, GR-26500, Patras, Greece

A. CHARALAMBOPOULOS
Department of Materials Science and Engineering
University of Ioannina, GR-45110, Ioannina, Greece

D. E. BESKOS
Department of Civil Engineering
University of Patras, GR-26500, Patras, Greece

An advanced Boundary Element Method (BEM) for solving two-dimensional (2D), frequency domain, elastodynamic problems in materials with microstructural effects is presented. The analysis is performed in the context of Mindlin's Form II gradient elastic theory. The frequency domain Form II gradient elastodynamic fundamental solution of the equation of motion is employed, while the integral representation of the problem, consisting of two boundary integral equations, one for displacements and the other for its normal derivative is utilized. The global boundary of the analyzed domain is considered smooth without corners, discetized into quadratic line elements. A representative 2D numerical example is provided to illustrate the method, demonstrate its accuracy and assess the gradient effect on the response.

1. Introduction

In elastic materials with microstructure, such as bones, polymers, polycrystals, granular or composite materials, microstructural effects are important and have to be included in their constitutive equations. These microstructural effects can be successfully modeled in a macroscopic level by employing the general higher order gradient elastic theory of Mindlin [1–3]. According to this theory, beyond the classical Lame constants, new internal length scale parameters that correlate the stiffness and the inertia of the microstructure with the macrostructure are introduced. However, due to its complexity and the lack of information about the size of the new intrinsic parameters, only simplified versions of that theory have been exploited so far for the analytical solution of elastodynamic boundary value problems with microstructural effects [4-11].

For realistic engineering problems characterized by complicated geometry and boundary conditions, analytical methods of solution are inadequate and resort has to be made to numerical methods, such as the finite element method (FEM) or the boundary element method (BEM). Among the efforts made for FEM and BEM solutions of boundary value problems in gradient elastodynamics, one can mention the works of Askes et al. [12], Polyzos et al. [13, 14], Polyzos [15] and Tsepoura and Polyzos [16].

In the present work the BEM in its direct form is employed for the solution of two-dimensional frequency domain dynamic problems in the framework of a simple gradient elastic theory obtained as the simplest possible special case of the general elastodynamic one of Mindlin [2, 15].

2. Mindlin's Form-II Simplified Gradient Elastic Theory and Integral Representation of the Problem

Mindlin in Form II version of his general gradient elastic theory [1], considering the potential energy density to be a quadratic form of the strains and gradient of strains and the kinetic energy density a function of the velocities and gradient of velocities and taking their variation, obtained an equation of motion which in frequency domain is written as

$$\partial_j(\hat{\tau}_{jk} - \partial_i \hat{\mu}_{ijk}) + \rho\omega^2(u_k - h^2 \partial_i \partial_i u_k) = 0 \tag{1}$$

where $\partial_i, \hat{\tau}_{jk}, \partial_i \hat{\mu}_{ijk}$ represent the gradient operator, the symmetric Cauchy-like tensor and the symmetric relative stress tensor, respectively, u_k stands for displacement vector, ρ, ω are the mass density and frequency, respectively, while h^2 represents the velocity gradient coefficient (units m^2), which is the intrinsic parameter correlating the inertia of the microstructure with the macrostructural dynamic behavior of the considered materials and/or structures. Eq. (1) is accompanied by the classical boundary conditions

$$u_k = \overline{u} \quad \text{and/or} \quad \hat{p}_k = \overline{p} \tag{2}$$

and the non-classical ones

$$q_k = n_l \partial_l u_k = \overline{q} \quad and/or \quad \hat{R}_k = \overline{R} \tag{3}$$

$$\text{and } \hat{E}_k = \overline{E} \tag{4}$$

where $\overline{u}, \overline{p}, \overline{q}, \overline{R}, \overline{E}$ represent prescribed vectors, n_l is the unit vector normal to the boundary and $\hat{p}_k, \hat{R}_k, \hat{E}_k$ are the traction, double traction and jump traction vectors, respectively, defined as

$$\hat{p}_k = n_j \hat{\tau}_{jk} - n_i n_j D \hat{\mu}_{ijk} - (n_j D_i + n_i D_j)\hat{\mu}_{ijk} + (n_i n_j D_l n_l - D_j n_i)\hat{\mu}_{ijk} \tag{5}$$

$$\hat{R}_k = n_i n_j \hat{\mu}_{ijk} \tag{6}$$

$$\hat{E}_k = \left\| n_i m_j \hat{\mu}_{ijk} \right\| \tag{7}$$

with $D \equiv n_l \partial_l$ and $D_j \equiv (\delta_{jl} - n_j n_l)\partial_l$.

The non-classical boundary condition described in Eq. (7) exists only when non-smooth boundaries are considered. Double brackets indicate that the enclosed quantity is the difference between its values taken on the two sides of a corner while m_j is a vector being tangential to the lines forming the corner.

In the simplest possible case the Cauchy-like stress tensor $\hat{\tau}_{ij}$ and the third order double stress tensors $\hat{\mu}_{ijk}$ are written as

$$\hat{\tau}_{ij} = 2\mu e_{ij} + \frac{2\mu\nu}{1-2\nu} e_{kk}\delta_{ij} \tag{8}$$

$$\hat{\mu}_{ijk} = g^2 \partial_i \hat{\tau}_{jk} \tag{9}$$

where μ, ν is the shear modulus and Poisson ratio, respectively, $e_{ij} = (\partial_i u_j + \partial_j u_i)/2$ the strain tensor and g^2 is the volumetric strain gradient energy coefficient (units m^2), the constant which in conjunction with velocity gradient coefficient h^2 relates the microstructure with the macrostructure.
Following the methodology of [15], it is easy to find one that the 2D fundamental solution of Eq. (1) is of the form

$$u_{ij}^* = \frac{1}{2\pi\rho\omega^2}\left[\Psi(r)\delta_{ij} - \mathrm{X}(r)\partial_i r \partial_j r\right] \tag{10}$$

where $r = |\mathbf{x} - \mathbf{y}|$, $\mathbf{x}$ is the point where the displacement field u_{ij}^* due to a unit force applied at a point $\mathbf{y}$ should be obtained and

$$X(r) = -w_1 K_2(ik_1 r) + w_2 K_2(ik_2 r) + w_1 K_2(a_1 r) - w_2 K_2(a_2 r) \tag{11a}$$

$$\begin{aligned}\Psi(r) = &-w_1\left[K_2(ik_1 r) - K_0(ik_1 r)\right]/2 + w_2\left[K_2(ik_2 r) + K_0(ik_2 r)\right]/2 - \\ &- w_1\left[K_0(a_1 r) - K_2(a_1 r)\right]/2 - w_2\left[K_0(a_2 r) + K_2(a_2 r)\right]/2\end{aligned} \tag{11b}$$

with $w_{1,2} = k_{1,2}^2(1 + L_{1,2}^2 k_{1,2}^2)/(1 + 2L_{1,2}^2 k_{1,2}^2)$, $k_1^2\left(1 + L_1^2 k_1^2\right) = k_P^2$, $\mathrm{k}_2^2\left(1 + L_2^2 k_2^2\right) = k_S^2$, $\omega / k_{p,s} = g c_{p,s} / L_1 = \sqrt{c_{p,s}^2 - \omega^2 h^2}$ and $a_{1,2} = \sqrt{1/L_{1,2}^2 + k_{1,2}^2}$. $c_{p,s}$ are the phase velocities corresponding to longitudinal (p) and shear (s) waves of the classical theory of elasticity and K_n modified Bessel functions of second kind and n-th order.

Considering a 2D gradient elastic material of volume V surrounded by a smooth surface S and characterized by the material constants μ, ν, g^2, h^2 and inserting the fundamental solutions (11) into a reciprocity identity provided in [14, 15], one can show that the frequency domain gradient elastic boundary value problem (Eqs. 1-3) admits an integral representation of the form

$$\begin{aligned} &\frac{1}{2}u_k(\mathbf{x}) + \int_S p^*_{kj}(\mathbf{x},\mathbf{y})\cdot u_j(\mathbf{y})dS_\mathbf{y} + \int_S R^*_{kj}(\mathbf{x},\mathbf{y})\cdot q_j(\mathbf{y})dS_\mathbf{y} = \\ &= \int_S u^*_{kj}(\mathbf{x},\mathbf{y})\cdot p_j(\mathbf{y})dS_\mathbf{y} + \int_S q^*_{kj}(\mathbf{x},\mathbf{y})\cdot R_j(\mathbf{y})dS_\mathbf{y} \end{aligned} \tag{12}$$

where $\mathbf{u}^*_{kj}, \mathbf{q}^*_{kj}, \mathbf{p}^*_{kj}, \mathbf{R}^*_{kj}$ are the fundamental solution of Eq. (1), its normal derivative, the fundamental traction and double stress traction tensors, respectively.

Observing Eq. (12), one realizes that it contains four unknown vector fields, $u_k(\mathbf{x}), p_k(\mathbf{x}), R_k(\mathbf{x})$ and $q_k(\mathbf{x})$. Thus, the evaluation of the unknown fields requires the existence of one more integral equation. This integral equation is obtained by applying the operator $\partial / \partial n_x$ on Eq. (12), i.e.

$$\begin{aligned} &\frac{1}{2}q_k(\mathbf{x}) + \int_S \frac{\partial p^*_{kj}(\mathbf{x},\mathbf{y})}{\partial n_\mathbf{x}}\cdot u_j(\mathbf{y})dS_\mathbf{y} + \int_S \frac{\partial R^*_{kj}(\mathbf{x},\mathbf{y})}{\partial n_\mathbf{x}}\cdot q_j(\mathbf{y})dS_\mathbf{y} = \\ &\int_S \frac{\partial u^*_{kj}(\mathbf{x},\mathbf{y})}{\partial n_\mathbf{x}}\cdot p_j(\mathbf{y})dS_\mathbf{y} + \int_S \frac{\partial q^*_{kj}(\mathbf{x},\mathbf{y})}{\partial n_\mathbf{x}}\cdot R_j(\mathbf{y})dS_\mathbf{y} \end{aligned} \tag{13}$$

Integral equations (12) and (13) accompanied by the classical and non-classical boundary conditions (Eqs. (2) and (3)) form the integral representation of a well-posed 2D gradient elastodynamic boundary value problem which, can be solved via the standard BEM machinery.

3. BEM Formulation and Numerical Results

The goal of the BEM is to solve numerically the boundary value problem constituted by the system of two integral equations (Eqs. (12) and (13)) and the boundary conditions (2), (3) since only smooth boundaries are considered. To this end the global boundary S is discretized into e three-noded, linear, quadratic, continuous and discontinuous isoparametric boundary elements. Then, all the nodal fields of the corresponding fields are expressed to the local co-ordinate system ξ with the aid of the shape functions N_α (α = 1, 2, 3). Adopting a global numbering for the total L nodes (each pair (e, α) is associated to a number β) the integral equations (12) and (13) for the node k are written as

$$\frac{1}{2}\mathbf{u}^k + \sum_{\beta=1}^{L} \tilde{\mathbf{H}}_\beta^k \cdot \mathbf{u}^\beta + \sum_{\beta=1}^{L} \tilde{\mathbf{K}}_\beta^k \cdot \mathbf{q}^\beta = \sum_{\beta=1}^{L} \tilde{\mathbf{G}}_\beta^k \cdot \mathbf{p}^\beta + \sum_{\beta=1}^{L} \tilde{\mathbf{L}}_\beta^k \cdot \mathbf{R}^\beta \tag{14}$$

$$\frac{1}{2}\mathbf{q}^k + \sum_{\beta=1}^{L} \tilde{\mathbf{S}}_\beta^k \cdot \mathbf{u}^\beta + \sum_{\beta=1}^{L} \tilde{\mathbf{T}}_\beta^k \cdot \mathbf{q}^\beta = \sum_{\beta=1}^{L} \tilde{\mathbf{V}}_\beta^k \cdot \mathbf{p}^\beta + \sum_{\beta=1}^{L} \tilde{\mathbf{W}}_\beta^k \cdot \mathbf{R}^\beta \tag{15}$$

where the vectors $\mathbf{b}^k, \mathbf{b}^\beta$ represent the two components of the parameter vector **b** at the nodes k and β, respectively.

Collocating Eqs. (14) and (15) at all nodal points L one obtains the following linear system of algebraic equations

$$\begin{bmatrix} \frac{1}{2}\tilde{\mathbf{I}} + \tilde{\mathbf{H}} & \tilde{\mathbf{K}} \\ \tilde{\mathbf{S}} & \frac{1}{2}\tilde{\mathbf{I}} + \tilde{\mathbf{T}} \end{bmatrix} \cdot \begin{Bmatrix} \mathbf{u} \\ \mathbf{q} \end{Bmatrix} = \begin{bmatrix} \tilde{\mathbf{G}} & \tilde{\mathbf{L}} \\ \tilde{\mathbf{V}} & \tilde{\mathbf{W}} \end{bmatrix} \cdot \begin{Bmatrix} \mathbf{p} \\ \mathbf{R} \end{Bmatrix} \tag{16}$$

Satisfying the boundary conditions (Eqs. (2) and (3)) and rearranging Eq. (16) one produces the final linear system of algebraic equations of the form

$$\tilde{\mathbf{A}} \cdot \mathbf{X} = \mathbf{B} \tag{17}$$

where the vectors **X** and **B** contain all the unknown and known nodal components of the boundary fields, respectively. The singular and hypersingular integrals involved in Eqs. (16) are evaluated with high accuracy applying a methodology for direct treatment in unified manner of CPV and hypersingular integrals [14]. Finally, the system of Eq. (17) is solved via a typical LU-decomposition algorithm and the vector **B** comprising all the unknown nodal values of **u, p, R, q** is evaluated.

In order to demonstrate the accuracy of the above describe BEM, a benchmark dealing with a cylindrical cavity embedded in an infinitely extended gradient elastic space and subjected to an internal harmonic displacement $u_0 e^{i\omega t}$ at ρ=D, is numerically solved. For the classical boundary condition $u(D) = u_0$ and the non-classical one $q(D)=0$, the analytical solution of this problem (derived by the authors) is:

$$u(\rho) = A_1 H_1^{(1)}(k_1\rho) + A_2 H_1^{(1)}(k_2\rho) \tag{18}$$

where $H_n^{(1)}(\cdot)$ is the Hankel function of first type and n-th order, and

$A_1 = -k_2 u^{(c)}\left(-H_0^{(1)}(k_2 D) + H_2^{(1)}(k_2 D)\right)/B_1$,

$A_2 = -k_1 u^{(c)}\left(-H_0^{(1)}(k_1 D) + H_2^{(1)}(k_1 D)\right)/B_2$

$$B_1 = k_2 H_0^{(1)}(k_2 D) H_1^{(1)}(k_1 D) - k_1 H_0^{(1)}(k_1 D) H_1^{(1)}(k_2 D) +$$
$$- k_1 H_1^{(1)}(k_2 D) H_2^{(1)}(k_1 D) - k_2 H_1^{(1)}(k_1 D) H_2^{(1)}(k_2 D)$$
$$B_2 = -k_2 H_0^{(1)}(k_2 D) H_1^{(1)}(k_1 D) + k_1 H_0^{(1)}(k_1 D) H_1^{(1)}(k_2 D) -$$
$$- k_1 H_1^{(1)}(k_2 D) H_2^{(1)}(k_1 D) + k_2 H_1^{(1)}(k_1 D) H_2^{(1)}(k_2 D)$$

The above described problem has been solved numerically with the aid of the 2D model depicted in Figure 1 (D=0.287, $u_0=3x10^{-4}$m). Both classical and non-classical boundary conditions are presented in detail in the same Figure. Due to the symmetry, only the quarter of the cylinder has been descretized and the radial displacements have been evaluated for the frequency of $\omega=2\pi n$, n=20kHz. The obtained results are compared to analytical ones for the pairs (g=0, h=0) , ($g=5.10^{-4}$m, $h=10^{-4}$m) and ($g=10^{-4}$m, $h=5.10^{-4}$m) and depicted in Figure 2. For all cases the Young modulus, the Poisson ratio and the mass density are: $E=14.10^9$ Pa, $\nu=0$ and $\rho=1500$ Kg/m^3, respectively .As it is apparent from Figure 2 the agreement between analytical and numerical results is excellent.

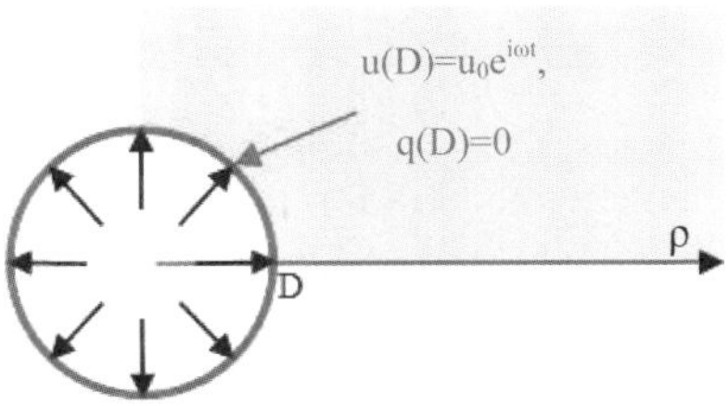

Figure 1: Cylindrical cavity embedded in a gradient elastic space. Geometry and boundary conditions

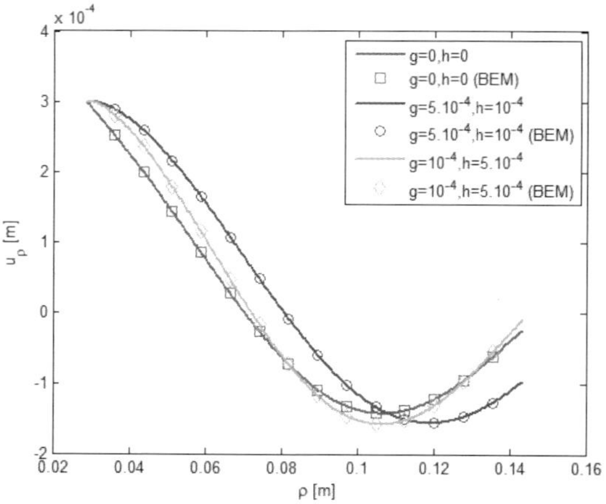

Figure 2: Radial displacements for the space outside the cavity depicted in Figure 1. Comparison between BEM (dots) and analytical (lines) solutions for different values of g and h.

4. Conclusions

A boundary element method for the solution of two dimensional frequency domain dynamic problems concerning linear elastic materials with microstructural effects has been developed. A simple strain gradient elastic theory, proposed by Mindlin, has been exploited. The boundary integral equations of the problem consist of one equation for the displacement and another one for its normal derivative. Their numerical implementation is accomplished with the aid of quadratic quadrilateral elements and advanced integration algorithms for the highly accurate evaluation of singular integrals have been utilized. A representative numerical example has been solved and the high accuracy of the proposed BEM has been assessed.

References

1. Mindlin RD. Microstructure in linear elasticity. Arch Rat Mech Anal 1964; 10: 51-78.
2. Mindlin RD. Second gradient of strain and surface tension in linear elasticity. Int J Solids Struct 1965; 1: 417-438.
3. Charalambopoulos A, Gergidis LN. On the dyadic scattering problem in three-dimensional gradient elasticity: an analytic approach. Journal of Physics A: Mathematical and Theoretical 2008; 41: 1-33.
4. Altan BS, Evensen HA, Aifantis EC. Longitudinal vibrations of a beam: A gradient elastic approach. Mech Res Comm 1996; 23 (1): 35-40.

5. Tsepoura KG, Papargyri-Beskou S, Polyzos D, Beskos DE. Static and dynamic analysis of a gradient elastic bar in tension. Arch Appl Mech 2002; 72: 483-497.
6. Papargyri-Beskou S, Polyzos D, Beskos DE. Dynamic analysis of gradient elastic flexural beams. Struct Engng & Mech 2003; 15: 705-716.
7. Papargyri-Beskou S, Polyzos D, Beskos DE. Wave dispersion in gradient elastic solids and structures: A unified treatment. International Journal of Solids and Structures 2009; 46: 3751–3759.
8. Papargyri-Beskou S, Mylonakis G. Wave dispersion studies in granular media by analytical and analytical–numerical methods. Soil Dynamics and Earthquake Engineering 2009; 29: 883–887.
9. Georgiadis HG, Vardoulakis I. Anti-plane shear Lamb's problem treated by gradient elasticity with surface energy. Wave Motion 1998; 28 (4): 353-366.
10. Georgiadis HG. The mode III crack problem in micro-structured solids governed by dipolar gradient elasticity: static and dynamic analysis. Journal of Applied Mechanics – Transactions of the ASME 2003; 70 (4): 517-530.
11. Vavva MG, Protopappas VC, Gergidis LN, Charalambopoulos A, Fotiadis DI, Polyzos D, 2009. Velocity dispersion of guided waves propagating in a free gradient elastic plate: Application to cortical bone. Journal of the Acoustical Society of America; 125(5): 3414-3427.
12. Askes H, Bennett T, Aifantis EC. A new formulation and C^0-implementation of dynamically consistent gradient elasticity. Int. J. Numer. Meth. Engng 2007; 72:111–126.
13. Polyzos D, Tsepoura KG, Beskos DE. BEM solutions of frequency domain gradient elastodynamic 3-D problems. Electronic Journal of Boundary Elements 2003; 1(2): 174-200.
14. Polyzos D, Tsepoura KG, Beskos DE. Transient dynamic analysis of 3-D gradient elastic solids by BEM. Computers and Structures 2005; 83: 783–792.
15. Polyzos D. 3D frequency domain BEM for solving dipolar gradient elastic problems. Computational Mechanics 2005; 35: 292–304.
16. Tsepoura KG, Polyzos D (2003) Static and harmonic BEM solutions of gradient elasticity problems with axisymmetry. Computational Mechanics 32: 89–103.

WAVE DISPERSION IN GRANULAR SOIL BY DEM ANALYSIS IN TWO AND THREE DIMENSIONS

C.N. THOMAS
Department of Civil Engineering, University of Patras
Rio 26500, Greece
e-mail: kthomas@upatras.gr

G. MYLONAKIS
Department of Civil Engineering, University of Patras
Rio 26500, Greece
e-mail: mylo@upatras.gr

S. PAPARGYRI-BESKOY
Department of Civil Engineering, Aristotle University of Thessaloniki
Thessaloniki 54124, Greece
e-mail: spapargy@civil.auth.gr

Numerical studies are conducted to investigate the existence of wave dispersion in resonant column tests on dry granular soil. The distinct element method (DEM) in the time domain is employed in two and three-dimensions to simulate the laboratory apparatus. The investigations focus on the effect of sample width, voids ratio, viscous damping and wavelength, on propagation velocities of longitudinal harmonic waves in rectangular samples consisting of uniform grains. It is shown that granular materials may exhibit anomalous dispersion that is, wave velocities that increase with increasing excitation frequency. This increase may be significant for squatty samples, yet becomes less pronounced for slender samples. The dispersion effects tend to be more pronounced in 2D over 3D analyses which suggests waveguide dispersion. Similar findings have been reported in some experiments on granular materials, but have not been systematically explored by numerical means. Results are presented in the form of dimensionless graphs and charts that elucidate the salient features of the problem.

Keywords: Numerical analyses; Wave dispersion; Distinct element method.

1. Introduction

Waves in real materials propagate at velocities which depend on frequency. Such media are called *"dispersive"*, to signify that traveling disturbances tend to spread out spatially causing significant distortion of signals even at small distances

While in non-dispersive media only one propagation velocity per wave mode can be defined, in dispersive media one must distinguish between two distinct quantities: *phase velocity* and *group velocity*. The former can be thought of as the velocity at which the phase of a particular frequency component of the wave propagates. The latter is associated with the propagation of wave amplitude. In case where phase velocity decreases with frequency (or wavenumber), the medium is said to exhibit *normal dispersion*. On the other hand, if phase velocity increases with frequency or wavenumber, the medium is said to exhibit *anomalous dispersion*.

The issue is controversial among experimentalists; as some report no wave dispersion [2, 3], whereas others have clearly noticed wave dispersion on their measurements [4, 5]. It is interesting to know that some of them [3] state that propagation velocities are independent of frequency (no dispersion), although dispersion is clearly inferred from their reports, while some others [4] attribute the observed dispersion to the excitation of the body modes of the specimen. Wave dispersion has been clearly measured (and theoretically proved) in the works of investigators dealing with propagation of surface waves in stratified media [2–5]. It has also been observed in some experimental works in geo-synthetic materials [7].

Most theoretical studies concentrate in unbounded media, by employing analytical techniques to explore the dispersion relations under different assumptions for geometric and material behavior. On the other hand, only a handful of numerical studies on bounded granular models have been reported [6]. These research findings are discussed in [6, 13].

Notwithstanding the theoretical significance of the aforementioned studies, it is evident that there is no consensus as to the nature of wave dispersion (i.e., normal or anomalous) in granular materials. The lack of knowledge and some contradictory findings among the various experimental and analytical studies provided the initial motivation for this work.

The object of this study is to investigate wave dispersion effects in granular media using the distinct element method (DEM), which is employed directly into the time domain using the computer platforms PFC2D and PFC3D [9]. The basic aspects of the method are well known and will not be repeated here [10]. However, the representative applications of the method in soil mechanics under static or dynamic load presented in references [11-12] should be mentioned.

Following a brief review of the fundamentals of wave dispersion, the dependence of longitudinal wave velocity will be investigated on: (1) wavenumber (frequency), (2) specimen aspect ratio, and (3) particle size.

2. Review of Wave Dispersion Fundamentals

Consider the pair of complex harmonic waves (Fig. 1a)

$$u_1 = A \exp i \left[\left(\kappa + \frac{\Delta\kappa}{2} \right) x - \left(\omega + \frac{\Delta\omega}{2} \right) t \right] \quad (1)$$

$$u_2 = A\exp i\left[\left(\kappa - \frac{\Delta\kappa}{2}\right)x - \left(\omega - \frac{\Delta\omega}{2}\right)t\right] \tag{2}$$

which have same amplitude A and direction of propagation, but have wavenumbers and cyclic frequencies which differ by $\Delta\kappa$ and $\Delta\omega$, respectively.

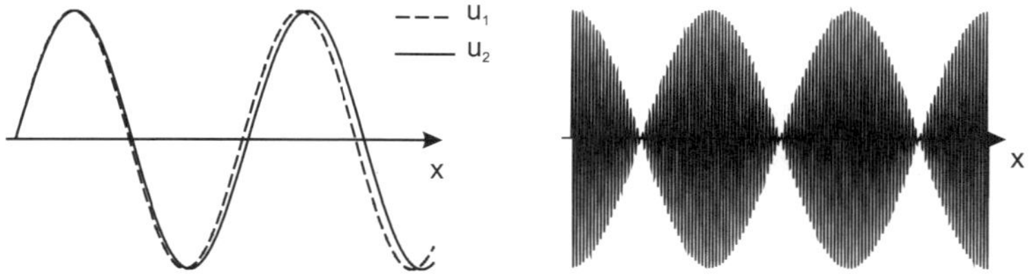

Fig. 1. Superposition of two sinusoidal waves having equal amplitude and slightly different frequencies: (a) the two waves and (b) superposition pattern showing a beat phenomenon in space.

The superposition of the waves can be cast in the real-valued form

$$u(x,t) = \mathrm{Re}\left(u_1 + u_2\right) = 2A\cos\left(\frac{\Delta\kappa}{2}x - \frac{\Delta\omega}{2}t\right)\cos\left(\kappa x - \omega t\right) \tag{3}$$

which describes two propagating waves: a high-frequency wave of wavenumber κ (i.e., the average wavenumber of the initial waves) and wavelength $\lambda=2\pi/\kappa$, and a low-frequency modulating wave, with wavenumber $\Delta\kappa/2$ and wavelength $4\pi/\Delta\kappa$. Equation (3) is plotted in Fig. 1b. Evidently, this superposition results in a beat-like pattern which has little to do with the original wave components.

It can be readily shown that the low-frequency wave of Fig. (1b) propagates towards positive x with a velocity V_{group} which is known as *group velocity*. The high-frequency wave propagates with a velocity V_{phase} which is called *phase velocity*. These velocities are defined by [6, 13]

$$V_{group} = \lim_{\Delta\kappa\to\infty}\frac{\Delta\omega}{\Delta\kappa} = \frac{d\omega}{d\kappa}\,, \qquad V_{phase} = \frac{\omega}{\kappa} \tag{4a,b}$$

Combining Eqs. (1-2), (4a,b), the following pair of relations can be derived:

$$V_{group} = V_{phase} + \kappa\frac{dV_{phase}}{d\kappa}\,, \qquad V_{group} = V_{phase} - \lambda\frac{dV_{phase}}{d\lambda} \tag{5a,b}$$

which define the so-called forward problem i.e., finding group velocity from phase velocity. Evidently, if the derivative $dV_{phase}/d\kappa$ is positive (or equivalently if $dV_{phase}/d\lambda$ is negative) that is, if phase velocity is an increasing function of frequency, group velocity is higher than phase velocity. The opposite is true when the derivatives have opposite signs. Note that these attributes have nothing to do with the nominal wave velocity at zero frequency V_o (Fig. 2) or with the ratio V_{phase}/V_o. In other words, the nature of wave dispersion at a certain frequency relates to the *derivative* of phase velocity with respect to the spectral parameters, not to phase velocity itself. Note that in the limit $\omega=0$, group velocity and phase velocity coincide, as evident from Eq. (5a).

The inverse problem i.e., finding phase velocity from group velocity at a specific frequency (not wavenumber) is more involved as discussed in [6].

3. Numerical studies

The problem addressed in this article is shown in Fig. 2: a rectangular specimen consisting of uniform circular or spherical grains is excited at one end by a harmonic velocity $V_y=V_{ymax}\,exp(i\omega t)$. The grains are considered elastic disks or spheres (in 2D and 3D respectively), connected with linear springs and dampers in both radial and tangential directions [9, 10]. The stiffness of the radial springs, k_n, is determined from the pair of equations [6, 9]:

$$k_n = 2V_o^2\rho \,, \qquad \rho_p = \chi\,\rho \qquad \text{(6a,b)}$$

where V_o is the longitudinal wave velocity at zero frequency, ρ and ρ_p are the average density and particle density, respectively and χ is a dimensionless constant equal to $(4/\pi)$ and $(3/2)$ in 2D and 3D models, respectively.

The stiffness of tangential springs k_s can be defined in an approximate manner through the dimensionless ratio

$$\frac{k_s}{k_n} = \frac{3(1-2\nu)}{2(1+\nu)} \qquad (7)$$

which defines the ratio of shear modulus to bulk modulus in a linear isotropic elastic medium of Poisson's ratio ν.

From Fig. 2, it is easy to show [6, 13] that the number of dimensional parameters governing the response is six. Given that these parameters involve three fundamental units (mass, time and length), Buckingham's theorem specifies that the number of dimensionless parameters required to describe wave propagation in the sample is three (6 3=3). If one adds the dimensionless parameters (e, ν, β) to this value, the total number of controlling variables is six (6).

In light of the above, the following dimensionless variables are introduced to describe the results [6, 13]: slenderness ratio H/B, width to particle size ratio B/d, particle size to wavelength (or dimensionless wavenumber) ratio d/λ, Poisson's ratio ν, voids ratio (e_{2D} or e_{3D}) and viscous damping ratio β. Thirty seven different configurations were investigated in total, with H/B varying from

2 to 16, B/d from 25 to 100, e_{2D} from 0.1 to 0.67 and β from 0% to 3%. The whole set of parameters are provided in the Table accompanying Fig. 2.

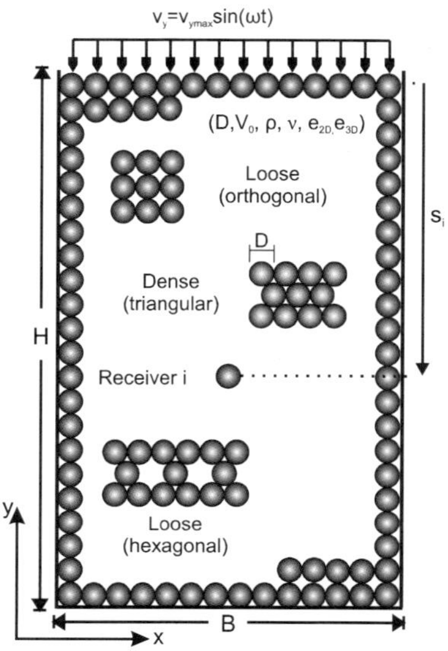

Num	Name	H/B	B/d	e_{2D}	e_{3D}	V_0(m/sec)	Damping
1	Reg-Bspec1	2	25	0.27	0.91	100	-
2	Reg-Bspec2	2	25	0.27	0.91	50	-
3	Spec3	2	25	0.27	0.91	25	-
4	Reg-Bspec3	4	25	0.27	0.91	100	-
5	Reg-Bspec4	3	25	0.27	0.91	100	-
6	Reg-Bspec5	1	25	0.27	0.91	100	-
7	Reg-Bspec6	2	50	0.27	0.91	100	-
8	Reg-Bspec7	2	100	0.27	0.91	100	-
9	Reg-Bspec8	8	25	0.27	0.91	100	-
10	Reg-Bspec9	1	50	0.27	0.91	100	-
11	Reg-Bspec10	1	100	0.27	0.91	100	-
12	Reg-Bspec11	3	50	0.27	0.91	100	-
13	Reg-Bspec12	3	100	0.27	0.91	100	-
14	Reg-Bspec13	4	50	0.27	0.91	100	-
15	Reg-Bspec14	4	100	0.27	0.91	100	-
16	Reg-Bspec15	8	50	0.27	0.91	100	-
17	Reg-Bspec16	8	100	0.27	0.91	100	-
18	Reg-Bspec17	16	25	0.27	0.91	100	-
19	Reg-Vis1-Bspec1	2	25	0.27	0.91	100	1% Viscous
20	Reg-Vis2-Bspec1	2	25	0.27	0.91	100	2% (Viscous)
21	Reg-Vis3-Bspec1	2	25	0.27	0.91	100	3% (Viscous)
22	Hex-Bspec1	2	25	0.1	0.65	100	-
23	Hex-Bspec2	1	25	0.1	0.65	100	-
24	Hex-Bspec3	1	50	0.1	0.65	100	-
25	Hex-Bspec4	1	100	0.1	0.65	100	-
26	Hex-Bspec5	2	50	0.1	0.65	100	-
27	Hex-Bspec6	2	100	0.1	0.65	100	-
28	Hex-Bspec7	3	25	0.1	0.65	100	-
29	Hex-Bspec8	3	50	0.1	0.65	100	-
30	Hex-Bspec9	3	100	0.1	0.65	100	-
31	Hex-Bspec10	4	25	0.1	0.65	100	-
32	Hex-Bspec11	4	50	0.1	0.65	100	-
33	Hex-Bspec12	4	100	0.1	0.65	100	-
34	Hex-Bspec13	8	25	0.1	0.65	100	-
35	Hex-Bspec14	8	50	0.1	0.65	100	-
36	Hex-Bspec15	8	100	0.1	0.65	100	-
37	Hex2-Bspec1	2	25	0.65	1.48	100	-

Fig. 2. Problem and cases under consideration. Specimen geometry, material properties V_0, ρ, ν, e_{2D}, e_{3D} and particle configuration.

4. Determination of time of first arrival

Essential to the analyses presented herein is the determination of the time of first wave arrival at specific locations in the sample (termed "receiver" points). Determining this time is not trivial, as the wave arrival does not manifest itself in the form of a sudden departure of particle velocity from zero. Instead, the velocity increases in a gradual manner (apparently as a result of wave dispersion) and is available in the form of distinct data points. To achieve this goal, alternative techniques were examined [6, 13]. Following a careful analysis of the numerical data, the arrival time t_a in Fig. 3 was chosen, corresponding to the time required for particle velocity to reach a specific value ($V_{par,\,a}$), which is taken as a percentage of peak excitation velocity atop the sample (V_{ymax}).

Alternative values of the threshold ratio $N= V_{par,\,a}/V_{ymax}$ were examined, ranging from 10^{-2} to 10^{-4}. Different choices of this ratio generally lead to different results, which is clearly unacceptable. A more consistent solution was obtained on the following physical requirement

$$V / V_0(\omega \to 0) = 1 \qquad (8)$$

which was imposed on the numerical data of Fig. 4 (indicated by "with adjustment"). In what follows, the velocity obtained from the above procedure is assumed to be a group velocity for the specific excitation frequency.

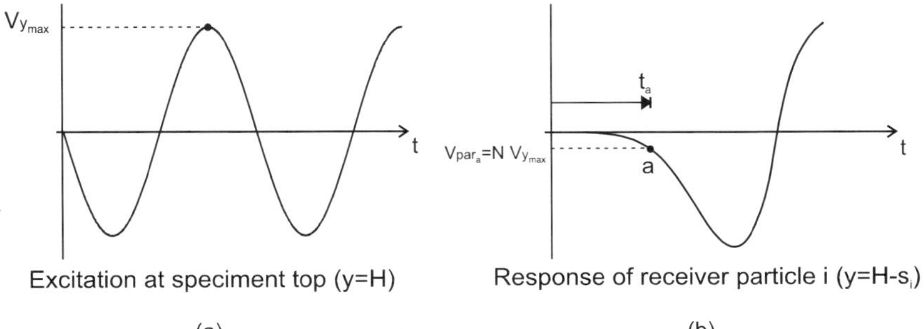

Fig. 3. Definition of time of first wave arrival in the current approach

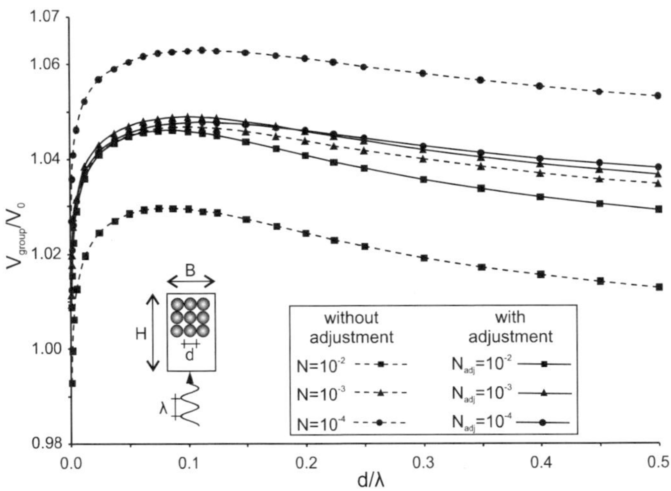

Fig. 4. Adjustment of numerical results for different values of N

5. Results

The influence of some of the aforementioned parameters on P-wave propagation velocity is examined in Figs. 5 and 6, plotted in terms of group velocity. Figure 5 depicts group velocities as function of wavenumber for undamped samples of different slenderness ratios H/B. Evidently, group velocity increases with wavenumber by as much as 8% compared to the low-frequency value V_o. This suggests anomalous dispersion, which is in agreement with earlier findings [6, 13]. Note that wave dispersion is less pronounced in the 3D case.

The effect of particle size on group velocity is presented in Fig. 6, as function of frequency and wavenumber, respectively. It is observed that V_{group} decreases with decreasing particle size - an anticipated trend since dispersion effects are expected to diminish as the discrete medium tends to a continuum.

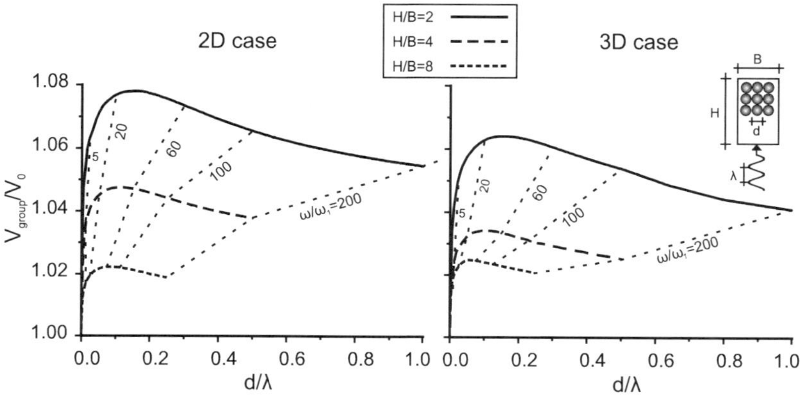

Fig. 5. Group P-velocity versus d/λ for different H/B ratios and loosest possible particle configuration; e2D=0.27, e3D=0.91, B/d=25 and β=0.

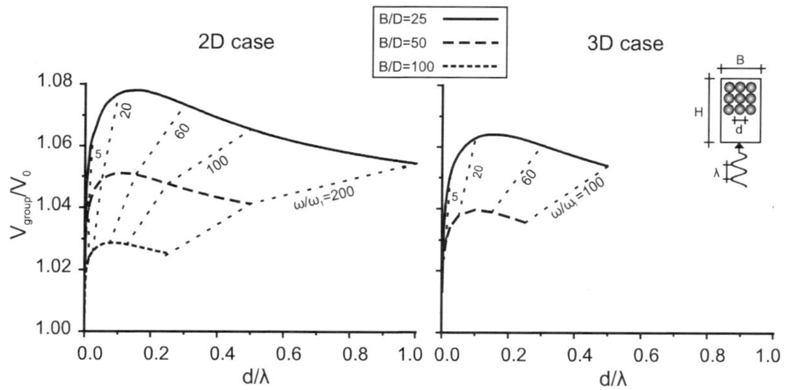

Fig. 6. Group P-velocity versus d/λ for different B/D ratios and loosest possible particle configuration; e2D=0.27, e3D=0.91, H/B=2 and β=0.

6. Conclusions

The following conclusions were drawn from this study:

(1) Anomalous dispersion was observed for wavelengths as large as 10 times the particle diameter.

(2) In squatty specimens involving coarse materials, group velocity may exceed the low-frequency wave velocity by 10% or so. For slender specimens involving finer materials, the phenomenon is less pronounced. This suggests that dispersion may be primarily of geometric as opposed to material origin.

(3) Dispersion effects are more pronounced in 2D over 3D analyses.

It is fair to mention that for wavelengths comparable to the particle diameter, the method may loose accuracy, as wave propagation within the particles is not modeled.

References

1. W.C. Elmore and M.A. Heald, Physics of waves, Dover Publ., New York (1969).
2. S. Yamashita and T. Suzuki, Small strain stiffness on anisotropic consolidated state of sands by bender elements and cyclic loading tests, Proceedings of the 15th international conference on soil mechanics and geotechnical engineering, Instabull Vol. 1, A.A. Balkema, Lisse (2001), pp. 325–328.
3. L.Q. AnnDan, J. Koseki and T. Sato, Comparison of Young's moduli of dense sand and gravel measured by dynamic and static methods, Geotech Test J 25 (4) (2002), 1–20.
4. P.D. Greening and D.F.T. Nash, Frequency domain determination of G0 using bender elements, Geotech Test J 27 (3) (2004), pp. 1–7.
5. J.A. Donald and S.D. Butt, Experimental technique for measuring phase velocities during triaxial compression tests, Int J Rock Mech Min Sci 42 (2005), pp. 307–314.
6. Thomas CN. Dispersion of longitudinal waves in granular soil by distinct element method, MSc Thesis, Department of Civil Engineering, University of Patras, Greece, 2008 [in Greek].
7. Xenaki B. Experimental investigations of mechanical behavior of geofoam under static dynamic/cycling loading, PhD dissertation, Department of Civil Engineering, University of Patras, Greece, 2005 [in Greek].
8. Wichtmann T, Sellountos EI, Tsinopoulos SV, Niemunis A, Papargyri-Beskou S, Polyzos D, et al. Wave dispersion in dry sand by experimental, analytical and numerical methods. In: Georgiou G, et al., editors., Proceedings of fifth international congress on computational mechanics GRACM05, Limassol, Cyprus, Nicosia, 2005. p. 487–93.
9. Itasca Consulting Group, Inc. PFC2D-PFC3D (Particle Flow Code in 2-3 Dimensions), Version 3.1, (2004), Theory and background, Minneapolis.
10. P.A. Cundall and O.D.L. Strack, A discrete numerical model for granular assemblies, Geotechnique 29 (1979), pp. 47–65.
11. C. O'Sullivan, J.D. Bray and M. Riemer, Examination of the response of regularly packed specimens of spherical particles using physical tests and discrete element simulations, Eng Mech J. ASCE 130 (2004), pp. 1140–1150.
12. J. Fortin, O. Millet and G. de Saxcé, Numerical simulation of granular materials by an improved discrete element method, Int J Numer Methods Eng 62 (2005), pp. 639–663.
13. Thomas CN, Papargyri-Beskou S, Mylonakis G. Wave dispersion in dry granular materials by the distinct element method. Soil Dyn. Erthq. Eng., 29 (5),pp 888-897 2009

FINITE ELEMENT MODELS FOR GENERALIZED COUPLED THERMOELASTIC PROBLEMS WITH MICRO INERTIA

S. P. FILOPOULOS, S. I. MARKOLEFAS, T. K. PAPATHANASIOU AND G. J. TSAMASPHYROS

National Technical University of Athens,
School of Applied Mathematical and Physical Sciences,
9 Iroon Polytechniou Street,
Athens 15773, Greece
E-mail: sfilop@mail.ntua.gr

In this paper we perform finite element analysis of an elastic bar within the context of generalized coupled thermoelasticity. For the description of the elastic behavior of the bar, we adopt the dipolar strain gradient model (Mindlin Form II), including micro-inertia terms. Furthermore, the thermoelastic model used in the present work is consistent with Lord - Shulman theory, i.e. heat conduction with one relaxation time. For the representation of the displacement field we use C^1 interpolation, whereas for the temperature field, C^0 interpolation is adequate. The semi-discrete system of equations is integrated in time, using the Newmark method. The characteristic case of an Ultra-Short-Laser pulse load is solved numerically and the temperature, displacement and stresses are calculated. These results are compared with the uncoupled case.

1. Introduction

According to Mindlin[1], material points of continua can be thought of as micro-continua, i.e. possessing their own internal displacement field, described in reference to a local coordinate system. In each material point, this internal displacement field is expanded in Taylor series. Retaining only the linear terms of the expansion, yields the dipolar theory, which is the simplest possible in the context of higher order continuum theories. The continuum under consideration is supposedly consistent of structural units (micro-media) in the form of cubes with edges of size $2h$ [9]. In this way, h becomes an intrinsic length of the material structure. Furthermore, the continuum is assumed homogeneous in the sense that the relative deformation, that is the difference between the gradient of the macroscopic displacement and the micro-strain, is zero.

Regarding the heat transfer phenomenon, in the context of classical theory of thermoelasticity, the differential equation for temperature is parabolic and thus thermal disturbances appear to propagate at infinite speed through the continuum. This result of the Fourier-Duhamel law is completely unrealistic, especially for cases involving extremely short transient durations or for very low temperatures near absolute zero. As a consequence, modified theories have been proposed to account for the finite speed of thermal stress wave propagation[10,11,12]. Most of these approaches are based on the general notion of relaxing the heat flux in the classical Fourier heat conduction equation, thereby, introducing a non-Fourier effect.

In the first section of this paper we present the governing and constitutive equations for a single-parameter dipolar strain gradient homogenous linear isotropic and centro-symmetric thermoelastic continuum with Vernotte-Cattaneo law for heat conduction, which is a partial result of the general theory of dipolar strain gradient thermoelastic continua with second sound effects developed in Ref. 6. In the sequel, we consider the specific case of a 1-D problem, i.e. a fixed-free bar, thermally insulated at both ends and subjected to thermal loading. In section 3 we derive the weak formulation of the problem and the finite element semi-discrete form using the standard Galerkin finite element method. Finally in section 3 we present the numerical results for the case of loading by an Ultra-Short laser pulse applied at the free end.

2. Governing Equations

The governing equations for a linear isotropic, centro-symmetric thermoelastic continuum with microstructure, which is modeled using simplified, one-parameter dipolar strain gradient theory (Mindlin Form II) and a proper generalization of Lord-Shulman theory, are[1,2,6,7]

$$\left(-1+l^2\partial_i^2\right)\left[(\lambda+\mu)\,u_{j,jk}+\mu u_{k,jj}\right]=F_k-\Phi_{jk,j}-\rho\ddot{u}_k+\frac{1}{3}\rho h^2\ddot{u}_{k,jj}-(3\lambda+2\mu)\,\alpha_t\vartheta_{,k} \tag{1}$$

$$\left(1+\tau_0\partial_t\right)\left[\rho c_\varepsilon\dot{\vartheta}+(3\lambda+2\mu)\,\alpha_t\theta_0\dot{u}_{j,j}\right]-k\vartheta_{,jj}=\rho\left(r+\tau_0\dot{r}\right) \tag{2}$$

where λ, μ are the Lamè constants, l is a characteristic material length related to volumetric strain energy, u_k are the components of the displacement field, F_k are the components of the body force vector, Φ_{jk} the components of the body double force tensor, h is the characteristic size of the micro-media, θ_0 is the absolute reference temperature, $\vartheta := \theta - \theta_0$, α_t is the linear expansion coefficient and τ_0 is the Lord-Shulman relaxation time. The traction boundary conditions for (1) are

$$n_j\left(\tau_{jk} - \mu_{ijk,i}\right) - D_j\left(n_i\mu_{ijk}\right) + \left(D_l n_l\right) n_i n_j \mu_{ijk} + \frac{1}{3}\rho h^2 n_j \ddot{u}_{k,j} = P_k \quad (3a)$$

$$n_i n_j \mu_{ijk} = R_k \quad (3b)$$

$$\left[n_i m_j \mu_{ijk}\right] = E_k \quad (3c)$$

where $m_j := e_{mlj} s_m n_l$, e_{mlj} is the Levi-Civita symbol, s_m are the components of the tangent vector on a boundary curve C across which the unit normal vector is discontinuous, $[*]$ denotes the jump of the enclosed quantity, P_k is the auxiliary surface traction, R_k and E_k are the first and second auxiliary dipolar surface tractions, i.e.

$$P_k := t_k + n_j \Phi_{jk} + \left(D_l n_l\right) n_j T_{jk} - D_j T_{jk} \quad (4a)$$

$$R_k := n_j T_{jk} \quad (4b)$$

$$E_k := \left[m_j T_{jk}\right] \quad (4c)$$

and $D_j := \left(\delta_{jk} - n_j n_k\right) \partial/\partial x_k$. The relevant constitutive equations read[6]

$$\tau_{ij} = \lambda\delta_{ij}\varepsilon_{kk} + 2\mu\varepsilon_{ij} - \left(3\lambda + 2\mu\right)\alpha_t\vartheta\delta_{ij} \quad (5a)$$

$$\mu_{ijk} = l^2\left(2\mu\kappa_{ijk} + \lambda\kappa_{irr}\delta_{jk}\right) \quad (5b)$$

$$\rho\eta = \frac{\rho c_\varepsilon}{\theta_0}\vartheta - \left(3\lambda + 2\mu\right)\alpha_t\varepsilon_{kk} \quad (5c)$$

$$q_i + \tau_0\dot{q}_i = -k\vartheta_{,i} \quad (5d)$$

where τ_{ij}, μ_{ijk}, ε_{ij}, κ_{ijk} and η are the Cauchy stress, the dipolar (double) stress, the infinitesimal strain, the gradient of strain and the specific entropy, respectively.

Now let us consider the case of a slender bar of rectangular cross-sectional area, with adiabatic ends, having thickness b, width W and length L, where $b << W$, which is free of body forces and body double forces and is subjected to convection losses from its surface to a surrounding medium of uniform temperature T_∞. The bar material is assumed thermally and elastically isotropic. The corresponding coupled thermoelastic system of equations can be derived either from (1) and (2) or from equilibrium considerations for the specific case:

$$-l^2\partial_x^4 u + \partial_x^2 u - c_{el}^{-2}\ddot{u} + \frac{h^2}{3c_{el}^2}\partial_x^2\ddot{u} = \alpha_t\partial_x\vartheta \quad (6)$$

$$c_{th}^2 \partial_x^2 \vartheta - \frac{2h_c}{b\tau_0 \rho c_\epsilon} \vartheta - \ddot{\vartheta} - \left(\frac{1}{\tau_0} + \frac{2h_c}{b\rho c_\epsilon} \right) \dot{\vartheta} -$$
$$\frac{c_{el}^2 \alpha_t T_\infty}{\tau_0 c_\epsilon} \left(\partial_x \dot{u} + \tau_0 \partial_x \ddot{u} \right) = - \frac{1}{\tau_0 \rho c_\varepsilon} \left[Q_0 \left(x, t \right) + \tau_0 \dot{Q}_0 \left(x, t \right) \right] \tag{7}$$

where h_c is the convection coefficient, $c_{el} = \sqrt{E/\rho}$, $c_{th} = \sqrt{k/\left(\rho c_\varepsilon \tau_0\right)}$ and $Q_0\left(x,t\right)$ is the thermal loading function. Equations (6) and (7) must be supplemented by initial and boundary conditions. For the temperature field $\vartheta\left(x,t\right)$ the conditions are:

$$\vartheta\left(x,0\right) = \dot{\vartheta}\left(x,0\right) = 0 \tag{8a}$$
$$\partial_x \vartheta\left(0,t\right) = \partial_x \vartheta\left(L,t\right) = 0 \tag{8b}$$

We consider the case of the fixed-free bar. Therefore, for he displacement field $u\left(x,t\right)$ we have the following initial and boundary conditions

$$u\left(x,0\right) = \dot{u}\left(x,0\right) = 0 \tag{9}$$
$$u\left(0,t\right) = 0 \tag{10}$$
$$R\left(0,t\right) = R\left(L,t\right) = P\left(L,t\right) = 0 \tag{11}$$

where the surface traction P with dimensions of stress (force/area) and dipolar surface traction R with dimensions of (force/length) assume the following specific form:

$$R = l^2 E \partial_x^2 u \tag{12}$$
$$P = E\partial_x u - l^2 \partial_x^3 u + \left(\rho h^2/3\right) \partial_x \ddot{u} \tag{13}$$

We now introduce the following non-dimensional variables:

$$\xi = \frac{x}{\tau_0 c_{th}}, \quad \eta = \frac{t}{\tau_0}, \quad U = \frac{c_{el}}{\alpha_t \kappa T_\infty} u, \quad \theta = \frac{\vartheta}{T_\infty} \tag{14}$$

On defining the quantities

$$\beta = 2h_c\tau_0/(b\rho c_\varepsilon) \qquad \gamma = \tau_0/(\rho c_\varepsilon T_\infty) \qquad \zeta = c_{el}\alpha_t^2 T_\infty c_{th}/c_\varepsilon$$
$$\lambda_1 = l^2/(\tau_0 c_{th})^2 \qquad \lambda_2 = (c_{th}/c_{el})^2 \qquad \lambda_3 = [h/(\sqrt{3}\tau_0 c_{el})]^2$$

equations (6) and (7) along with the initial and boundary conditions become:

$$-\lambda_1 \partial_\xi^4 U + \partial_\xi^2 U - \lambda_2 \partial_\eta^2 U + \lambda_3 \partial_\xi^2 \partial_\eta^2 U = \lambda_2^{-1/2} \partial_\xi \theta \tag{15}$$

$$\partial_\xi^2 \theta - \partial_\eta^2 \theta - (1+\beta)\partial_\eta \theta - \zeta(\partial_\xi \partial_\eta U + \partial_\xi \partial_\eta^2 U) = -\gamma f(\xi, \eta) \tag{16}$$

$$\theta(\xi,0)=\theta_{,\eta}(\xi,0)=U(\xi,0)=U_{,\eta}(\xi,0)=0 \tag{17}$$

$$\theta_{,\xi}(0,\eta)=\theta_{,\xi}(\xi_L,\eta)=0 \tag{18}$$

$$U(0,\eta)=0 \tag{19}$$

$$U_{,\xi\xi}(0,\eta)=U_{,\xi\xi}(\xi_L,\eta)=0 \tag{20}$$

$$\left(U_{,\xi}-\lambda_1 U_{,\xi\xi\xi}+\lambda_3 U_{,\xi\eta\eta}\right)|_{(\xi_L,\eta)}=0 \tag{21}$$

3. Weak formulation and finite element discretization

The weak form of the system (15) and (16) in a typical element e reads

$$\begin{aligned}\lambda_2\int_e \delta U\,U_{,\eta\eta}+\lambda_3\int_e(\delta U)_{,\xi}U_{,\eta\eta\xi}+\lambda_1\int_e(\delta U)_{,\xi\xi}U_{,\xi\xi}+\\ \int_e(\delta U)_{,\xi}U_{,\xi}+\lambda_2^{-1/2}\int_e\delta U\,\theta_{,\xi}=\Big[(\delta U)_{,\xi}\lambda_1 U_{,\xi\xi}\Big]_e+\\ \Big[\delta U\left(U_{,\xi}-\lambda_1 U_{,\xi\xi\xi}+\lambda_3 U_{,\eta\eta\xi}\right)\Big]_e\end{aligned} \tag{22}$$

$$\begin{aligned}\int_e\delta\theta\,\theta_{,\eta\eta}+(1+\beta)\int_e\delta\theta\,\theta_{,\eta}+\int_e(\delta\theta)_{,\xi}\theta_{,\xi}+\beta\int_e\delta\theta\,\theta-\\ \zeta\int_e(\delta\theta)_{,\xi}U_{,\eta\eta}-\zeta\int_e(\delta\theta)_{,\xi}U_{,\eta}=\gamma\int_e\delta\theta\,f(\xi,\eta)+\\ \Big[\delta\theta\,\theta_{,\xi}\Big]_e-\zeta\Big[\delta\theta\,U_{,\eta}\Big]_e-\zeta\Big[\delta\theta\,U_{,\eta\eta}\Big]_e\end{aligned} \tag{23}$$

where $[*]_e$ denotes the difference of the values of the enclosed quantity at the element boundary nodes. From (22) and (23) it is obvious that the shape functions for U and θ belong to $H^2(e)$ and $H^1(e)$ respectively. Since the problem is 1-dimensional, this translates to C^1 interpolation for U and C^0 for θ. Setting $\theta_e=\mathbf{v}_\theta\,\mathbf{q}_e$, $U_e=\mathbf{v}_u\,\mathbf{d}_e$, where $\mathbf{v}_\theta$, $\mathbf{v}_u$ are the matrices of the element shape functions for temperature and displacement respectively and $\mathbf{q}_e$, $\mathbf{d}_e$ are the vectors of the unknown nodal temperatures and displacements, we obtain the following semi-discrete form of the system (22) - (23)

$$\mathbf{m}_\theta^e\frac{\mathrm{d}^2}{\mathrm{d}\eta^2}\mathbf{q}_e+\mathbf{c}_\theta^e\frac{\mathrm{d}}{\mathrm{d}\eta}\mathbf{q}_e+\mathbf{k}_\theta^e\mathbf{q}_e+\mathbf{m}_{\theta u}^e\frac{\mathrm{d}^2}{\mathrm{d}\eta^2}\mathbf{d}_e+\mathbf{c}_{\theta u}^e\frac{\mathrm{d}}{\mathrm{d}\eta}\mathbf{d}_e=\mathbf{f}_{\theta\mathrm{v}}^e+\mathbf{f}_{\theta\mathrm{b}}^e \tag{24}$$

$$\mathbf{m}_u^e\frac{\mathrm{d}^2}{\mathrm{d}\eta^2}\mathbf{d}_e+\mathbf{k}_u^e\mathbf{d}_e+\mathbf{k}_{u\theta}^e\mathbf{q}_e=\mathbf{f}_{u\mathrm{b}}^e \tag{25}$$

where we have set

$$\mathbf{m}_\theta^e = \int_e \mathbf{v}_\theta^{\mathrm{T}} \mathbf{v}_\theta \mathrm{d}\xi, \quad \mathbf{k}_\theta^e = \int_e \frac{\mathrm{d}}{\mathrm{d}\xi}\mathbf{v}_\theta^{\mathrm{T}} \frac{\mathrm{d}}{\mathrm{d}\xi}\mathbf{v}_\theta \mathrm{d}\xi + \beta \mathbf{m}_\theta^e$$

$$\mathbf{m}_{\theta u}^e = -\zeta \int_e \frac{\mathrm{d}}{\mathrm{d}\xi}\mathbf{v}_\theta^{\mathrm{T}} \mathbf{v}_u \mathrm{d}\xi, \quad \mathbf{c}_{\theta u}^e = \mathbf{m}_{\theta u}^e, \quad \mathbf{c}_\theta^e = (1+\beta)\,\mathbf{m}_\theta^e$$

$$\mathbf{m}_u^e = \lambda_2 \int_e \mathbf{v}_u^{\mathrm{T}} \mathbf{v}_u \mathrm{d}\xi + \lambda_3 \int_e \frac{\mathrm{d}}{\mathrm{d}\xi}\mathbf{v}_u^{\mathrm{T}} \frac{\mathrm{d}}{\mathrm{d}\xi}\mathbf{v}_u \mathrm{d}\xi$$

$$\mathbf{k}_u^e = \lambda_1 \int_e \frac{\mathrm{d}^2}{\mathrm{d}\xi^2}\mathbf{v}_u^{\mathrm{T}} \frac{\mathrm{d}^2}{\mathrm{d}\xi^2}\mathbf{v}_u \mathrm{d}\xi + \int_e \frac{\mathrm{d}}{\mathrm{d}\xi}\mathbf{v}_u^{\mathrm{T}} \frac{\mathrm{d}}{\mathrm{d}\xi}\mathbf{v}_u \mathrm{d}\xi$$

$$\mathbf{k}_{u\theta}^e = \lambda_2^{-1/2} \int_e \mathbf{v}_u^{\mathrm{T}} \frac{\mathrm{d}}{\mathrm{d}\xi}\mathbf{v}_\theta \mathrm{d}\xi, \quad \mathbf{f}_{\theta \mathrm{v}}^e = \gamma \int_e \mathbf{v}_\theta^{\mathrm{T}} f(\xi,\eta)\, \mathrm{d}\xi$$

$$\mathbf{f}_{\theta \mathrm{b}}^e = \{-\partial_\xi \theta|_{\xi_1^e} - \zeta\left(\partial_\eta U + \partial_\eta^2 U\right)|_{\xi_1^e} \quad \partial_\xi \theta|_{\xi_2^e} + \zeta\left(\partial_\eta U + \partial_\eta^2 U\right)|_{\xi_2^e}\}^{\mathrm{T}}$$

$$\mathbf{f}_{u\mathrm{b}}^e = \{-\tilde{P}_1^e \quad -\tilde{R}_1^e \quad \tilde{P}_2^e \quad \tilde{R}_2^e\}^{\mathrm{T}}$$

$$\tilde{P}_k^e = (U_{,\xi} - \lambda_1 U_{,\xi\xi\xi} + \lambda_3 U_{,\eta\eta\xi})|_{\xi_k^e}, \quad \tilde{R}_k^e = (\lambda_1 U_{,\xi\xi})|_{\xi_k^e}, \quad k=1,2$$

Using the above results, the system of element equations is written in a compact form

$$\mathbf{M}_e \ddot{\mathbf{r}}_e + \mathbf{C}_e \dot{\mathbf{r}}_e + \mathbf{K}_e \mathbf{r}_e = \mathbf{f}_e$$

where

$$\mathbf{r}_e := \begin{Bmatrix} \mathbf{q}_e \\ \mathbf{d}_e \end{Bmatrix}, \mathbf{M}_e := \begin{bmatrix} \mathbf{m}_\theta^e & \mathbf{m}_{\theta u}^e \\ \mathbf{0} & \mathbf{m}_u^e \end{bmatrix}, \mathbf{C}_e := \begin{bmatrix} \mathbf{c}_\theta^e & \mathbf{c}_{\theta u}^e \\ \mathbf{0} & \mathbf{0} \end{bmatrix}$$

$$\mathbf{K}_e := \begin{bmatrix} \mathbf{k}_\theta^e & \mathbf{0} \\ \mathbf{k}_{u\theta}^e & \mathbf{k}_u^e \end{bmatrix}, \mathbf{f}_e := \begin{Bmatrix} \mathbf{f}_{\theta \mathrm{v}}^e + \mathbf{f}_{\theta \mathrm{b}}^e \\ \mathbf{f}_{u\mathrm{b}}^e \end{Bmatrix}$$

We realize the interpolation for θ and U using polynomials of degrees $p_\theta = 1$ and $p_u = 3$ or $p_\theta = 2$ and $p_u = 5$ respectively.

4. Results and Discussion

The bar is subjected to an ultra-short laser pulse at $x = x_0$ which has duration t_p and affects a zone of length Δx. The pulse load function Q_0 is expressed as

$$Q_0(x,t) = \frac{Pe^{-t/t_p}}{\Delta x}\frac{t}{t_p^2}\left[H(x-x_0) - H(x-x_0-\Delta x)\right]$$

Accordingly, the loading function $f(\xi,\eta)$ is given by

$$f(\xi,\eta) = \frac{Pe^{-\eta/\eta_1}}{\Delta\xi \eta_1^2 \sqrt{\kappa \tau_0^3}}\left[1 + \eta\left(1 - \frac{1}{\eta_1}\right)\right]\left[H(\xi-\xi_0) - H(\xi-\xi_0-\Delta\xi)\right]$$

with $\eta_1 = t_p/\tau_0$ and $H(x)$ the Heaviside generalized function. The normalized Cauchy, total and double stresses (S, S_t, M) are defined as follows:

$$S := \tau / (E\alpha_t T_\infty) = \sqrt{\lambda_2} U_{,\xi} - \theta$$

$$S_t := \sigma_t / (E\alpha_t T_\infty) = \sqrt{\lambda_2} (U_{,\xi} - \lambda_1 U_{,\xi\xi\xi}) - \theta$$

$$M := \mu \left(E\alpha_t T_\infty \sqrt{\kappa\tau_0}\right)^{-1} = \lambda_1 \sqrt{\lambda_2} U_{,\xi\xi}$$

where σ_t is the total stress $\sigma_t := \tau - \partial\mu/\partial_x$ and κ is the thermal diffusivity $\kappa := k/(\rho c_\varepsilon)$. For the numerical experiments, the following material data and parameter values were used

$\rho = 8960\ \mathrm{kg/m^3}$	$\alpha_t = 16.5 \times 10^{-6}\mathrm{grad}^{-1}$	$\eta_1 = 0.2$
$c_\varepsilon = 385\ \mathrm{J/kg}$ - grad	$T_\infty = 300\,\mathrm{K}$	$E = 1.1 \times 10^{11}\,\mathrm{Pa}$
$\tau_0 = 4.348 \times 10^{-12}\,\mathrm{s}$	$P = 1$	$\xi_L = 4$
$k = 401\ \mathrm{W/m}$ - grad	$\xi_0 = 0$	$\eta_t = 40$

The above values correspond to the elastic and thermal properties of copper. The value for the relaxation time was taken from Ref. 13 and is typical for metals. Further, convection losses were not taken into account. The interpolation degrees chosen are $p_\theta = 2$ and $p_u = 5$. For the numerical integration in time we employ the Newmark method. For obtaining the results we use a uniform grid in space of 100 elements and 2000 time steps. Numerical experiments with 400 elements and 4000 time steps result in deviations of less than 2 percent with substantial increase in computation time.

In Figures 1 - 4 below we show some results for the fixed-free configuration. It is observed that, for the selected material, the coupled and uncoupled models yield practically the same results. It is also evident that the microstructural model is substantially different from that of classical elasticity. Furthermore, the consideration of micro-inertia is also important. Actually, as it has been shown in Refs. 14 and 15, neglecting micro-inertia leads to incorrect results for the dispersion of elastic waves in micro-structured solids.

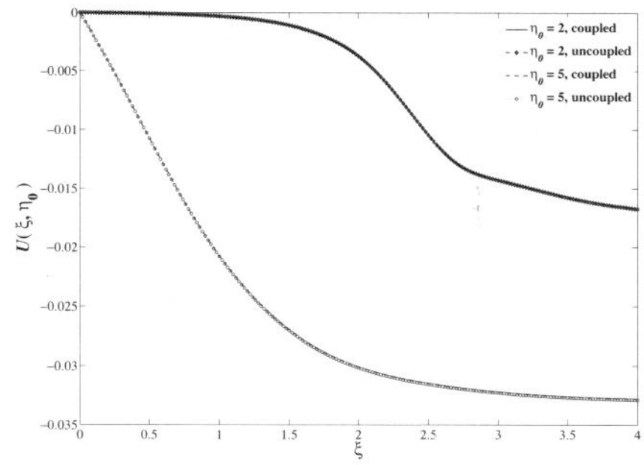

Figure 1. Distribution of normalized displacement at $\eta = 2$ and $\eta = 5$ for systems with and without coupling.

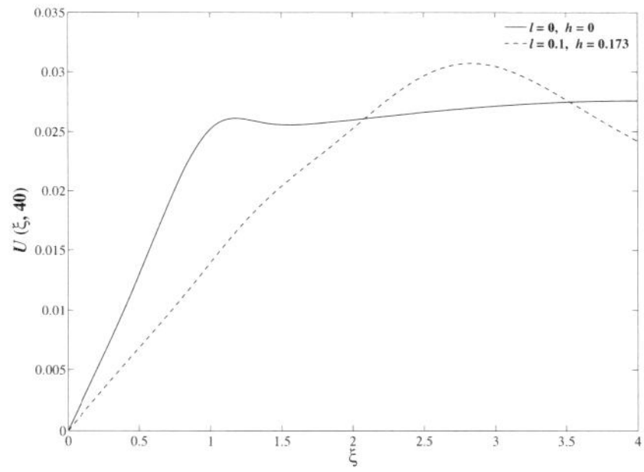

Figure 2. Coupled System. Distribution of normalized displacement at $\eta = 40$ for systems with and without ($l = h = 0$) microstructure.

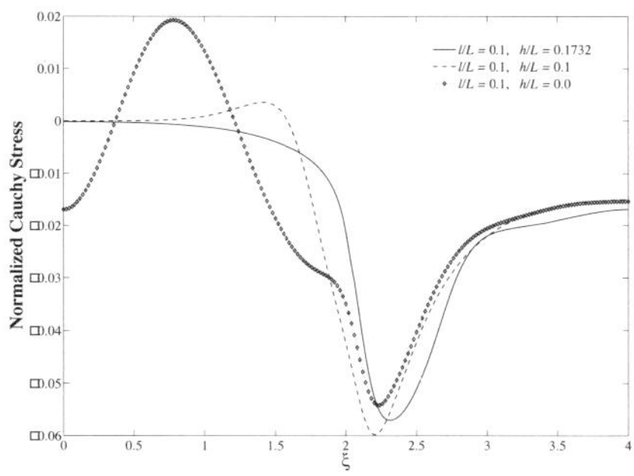

Figure 3. Coupled System. Distribution of normalized Cauchy stress at $\eta = 2$ for $l/L = 0.1$ and different values of h/L.

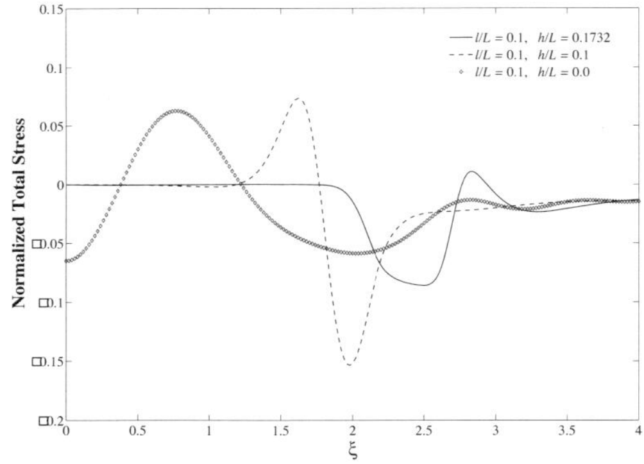

Figure 4. Coupled System. Distribution of normalized total stress at $\eta = 2$ for $l/L = 0.1$ and different values of h/L.

References

1. R. D. Mindlin, *Arch. Rat. Mech. Anal.* **16**, 51 (1964).
2. H. W. Lord and Y. Shulman, *J. Mech. Phys. Solids* **15**, 299 (1967).
3. D. S. Chandrasekharaiah, *Int. J. Engng Sci.* **24**, 1389 (1986).
4. D. Iesan, *J. Therm. Stresses* **6**, 167 (1983).
5. M. Ciarletta, *J. Therm. Stresses* **19**, 731 (1996).
6. S. P. Filopoulos, *Ph. D. Thesis* N.T.U.A. (2009).
7. S. P. Filopoulos, T. K. Papathanasiou and S. I. Markolefas *in preparation.*
8. S. P. Filopoulos, Th. K. Papathanassiou and G. J. Tsamasphyros, *J. Therm. Stresses* **32**, 905 (2009).
9. H. G. Georgiadis, I. Vardoulakis and E. G. Velgaki, *J. Elasticity* **74**, 17 (2004).
10. P. Vernotte, *C.R. Acad. Sci.*, **246**, 3154 (1958).
11. C. Cattaneo, *C.R. Acad. Sci.*, **247**, 431 (1958).
12. K. K. Tamma and R. R. Namburu, *Comput. Mech.*, **9**, 73 (1992).
13. D. W. Tang and N. Araki, *Heat Mass Transfer*, **31**, 359 (1996)
14. S. Papargyri-Beskou, D. Polyzos and D. E. Beskos, *Int. J. Solids Struct.*, **46**, 3751 (2009)
15. S. P. Filopoulos, T. K. Papathanasiou, S. I. Markolefas and G. J. Tsamasphyros, *Comput. Mech.*, *submitted*

SIGNIFICANCE OF SIZE AND LOCATION OF THE LESION FOR PROGNOSIS OF COLLAPSE OF OSTEONECROSIS OF THE FEMORAL HEAD FOR PATIENTS TREATED WITH TANTALUM ROD

S.D. Koumpoyiannis
Department of Computer and Communication Engineering,
University of Thessaly,
Volos, Greece

N. Aravas
Department of Mechanical Engineering Engineering,
University of Thessaly,
Volos, Greece

S. Varitimidis
Orthopaedic Department,
University Hospital of Larissa,
Larissa Greece

E.N. Houstis
Department of Computer and Communication Engineering,
University of Thessaly,
Volos, Greece

V. Bagiatis
Department of Biochemistry and Biotechnology,
University of Thessaly,
Larissa, Greece

K. Malizos
Orthopaedic Department,
University Hospital of Larissa,
Larissa Greece

The purpose of this study is to predict osteonecrosis of the femoral head by examining location and size of the lesion of the femoral head. A triangulation approximation algorithm is used for the computation of the geometric center of the lesion based on MRI scans for patients treated with a tantalum rod.

Twenty nine patients (thirty four hips) with osteonecrosis of the femoral head were treated with the tantalum rod procedure. The mean age was 39,21 years (14 - 59). All necrotic lesions were confirmed by magnetic resonance imaging. A customized triangulation approximation algorithm was developed in order to compute the geometric center of the lesion of the reconstructed model. Lesion volume, the distance from the origin (ρ or R), latitude (φ) and longitude (θ) of the geometric center of the lesion within the femoral head were calculated.

The mean follow-up was 3 (1-7) years. Ten hips (29.4%) showed radiographic progression. Statistical method Cox regression analysis showed that lesion volume and longitude angle (θ) were important factors to predict radiographic progression. Lesions having an angle (θ) of [15° - 35°] and [50 °- 60 °] and size c in Steinberg scale with lesion size greater than 27% of the femoral head were more likely to develop radiographic progression. For hips that developed radiographic progression out of this range, lesion size was the only significant factor.

The results suggest that lesion volume is an important factor strongly correlated with the risk of collapse of the femoral head. The location of the lesion of the necrotic area is an important supplemental parameter in order to predict with accuracy the progression of the disease. However further studies with the current algorithm that computes the geometric center of the lesion, larger study groups and long-term follow-up are needed in order to predict accurately the progression of the disease and the collapse of the femoral head.

1. Introduction

Osteonecrosis of the femoral head, also known as avascular necrosis, is a disease that occurs when blood supply to the femoral head is disrupted. The disruption of the blood supply leads in the avascular necrosis of the bone and eventually in the collapse of the diseased bone. In most of the cases osteonecrosis affects articulation areas and results in collapse of the articular surface and eventually in the damage of the joint. Osteonecrosis can occur in any bone but mostly affects the femoral head. The disease has been identified after trauma of femoral head fracture, blood disorders, e.g., sickle cell anemia, or as a consequence of pathologic situations such as use of corticosteroids, alcohol, cigarette smoking. Women in pregnancy (14, 15) and people in radiation therapy are in a risk of developing the disease.

Osteonecrosis is responsible for a large number arthroplasties in young and energetic patients (10). In the early stages of osteonecrosis patients can be treated with different joint preserving procedures such as free vascularized fibular graft, tantalum rod or core decompression (11, 12, 13, 14). Following collapse of the femoral head and as osteoarthritis progresses patients are treated with hip arthroplasty. In the early stages of osteonecrosis it is important to predict the collapse of the femoral head and focus on those patients who can benefit from the treatment mentioned before. In this study we are interested in patients treated with tantalum rod as a joint-preserving treatment. Tantalum is a foam-like metal that is used in several orthopaedic procedures, for hip and knee arthroplasty, spine surgery, and as bone graft substitute (Figures 1, 2). This metal has excellent biocompatibility, is safe, and has a high volumetric porosity. It is also corrosion resistant and its elasticity is similar to that of subchondral bone. Its strength, fatigue properties, and duration limits make it superior to natural bone grafts (11, 12, 13). Another study shows that tantalum implants in a simulation of a necrotic femoral head, reduced subchondral plate variation and its strength was 9.3 times greater than the maximum force it sustains when placed in the femoral head (21).

Searching for the factors that cause collapse of the femoral head many authors have noted the relation of size and the location of the necrotic lesion (15, 16, 17, 18). The first techniques that were proposed used exclusively radiographs for estimation of size of the necrotic area, whereas more recent techniques are based on the magnetic resonance imaging (MRI) that allows 3D representation of the lesion and the precise estimate of its size. Lately volumetric and quantitative methods have been developed by processing multiple MRI slices to calculate accurately the area (18) and the center of gravity of the lesion of the femoral head (2). The newer systems of classification of osteonecrosis are based on the estimate of extent of necrotic lesion and incorporate this information in their stages. Due to anatomic limitations and significant variations in individuals, the sphere equivalent of the femoral head, defined as the sphere with the smallest radius enclosing the femoral head, has been introduced by Malizos et al. (4). The aforementioned sphere has a larger volume than the corresponding anatomic femoral head. A method that calculates the latitude and longitude using the geometric center of the lesion has not yet being developed to our knowledge. In this study, we propose a new quantitative method that determines the geometric center of the necrotic lesion of the femoral head and predicts the incidence of collapse.

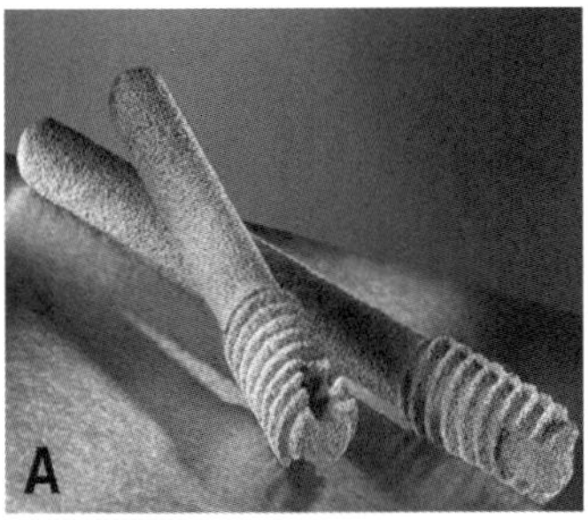

Figure 1 Porus tantalum implant. Retrieved from http://www.zimmer.com.

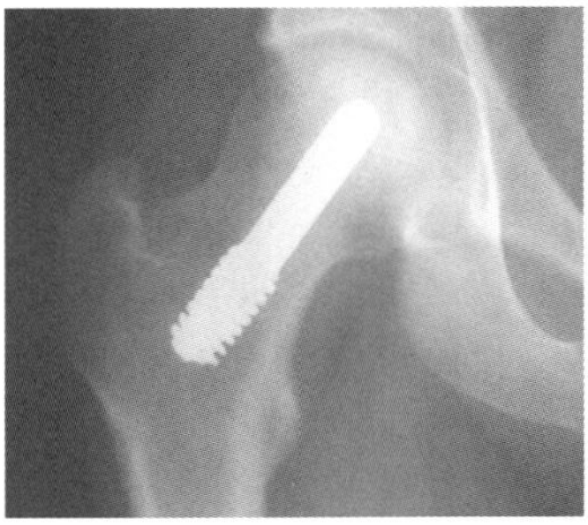

Figure 2 Porous tantalum implant in a Arco stage II osteonecrotic hip, a 4 years post-operatively.

1.1. *Materials and Methods*

Osteonecrosis of the femoral head was evaluated in 28 patients of whom 23 were male and 5 were female with a mean age of 38.43 years. Preoperative MRI revealed 34 hips with osteonecrosis. The disease was attributed to several etiologic factors, such as overuse of alcohol in 3 patients (3 hips), previous temporary use of corticosteroids in 15 patients (19 hips), sickle cell disease in 2 patients (2 hips) and idiopathic in 9 patients (10 hips). There was no significant difference in the mean age of patients with different aetiologies.

All patients were subjected to MR imaging using Philips Intera 1 Tesla superconducting unit. All MR images were obtained under non weight bearing conditions. The field of view was 320 mm, the matrix 256 x 256, and the slice thickness was 5mm without interslice gaps. T1- and T2-weighted images taken in coronal sections were used for the creation of three dimensional images of the hips. MR images were digitized with a VIDAR VXR 12 scanner (Vidar Systems Corporation, Va., USA)

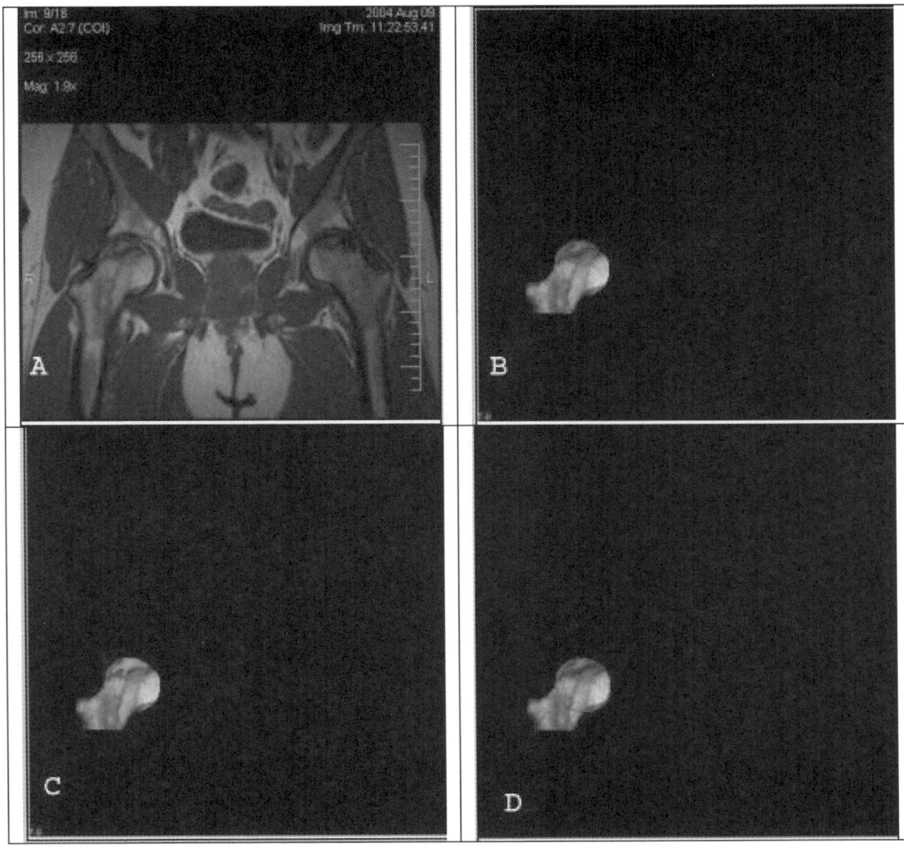

Figure 3 A shows one of the MR images. B with the software analyse the necrotic lesion is marked manually. C with the same software the area of the specific necrotic lesion it has been calculated and the boundary 2d coordinates are created. D this is the smallest circle that circumscribes the femoral head and the radius is calculated.

(Fig. 3) or created in a dicom medical format. All data were processed on a Pentium PC workstation in Windows XP environment using the ANALYZE image analysis software Version 7.5 (Mayo Foundation, Minn., USA). The affected region, defined as the region with an abnormal signal, was outlined for each MRI slice using the tracing tool of ANALYZE. A "reference circle" was identified on each MRI slice as the smallest circle enclosing the affected area of the slice. The sphere equivalent of the femoral head was then determined, an appropriate Cartesian coordinate system $x-y-z$ was introduced, and the geometric center of the legion was found as described in the following.

The MRI slice with the largest reference circle was used to define the vertical plane $x=0$, and the positive direction of the z-axis was taken upwards. The center of this circle was chosen as the origin O of the global Cartesian coordinate system and the center of the sphere equivalent to the femoral head. The positive x-axis was directed ahead of the patient, and the y-axis was chosen so that the resulting $x-y-z$ Cartesian coordinate system was right-handed (see Figure 4). The front half of the sphere $(x>0)$ is referred to as anterior and the region $x<0$ for as posterior. The MR images were used to determine the size of the femoral head in the x-direction, say D_x. Let D_{yz} be the size of the diameter of the maximum reference circle on the MRI slices. Then the diameter D of the sphere equivalent to the femoral head was determined as the maximum of D_x and D_{yz}, i.e., $D=\max(D_x,D_{yz})$.

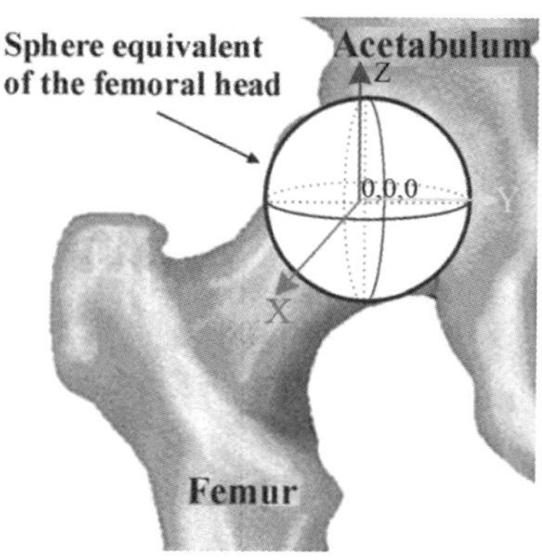

Figure 4 3D using spherical coordinates. For $x>0$ the front half of the sphere is referred to as anterior and for $x<0$ as posterior.

We introduce also spherical coordinates (ρ,φ,θ), where ρ (or R) is the distance from the origin (center of the sphere), φ is the angle from the z-axis, known as the colatitude or zenith and takes values in the range 0° to 180°, and θ is the angle from the x-axis $\left(0^{\circ} \leq \theta \leq 360^{\circ}\right)$ (see Figure 5). These coordinates are similar to the "latitude" (φ) and "longitude" (θ) system used in geography, with the origin at the center of the Earth.

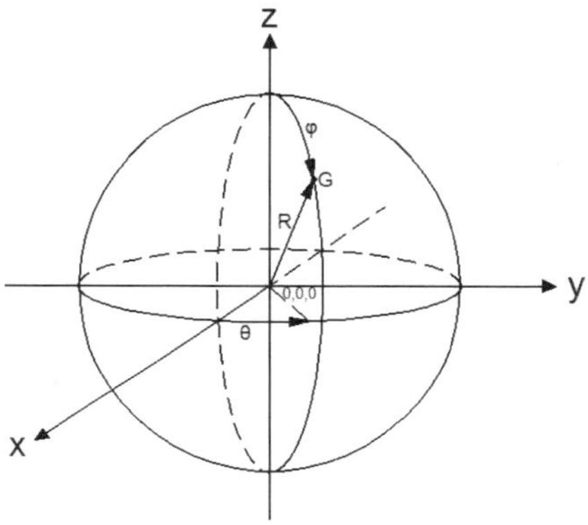

Figure 5 Spherical coordinate system.

The Cartesian and polar coordinates are related through the following equations

$$R=\sqrt{x^2+y^2+z^2}, \qquad \theta=\tan^{-1}\frac{y}{x}, \qquad \phi=\tan^{-1}\frac{\sqrt{x^2+y^2}}{z},$$

where $0 \leq \theta \leq 2\pi$ and $0 \leq \phi \leq \pi$, and

$$x=R\sin\phi\cos\theta, \qquad y=R\sin\phi\sin\theta, \qquad z=R\cos\phi.$$

The volume of the sphere was then calculated using the formula

$$V_{femoral}=\frac{\pi}{6}D^3.$$

The coordinates of the perimeter of necrotic lesion for each image that contains a necrotic lesion were obtained by the ANALYZE software and stored in file which was then input to our customized software in order to compute the volume V_{lesion} and the position of the geometric center of the necrotic lesion. Both the Cartesian coordinates of the center and the corresponding geographical latitude and longitude were determined. The volume fraction of the lesion in the femoral head was determined also as the ratio $\frac{V_{lesion}}{V_{femoral}}$.

The customized software was written in Microsoft C++ version 6.0. The divisions (octants) and abbreviations introduced by Malizos et al. (4) for the description of the femoral head were used (Table 1):

ASM Anterior superior medial $\left(0<\theta<\frac{\pi}{2}, 0<\phi<\frac{\pi}{2}\right)$ $(x>0, y>0, z>0)$	PSM Posterior superior medial $\left(\frac{\pi}{2}<\theta<\pi, 0<\phi<\frac{\pi}{2}\right)$ $(x<0, y>0, z>0)$
ASL Anterior superior lateral $\left(\frac{3\pi}{2}<\theta<2\pi, 0<\phi<\frac{\pi}{2}\right)$ $(x>0, y<0, z>0)$	PSL Posterior superior lateral $\left(\pi<\theta<\frac{3\pi}{2}, 0<\phi<\frac{\pi}{2}\right)$ $(x<0, y<0, z>0)$
AIM Anterior inferior medial $\left(0<\theta<\frac{\pi}{2}, \frac{\pi}{2}<\phi<\pi\right)$ $(x>0, y>0, z<0)$	PIM Posterior inferior medial $\left(\frac{\pi}{2}<\theta<\pi, \frac{\pi}{2}<\phi<\pi\right)$ $(x<0, y>0, z<0)$
AIL Anterior inferior lateral $\left(\frac{3\pi}{2}<\theta<2\pi, \frac{\pi}{2}<\phi<\pi\right)$ $(x>0, y<0, z<0)$	PIL Posterior inferior lateral $\left(\pi<\theta<\frac{3\pi}{2}, \frac{\pi}{2}<\phi<\pi\right)$ $(x<0, y<0, z<0)$

Table 1

The methodology developed for the calculation of V_{lesion} and the location of the geometric center of the lesion is described in the following. Let N be the total number of MRI slices perpendicular to the x-axis. On each such slice the perimeter of the lesion is defined by the (y_i, z_i) coordinates a total of M discrete points. Then area of the lesion on the k-th MRI slice is determined from

$$A_k = \frac{1}{2}\sum_{i=1}^{N}\left(y_i^k\, z_{i+1}^k - y_{i+1}^k\, z_i^k\right),$$

where the numbering of the points is such that i increases counterclockwise, and $(y_{M+1}, z_{M+1}) \equiv (y_1, z_1)$. The coordinates (y_G^k, z_G^k) of the center G^k of the area of the legion on the k-th slice are determined from

$$y_G^k = \frac{1}{6A_k} \sum_{i=1}^{M} \left(y_i^k \, z_{i+1}^k - y_{i+1}^k \, z_i^k \right) \left(y_i^k + y_{i+1}^k \right),$$

$$z_G^k = \frac{1}{6A_k} \sum_{i=1}^{M} \left(y_i^k \, z_{i+1}^k - y_{i+1}^k \, z_i^k \right) \left(z_i^k + z_{i+1}^k \right).$$

The total volume of the legion is determined as

$$V_{lesion} = \sum_{k=1}^{N} A_k \, \Delta x_k \ ,$$

where Δx_k is the thickness of the k-th MRI slice. Finally, the Cartesian coordinates of the center G of the lesion are determined from

$$x_G = \frac{1}{V_{lesion}} \sum_{k=1}^{M} x_k \, A_k \, \Delta x_k \ , \qquad y_G = \frac{1}{V_{lesion}} \sum_{k=1}^{M} y_G^k \, A_k \, \Delta x_k \ ,$$

$$z_G = \frac{1}{V_{lesion}} \sum_{k=1}^{M} z_G^k \, A_k \, \Delta x_k \ ,$$

where x_k is the x-coordinate of the k-th MRI slice. The proof of the formulae listed above is given in the Appendix.

Lesion size was expressed as a percentage of the volume of the entire femoral head, and location was expressed as the latitude and longitude of a vector of the necrotic lesion. Our software determines in which octant of the femoral head is the calculated Geometric Center of the lesion Figure 6.

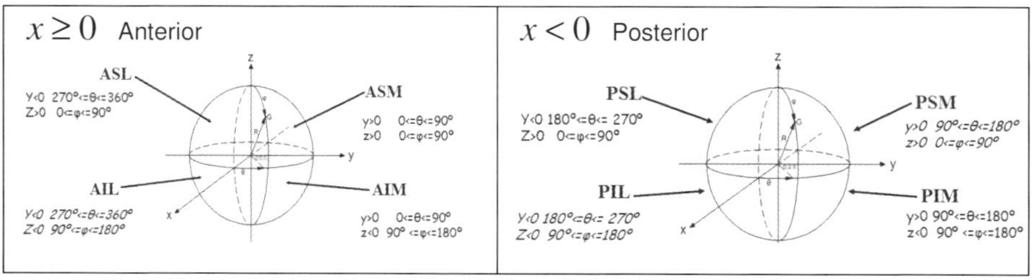

Figure 6

Volume of the necrotic area ranged from 11.4% to 63.3%, latitude (φ) ranged from 8° to 106° degrees, and longitude (θ) ranged from 6 ° to 83 ° degrees and the R vector value from 4 to 23 mm. To be confident with our results the evaluation of collapse was done by one of the authors and another evaluation by an independent surgeon. The evaluators were close in their estimates in 34 cases but in only 2 cases were not agreed 5.8%. To determine the influence of the calculated lesion morphological parameters on the risk of progression of collapse, a statistical analysis with a non-parametric Mann–Whitney U-test was performed. Fisher's exact probability test was applied in the percentage of lesion size according to Steinberg classification. Cox regression test with parameters such as patient age, lesion size, R, φ°, θ° was performed. Values of $P<0.05$, were regarded as statistically significant.

1.2. *Results*

During the study period 10 out of 34 hips (29.4%) showed radiographic progression (Table I) and 5 of them were treated with a total hip arthroplasty. Man-Whitney non parametric statistical analysis showed that examining only the lesions in our patient data and the radiographic progression of the femoral head the value of $p = 0$ which is <0.05 and that means acceptable data.

Lesions were classified according to their percentage in the femoral head according to Steinberg scale (9) as follows: Group a for lesions having necrotic volume less than 15%, Group b for lesions having necrotic volume between 15% and 30% and Group c for lesions having necrotic volume greater than 30%. Of the 2 hips in group a, 0 (0%) collapsed, of the 19 hips in group b, 1 (5%) collapsed and of the 13 hips in group c 9 (75%) collapsed. Fisher's Exact Probability showed that there was no significant association between Lesions in group a and Lesions in group b. On the other hand groups b, c showed statistically important and acceptable $p=0.002$ which is <0.05 result which means group c has chance to collapse than group b or a. Table 2. Most of the calculated Geometric Center occurred in the ASM octant with 19 no collapse femoral heads (55,88%) and 9 collapsed femoral heads (26,47%), 5 Non collapsed femoral heads in the ASL octant (14,71%) and 1 collapsed femoral head in the AIM octant (2.9%).

Table 2 Associatin between Steinberg & Collapse : Fisher's exact test

Steinberg scales	p
a-b	1,00
b-c	0,02

Cox proportional hazards models with parameters such as patient age, lesion size, follow up period, R φ°, θ°, were performed in enter method and showed that lesion (p:0.001) and longitude angle (θ) (p:0.025) could be important factors to predict radiographic progression. The results of the cox regression analysis with the two factors are presented in (table 3). We can observe the scatter with the parameters lesion size (%) and the angle θ (longitude) as shown in (Fig.7).

Table 3 Associations in Cox Regression Analysis of Osteonecrosis patients

Patient characteristic

	p	(95% CI)
Lesion	0,001	287488,4 (208,4 – 4E+008)
Thita	0,025	0,941(0,892 – 0,992)

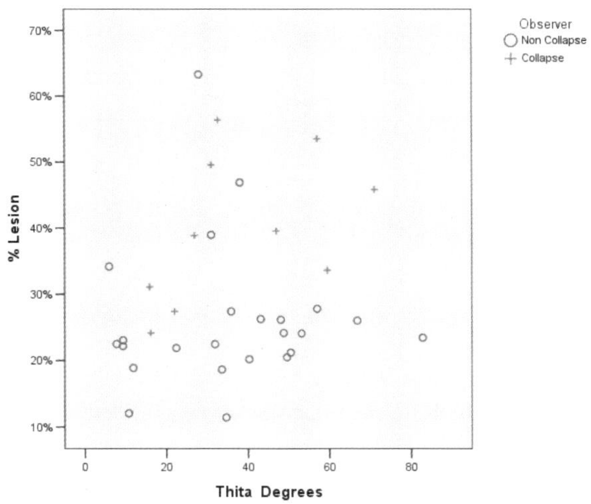

Figure 7 We can see that femoral heads having longitude angle (θ°) of 15° - 35° and 50 °- 60 ° and being in Group c – except 2 femoral heads with Lesion volume 27% of the total volume- were more likely to develop radiographic progression with p = 0.025. For hips that developed radiographic progression out of this range, lesion size was the only significant factor.

Table 4 Patient description according to etiology. Number of patients: *n*=34, number of hips: *n*=28

N=34 Hips 28 Patients

	No Collapse (n=24)	Collapse of FH (n=10)
Male : female 23:5	18 : 6	8 : 2
Bilaterality : unilaterality 12 : 22	10 : 14	2 : 8
Etiology		
Alcohol 2 (3)	2	1
Steroid 15 (19)	13	6
Primary 9 (10)	9	1
Sicle Thal. 2	0	2
ARCO stage		
I 3	2	1
II 17	15	2
III 14	7	7
Parameters		
Volume (%)a	24.6±9.46	39.39±10.81
Latitude φ (_)a	33.875±11.646	40.253±24.029
Longitude θ (_)a	30.108±17.697	27.68±17.968

a Values are mean _ S.D.
b P < 0:05 between hips with and without collapse (Fisher's exact probability test).
c P < 0:001 between hips with and without collapse (Mann–Whitney U-test).

1.3. *Discussion*

Since the late 1960s, several studies have evaluated effectiveness of potential techniques to treat ONFH, such as core decompression, vascularized and nonvascularized fibular bone grafts, and angular or rotational osteotomies.

The research so far indicates that mechanical assistance can be useful for preventing progression of osteonecrosis of the femoral head. Buckley et al. in a long-term study of twenty hips has shown that compression and cortical structural grafting was successful for the treatment of stage-I or II atraumatic due to use of mechanical support of subchondral plate during the healing proccess of the lession. The results of research show that the elastic modulus of porus tandalum is similar to that of the bone and can carry loads (19,20,21,22).

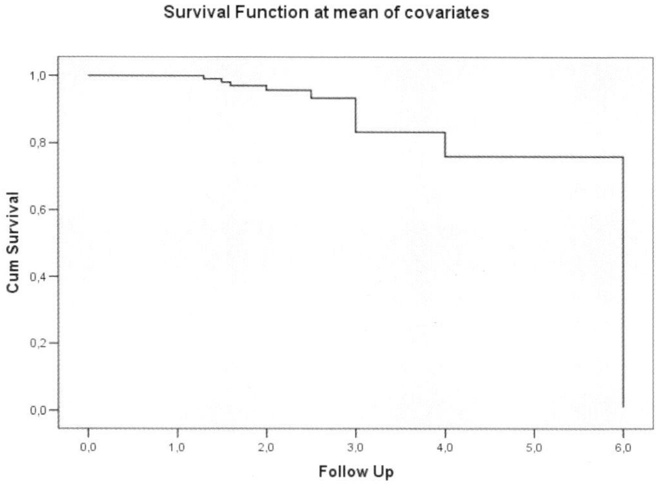

Figure 8 Cox regression with the parameter follow up as a time variable, shows that after the 6th year there is no patient without collapse. The cumulative expected probability is 70% for the patients to survive without collapse till the 4th year of the follow up period.

During the follow-up period, detection of collapse was based on radiographs in a bidirectional view. Other imaging modalities, such as MR imaging or CT scanning, may enable a more sensitive detection of early-focal collapse of the femoral head than plain radiographs. Considering the high cost of repeated MR imaging or CT scanning examinations, we propose the use of sequential radiographic examinations.

The X-ray was the first imaging technique used to estimate the extent of damage in osteonecrosis of the femoral head because of its availability and low cost. The use of X-ray gives an overview of the injury, but does not allow accurate estimation of actual size of the Lesion. Moreover, the X-ray can determine accurately the sphericity of the head. The techniques use a combination anteriosuperior and lateral radiographs give a better description of the extent of damage, but also to obtain accurate estimate of its size. Kerboul et al. (26) used a mixture (AP - anteroposterior) and lateral by measuring the angle (arc) of AP necrosis in X-ray and the corresponding arc in lateral radiography. The sum (in degrees) of the two angles is the ratio Kerboul. The extent of damage is considered large for values of index greater than the 200 for intermediate values between 160th to 200th and low values for less than

160th. Mont et al. (27) used the ratio Kerboul found that the main factor associated with disease progression and ultimately the collapse of the femoral head is the advanced stage of necrosis.

Later researches found that the prognosis of osteonecrosis requires precision in measuring the extent and location of damage, obtained only by using MR imaging. The MRI combines three-dimensional analysis of potential harm to the sensitivity in the early diagnosis of osteonecrosis. Furthermore, no radiation burden to the question (such as computer tomography examination). These reasons have direct interest in MRI as the most suitable technique for quantitative analysis of necrotic damage. The first methods presented make two-dimensional description of the injury, without exploiting the three-dimensional magnetic character. Beltran et al. (28) first used MR image and found 4 types of osteonecrosis according to lesion volume Type A: hips without necrosis. • Type B: hips collapse up to 25% • Type C: hips with collapse 25% to 50%. • Type D: hips with collapse greater than 50%. Recent techniques use a combination of sagital and coronal sections, giving more precise information on the form, but do not calculate the actual amount of damage. Numerous publications indicate that the prognosis of collapse the femoral head suggests that lesion size is of great importance (23, 24).Koo et al. (25) showed accurate prediction using an index calculated by multiplying the arc of the lesion in the mid-coronal MR image by that in the mid-sagital MR image and then calculates using the formula: (A/180) x (B/180) x 100. This formula is called necrosis index and describes the extent of damage of the femoral head. Location of the necrotic area of the femoral head has also been proposed for prognosis of the risk of collapse (28, 29). Nishii et al. (2) studied the importance of size and location for the prediction of risk of collapse of the femoral head. To assess the size and location of damage suggested a new method based on MRI. The method applies identification of the necrosis by hand using a computer. After the identification of necrosis in the area, three-dimensional reconstruction of the damage is applied. Correspondingly, the head is considered a sphere with a radius specified in the central frontal incision MRI. The method calculates the percentage of necrosis and the relative position of the head (distance of center of gravity of the harm from the center of the sphere). It should be noted that this technique requires a «fine» section (MRI thickness 1mm). The authors suggest that lesion size should be determined, and when the lesion is small, its location, especially at the anterosuperior portion of the femoral head, should be considered.

In this paper, we found that lesion size was the most important factor among various related factors that predicted collapse. Lesion volume showed a strong correlation with risk of collapse according to cox regression analysis, and hips with lesion volumes greater than 30% of the femoral head (7) showed a high incidence of collapse (75%). Lesions with geometric center in the anterior superior medial portion of the femoral head having an angle (θ) of 15° - 35° and 50° - 60° degrees and lesion size greater than 27% of the femoral head were more likely to develop radiographic progression. Only one patient with necrotic lesion of 46% and geometric center in the anterior inferior medial part of the femoral head collapsed in a follow up of a 6 years with (θ) 71° degrees. Femoral heads with small necrotic lesions (0 to 27%) did not develop any radiographic progression. Only one hip with lesion size of 27% and (θ) of 25° degrees collapsed in a follow up period of 5 years. In our study we used 10 MR slices per patient with slice thickness 5 mm without interslice gaps. In our pilot preliminary study when we used 5 MR slices per hip the results we obtained varied dramatically and we could not draw any conclusions. Subsequently, the number of MR slices per hip was increased.

The technique described might be applicable in clinical orthopaedic settings since it requires MR slice thickness of 5mm. Any CAD Software that reads MR images and produces 2d x and y coordinates will be appropriate to produce input to our customized software. However, in the near future, the development of sophisticated, widely available software of automatic image processing is expected to reduce the time and effort required for the clinical use of three-dimensional morphological analysis. Until then, additional research for developing newer algorithms for calculating three-dimensional lesion size and location, mid-coronal and mid-sagittal MR images, or appropriate directions and positions of MR imaging within the femoral head, is necessary to meet clinical requirements.

The methodology described above uses tracing by hand to determine the limits of the damage. It is quite difficult to be repeated for each of the slices separately with accuracy because it is semi-automatic procedure.

The method described in this paper predicts with accuracy the progression of the disease. However, further investigation with larger data sets and other algorithms and criteria are still needed.

Appendix

Let N be the total number of MRI slices perpendicular to the x-axis. On each such slice the perimeter of the lesion is defined by the (y_i, z_i) coordinates a total of M discrete points as shown in Fig. 9. The numbering of the points is such that i increases counter-clockwise.

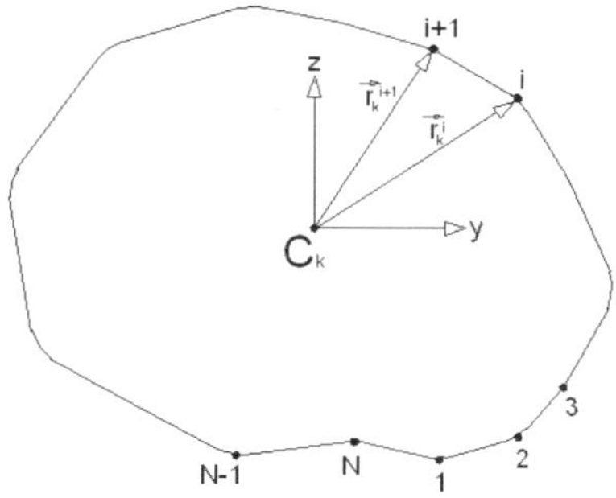

Figure 9

On the k -th slice $(k=1,2,\ldots,M)$, let

$$\mathbf{r}_k^{(i)} = y_i^k\,\mathbf{e}_y + z_i^k\,\mathbf{e}_z \qquad \text{and} \qquad \mathbf{r}_k^{(i+1)} = y_{i+1}^k\,\mathbf{e}_y + z_{i+1}^k\,\mathbf{e}_z$$

$$(i=1,2,\ldots,N)$$

be the position vectors on the plane of two consecutive points of the perimeter with respect to C_k, where C_k the point of intersection of the x -axis and the plane of the slice. The area of the triangle defined by C and point i and $i+1$ on the perimeter is

$$A_k^{(i)} = \frac{1}{2}\left|\mathbf{r}_k^{(i)} \times \mathbf{r}_k^{(i+1)}\right| = \frac{1}{2}\left(y_i^k\, z_{i+1}^k - y_{i+1}^k\, z_i^k\right) \qquad (i=1,2,\ldots,N),$$

with $\left(y_{M+1}^k, z_{M+1}^k\right) \equiv \left(y_1^k, z_1^k\right)$. The position of the centroid of the aforementioned triangle is defined by

$$\mathbf{r}_{G_k}^{(i)} = \frac{1}{3}\left(\mathbf{r}_k^{(i)} + \mathbf{r}_k^{(i+1)}\right) = \frac{1}{3}\left(y_i^k + y_{i+1}^k\right)\mathbf{e}_y + \frac{1}{3}\left(z_i^k + z_{i+1}^k\right)\mathbf{e}_z\,.$$

The total area is determined as

$$A_k = \sum_{i=1}^{M} A_k^{(i)} = \frac{1}{2}\sum_{i=1}^{M}\left(y_i^k\, z_{i+1}^k - y_{i+1}^k\, z_i^k\right).$$

The position of the center G_k of A_k is determined from

$$\mathbf{r}_G^k = \frac{1}{A_k}\int_{A_k}\mathbf{r}\, dA = \frac{1}{A_k}\sum_{i=1}^{M}\int_{A_k^{(i)}}\mathbf{r}\, dA = \frac{1}{A_k}\sum_{i=1}^{M}\mathbf{r}_{G_k}^{(i)}\, A_k^{(i)} ,$$

so that the (y,z) coordinates of the center of the k -th slice are

$$x_G^k = \frac{1}{6A}\sum_{i=1}^{M}\left(y_i^k + y_{i+1}^k\right)\left(y_i^k\, z_{i+1}^k - y_{i+1}^k\, z_i^k\right) \quad \text{and}$$

$$y_G^k = \frac{1}{6A}\sum_{i=1}^{M}\left(z_i^k + z_{i+1}^k\right)\left(y_i^k\, z_{i+1}^k - y_{i+1}^k\, z_i^k\right).$$

The total volume of the lesion is determined as

$$V_{lesion} = \sum_{k=1}^{N} A_k\, \Delta x_k \ .$$

The Cartesian coordinates of the center G of the lesion is determined from

$$\mathbf{r}_G = \frac{1}{V_{lesion}}\int_{V_{lesion}}\mathbf{r}\, dV = \frac{1}{V_{lesion}}\sum_{k=1}^{M}\left(\int_{A_k}\mathbf{r}\, dA\right)\Delta x_k = \frac{1}{V_{lesion}}\sum_{k=1}^{M}\mathbf{r}_G^k\, A_k\, \Delta x_k\ ,$$

so that

$$x_G = \frac{1}{V_{lesion}} \sum_{k=1}^{M} x_k \, A_k \, \Delta x_k \, , \qquad y_G = \frac{1}{V_{lesion}} \sum_{k=1}^{M} y_G^k \, A_k \, \Delta x_k \, ,$$

$$z_G = \frac{1}{V_{lesion}} \sum_{k=1}^{M} z_G^k \, A_k \, \Delta x_k \, ,$$

where x_k is the x-coordinate of the k-th MRI slice.

References

1. Nishii T, Sugano N, Ohzono K, Sakai T, Haarguchi K, Yoshikawa H. Clin Orthop 2002; 400:149-157.
2. Nishii T, Sugano N, Ohzono K, Sakai T, Sato Y, Yoshikawa H. J Orthopaed Res 2002; 20:130-136.
3. Zoroofi RA, Nishii T, Sato Y, Sugano N, Yoshikawa H, Tamura S. Comput Med Imaging Graph 2001; 25:511-521.
4. Malizos KN, Siafakas MS, Fotiadis DI, Karachalios TS, Soukakos PN.. Skeletal Radiol 2001; 30:686-693.
5. Ito H, Matsuno T, Kaneda K. Clin Orthop 1999;358:149–57.
6. Shimizu K, Moriya H, Akita T, Sakamoto M, Suguro T. J Bone Jt Surg [Am] 1994;76:215–23.
7. Steinberg ME, Bands RE, Parry S, Hoffman E, Chan T, Hartman KM. Clin Orthop 1999;367:262–71.
8. Sugano N, Takaoka K, Ohzono K, Matsui M, Masuhara K, Ono K. Clin Orthop 1994;303:155–64.
9. Steinberg ME, Hayken GD, Steinberg DR. A quantitative system for staging avascular necrosis. J Bone Jt Surg [Br] 1995; 77:34–41.
10. Mont MA, Hungerford DS. J Bone Jt Surg [Am] 1995;77:469–74.
11. Mont MA, Carbone JJ, Fairbank AC. Clin Orthop Relat Res. 1996; 324:169-78.
12. Steinberg ME, Larcom PG, Strafford B, et al. Clin Orthop Relat Res. 2001; (386):71-8.
13. Soucacos PN, Beris AE, Malizos K, Koropilias A, Zalavras H, Dailiana Z. Clin Orthop Relat Res. 2001; (386):120-30.
14. Tsao AK, Roberson JR, Christie MJ. J Bone Joint Surg Am. 2005; 87 Suppl 2:22-7.
15. Sugano N, Takaoka K, Ohzono K, Matsui M, Masuhara K, Ono K. Clin Orthop 1994;303:155–64.
16. Ito H, Matsuno T, Kaneda K. Clin Orthop 1999;358:149–57.
17. Koo KH, Kim R. J Bone Jt Surg [Br] 1995;77:875–80.
18. Steinberg ME, Bands RE, Parry S, Hoffman E, Chan T, Hartman KM. Clin Orthop 1999;367:262–71.

19. Bobyn JD, Stackpool GJ, Hacking SA, Tanzer M, Krygier JJ. Jbone Joint Surg BR. 1999;81:907-14.
20. Bobyn JD, Poggie RA, Krygier JJ, Lewallen DG, Hanssen AD, Lewis RJ, Unger AS, O'Keefe TJ, Christie MJ, Nasser S,Wood JE, Stulberg D, Tanzer M. Jbone Joint Surg Am. 2004;86 Suppl 2:123-9.
21. Heiner AD, Brown TD, Poggie RA. Trans Orthop Res Soc. 2001;26:480.
22. Tsao AK, Roberson JR, Christie MJ. J Bone Joint Surg Am. 2005; 87 Suppl 2:22-7.
23. Steinberg ME, Bands RE, Parry S, Hoffman E, Chan T, Hartman KM. Clin Orthop 1999;367:262–71.
24. Mazieres B, Marin F, Chiron P, Moulinier L, Amigues J-M,Laroche M, Cantagrel A. Ann Rheum Dis 1997;56:747–50.
25. Koo KH, Kim R. J Bone Jt Surg [Br] 1995;77:875–80.
26. Kerboul M, Thomine J, Postel M, Merle D'Aubigne R. J Bone Joint Surg Br 1974; 56: 291-296.
27. Mont MA, Fairbank AC, Petri M, Hungerford DS. Clin Orthop 1997; 334:91-97.
28. Shimizu K, Moriya H, Akita T, Sakamoto M, Suguro T. J Bone Jt Surg [Am] 1994;76:215–23.
29. Sugano N, Ohzono K, Masuhara K, Takaoka K, Ono K. Clin Orthop 1994;305: 190–9.

TWO-DIMENSIONAL BOUNDARY ELEMENT MODELING OF WAVE DISPERSION IN BONE-MIMICKING PLATES WITH MICROSTRUSTURAL EFFECTS

M. G. VAVVA, D. I. FOTIADIS
Unit of Medical Technology and Intelligent Information Systems, Dept. of Material Science and Engineering, University of Ioannina, GR 45110 Ioannina, Greece

A. PAPACHARALAMBOPOULOS, V. C. PROTOPAPPAS, D. POLYZOS
Dept. of Mechanical Engineering and Aeronautics University of Patras, GR 265 00, Greece

The study of ultrasound wave propagation in bones is mostly based on the classical theory of linear elasticity. However, bone is a material with microstructure and thus its mechanical behaviour can be successfully modeled macroscopically using enhanced theories of elasticity. In this study we use the Boundary Element Method (BEM) to numerically investigate guided wave propagation in 2D bone models. The bone is modeled as an isotropic elastic material plate and its microstructure is taken into account by introducing the gradient coefficient g and the micro-inertia term h in the constitutive equations. Numerical results are derived for three different combinations between g and h. In all cases the values of g and h are of the order of the osteons' size. The distance between the transmitter and the receiver is 59 mm. The bone is considered free of stresses and double stresses. The received signals are analyzed in the time-frequency (t-f) domain. The propagating guided modes are characterized in the t-f representations by incorporating the theoretical velocity dispersion curves derived from the gradient theory of elasticity. The t-f representations of the signals obtained from the 2D plates with microstructure are found significantly different from those in the case of linear elastic plates. For the smallest value of h, i.e., $h = 10^{-5}\ m$, the lowest order modes in the gradient elastic theory could be well characterized in the t-f domain.

1. Introduction

Bone is a natural composite with complex structure and microstructural effects (Rho, 1998). However, previous computational and experimental studies aiming at the assessment of osteoporosis (Bossy, 2004, Moilanen, 2006) and the evaluation of the fracture healing process (Protopappas, 2006, 2007, 2008, Dodd, 2007, Vavva, 2008) investigate ultrasound guided wave propagation by assuming the bone to be a homogeneous material. Furthermore, the analysis is performed in the context of the linear theory of classical elasticity, which however cannot adequately describe the mechanical behavior of materials with microstructure. Bone's microstructural effects can be successfully modeled in a macroscopic manner by using enhanced theories of elasticity, such as the couple stresses elastic theories (Cosserat, 1909), higher order gradient theories (Mindlin, 1964), non-local and micropolar elastic theories (Eringen 2002). The most general gradient elastic theory has been proposed by Mindlin in three versions i.e., Form I, II and III. Mindlin's Form-II leads to a total stress tensor, which is symmetric as in the case of classical elasticity.

In a recent study, by using the gradient theory of elasticity, we showed that microstructure significantly affects guided wave propagation in two dimensional bone mimicking plates even at medium frequencies, i.e. 0.7-1 MHz, commonly used in ultrasonic bone studies (Vavva, 2009). The group velocity dispersion curves of the guided modes propagating in two-dimensional (2D) isotropic bone-mimicking plates with microstructural effects were derived and were found modified from the Lamb modes. Furthermore, analysis of *ex-vivo* measurements on an intact tibial bone showed that the gradient theory could describe the dispersion of the propagating guided modes better than the classical Lamb wave theory.

In this study we proceed in the computational investigation of microstructural effects on guided wave propagation using 2D Boundary Element models of bone. Numerical simulations of wave propagation are also performed for an isotropic linear elastic plate for comparison purposes. The Boundary Element Method was selected as the most appropriate, since it can solve elasto-dynamic problems with low computational effort. The obtained signals are analyzed in the time-frequency (t-f) domain. The propagating guided modes are characterized in the t-f representations of the obtained signals by incorporating the theoretical velocity dispersion curves derived from the gradient theory of elasticity.

2. Materials and Methods

2.1. *Mindlin's Form II simplified gradient elastic theory and wave dispersion*

In Mindlin Form-II, the strain energy density function is assumed to be a quadratic form of the classical strains and the gradient of strains. The microstructural characteristics are introduced in the stress analysis by considering two additional elastic constants, representing the characteristic length in bone: (a) the gradient coefficient g, introduced in the strain energy and (b) the micro-inertia term h, in the kinetic energy.

The equation of motion in a gradient elastic continuum is given by (Vavva, 2009):

$$(1-g^2\nabla^2)\left[\mu\nabla^2\mathbf{u}+(\lambda+\mu)\nabla\nabla\cdot\mathbf{u}\right]=\rho(\ddot{\mathbf{u}}-h^2\nabla^2\ddot{\mathbf{u}}), \tag{1}$$

where $g^2\nabla^2\left[\mu\nabla^2\mathbf{u}+(\lambda+\mu)\nabla\nabla\cdot\mathbf{u}\right]$ and $\rho h^2\nabla^2\ddot{\mathbf{u}}$ are the micro-structural and the micro-inertia terms, respectively, and the operator ∇ is the Laplacian. It can be observed that when the volumetric strain gradient coefficient g^2 and the inertia term h^2 become zero the equation of motion in the gradient elastic theory becomes identical to that obtained in the classical elastic case.

Consider a free two-dimensional (2D) plate and a Cartesian co-ordinate system Ox_1x_2 with the axis Ox_1 being the axis of symmetry of the plate. In the gradient theory the plate is assumed not only free from stresses, as in the classical case but also free from double stresses i.e.,

$$\mathbf{P}|_{x_2=d/2}=\mathbf{P}|_{x_2=-d/2}=0\,,\ \mathbf{R}|_{x_2=d/2}=\mathbf{R}|_{x_2=-d/2}=0\,, \tag{2}$$

where **P** and **R** are the traction vector and the double traction vector, respectively, and d is the plate thickness. Solution to Eq. (1) and satisfaction of the boundary conditions result in two systems of four unknown constants (Vavva, 2009). Vanishing of each determinant yields the characteristic dispersion equations for the symmetric and antisymmetric modes propagating in a gradient elastic plate.

2.2. *Boundary Element Method (BEM)*

Numerical simulation of guided wave propagation in 2D bone mimicking plates with microstructure is performed using the advanced Boundary Element Method/Fast Fourier transform code which is able to solve dynamic gradient elastic problems. As it is shown in (Polyzos, 2005), Eq. (1) in the frequency domain is transformed into an integral representation, along the boundary of the geometry. The Fast Fourier Transform then is used to solve a transient problem.

The bone is modeled as a 2D plate with microstructural effects and mechanical properties equal to those of bone, i.e. Young's modulus E_{bone} = 14 GPa, Poisson's ratio v_{bone} = 0.37 and density ρ = 1500 Kg/m^3 (Protopappas, 2006). The plate thickness is 4 mm, and the resulting classical bulk longitudinal and shear velocities are 4063 m/s and 1846 m/s, respectively. The plate is assumed to have rounded corners, due to the fact that the gradient elastic continuum needs extra boundary conditions at non-smooth boundaries.

The distance between the transmitter and receiver is 59 mm. The bone is considered free of stresses and double stresses. The length of the plate is 180 mm. In order to avoid the reflected waves from the side ends to arrive at the receiver the duration of the received waveform is truncated accordingly.

The plate was assumed to have fixed ends (**u**=0). The rest of the plate is free of stresses and double-stresses. The transmitter creates a stress-excitation of the form $P_y=P_0\delta(t)$, where $\delta(t)$ is the Dirac-distribution (Fig. 1).

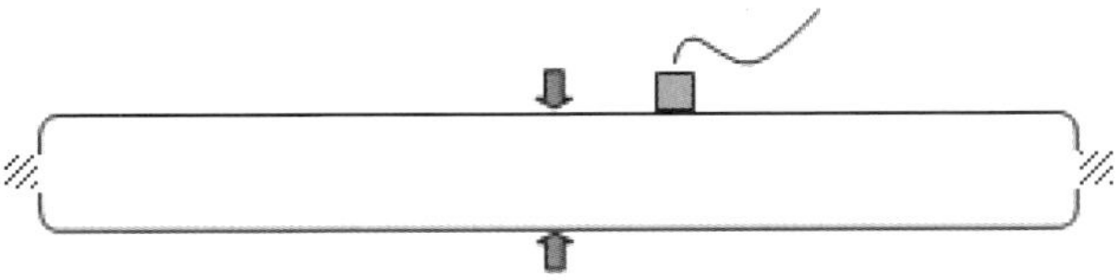

Figure 1: Geometry of the 2D bone mimicking plate. The arrow stands for the excitation and the blue square for the receiver.

Simulations are performed for three different combinations between g and h, i.e., $h=10^{-4}$, $g=10^{-5}$ in Case1, $h=10^{-4}$, $g=5\cdot10^{-4}$ in Case2 and $h=10^{-5}$, $g=10^{-4}$ in Case3. It is obvious that in Case1, $g<h$ whereas in Cases2 and 3, $g>h$. However, in all cases the values of g and h are comparable to bone's microstructure (Harvesian systems, osteons), i.e. from 10-500 μm (Rho, 1998).

For all the examined cases of g and h we also calculate the corresponding theoretical velocity dispersion curves for a free 2D gradient elastic plate. The symmetric (dashed red lines) and antisymmetric modes (dotted red lines) propagating in the bone-mimicking plate with microstructure in Case1, and Case2 are presented in Figs. 2(a) and (b), respectively, in the form of frequency-group (f, c_g) velocity dispersion curves (denoted herein as g_Sn and g_An, where n is the mode number). The group velocity dispersion curves of the Lamb modes calculated for a classical elastic plate with the same geometrical and mechanical properties are also presented in Figs.2(a) and (b) (black lines) for comparison purposes. The Lamb modes are denoted as Sn (dashed lines) and An (red lines), where $n=0,1,2...$

The obtained signals are analyzed in the time-frequency domain. We use the reassigned Smooth-Pseudo Wigner-Ville (RSPWV) energy distribution function because it accurately localizes guided waves (Protopappas, 2006). For the given transmitter-receiver distance, the theoretical dispersion curves derived from both gradient elastic theory are converted to t-f curves and are superimposed on the t-f plane of the corresponding signals obtained from the BEM simulations.

2 Results

The t-f representations of the signals obtained from the linear isotropic and the gradient elastic models of bone are shown in Figs 3and 4. In each figure, the superimposed constant lines represent the corresponding dispersion curves of the theoretical symmetric (dashed lines) and antisymmetric modes (dotted lines).

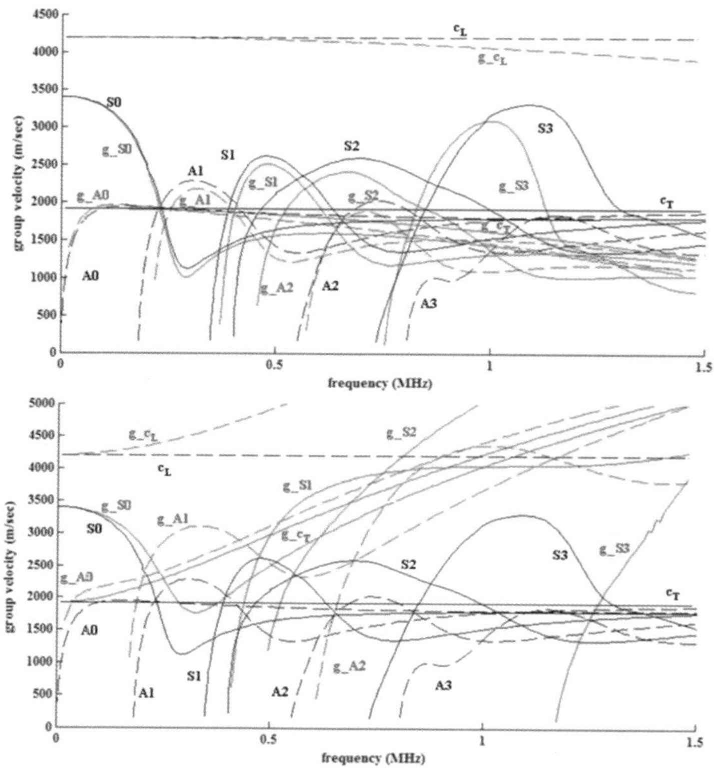

Figure 2: Group velocity dispersion curves for a free bone-mimicking plate for: (a) Case1 and (b) Case2. The Lamb modes are also presented (black lines).

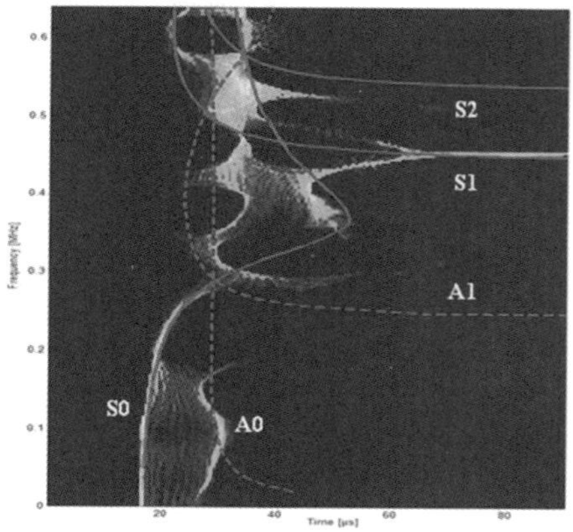

Figure 3: T-f representation of the signal from the isotropic linear elastic plate. The corresponding symmetric (dashed lines) and antisymmetric (dotted lines) Lamb modes are shown.

It can be shown that for very low frequencies microstructure does not influence the propagation of the lowest order mode (S0 and g_S0) which is clearly detected in most of the examined cases. For the Case1 ($g<h$) the g_S0 and g_S1 the modes are described in the 0.01- 0.3 MHz frequency range and in the 0.45 - 0.46 frequency range, respectively (Fig. 4(a)). However, in that case the propagation of the g_A0 is poorly supported. For the Case2 ($g>h$) only the g_S0 mode could be well detected in the 0.01 - 0.31 frequency range (Fig. 4(b)). Finally in Fig. 4(c), i.e. for $h=10^{-5}$ m again the g_S0 and g_S1 modes are dominant and can be identified in the 0.01- 0.31 MHz frequency range and in the 0.454 - 0.45 frequency range, respectively. It can be observed that in all cases the propagation of the antisymmetric modes is not supported.

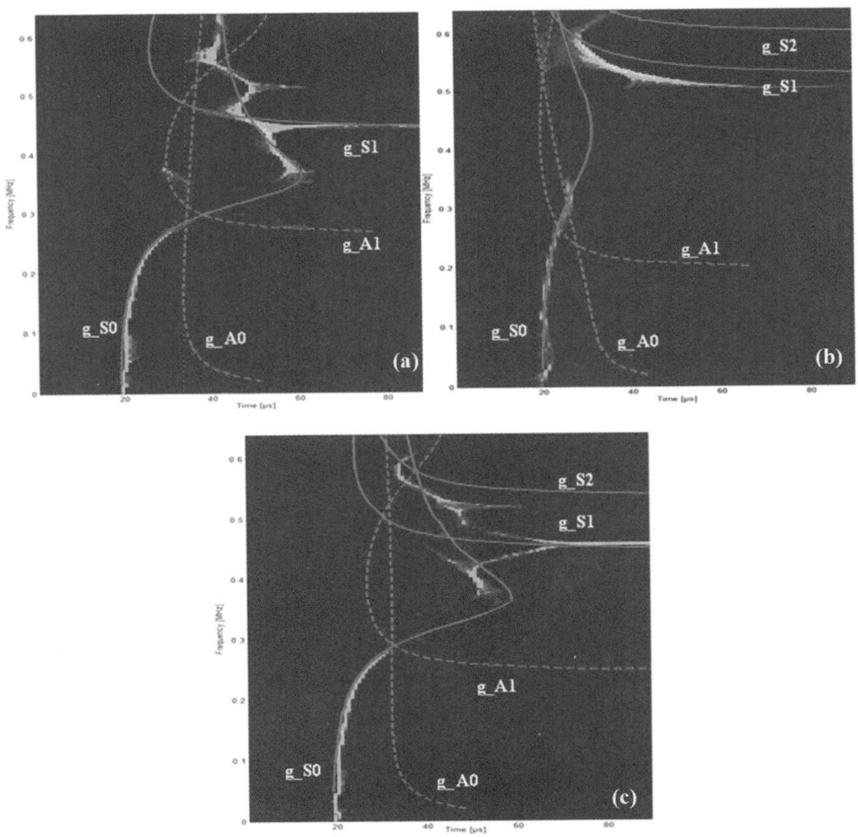

Figure 4: T-f representation of the signal from gradient elastic plate in (a) Case1 ($g=10^{-5}$, $h=10^{-4}$), (b) Case2 ($g=5\cdot10^{-4}$, $h=10^{-4}$) and (c) Case3 ($g=10^{-4}$, $h=10^{-5}$). The corresponiding theoretical dispersion curves of the symmetric (dashed lines) and the antisymmetric (dotted lines) modes derived from the gradient elastic theory are superimposed.

3 Discussion

In this work we presented a BEM study on ultrasound wave propagation in 2D bone mimicking plates with microstructural effects using Mindlin Form II gradient theory. The main advantages of this theory as compared to other enhanced theories are its simplicity and the symmetry of all stress tensors involved. The elastic constants g and h are incorporated in the stress analysis to account for bone's microstructure. For each value of h the g is considered to have an either higher or smaller value as in [11]. The case $g = h$ is not examined here since it has been both analytically and computationally proved in (Vavva, 2009) that when g and h are equal, no differences exist in mode dispersion between the gradient and the classical theories.

For all the examined cases, the t-f representations of the signals obtained from the 2D gradient elastic plates are different from those in the case of linear elastic plate. This is further supported by the theoretical results since the dispersion curves of the modes derived from the gradient elastic theory are modified from those predicted by the classical Lamb wave theory (Vavva, 2009). These findings demonstrate that bone's microstructural characteristics give rise to major changes in the dispersion of the propagating guided modes by inducing material and geometrical dispersion.

In all the examined cases the g_S0 and g_S1 modes can be clearly detected in the t-f representations, whereas the propagation of the antisymmetric modes is not supported. This can be justified by the symmetric excitation form (Fig.(1)).

Overall the modes propagating in the 2D bone models with microstructural effects can be successfully characterized from the theoretical dispersion curves derived from the gradient theory of elasticity. Our results show that BEM simulations can be further used for the assessment of pathological situations of bone such as bone fracture and osteoporosis. However, 3D bone models with microstructure accounting for bone's irregular geometry and anisotropy need to be developed to better interpret measurements on real bones.

References

1. Rho, J.Y. (1998). Mechanical properties and the hierarchical structure of bone. Med. Eng. Phys., 20(2), 92-102.
2. Bossy, E., Talmant, M., & Laugier, P. (2004). Three-dimensional simulations of ultrasonic axial transmission velocity measurement on cortical bone models. J. Acoust. Soc. Am., 115(5), 2314–2324.
3. Moilanen, P., Nicholson, P.H.F., Kilappa, V., Cheng, S., & Timonen, J. (2006). Measuring guided waves in long bones: Modeling and experiments in free and immersed plates. Ultrasound Med. Biol., 32(5), 709-719.

4. Protopappas, V.C., Vavva, M.G., Fotiadis, D.I., & Malizos, K.N. (2008). Ultrasonic Monitoring of Bone Fracture Healing. IEEE Trans. on Ultr. Ferroel. Freq. Contr., 55(6), 1243-1254.
5. Protopappas, V.C., Fotiadis, D.I., & Malizos, K.N. (2006). Guided ultrasound wave propagation in intact and healing long bones. Ultrasound Med. Biol. 32(5), 693-708,.
6. Protopappas V.C., Kourtis, I.C., Kourtis, L.C., Malizos, K.N., Massalas C.V. & Fotiadis, D.I. Three-Dimensional Finite Element Modelling of Guided Ultrasound Wave Propagation in Intact and Healing Long Bones.(2007). J. Acoust. Soc. Am., 121(6), 3907-3921.
7. Dodd, S.P., Cummingham, J.L., Miles, A.W., Gheduzzi, S., Humphrey, V.F. An in vitro study of ultrasound signal loss across simple fractures in cortical bone mimics and bovine cortical bone samples. (2007). Bone 40(3), 656-661.
8. Vavva, M.G., Protopappas, V. C., Gergidis, L. N., Charalampopoulos, A., Fotiadis D.I., & Polyzos, D. The Effect of Boundary Conditions on Guided Wave Propagation in Two-Dimensional Models of Healing Bone. Ultrasonics (2008). 48 598-606.
9. Cosserat, E. & Cosserat, F. Théorie des Corps Déformables (1909). Paris: Hermann et Fils, Paris.
10. Mindlin, R.D. Micro-structure in linear elasticity (1964). Arch. Rat. Mech. Anal., 16, 51-78.
11. Eringen, A.C. Nonlocal Continuum Field Theories, (2002) NY: Springer, New York.
12. Vavva, M.G., Protopappas, V.C., Fotiadis, D.I., & Polyzos, D. Velocity dispersion Curves of Guided Waves Propagating in a Free Gradient Elastic Plate: Application to Cortical Bone. (2009). 125, 3414-3427.
13. Polyzos, D. Tsepoura K.G. & Beskos, D.E. Transient dynamic analysis of 3D gradient elastic solids by BEM. (2005). Computers and Structures. 83(10-11), 783-790.

Mathematical Methods in Mechanics and Biomedical Engineering

NOVEL INTEGRAL REPRESENTATIONS FOR IRROTATIONAL STOKES' FLOW IN THE EXTERIOR OF A SPHERE

G. DASSIOS AND M. DOSCHORIS

Department of Chemical Engineering, Division of Applied Mathematics, University of Patras and ICE-HT/FORTH, Greece

In the present work, solutions of the Stokes' operator E^2, which governs the irrotational axisymmetric Stokes' flow, are obtained in terms of novel integral representations valid in the exterior of a sphere. The formulae obtained, by means of the generalized transform introduced recently by Fokas and his collaborators, have the advantage of being uniformly convergent at the boundary. Furthermore, it is possible to solve problems with more complicated boundary conditions which demand the solution of a Riemann-Hilbert problem. This is achieved via the so called *global relation*, i.e. an integral relation connecting the Dirichlet boundary values with the Neumann boundary values. On the other hand, the Dirichlet-to-Neumann correspondence, needed to evaluate the unknown boundary data, for simple boundary data is evaluated by algebraic manipulations of the global relation alone. In the present work we only consider simple boundary conditions, corresponding to the Dirichlet and Neumann problems.

1. Introduction

Within the last decade a generalized transform has been developed by Fokas and his collaborators [2]. The novelty of this transformation is focused on the fact that it is a transform that meets the particular analytical and geometrical characteristics of the problem at hand. A crucial part of the theory concerns the manipulation of the so-called *global relation*, which is an integral relation connecting the boundary values of the solution (Dirichlet data) with the normal derivative of the solution on the boundary (Neumann data).

We consider the motion of a solid sphere of radius a, moving with constant velocity U through an unbounded fluid which otherwise is at rest. Aim of the present paper is to obtain solutions for the irrotational Stokes' flow Ψ_D, Ψ_N that solve the exterior Dirichlet and Neumann boundary value problems, respectively. Denote the Dirichlet data on the boundary by

$g_D(U,\zeta)$ and the Neumann data on the boundary by $g_N(U,\zeta)$. Furthermore, for the well-posedness of the exterior problem we demand that the solution should satisfy the asymptotic condition

$$\Psi(\mathbf{r}) \to 0, \quad r \to \infty \,. \tag{1.1}$$

Introducing the variable $\zeta = \cos\theta$, $\theta \in (0,\pi)$ the equation which governs the irrotational Stokes' flow reads

$$\left(\frac{\partial^2}{\partial r^2} + \frac{1-\zeta^2}{r^2}\frac{\partial^2}{\partial \zeta^2}\right)\Psi(r,\zeta) = 0 \,. \tag{1.2}$$

2. The Global Relation

Let the stream function $\Psi(r,\zeta)$ satisfy (1.2) and suppose $\overline{q}(r,\zeta)$ is a solution to the formal adjoint equation of (1.2),

$$\left(\frac{\partial^2}{\partial r^2} + \frac{\partial^2}{\partial \zeta^2}\frac{1-\zeta^2}{r^2}\right)\overline{q}(r,\zeta) = 0 \,. \tag{2.1}$$

Multiplying (1.2) by $\overline{q}(r,\zeta)$, (2.1) by $\Psi(r,\zeta)$ and using simple algebraic manipulations we arrive at the divergence form

$$\frac{\partial}{\partial r}\left(\overline{q}\frac{\partial \Psi}{\partial r} - \Psi\frac{\partial \overline{q}}{\partial r}\right) + \frac{\partial}{\partial \zeta}\left[\frac{1-\zeta^2}{r^2}\left(\overline{q}\frac{\partial \Psi}{\partial \zeta} - \Psi\frac{\partial \overline{q}}{\partial \zeta}\right) + \frac{2\zeta}{r^2}\Psi\overline{q}\right] = 0 \,. \tag{2.2}$$

Applying Green's theorem to a closed subdomain of Ω, we obtain

$$\int_{\partial\Omega}\left\{\left(\overline{q}\frac{\partial \Psi}{\partial r} - \Psi\frac{\partial \overline{q}}{\partial r}\right)d\zeta - \left[\frac{1-\zeta^2}{r^2}\left(\overline{q}\frac{\partial \Psi}{\partial \zeta} - \Psi\frac{\partial \overline{q}}{\partial \zeta}\right) + \frac{2\zeta}{r^2}\Psi\overline{q}\right]dr\right\} = 0 \,, \tag{2.3}$$

where $\partial\Omega$ is the boundary of the subdomain.

3. The Dirichlet-to-Neumann Correspondence

Separating variables in (2.1) we observe that the $r-$dependency is described by

$$\overline{R}_1(r;\nu) = r^{\nu+2}, \qquad \overline{R}_2(r;\nu) = r^{-\nu-1}, \quad \nu \in \mathbb{C} \,, \tag{3.1}$$

where the $\zeta-$dependence is given by the Gegenbauer functions of order $\frac{3}{2}$. Therefore, (2.1) accepts solutions of the form $\overline{q}(r,\zeta;\nu) = \overline{R}(r;\nu)\,G_\nu(\zeta)$, where $\overline{R}(r;\nu)$ are given by (3.1) and $G_\nu(\zeta)$ stands for any Gegenbauer function of order $\frac{3}{2}$. The solution $\overline{q}(r,\zeta;\nu)$ remains bounded as $r \to \infty$ for

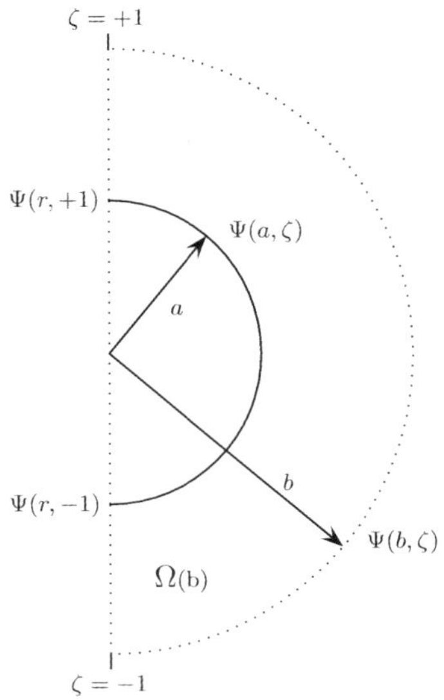

Figure 1. The exterior shell domain $\Omega(b) = \{a \le r \le b,\ -1 < \zeta < +1\}$ which tends to $\Omega = \{a \le r < +\infty,\ -1 < \zeta < 1\}$ as $b \to \infty$.

$\operatorname{Re}\nu \in (-\infty, -2]$ when the functions $\overline{R}_1(r;\nu)$ are chosen, and for $\operatorname{Re}\nu \in [-1, +\infty)$ when the functions $\overline{R}_2(r;\nu)$ are chosen. Thus, applying (2.3) in the domain Ω, depicted in Figure 1, it is straightforward to show that

$$a\mathfrak{N}_\nu - (\nu+2)\mathfrak{D}_\nu = \frac{4}{\pi}\sin\nu\pi\frac{1}{a}\int_a^\infty \tilde{\Psi}(\tau,-1)\left(\frac{r}{a}\right)^\nu dr, \quad \operatorname{Re}\nu \le -1\,, \tag{3.2}$$

$$a\mathfrak{N}_\nu + (\nu+1)\mathfrak{D}_\nu = \frac{4}{\pi}\sin\nu\pi\frac{1}{a}\int_a^\infty \tilde{\Psi}(\tau,-1)\left(\frac{a}{r}\right)^{\nu+3} dr, \quad \operatorname{Re}\nu \ge -2\,, \tag{3.3}$$

where

$$\mathfrak{D}_\nu = \int_{-1}^{+1} g_D(\zeta) C_\nu^{\frac{3}{2}}(\zeta)\,d\zeta, \qquad \mathfrak{N}_\nu = \int_{-1}^{+1} g_N(\zeta) C_\nu^{\frac{3}{2}}(\zeta)\,d\zeta\,. \tag{3.4}$$

4. A Novel Integral Representation

Applying the global relation (2.3) in the subdomain $\Omega_1(\zeta) = \{0 \le r \le b, \zeta \le t < +1\}$, depicted in Figure 2, with $\overline{q}(r,\zeta;\nu)$ replaced by $\overline{R}(r;\nu)\,C_\nu^{\frac{3}{2}}(\zeta)$, we obtain

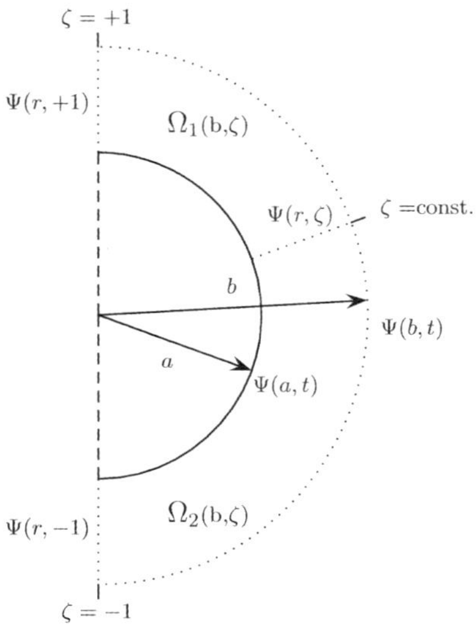

Figure 2. The subdomains $\Omega_1(\zeta) = \{a \le r \le b,\ \zeta \le t < +1\}$ and $\Omega_2(\zeta) = \{a \le r \le b,\ -1 < t \le \zeta\}$.

$$\int_a^\infty \left[(1-\zeta^2)\left(C_\nu^{\frac{3}{2}}(\zeta)\,\frac{\partial\Psi}{\partial\zeta} - \frac{dC_\nu^{\frac{3}{2}}(\zeta)}{d\zeta}\,\Psi\right) + 2\zeta C_\nu^{\frac{3}{2}}(\zeta)\Psi\right]\overline{R}(r;\nu)\,\frac{dr}{r^2}$$
$$= -\int_\zeta^{+1}\left[\overline{R}(a;\nu)\,g_N(t) - \frac{d\overline{R}(a;\nu)}{dr}\,g_D(t)\right]C_\nu^{\frac{3}{2}}(t)\,dt\,. \tag{4.1}$$

Similarly, applying the global relation (2.3) in the subdomain $\Omega_2(\zeta) = \{a \le r \le b,\ -1 < t \le \zeta\}$, depicted in Figure 2, with $\overline{q}(r,\zeta;\nu)$ replaced by $\overline{R}(r;\nu)\,D_\nu^{\frac{3}{2}}(\zeta)$, we obtain

$$\int_a^\infty \left[(1-\zeta^2)\left(D_\nu^{\frac{3}{2}}(\zeta)\,\frac{\partial\Psi}{\partial\zeta} - \frac{dD_\nu^{\frac{3}{2}}(\zeta)}{d\zeta}\,\Psi\right) + 2\zeta D_\nu^{\frac{3}{2}}(\zeta)\Psi\right]\overline{R}(r;\nu)\,\frac{dr}{r^2}$$
$$= \int_{-1}^{\zeta}\left[\overline{R}(a;\nu)\,g_N(t) - \frac{d\overline{R}(a;\nu)}{dr}\,g_D(t)\right]D_\nu^{\frac{3}{2}}(t)\,dt - 2\cos\nu\pi\int_a^\infty \tilde{\Psi}(\tau,-1)\,\overline{R}(a;\nu) \tag{4.2}$$

Multiplying (4.1) by $D_\nu^{\frac{3}{2}}(\zeta)$, (4.2) by $C_\nu^{\frac{3}{2}}(\zeta)$ and subtracting them in order to eliminate the unknown function $\frac{\partial\Psi(r,\zeta)}{\partial\zeta}$, we obtain

$$\begin{aligned}
&\int_a^\infty \Psi(r,\zeta)\,\overline{R}(r;\nu)\,\frac{dr}{r^2} \\
&= (1-\zeta^2)\,\frac{\Gamma(\nu+1)}{\Gamma(\nu+3)}\,\mathfrak{G}_\nu^{\frac{3}{2}}(\zeta)\left[\overline{R}(a;\nu)\,g_N(t) - \frac{d\overline{R}(a;\nu)}{dr}\,g_D(t)\right]dt, \\
&- 2\cos\nu\pi\,C_\nu^{\frac{3}{2}}(\zeta)\int_a^\infty \tilde{\Psi}(\tau,-1)\,\overline{R}(a;\nu)\,\frac{dr}{r^2}, \quad \nu\in\mathbb{C}.
\end{aligned} \tag{4.3}$$

where the operator

$$\mathfrak{G}_\nu^{\frac{3}{2}}(\zeta) \equiv C_\nu^{\frac{3}{2}}(\zeta)\int_{-1}^{\zeta} dt\,D_\nu^{\frac{3}{2}}(t) + D_\nu^{\frac{3}{2}}(\zeta)\int_\zeta^1 dt\,C_\nu^{\frac{3}{2}}(t), \quad \nu\in\mathbb{C} \tag{4.4}$$

is an integral operator, which we will refer to as the *Gegenbauer Integral Operator*.

4.1. *The half-plane* $Re\,\nu \in (-\infty, -2]$

In the half-plane $\mathrm{Re}\,\nu \in (-\infty,-2]$, $\overline{q}(r,\zeta;\nu)$ remains bounded as r tends to infinity if $\overline{R}(r;\nu)$ is replaced by $r^{\nu+2}$. Thus, (4.3) becomes the Mellin transform for the function $r\,\Psi(r,\zeta)$,

$$\begin{aligned}
&\int_a^\infty r\,\Psi(r,\zeta)\,r^{\nu-1}\,dr = (1-\zeta^2)\,a^{\nu+1}\,\frac{\Gamma(\nu+1)}{\Gamma(\nu+3)}\,\mathfrak{G}_\nu^{\frac{3}{2}}(\zeta)\Big(a\,g_N(t) - (\nu+2)\,g_D(t)\Big) \\
&\quad - 2(1-\zeta^2)\,\frac{\Gamma(\nu+1)}{\Gamma(\nu+3)}\cos\nu\pi\,C_\nu^{\frac{3}{2}}(\zeta)\int_a^\infty \tilde{\Psi}(\tau,-1)\,\overline{R}(a;\nu)\,\frac{dr}{r^2} \\
&\quad \mathrm{Re}\,\nu < 0, \quad \nu+3 \neq -1,-2,\dots\,.
\end{aligned} \tag{4.5}$$

The inverse transform formula then implies

$$\begin{aligned}
\Psi(r,\zeta) &= \frac{1-\zeta^2}{2\pi i}\int_{c-i\infty}^{c+i\infty}\left(\frac{a}{r}\right)^{\nu+1}\frac{\Gamma(\nu+1)}{\Gamma(\nu+3)}\,\mathfrak{G}_\nu^{\frac{3}{2}}(\zeta)\Big(a\,g_N(t) - (\nu+2)\,g_D(t)\Big)d\nu \\
&- \frac{1-\zeta^2}{\pi i}\int_{c-i\infty}^{c+i\infty} r^{-\nu-1}\,\frac{\Gamma(\nu+1)}{\Gamma(\nu+3)}\cos\nu\pi\,C_\nu^{\frac{3}{2}}(\zeta)\left[\int_a^\infty \tilde{\Psi}(\tau,-1)\,\overline{R}(a;\nu)\,\frac{dr}{r^2}\right]d\nu \\
&.
\end{aligned} \tag{4.6}$$

It can be shown that [1]

$$C_{-\nu-3}^{\frac{3}{2}}(\zeta) = C_\nu^{\frac{3}{2}}(\zeta) \tag{4.7}$$

for all values of ν and

$$D_{-\nu-3}^{\frac{3}{2}}(\zeta) = D_\nu^{\frac{3}{2}}(\zeta) - \pi\cot(\pi\nu)\,C_\nu^{\frac{3}{2}}(\zeta) \tag{4.8}$$

for every $\nu \in \mathbb{C}$ except integral values of ν. Utilizing the formula which is obtained by replacing ν with $-\nu - 3$ in (4.5) where the expressions (3.3), (4.4) together with (4.7) and (4.8), are used, to eliminate the unknown boundary data from (4.5), we obtain the following equations

$$\Psi(r,\zeta) = -\frac{1-\zeta^2}{2\pi i}\int_{\mathcal{R}} \left(\frac{a}{r}\right)^{\nu+1} \frac{\Gamma(\nu+1)}{\Gamma(\nu+3)}\,(2\nu+3)\left(\mathfrak{G}_{\nu}^{\frac{3}{2}}(\zeta)\, g_{_D}(t)\right) d\nu \quad (4.9)$$

if Dirichlet boundary values are described, or

$$\Psi(r,\zeta) = a\frac{1-\zeta^2}{2\pi i}\int_{\mathcal{R}} \left(\frac{a}{r}\right)^{\nu+1} \frac{\Gamma(\nu+1)}{\Gamma(\nu+3)}\,\frac{2\nu+3}{\nu+1}\left(\mathfrak{G}_{\nu}^{\frac{3}{2}}(\zeta)\, g_{_N}(t)\right) d\nu \quad (4.10)$$

for the Neumann case.

They are derived by a deformation of the contour to the right complex ν−plane. Note that the deformed contour $\mathcal{R}$, depicted in Figure 3, begins

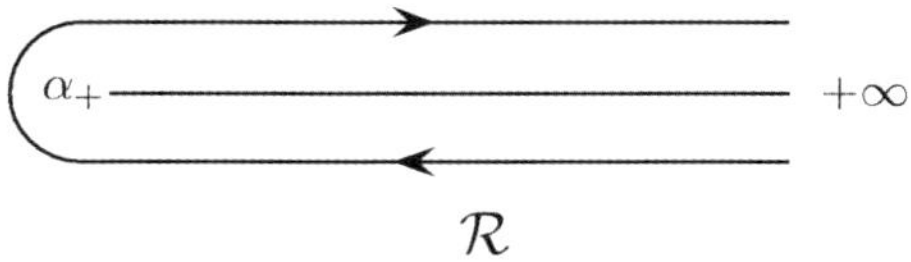

Figure 3. The contour $\mathcal{R}$.

and ends in the right complex ν−plane, such that $\mathrm{Re}\,\nu \to +\infty$ at each end.

4.2. *The half-plane* $\mathrm{Re}\,\nu \in [-1, +\infty)$

In the half-plane $\mathrm{Re}\,\nu \in [-1, +\infty)$, $\overline{q}(r, \zeta; \nu)$ remains bounded as r tends to infinity if $\overline{R}(r; \nu)$ is replaced by $r^{-\nu-1}$. Thus, following a procedure similar to the aforementioned, we find

$$\Psi(r,\zeta) = -\frac{1-\zeta^2}{2\pi i}\int_{\mathcal{L}} \left(\frac{r}{a}\right)^{\nu+2} \frac{\Gamma(\nu+1)}{\Gamma(\nu+3)}\,(2\nu+3)\left(\mathfrak{G}_{\nu}^{\frac{3}{2}}(\zeta)\, g_{_D}(t)\right) d\nu, \quad (4.11)$$

for the Dirichlet, and

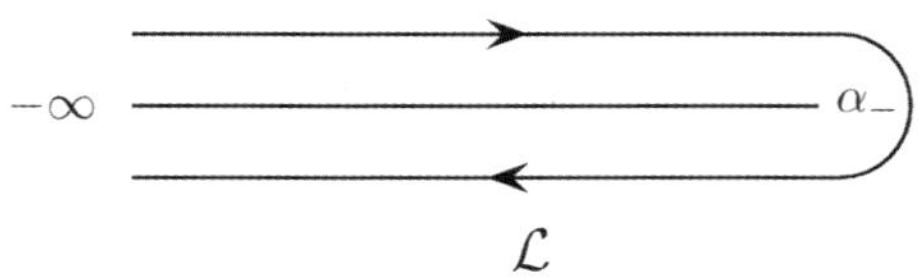

Figure 4. The contour $\mathcal{L}$.

$$\Psi(r,\zeta) = -a\frac{1-\zeta^2}{2\pi i}\int_{\mathcal{L}}\left(\frac{r}{a}\right)^{\nu+2}\frac{\Gamma(\nu+1)}{\Gamma(\nu+3)}\frac{2\nu+3}{\nu+2}\left(\mathfrak{G}_{\nu}^{\frac{3}{2}}(\zeta)\,g_N(t)\right)d\nu, \quad (4.12)$$

for Neumann case, derived by a deformation of the contour to the left complex $\nu-$plane. Note that, the deformed contour $\mathcal{L}$, depicted in Figure 4, begins and ends in the left complex $\nu-$plane, such that $\mathrm{Re}\,\nu \to -\infty$ at each end.

References

1. A. Erdélyi, editor. *Higher Transcendental Functions*, volume 1. McGraw-Hill, 1953.
2. A.S. Fokas. *A Unified Approach To Boundary Value Problems*. SIAM, Philadelphia, Pennsylvania, 1st edition, 2008.

THE NEED FOR A FIRST-ORDER QUASI LORENTZ TRANSFORMATION

Dan Censor

Department of Electrical and Computer Engineering,
Ben–Gurion University of the Negev
84105 Beer–Sheva, Israel
E-mail: censor@ee.bgu.ac.il

Solving electromagnetic scattering problems involving non-uniformly moving objects requires an approximate but consistent extension of Einstein's Special Relativity theory, which originally is valid for constant velocities only. For moderately varying velocities a quasi Lorentz transformation is presented. The conditions for form-invariance of the Maxwell equations, the so-called "principle of relativity", are shown to hold for a broad class of motional modes and time scales. An example of scattering by an harmonically oscillating mirror is given elsewhere.

1. Introduction and Rationale

Einstein's Special Relativity (SR) theory [1] is based on the invariance of c, the speed of light in free space (vacuum), for all inertial observers. The other postulate so called "principle of relativity" (PR) for the Maxwell Equations (ME), is discussed below. From the invariance of c follow [1] the spatial and temporal Lorentz transformations (LT)

$$\mathbf{r}' = \tilde{\mathbf{U}} \cdot (\mathbf{r} - \mathbf{v}t) \tag{1a}$$

$$t' = \gamma(t - \mathbf{v} \cdot \mathbf{r} / c^2) \tag{1b}$$

$$\begin{gathered} \tilde{\mathbf{U}} = \tilde{\mathbf{I}} + (\gamma - 1)\hat{\mathbf{v}}\hat{\mathbf{v}},\ \gamma = (1 - \beta^2)^{-1/2} \\ \hat{\mathbf{v}} = \mathbf{v} / v,\ \beta = v / c,\ c = (\mu_0 \varepsilon_0)^{-1/2},\ v = |\mathbf{v}| \end{gathered} \tag{1c}$$

respectively, written in a coordinate-independent form, involving inertial systems of reference $\mathbf{\Gamma}$ and $\mathbf{\Gamma}'$ and their associated space-time coordinates $\mathbf{r}, t$, and $\mathbf{r}', t'$, respectively. The corresponding inverse transformations are given by

$$\mathbf{r} = \tilde{\mathbf{U}} \cdot (\mathbf{r}' - \mathbf{v}'t'),\ \mathbf{v}' = -\mathbf{v} \tag{1a*}$$

$$t = \gamma(t' - \mathbf{v}' \cdot \mathbf{r}' / c^2) \tag{1b*}$$

Typically, inverse transformations, e.g., (1*), are derived by algebraically solving the initial formulas for the sought variables. Formally, such inverse transformations follow by interchanging primed and unprimed symbols where

$\mathbf{v}' = -\mathbf{v}$. Henceforth inverse formulas will be referred to by an asterisk on the initial equation number, even if the equation is not presented explicitly.

So far the LT (1) is restricted to constant velocities $\mathbf{v}$. If we try to formulate a systematic model for varying velocities, at least one reference frame ceases to be inertial. We cannot claim that such an approximate model is applicable to arbitrarily large velocities, therefore for varying velocities the model will be restricted to a first order approximation in the normalized velocity $\beta = v / c$. Furthermore, for the model to be simple and tractable, the effects of the varying velocities are viewed as kinematical only, i.e., no attempt will be made to incorporate the acceleration effects into the media properties and/or the ME. For sufficiently small accelerations this seems to provide a reasonable theoretical framework.

Such a model based on the d'Alembert solution of the one-dimensional wave equation has already been proposed in the literature [2]. This model was exploited for analyzing scattering by harmonically moving mirrors [3], and is therefore relevant to the present discourse. Other attempts of incorporating varying velocities, based on different modeling, have been made too, e.g., see [4] and references mentioned therein.

A very suggestive methodology for generalizing the problem to varying velocities is to approximate an inhomogeneously moving reference frame as a sequence of systems moving with different local/instantaneous velocities $\overline{\mathbf{v}} = \overline{\mathbf{v}}(\overline{\mathbf{r}}, \overline{t})$, instead of the continuously changing $\mathbf{v} = \mathbf{v}(\mathbf{r}, t)$. The velocity $\overline{\mathbf{v}}$ is assumed to be valid for a limited span of space-time in the vicinity of the world event $\overline{\mathbf{r}}, \overline{t}$. When the discrepancy becomes too large the velocity is updated. Thusly the continuous change of $\mathbf{v} = \mathbf{v}(\mathbf{r}, t)$ is replaced by a sequence of discrete constant velocities, figuratively like a sequence discrete frames in a movie. Accordingly (1) would be replaced by

$$\mathbf{r}' = \tilde{\mathbf{U}}(\overline{\mathbf{v}}) \cdot (\mathbf{r} - \overline{\mathbf{v}}t) \tag{2a}$$

$$t' = \gamma(\overline{\mathbf{v}})(t - \overline{\mathbf{v}} \cdot \mathbf{r} / c^2) \tag{2b}$$

As a multi-scale method (2) is plausible for systems where the velocity changes slowly and monotonically over space and time regions large in comparison to other system parameters. However this assumption holds only for a restricted class of configurations. It does not hold, for example, in the case of a wave emitted by an oscillating antenna. Intuitively one expects the antenna motion to cause a Doppler effect which during the mechanical cycle creates higher and lower frequencies. The results would then be akin to a frequency modulated carrier wave of frequency ω, creating sidebands of frequencies $\omega \pm n\Omega$, with Ω corresponding to the mechanical frequency. This is a result that cannot be displayed when the local instantaneous velocity concept is employed, because for each incremental change of $\overline{\mathbf{v}}$ a different Doppler frequency shift is created, and the resulting continuous spectrum does not merge into the expected discrete sideband frequencies.

Our goal is to find a first order Quasi Lorentz Transformation (QLT) that will satisfactorily deal with scattering by objects moving at varying velocities[a].

2. First Order Special Relativity; Quasi Lorentz Transformation

First order SR is derived by keeping in the LT and the ME only first order terms in the normalized dimensionless velocity β, and neglecting terms involving higher powers. Due care is called for, to guarantee that the ensuing approximations are mathematically consistent in the quantity β, prescribes for (1). This results in

$$\mathbf{r}' = \mathbf{r} - \mathbf{v}t \tag{3a}$$

$$t' = t - \mathbf{v}\cdot\mathbf{r} / c^2 \tag{3b}$$

and its associated inverse transformation

$$\mathbf{r} = \mathbf{r}' - \mathbf{v}'t' \tag{3a*}$$

$$t = t' - \mathbf{v}'\cdot\mathbf{r}' / c^2 \tag{3b*}$$

It is inconsistent to simply neglect terms involving c^2, arguing that these are negligible for low velocities. This becomes clear upon rewriting (3) in terms of normalized time ct and dimensionless β, $\hat{\mathbf{v}}$

$$\mathbf{r}' = \mathbf{r} - \beta\hat{\mathbf{v}}ct \tag{4a}$$

$$ct' = ct - \beta\hat{\mathbf{v}}\cdot\mathbf{r} \tag{4b}$$

showing that a consistent approximation leads to (3). Consequently the Galilean transformation (GT)

$$\mathbf{r}' = \mathbf{r} - \mathbf{v}t \tag{5a}$$

$$t' = t \tag{5b}$$

is *cannot* be considered as the limiting case of the LT (1) for small β, it rather constitutes a limiting case for $c \to \infty$ [5]. For finite c the ME predict the existence of electromagnetic waves propagating at a finite speed, thus encompassing lightwave phenomena which are supported by numerous empirical observations. From this point of view, the GT is incompatible with the ME and inadequate for describing global space-time phenomena.

The differential representations

$$d\mathbf{r}' = \tilde{\mathbf{U}}\cdot(d\mathbf{r} - \mathbf{v}dt) \tag{6a}$$

$$dt' = \gamma(dt - \mathbf{v}\cdot d\mathbf{r} / c^2) \tag{6b}$$

differ from (1), (2), only within a trivial constant of integration. Similarly the first order representations

---

[a] See also http://www.ee.bgu.ac.il/~censor/quasi-lorentz.pdf

$$d\mathbf{r}' = d\mathbf{r} - \mathbf{v}dt \tag{7a}$$

$$dt' = dt - \mathbf{v} \cdot d\mathbf{r} / c^2 \tag{7b}$$

can replace (3) for all purposes.

For varying velocities we can rewrite (7), using a space-time dependent velocity $\mathbf{v}(\mathbf{r}, t)$, without incurring the inconsistencies inherent in (3) when $\mathbf{v}$ is arbitrarily replaced by $\mathbf{v}(\mathbf{r}, t)$. Thus the differential first order QLT is *postulated* as

$$d\mathbf{r}' = d\mathbf{r} - \mathbf{v}(\mathbf{r}, t)dt \tag{8a}$$

$$dt' = dt - \mathbf{v}(\mathbf{r}, t) \cdot d\mathbf{r} / c^2 \tag{8b}$$

and its corresponding inverse formulas

$$d\mathbf{r} = d\mathbf{r}' - \mathbf{v}'(\mathbf{r}', t')dt' \tag{8a*}$$

$$dt = dt' - \mathbf{v}'(\mathbf{r}', t') \cdot d\mathbf{r}' / c^2 \tag{8b*}$$

In computing the inverse transformations (8*), to the first order in β, it is noted that $\mathbf{v}$ is already of first order in the velocity hence

$$\mathbf{v}'(\mathbf{r}', t') = -\mathbf{v}(\mathbf{r}, t) \tag{9}$$

Integration of the local differential expressions (8) leads to global space-time formulas. From (8a) we obtain for the space transformation

$$\mathbf{r}' = \mathbf{r} - \int_{t_0}^{t} \mathbf{v}(\mathbf{r}, \underset{\sim}{t})d\underset{\sim}{t} \tag{10a}$$

with $\underset{\sim}{t}$ denoting the integration variable. Differentiating (10a) with respect to t, invoking the Leibniz rule for differentiation of an integral, leads back to (8a). Integrating the corresponding temporal transformation (8b) yields

$$t' = t - c^{-2} \int_{\mathbf{r}_0}^{\mathbf{r}} \mathbf{v}(\underset{\sim}{\mathbf{r}}, t) \cdot d\underset{\sim}{\mathbf{r}} \tag{10b}$$

with the integration variable denoted by $\underset{\sim}{\mathbf{r}}$.

Integrating (8*) and incorporating (9) yields the inverse of (10), given explicitly by

$$\mathbf{r} = \mathbf{r}' - \int_{t_0'}^{t'} \mathbf{v}'(\mathbf{r}', \underset{\sim}{t}')d\underset{\sim}{t}' \tag{10a*}$$

$$t = t' - c^{-2} \int_{\mathbf{r}_0'}^{\mathbf{r}'} \mathbf{v}'(\mathbf{r}', \underset{\sim}{t}') \cdot d\underset{\sim}{\mathbf{r}}' \tag{10b*}$$

The integral forms are only valid when it is possible to return to the initial differential formula. This is feasible only when the line integrals in (10b), (10b*) are determined by the integration end points only and independent of the integration path. Otherwise the Leibniz rule for differentiation of integrals is not applicable. Hence the velocity field must be a conservative, satisfying, respectively

$$\partial_{\mathbf{r}} \times \mathbf{v}(\mathbf{r}, t) = 0 \tag{11a}$$

$$\partial_{\mathbf{r}'} \times \mathbf{v}'(\mathbf{r}', t') = 0 \tag{11b}$$

A less restrictive case is afforded by time dependent velocities $\mathbf{v}(t)$, spatially independent of the position vectors $\mathbf{r}$ or $\mathbf{r}'$. For this case (10a) remains unchanged, but (10b) trivially integrates into

$$t' = t - \mathbf{v}(t) \cdot \mathbf{r} / c^2 \tag{10c}$$

$$t = t' - \mathbf{v}'(t') \cdot \mathbf{r}' / c^2 \tag{10c*}$$

similar to (3b) for constant $\mathbf{v}$, and the obvious inverse (10c*) follows. With (11) automatically satisfied, $\mathbf{v}(t)$ is not restricted any more to laminar flow. This is very important for special cases like orbiting scatterers.

3. The Principle of Relativity

Einstein [1] postulated the PR, namely the form-invariance of the ME. For simplicity we confine the discussion to source-free regions. Thus in in $\mathbf{\Gamma}$ the ME are given by

$$\partial_{\mathbf{r}} \times \mathbf{E} = -\partial_t \mathbf{B} \tag{12a}$$

$$\partial_{\mathbf{r}} \times \mathbf{H} = \partial_t \mathbf{D} \tag{12b}$$

$$\partial_{\mathbf{r}} \cdot \mathbf{D} = 0 \tag{12c}$$

$$\partial_{\mathbf{r}} \cdot \mathbf{B} = 0 \tag{12d}$$

where $\partial_{\mathbf{r}}$ (the "Nabla" operator ∇) and ∂_t denote the space and time differential operators, respectively. In general all the fields are space and time dependent, e.g., $\mathbf{E} = \mathbf{E}(\mathbf{r}, t)$.

According to the PR, in another inertial system $\mathbf{\Gamma}'$, the form-invariance of the ME (12) is preserved, thusly

$$\partial_{\mathbf{r}'} \times \mathbf{E}' = -\partial_{t'} \mathbf{B}' \tag{12a*}$$

$$\partial_{\mathbf{r}'} \times \mathbf{H}' = \partial_{t'} \mathbf{D}' \tag{12b*}$$

$$\partial_{\mathbf{r}'} \cdot \mathbf{D}' = 0 \tag{12c*}$$

$$\partial_{\mathbf{r}'} \cdot \mathbf{B}' = 0 \tag{12d*}$$

where $\mathbf{E}' = \mathbf{E}'(\mathbf{r}', t')$ etc., in terms of the appropriate native coordinates in $\mathbf{\Gamma}'$.

The PR *statement* associating (12), (12*), is incomplete unless the LT relating the fields is provided, facilitating the derivation of the field transformation (FT) formulas. A full SR derivation is given for example by Kong [5]. Here we need the FT only to the first order in β, which is simpler to derive. The chain rule of calculus

$$\partial_{\mathbf{r}'} = (\partial_{\mathbf{r}'} \mathbf{r}) \cdot \partial_{\mathbf{r}} + (\partial_{\mathbf{r}'} t) \partial_t \tag{13a}$$

$$\partial_{t'} = (\partial_{t'} t)\partial_t + (\partial_{t'}\mathbf{r})\cdot\partial_{\mathbf{r}} \tag{13b}$$

is applied to (1) specifying a LT for the differential operators

$$\partial_{\mathbf{r}'} = \tilde{\mathbf{U}}\cdot(\partial_{\mathbf{r}} + \mathbf{v}\partial_t / c^2) \tag{14a}$$

$$\partial_{t'} = \gamma(\partial_t + \mathbf{v}\cdot\partial_{\mathbf{r}}) \tag{14b}$$

The approximation to first order in β follows as

$$\partial_{\mathbf{r}'} = \partial_{\mathbf{r}} + \mathbf{v}\partial_t / c^2 \tag{15a}$$

$$\partial_{t'} = \partial_t + \mathbf{v}\cdot\partial_{\mathbf{r}} \tag{15b}$$

The corresponding differential operators for the GT are obtained by applying (13) to (5), or equivalently, by taking the limit $c \to \infty$ in (15).

Substituting (14) in (12*) and collecting terms, (12) is obtained subject to

$$\mathbf{E}' = \tilde{\mathbf{V}}\cdot(\mathbf{E} + \mathbf{v}\times\mathbf{B}) \tag{16a}$$

$$\mathbf{B}' = \tilde{\mathbf{V}}\cdot(\mathbf{B} - \mathbf{v}\times\mathbf{E} / c^2) \tag{16b}$$

$$\mathbf{D}' = \tilde{\mathbf{V}}\cdot(\mathbf{D} + \mathbf{v}\times\mathbf{H} / c^2) \tag{16c}$$

$$\mathbf{H}' = \tilde{\mathbf{V}}\cdot(\mathbf{H} - \mathbf{v}\times\mathbf{D}) \tag{16d}$$

$$\tilde{\mathbf{V}} = \gamma\tilde{\mathbf{I}} + (1-\gamma)\hat{\mathbf{v}}\hat{\mathbf{v}} \tag{16e}$$

The GT analog of (16) is obtained by effecting the limit $c \to \infty$

$$\mathbf{E}' = \mathbf{E} + \mathbf{v}\times\mathbf{B} \tag{17a}$$

$$\mathbf{B}' = \mathbf{B} \tag{17b}$$

$$\mathbf{D}' = \mathbf{D} \tag{17c}$$

$$\mathbf{H}' = \mathbf{H} - \mathbf{v}\times\mathbf{D} \tag{17d}$$

In free space (vacuum), in both $\mathbf{\Gamma}$, $\mathbf{\Gamma}'$, the constitutive relations are given by

$$\mathbf{D} = \varepsilon_0\mathbf{E} \tag{18a}$$

$$\mathbf{B} = \mu_0\mathbf{H} \tag{18b}$$

$$\mathbf{D}' = \varepsilon_0\mathbf{E}' \tag{18c}$$

$$\mathbf{B}' = \mu_0\mathbf{H}' \tag{18d}$$

Substitution of (18c), (18d), in (16) yields (18a), (18b), demonstrating that in free space the same constitutive relations apply to all inertial observers.

On the other hand, substitution of (18c), (18d), into (17) implies in $\mathbf{\Gamma}$

$$\mathbf{D} = \varepsilon_0(\mathbf{E} + \mathbf{v}\times\mathbf{B}) \tag{19a}$$

$$\mathbf{B} = \mu_0(\mathbf{H} - \mathbf{v}\times\mathbf{D}) \tag{19b}$$

amounting to a statement that while in $\mathbf{\Gamma}'$, *in free space*, the constitutive relations (18c), (18d), are velocity independent, in all other inertial, moving with relative velocity $\mathbf{v}$, the constitutive relations (19) do depend on $\mathbf{v}$. This sets aside $\mathbf{\Gamma}'$ as a preferred reference system, and allows observers in all other reference systems $\mathbf{\Gamma}$ to compute their absolute velocity relative to $\mathbf{\Gamma}'$ solely by measuring fields in the system $\mathbf{\Gamma}$ where they reside. Obviously (19) takes us back to Newtonian absolutism and the luminiferous aether theory, nowadays superseded by SR. See related work [6, 7].

Keeping in (16) only first order β terms yields

$$\mathbf{E}' = \mathbf{E} + \mathbf{v} \times \mathbf{B} \tag{20a}$$

$$\mathbf{B}' = \mathbf{B} - \mathbf{v} \times \mathbf{E} / c^2 \tag{20b}$$

$$\mathbf{D}' = \mathbf{D} + \mathbf{v} \times \mathbf{H} / c^2 \tag{20c}$$

$$\mathbf{H}' = \mathbf{H} - \mathbf{v} \times \mathbf{D} \tag{20d}$$

In (16), (20), the primed, unprimed, fields are functions of the appropriate native coordinates $\mathbf{r}', t'$, and $\mathbf{r}, t$, respectively. Because both coordinate sets appear in the same formula, the LT is required to mediate between the coordinates.

4. Maxwell Equations Form-Invariance for Varying Velocities

So far in (14)-(20), and the corresponding inverse expressions, constant $\mathbf{v}$ was assumed. Consider now the application of the chain rule (13) to (10*). With the proviso that (11) is satisfied, the line integral (10b*) is independent of the path, hence the Leibniz rule for differentiation of integrals applies. Consequently we now have (cf. (15))

$$\partial_{\mathbf{r}'} = \partial_{\mathbf{r}} + \mathbf{v}(\mathbf{r}, t)\partial_t / c^2 \tag{21a}$$

$$\partial_{t'} = \partial_t + \mathbf{v}(\mathbf{r}, t) \cdot \partial_{\mathbf{r}} \tag{21b}$$

Application of (21) to (12d*) yields

$$(\partial_{\mathbf{r}} + \mathbf{v}(\mathbf{r}, t)\partial_t / c^2) \cdot \mathbf{B}' = \partial_{\mathbf{r}} \cdot \mathbf{B}' + \mathbf{v}(\mathbf{r}, t) \cdot \partial_t \mathbf{B}' / c^2 = 0 \tag{22a}$$

Anticipating (12d), recast (22a) as

$$\partial_{\mathbf{r}} \cdot (\mathbf{B}' - \mathbf{B}) + \mathbf{v}(\mathbf{r}, t) \cdot \partial_t \mathbf{B}' / c^2 = 0 \tag{22b}$$

The last term (22b) is already of order β, hence the factor $\partial_t \mathbf{B}'$ can be approximated to order zero by $\partial_t \mathbf{B}$. Substituting from (12a) yields

$$\partial_{\mathbf{r}} \cdot (\mathbf{B}' - \mathbf{B}) - \mathbf{v}(\mathbf{r}, t) \cdot \partial_{\mathbf{r}} \times \mathbf{E} / c^2 = 0 \tag{22c}$$

or, using the vector formula for $\partial_{\mathbf{r}} \cdot (\mathbf{v} \times \mathbf{E})$

$$\partial_{\mathbf{r}} \cdot (\mathbf{B}' - \mathbf{B} + \mathbf{v}(\mathbf{r}, t) \times \mathbf{E} / c^2) - \mathbf{E} \cdot \partial_{\mathbf{r}} \times \mathbf{v}(\mathbf{r}, t) / c^2 = 0 \tag{22d}$$

Consequently (22d) subject to (11) reduces to the counterpart of (20b)

$$\mathbf{B}' = \mathbf{B} - \mathbf{v}(\mathbf{r}, t) \times \mathbf{E} / c^2 \tag{22e}$$

with the associated inverse

$$\mathbf{B} = \mathbf{B}' - \mathbf{v}'(\mathbf{r}', t') \times \mathbf{E}' / c^2 \tag{22e*}$$

We conclude that subject to (11) and (22e) or (22e*) the PR for the ME is valid for the pair (12d) and (12d*).

The analog of (22d) for (12c), (12c*), is

$$\partial_{\mathbf{r}} \cdot (\mathbf{D}' - \mathbf{D} - \mathbf{v}(\mathbf{r}, t) \times \mathbf{H} / c^2) + \mathbf{H} \cdot \partial_{\mathbf{r}} \times \mathbf{v}(\mathbf{r}, t) / c^2 = 0 \tag{23a}$$

Subject to (11) the analog of (20c) follows as

$$\mathbf{D}' = \mathbf{D} + \mathbf{v}(\mathbf{r}, t) \times \mathbf{H} / c^2 \tag{23b}$$

and the associated inverse is

$$\mathbf{D} = \mathbf{D}' + \mathbf{v}'(\mathbf{r}', t') \times \mathbf{H}' / c^2 \tag{23b*}$$

Hence subject to (11) and (23b) or (23b*) the PR for the ME is valid for the pair (12c) and (12c*).

We turn now to the vector ME. Application of (15) to (12a*) yields

$$\partial_{\mathbf{r}} \times \mathbf{E}' + \mathbf{v}(\mathbf{r}, t) \times \partial_t \mathbf{E}' / c^2 + \partial_t \mathbf{B}' + \mathbf{v}(\mathbf{r}, t) \cdot \partial_{\mathbf{r}} \mathbf{B}' = 0 \tag{24a}$$

Tentatively assuming (12a) and subtracting from (24a) yields

$$\partial_{\mathbf{r}} \times (\mathbf{E}' - \mathbf{E}) + \mathbf{v}(\mathbf{r}, t) \times \partial_t \mathbf{E} / c^2 + \partial_t (\mathbf{B}' - \mathbf{B}) + \mathbf{v}(\mathbf{r}, t) \cdot \partial_{\mathbf{r}} \mathbf{B} = 0 \tag{24b}$$

where the primes in the first order velocity terms (24b) have been dropped. Substituting for $\mathbf{B}' - \mathbf{B}$ from (22e) and using the rule for the derivative of a product leads to

$$\partial_{\mathbf{r}} \times (\mathbf{E}' - \mathbf{E}) + \mathbf{E} \times \partial_t \mathbf{v}(\mathbf{r}, t) / c^2 + \mathbf{v}(\mathbf{r}, t) \cdot \partial_{\mathbf{r}} \mathbf{B} = 0 \tag{24c}$$

Exploiting (12d) and the formula for $\partial_{\mathbf{r}} \times (\mathbf{v} \times \mathbf{B})$, (24c) is recast in the form

$$\begin{aligned} &\partial_{\mathbf{r}} \times (\mathbf{E}' - \mathbf{E} - \mathbf{v}(\mathbf{r}, t) \times \mathbf{B}) + \mathbf{E} \times \partial_t \mathbf{v}(\mathbf{r}, t) / c^2 \\ &+\mathbf{B} \cdot \partial_{\mathbf{r}} \mathbf{v}(\mathbf{r}, t) - \mathbf{B} \partial_{\mathbf{r}} \cdot \mathbf{v}(\mathbf{r}, t) = 0 \end{aligned} \tag{24d}$$

For constant $\mathbf{v}$ only the first term in (24d) is retained, reducing to (20a) as expected. For varying velocity the last three terms in (24d) must vanish in order to obtain the analog of (20a)

$$\mathbf{E}' = \mathbf{E} + \mathbf{v}(\mathbf{r}, t) \times \mathbf{B} \tag{24e}$$

with the associated inverse

$$\mathbf{E} = \mathbf{E}' + \mathbf{v}'(\mathbf{r}', t') \times \mathbf{B}' \tag{24e*}$$

Whether or not the transition from (24d) to (24e) is justified depends on the velocity field in question. For time dependent velocities $\mathbf{v}(t)$ (24d) becomes

$$\partial_{\mathbf{r}} \times (\mathbf{E}' - \mathbf{E} - \mathbf{v}(t) \times \mathbf{B}) + \mathbf{E} \times \partial_t \mathbf{v}(t) / c^2 = 0 \tag{24f}$$

For cases where the velocity variation is small compared to some representative frequency ω, the criterion for the validity of (24e) becomes

$$\partial_t \mathbf{v} \ll \omega \mathbf{v} \tag{25}$$

The discussion is now repeated for the ME (12b), (12b*). The counterpart of (24c) is now

$$\partial_{\mathbf{r}} \times (\mathbf{H}' - \mathbf{H}) + \mathbf{H} \times \partial_t (\mathbf{v}(\mathbf{r}, t) / c^2 - \mathbf{v}(\mathbf{r}, t) \cdot \partial_{\mathbf{r}} \mathbf{D} = 0 \tag{26a}$$

and like (24d), is recast as

$$\begin{aligned} &\partial_{\mathbf{r}} \times (\mathbf{H}' - \mathbf{H} + \mathbf{v}(\mathbf{r}, t) \times \mathbf{D}) + \mathbf{H} \times \partial_t (\mathbf{v}(\mathbf{r}, t) / c^2 \\ &-\mathbf{D} \cdot \partial_{\mathbf{r}} \mathbf{v}(\mathbf{r}, t) + \mathbf{D} \partial_{\mathbf{r}} \cdot \mathbf{v} = 0 \end{aligned} \tag{26b}$$

Once again, for constant $\mathbf{v}$ (26b) reduces to (20d), as expected. For time dependent velocities, if the last three terms in (26b) can be neglected, we obtain the analog of (20d)

$$\mathbf{H}' = \mathbf{H} - \mathbf{v}(\mathbf{r}, t) \times \mathbf{D} \tag{26c}$$

and the associated inverse is

$$\mathbf{H} = \mathbf{H}' - \mathbf{v}'(\mathbf{r}', t') \times \mathbf{D}' \tag{26c*}$$

For time dependent velocities $\mathbf{v}(t)$ (26b) becomes to the analog of (24f)

$$\partial_{\mathbf{r}} \times (\mathbf{H}' - \mathbf{H} + \mathbf{v}(t) \times \mathbf{D}) + \mathbf{H} \times \partial_t \mathbf{v}(t) / c^2 = 0 \tag{26d}$$

and the criterion (25) follows.

In free space (vacuum) the set (22e), (23b), (24e), (26c), or their inverses satisfy same constitutive relations (18).

References

1. A. Einstein, "Zur Elektrodynamik bewegter Körper", *Ann. Phys. (Lpz.)*, Vol. 17, 891-921, 1905; English translation: "On the Electrodynamics of moving bodies", *The Principle of Relativity*, Dover
2. D. Censor, "The generalized Doppler effect and applications", Journal of the Franklin Institute, vol. 295, pp. 103 - 115, 1973.
3. J. Van Bladel, *Relativity and Engineering*, Springer, 1984.
4. D. Censor, "Non-relativistic scattering: pulsating interfaces", PIER—Progress In Electromagnetic Research, vol. 54, pp. 263-281, 2005.
5. J.A. Kong, Electromagnetic Wave Theory, Wiley, 1986.
6. T. Shiozawa, "Phenomenological and Electron-Theoretical Study of the Electrodynamics of Rotating Systems", Proc. IEEE, vol. 61, pp. 1694-1702, 1973.
7. P. Hillion, "Maxwell's equations and accelerated frames". Phys. Rev. E, vol. 57, pp. 7239-7243, 1998.

COMPARING TWO BILINEAR EQUIVALENT MATRIX PENCILS *

A. A. PANTELOUS †

Department of Mathematical Sciences, University of Liverpool,
Mathematics and Oceanography Building, Peach Str.
L69 7ZL, Liverpool, U.K.
E-mail: A.Pantelous@liverpool.ac.uk

A. D. KARAGEORGOS AND G. I. KALOGEROPOULOS

Department of Mathematics, University of Athens,
Panepistimiopoulis, GR 15784, Greece
E-mail: athkar, gkaloger@math.uoa.gr

In this paper, an algorithmic technique for investigating the necessary and sufficient conditions for two homogeneous matrix pencils $sF_1 - \hat{s}G_1$ and $sF_2 - \hat{s}G_2$ to be bilinear equivalent is provided. The proposed problem is very interesting since the applications are merely numerous.

Keywords: Matrix Pencil; Bilinear Equivalence; Linear Systems.
AMS (classification): 15A22; 34D30; 14N05; 93D15

1. Introduction and Preliminaries

Properties of matrix pencils are studied for a long time and are related to a wide spectrum of different areas of system science and mathematical control theory. The pencils $sF - G$ and $F - \hat{s}G$, where $s, \hat{s} \in \mathbb{R}$ and $F, G \in \mathbb{F}^{m \times n}$ ($\mathbb{R}$ or $\mathbb{C}$) are related by the homogeneous pencil $sF - \hat{s}G$ (or equivalently by a pair $(F, -G)$), since the existence of the important notion of duality, the so-called *elementary divisor type duality or integrator-differentiator type duality*, has already been demonstrated, see [8] and [9].
We denote $(F, G) \in \mathbb{F}^{m \times n} \times \mathbb{F}^{m \times n}$, and $(s, \hat{s})$ to be a pair of indeterminates. The polynomial matrix $sF - \hat{s}G \in \mathbb{F}^{m \times n}[s, \hat{s}]$ is defined as the homogeneous matrix pencil of the pair (F, G). Clearly, $sF - \hat{s}G$ is a matrix over the ring

*This work is supported by etc, etc.
†corresponding author

$\mathbb{F}[s, \hat{s}]$. However, it may also be viewed as a matrix over $\mathbb{F}(s)[\hat{s}]$ or $\mathbb{F}(\hat{s})[s]$. Quite recently, in [6], for a given homogeneous pencil which is unstable, in the sense that it has roots in the closed right half-plane, a bilinear transformation is determined such that the new homogeneous bilinear-equivalent matrix pencil is stable. There, the notion of bilinear equivalence is introduced and a stabilization criterion of a homogeneous matrix pencil is finally derived.

In the present paper, we study and provide an algorithmic technique for investigating the necessary and sufficient conditions to be two matrix pencils bilinear equivalent. This problem is very interesting and many applications can be derived.

For instance, in the literature of economic modelling, one of the most famous generalized dynamic model is the Leontief's Multi Input-Multi Output (MIMO) financial model where the matrix of state vector represents the amount of the ith-commodity that j^{th}-sector has to produce, see for instance [1], [11]-[12], [14]-[15], [17]-[18], [24]-[25]. For the accuracy of our references, we would like to stress out that there are also some other macroeconomic models similar to Leontief, see characteristically [13] and [16].

In order to evaluate further the importance of our aspect, recently, a typical portfolio of different insurance products and the pricing process using the framework of a generalized linear discrete time model is investigated by [5] and [20]. Note that due to regulatory constraints, the resulting system is appeared to be singular (in the sense of system science). Furthermore, claims may be regarded as the input, the accumulated surplus as the state and the gross premiums as the output vector-variable of the designed system. Similarly to Leontief' model, as well as in many other applications, the matrix of state vector includes several important parameters (and not just numbers), i.e. the expense factor for the ith-insurance product; the interaction factor, which is the proportion of accumulated surplus of the ith-product transferred to the jth-insurance product and the profit sharing factor (feedback factor) for the ith-insurance product, which includes premium repayments and determines the percentage of accumulated surplus repaid to the policyholders.

Consequently, we are highly interested to investigate if some of the state vector parameters are changed, when the relative matrix pencil stays bilinear equivalent or not. Through this equivalence, we can affect even our managerial policy. For instance, by changing bilinear equivalently the expense factor, see previous application, we may keep our system unchanged. Now, we transfer the entire practical problems mentioned above into a mathematical framework. Thus, the study of them may be reduced to the study of linear differential (or difference) equations

$$F_1\underline{\dot{x}}(t) = G_1\underline{x}(t) \ (or \ F_1\underline{x}_{k+1} = G_1\underline{x}_k), \tag{1}$$

where $\underline{x} \in \mathbb{F}^n$ is the state vector and $F_1, G_1 \in \mathbb{F}^{m\times n}$ (mainly $\mathbb{F} = \mathbb{R}$) are its matrices. Consequently, if some of the parameters of the state matrices are changed, another system is finally constructed, i.e.

$$F_2\underline{\dot{x}}(t) = G_2\underline{x}(t) \ (or \ F_2\underline{x}_{k+1} = G_2\underline{x}_k), \tag{2}$$

where $F_2, G_2 \in \mathbb{F}^{m\times n}$ (mainly $\mathbb{F} = \mathbb{R}$). In those cases, the matrix pencil theory becomes a key tool for the solution of (1) and (2). Moreover, since the relative matrix pencils $sF_1 - G_1$ and $sF_2 - G_2$ for system (1) and (2) respectively, should be constructed, an intuitive question about their bilinear equivalence is derived. In the present paper, this problem is under consideration. However, in order to go a step further, the following preliminary results are presented; see for more details [23], [4] and [6].

We denote the set of ordered pairs of $m \times n$-matrices

$$\mathcal{L} \triangleq \left\{L : L = (F, -G)\,; \ F, G \in \mathbb{F}^{m\times n}\right\},$$

and the set of ordered pairs of indeterminates $\Theta \triangleq \{\theta : \theta = (s, \hat{s})\}$.

For every $L = (F, -G) \in \mathcal{L}$ and $\theta = (s, \hat{s}) \in \Theta$, the matrix $[L] = \begin{bmatrix} F & -G \end{bmatrix} \in \mathbb{F}^{m\times 2n}$ is called a *matrix representation* of L and the homogeneous polynomial matrix

$$L_\theta \triangleq L(s, \hat{s}) = \begin{bmatrix} F & -G \end{bmatrix} \begin{bmatrix} sI_n \\ \hat{s}I_n \end{bmatrix} = sF - \hat{s}G, \tag{3}$$

is referred as the θ-matrix pencil of L.

For the remainder, it is also necessary to define the following sets

$$\mathcal{L}_\theta \triangleq \{L_\theta : \ for \ a \ fixed \ \theta = (s, \hat{s}) \in \Theta \ and \ for \ every \ L = (F, -G) \in \mathcal{L}\},$$

$$\mathcal{K} \triangleq \left\{k : k = (M, N)\,; M \in \mathbb{F}^{m\times m}, N \in \mathbb{F}^{2n\times 2n}; \ \det M, \det N \neq 0\right\},$$

and a composition rule $(*)$ defined on $\mathcal{K}$ as follows,

$$* : \mathcal{K}\times\mathcal{K} \to \mathcal{K} : \ for \ every \ k_1 = (M_1, N_1) \in \mathcal{K}, and \ k_2 = (M_2, N_2) \in \mathcal{K}, \ then$$

$$k_1 * k_2 \triangleq (M_1, N_1) * (M_2, N_2) = (M_1M_2, N_2N_1) \tag{4}$$

The action $\circ$ of $\mathcal{K}$ on $\mathcal{L}$ is defined by

$$* : \mathcal{K} \times \mathcal{L} \to \mathcal{L} : \ for \ every \ k_1 = (M, N) \in \mathcal{K}, and \ L = (F, -G) \in \mathcal{L}, \ then$$

$$k \circ L \triangleq k \circ (F, -G) = L' = (F', -G') \in \mathcal{L} : [L'] = M\,[L]\,N \tag{5}$$

or equivalently, by

$$r\left[\, F' \; -G' \,\right] \triangleq M \left[\, F \; -G \,\right] N. \tag{6}$$

Thus, the action defines an equivalence relation $\mathcal{E}_{\mathcal{K}}$ on $\mathcal{L}$, and $\mathcal{E}_{\mathcal{K}}(L)$ denotes the equivalence class or orbit of $L \in \mathcal{L}$ under $\mathcal{K}$.
(***Bilinear Equivalence***): The subgroup $(\mathcal{B}, *)$ of $(\mathcal{K}, *)$, where

$$\mathcal{B} \triangleq \left\{ b : b = \left(I_m, \begin{bmatrix} \alpha I_n & \beta I_n \\ \gamma I_n & \delta I_n \end{bmatrix} \right) = (I_m, T_d)\,, d = \begin{bmatrix} \alpha & \beta \\ \gamma & \delta \end{bmatrix} \in \mathbb{R}^{2\times 2};\ \det d \neq 0 \right\} \tag{7}$$

is called the *Bilinear Equivalence Group* (BEG), see [10]. Every $b \in \mathcal{B}$ is generated by a projective transformation $d = \begin{bmatrix} \alpha & \beta \\ \gamma & \delta \end{bmatrix} \in \mathbb{R}^{2\times 2}$. The action of $\mathcal{B}$ on $\mathcal{L}$ is defined by

$$b \circ L \triangleq b \circ (F, -G) = L' = (F', -G') \in \mathcal{L} : [L'] = I_m\,[L]\,T_d \tag{8}$$

or equivalently, by

$$\left[\, F' \; -G' \,\right] \triangleq I_m \left[\, F \; -G \,\right] \begin{bmatrix} \alpha I_n & \beta I_n \\ \gamma I_n & \delta I_n \end{bmatrix} = \left[\, \alpha F - \gamma G \;\; \beta F - \delta G \,\right]. \tag{9}$$

The equivalence relation $\mathcal{E}_{\mathcal{B}}$ defined by (8) and (9) on $\mathcal{L}$ is called the *Bilinear Equivalence* (BE). Two pencils $L_\theta^1 = sF_1 - \hat{s}G_1 \in \mathcal{L}_\theta$, $L_{\theta'}^2 = \lambda F_2 - \hat{\lambda} G_2 \in \mathcal{L}_{\theta'}$, $\theta = (s, \hat{s}) \in \Theta$ and $\theta' = \left(\lambda, \hat{\lambda}\right) \in \Theta$ are said to be bilinearly equivalent, i.e. $L_\theta^1 \mathcal{E}_{\mathcal{H}} L_\theta^2$, if and only if there exists a transformation $d : (s, \hat{s}) \to \left(\lambda, \hat{\lambda}\right)$, thus a $b \in \mathcal{B}$ generated by d, such that $(F_2, -G_2) = b \circ (F_1, -G_1)$. By $\mathcal{E}_{\mathcal{B}}(F, G)$ is denoted the BE class or orbit of $L_\theta = sF - \hat{s}G$ or equivalently of $L = (F, -G)$.
The action of $b = (I_m, T_d)$ on $L_\theta = sF - \hat{s}G$ may be interpreted as

$$\begin{aligned} b \circ L(s, \hat{s}) &= I_m \left[\, F \; -G \,\right] \begin{bmatrix} \alpha I_n & \beta I_n \\ \gamma I_n & \delta I_n \end{bmatrix} \begin{bmatrix} \lambda I_n \\ \hat{\lambda} I_n \end{bmatrix} \\ &= \lambda(\alpha F - \gamma G) - \hat{\lambda}(-\beta F + \delta G) = \lambda F' - \hat{\lambda} G' = L'_{\theta'}, \end{aligned} \tag{10}$$

which clearly expresses a change in the indeterminates from $(s, \hat{s}) \xrightarrow{d} \left(\lambda, \hat{\lambda}\right)$ by the bilinear transformation

$$d : \begin{bmatrix} s \\ \hat{s} \end{bmatrix} = \begin{bmatrix} \alpha & \beta \\ \gamma & \delta \end{bmatrix} \begin{bmatrix} \lambda \\ \hat{\lambda} \end{bmatrix}, \tag{11}$$

where $\alpha, \beta, \gamma, \delta \in \mathbb{R}$ and $\det d = \alpha\delta - \beta\gamma \neq 0$.
It is worth noting that the bilinear transformation expresses a coordinate type transformation in the indeterminates, but not in the pair $(F, -G)$.

2. Main results

In this section, the bilinear equivalence of two matrix pencils

$$L_\theta^1 = sF_1 - \hat{s}G_1 \in \mathcal{L}_\theta \tag{12}$$

and

$$L_{\theta'}^2 = \lambda F_2 - \hat{\lambda}G_2 \in \mathcal{L}_{\theta'}, \tag{13}$$

where $\theta = (s, \hat{s}) \in \Theta$, $F_1, G_1 \in \mathbb{R}^{m\times n}$ and $\theta' = \left(\lambda, \hat{\lambda}\right) \in \Theta$, $F_2, G_2 \in \mathbb{R}^{m\times n}$ respectively, are investigated. By considering the expression (9), the following system is derived

$$\begin{bmatrix} F_2 & -G_2 \end{bmatrix} \triangleq I_m \begin{bmatrix} F_1 & -G_1 \end{bmatrix} \begin{bmatrix} \alpha I_n & \beta I_n \\ \gamma I_n & \delta I_n \end{bmatrix} = \begin{bmatrix} \alpha F_1 - \gamma G_1 & \beta F_1 - \delta G_1 \end{bmatrix}. \tag{14}$$

In what it follows, the following definitions are required. Analytically, for matrix $A = \begin{bmatrix} \underline{a}_1^t \\ \vdots \\ \underline{a}_m^t \end{bmatrix} \in \mathbb{R}^{m\times n}$ where $\underline{a}_i^t \in \mathbb{R}^{1\times n}$ is the ith row vector, we define $vec(A)$ as follows,

$$vec(A) \triangleq \begin{bmatrix} \underline{a}_1^t & \cdots & \underline{a}_m^t \end{bmatrix}^t \in \mathbb{R}^{nm} \tag{15}$$

Consequently, since $F_i = \begin{bmatrix} \left(\underline{f}_1^i\right)^t \\ \vdots \\ \left(\underline{f}_m^i\right)^t \end{bmatrix}$ where $\left(\underline{f}_j^i\right)^t \in \mathbb{R}^{1\times n}$ is the jth row vector of matrix F_i, $i = 1, 2$, and $G_i \begin{bmatrix} \left(\underline{g}_1^i\right)^t \\ \vdots \\ \left(\underline{g}_m^i\right)^t \end{bmatrix}$ where $\left(\underline{g}_j^i\right)^t \in \mathbb{R}^{1\times n}$ is the jth row vector of matrix G_i, $i = 1, 2$. The following vectors are derived, and

$$vec(F_i) \triangleq \left[\left(\underline{f}_1^i\right)^t \cdots \left(\underline{f}_m^i\right)^t \right]^t \text{ and } vec(G_i) \triangleq \left[\left(\underline{g}_1^i\right)^t \cdots \left(\underline{g}_m^i\right)^t \right]^t \tag{16}$$

for $i = 1, 2$.

Lemma 2.1. *There exists a transformation* $d : (s, \hat{s}) \to \left(\lambda, \hat{\lambda}\right)$, *such as* $(F_2, -G_2) = b \circ (F_1, -G_1)$ *if and only if systems (17) and (18) are solvable*

$$\left[vec\,(F_1)\; -vec\,(G_1)\right] \begin{bmatrix} \alpha \\ \gamma \end{bmatrix} = vec\,(F_2) \tag{17}$$

$$\left[vec\,(F_1)\; -vec\,(G_1)\right] \begin{bmatrix} \beta \\ \delta \end{bmatrix} = -vec\,(G_2)\,, \tag{18}$$

and

$$\alpha\delta - \beta\gamma \neq 0, \tag{19}$$

where $\left[vec\,(F_1)\; -vec\,(G_1)\right] \in \mathbb{R}^{mn\times 2}$.

Proof. . Combining (14) with expressions (16) and considering also basic linear algebra, the expressions (17)-(18) are derived straightforwardly. □

Taking into consideration the following lemma, our problem is divided into two interesting cases. Consequently, the first sufficient and necessary conditions are derived. In more details, these expressions are related to the exclusion of one possible direction.

Lemma 2.2. *In order to obtain a solution to our problem, the matrices*

$$\left[vec\,(F_1)\; -vec\,(G_1)\; vec\,(F_2)\right] \in \mathbb{R}^{mn\times 3} \tag{20}$$

and

$$\left[vec\,(F_1)\; -vec\,(G_1)\; -vec\,(G_2)\right] \in \mathbb{R}^{mn\times 3} \tag{21}$$

should have rank 1 or 2.

Proof. The rank of the matrices above can not take the value 0, since the known condition

$$rank\left[vec\,(F_1)\; -vec\,(G_1)\right] = rank\left[vec\,(F_1)\; -vec\,(G_1)\; vec\,(F_2)\right] = 0,$$

requires zero elements both for matrices F_1 and G_1. Moreover,

$$rank\left[vec\,(F_1)\; -vec\,(G_1)\right] = rank\left[vec\,(F_1)\; -vec\,(G_1)\; vec\,(F_2)\right] \neq 3,$$

otherwise (17), (18) are not solvable. □

Consequently, the only possible ranks for the Echelon form of linear systems (17) and (18) are 1 or 2. Now, in order to be benefited of the fact that the

ranks can not take the value of 3, the following definition should firstly be considered.
Thus, we denote the set of strictly increasing sequences of l integers, for $1 \leq l \leq m$, chosen from $1, 2, \ldots, m$, i.e. $Q_{2,3} = \{(1,2), (1,3), (2,3)\}$. Then the number of the sequences which belongs to $Q_{l,m}$ is $\binom{m}{l}$.
We suppose that $A = [a_{ij}] \in \mathcal{M}_{m,n}(\mathbb{R})$, where $\mathcal{M}_{m,n}(\mathbb{R})$ denotes the set of $m \times n$ matrices over $\mathbb{R}$. Let μ, ν be positive integers satisfying $1 \leq \mu \leq m$, $1 \leq \nu \leq n$, $\alpha = (i_1, \ldots, i_\mu) \in Q_{\mu,m}$ and $\beta = (j_1, \ldots j_\nu) \in Q_{\nu,n}$, then $A[\alpha|\beta] \in \mathcal{M}_{\mu,\nu}(\mathbb{R})$ denotes the sub-matrix of A which contains the rows $i_1, i_2, \ldots, i_\mu$ and the columns $j_1, j_2, \ldots, j_\nu$. Finally, for $1 \leq l \leq \min\{m, n\}$ the *l-compound matrix* or *l-adjugate* of A is the $\binom{m}{l} \times \binom{n}{l}$ matrix whose entries are $\det\{A[\alpha|\beta]\}$, $\alpha \in Q_{l,m}$, $\beta \in Q_{l,n}$ arranged lexicographically in α and β. This matrix will be designated by $C_l(A)$; see [2]-[3], [7], and [19] for many more details about that really interesting matrix.
Now, in Lemma 2.3, other necessary and sufficient conditions are derived.

Lemma 2.3. *In order to exist a bilinear equivalence between of pencils (12) and (13),*

$$C_3\left[vec(F_1)\ -vec(G_1)\ vec(F_2)\right] = \underline{0} \in \mathbb{R}^{\binom{mn}{3}}$$

and

$$C_3\left[vec(F_1)\ -vec(G_1)\ -vec(G_2)\right] = \underline{0} \in \mathbb{R}^{\binom{mn}{3}},$$

respectively, should be satisfied.

Proof. It is derived straightforwardly from Lemma 2.2 and the definition of compound matrices. □

Lemma 2.3 provides some necessary and sufficient conditions, but they are not the only ones. Some others are derived by considering the exact rank of matrices (20) and (21).

Lemma 2.4. *In order to obtain bilinear equivalent pencils, the following necessary and sufficient conditions are derived. Analytically,*
a) If the rank of matrices (20) and (21) is equal to 2, then for the rows-vectors $\begin{bmatrix} f_{ij}^1 & -g_{ij}^1 \\ f_{kl}^1 & -g_{kl}^1 \end{bmatrix}$ *which span the matrix* $\left[vec(F_1)\ -vec(G_1)\right] \in \mathbb{R}^{mn\times 2}$,

the relative rows-vectors of matrix $\left[vec(F_2)\; vec(G_2) \right] \in \mathbb{R}^{mn\times 2}$ *should satisfy the following condition,*

$$\det \begin{bmatrix} f_{ij}^2 & g_{ij}^2 \\ f_{kl}^2 & g_{kl}^2 \end{bmatrix} \neq 0,$$

where $i, k \in \{1, 2, \ldots, m\}$ *and* $j, l \in \{1, 2, \ldots, n\}$ *with* $i \neq j$ *and* $k \neq l$.
b) If the rank of matrices (20) and (21) is equal to 1, then

$$C_2 \left[vec(F_1)\; -vec(G_1)\; vec(F_2) \right] = \mathbb{O} \in \mathbb{R}^{\binom{mn}{2}\times 3}$$

and

$$C_2 \left[vec(F_1)\; -vec(G_1)\; -vec(G_2) \right] = \mathbb{O} \in \mathbb{R}^{\binom{mn}{2}\times 3},$$

respectively, should be satisfied. And,
if $f_{ij}^1 \neq 0$, *then the expression,*

$$f_{ij}^2 \delta + g_{ij}^2 \gamma \neq 0 \tag{22}$$

should be considered.
Otherwise, if $g_{ij}^1 \neq 0$, *the expression,*

$$g_{ij}^2 \alpha + f_{ij}^2 \beta \neq 0 \tag{23}$$

should be also considered.

Proof. a) Consider the case that rank of matrices (20) and (21) is equal to 2. Since,

$$rank \left[vec(F_1)\; -vec(G_1) \right] = rank \left[vec(F_1)\; -vec(G_1)\; vec(F_2) \right] = 2,$$

and equivalently,

$$rank \left[vec(F_1)\; -vec(G_1) \right] = rank \left[vec(F_1)\; -vec(G_1)\; -vec(G_2) \right] = 2,$$

then two linear independent row-vectors exist that span the matrix $\left[vec(F_1)\; -vec(G_1) \right] \in \mathbb{R}^{mn\times 2}$, i.e. $\begin{bmatrix} f_{ij}^1 & -g_{ij}^1 \\ f_{kl}^1 & -g_{kl}^1 \end{bmatrix}$ where $i, k \in \{1, 2, \ldots, m\}$ and $j, l \in \{1, 2, \ldots, n\}$ with $i \neq j$ and $k \neq l$. Moreover, considering the equalities above, we obtain two linear independent row-vectors,

$$\begin{bmatrix} f_{ij}^1 & -g_{ij}^1 & f_{ij}^2 \\ f_{kl}^1 & -g_{kl}^1 & f_{kl}^2 \end{bmatrix} \quad and \quad \begin{bmatrix} f_{ij}^1 & -g_{ij}^1 & -g_{ij}^2 \\ f_{kl}^1 & -g_{kl}^1 & -g_{kl}^2 \end{bmatrix},$$

that span the matrices (20) and (21), respectively.
Consequently, systems (17) and (18) are solvable and an expression of the

elements of the projective transformation $d = \begin{bmatrix} \alpha & \beta \\ \gamma & \delta \end{bmatrix} \in \mathbb{R}^{2\times 2}$ is derived.

Analytically, we obtain

$$\alpha = \frac{-f_{ij}^2 g_{kl}^1 + f_{kl}^2 g_{ij}^1}{-f_{ij}^1 g_{kl}^1 + f_{kl}^1 g_{ij}^1}, \beta = \frac{g_{ij}^2 g_{kl}^1 - g_{kl}^2 g_{ij}^1}{-f_{ij}^1 g_{kl}^1 + f_{kl}^1 g_{ij}^1},$$

$$\gamma = \frac{f_{ij}^1 f_{kl}^2 - f_{kl}^1 f_{ij}^2}{-f_{ij}^1 g_{kl}^1 + f_{kl}^1 g_{ij}^1} \ and \ \delta = \frac{-f_{ij}^1 g_{kl}^2 + f_{kl}^1 g_{ij}^2}{-f_{ij}^1 g_{kl}^1 + f_{kl}^1 g_{ij}^1}.$$

Now, through the expression (19), the restriction $\det \begin{bmatrix} f_{ij}^2 & g_{ij}^2 \\ f_{kl}^2 & g_{kl}^2 \end{bmatrix} \neq 0$ is derived.

b) Consider the alternative case. Since,

$$rank \left[vec(F_1) \ -vec(G_1) \right] = rank \left[vec(F_1) \ -vec(G_1) \ vec(F_2) \right] = 1,$$

and equivalently,

$$rank \left[vec(F_1) \ -vec(G_1) \right] = rank \left[vec(F_1) \ -vec(G_1) \ -vec(G_2) \right] = 1,$$

then a linear independent row-vector exists that span the matrix $\left[vec(F_1) \ -vec(G_1) \right] \in \mathbb{R}^{mn\times 2}$, for instance, $\left[f_{ij}^1 \ -g_{ij}^1 \right]$ where $i \in \{1, 2, \ldots, m\}$ and $j \in \{1, 2, \ldots, n\}$ with $i \neq j$.

Moreover, by also considering the equalities above, we obtain a linear independent row-vector, $\left[f_{ij}^1 \ -g_{ij}^1 \ f_{ij}^2 \right]$ and $\left[f_{ij}^1 \ -g_{ij}^1 \ -g_{ij}^2 \right]$, that span the matrices (20) and (21), respectively. Consequently, systems (17) and (18) are solvable and an expression of the elements of the projective transformation $d = \begin{bmatrix} \alpha & \beta \\ \gamma & \delta \end{bmatrix} \in \mathbb{R}^{2\times 2}$ is derived. Analytically, we obtain

$$f_{ij}^1 \alpha - g_{ij}^1 \gamma = f_{ij}^2, \tag{24}$$

and

$$f_{ij}^1 \beta - g_{ij}^1 \delta = -g_{ij}^2. \tag{25}$$

Since,$rank \left[vec(F_1) \ -vec(G_1) \right] \neq 0$, the elements f_{ij}^1, g_{ij}^1 can not be equal to zero, simultaneously. Thus, considering (14) and (15), then if $f_{ij}^1 \neq 0$, expression (12) should be considered. Otherwise, if $g_{ij}^1 \neq 0$, expression (2.12) should be considered. □

Remark 2.1. For the case where the rank of matrices (10) and (11) is equal to 1. Denote that the column vector $\begin{bmatrix} \alpha \\ c \end{bmatrix}$ and $\begin{bmatrix} \beta \\ \delta \end{bmatrix}$ spans the space

V_1 and V_2, respectively. Moreover, the column vector $\begin{bmatrix} f_{ij}^2 \\ g_{ij}^2 \end{bmatrix}$ spans the space V.

Then, by considering the results of Lemma 2.4 (b): (12)-(13), it is derived that

$$V_1 \nmid V \textit{ and } V_2 \nmid V.$$

The statement and the proof of Theorem 3.1 is a consequence of the whole procedure.

Theorem 2.1. *Two pencils are bilinear equivalent* if and only if *the following conditions are satisfied simultaneously,*
(a)

$$C_3\left[vec(F_1)\ -vec(G_1)\ vec(F_2)\right] = C_3\left[vec(F_1)\ -vec(G_1)\ -vec(G_2)\right] = \underline{0} \tag{26}$$

(b1) *For rank* $\left[vec(F_1)\ -vec(G_1)\right] = 2$, *the independent rows-vectors* $\begin{bmatrix} f_{ij}^1 & -g_{ij}^1 \\ f_{kl}^1 & -g_{kl}^1 \end{bmatrix}$ *span the matrix* $\left[vec(F_1)\ -vec(G_1)\right] \in \mathbb{R}^{mn\times 2}$, *and*

$$\det\begin{bmatrix} f_{ij}^2 & g_{ij}^2 \\ f_{kl}^2 & g_{kl}^2 \end{bmatrix} \neq 0 \tag{27}$$

should be satisfied, where rows-vector $\begin{bmatrix} f_{ij}^2 & -g_{ij}^2 \\ f_{kl}^2 & -g_{kl}^2 \end{bmatrix}$ *span the matrix* $\left[vec(F_2)\ vec(G_2)\right] \in \mathbb{R}^{mn\times 2}$, *for* $i,k \in \{1,2,\ldots,m\}$ *and* $j,l \in \{1,2,\ldots,n\}$ *with* $i \neq j$ *and* $k \neq l$. *Moreover,*

$$\alpha = \frac{-f_{ij}^2 g_{kl}^1 + f_{kl}^2 g_{ij}^1}{-f_{ij}^1 g_{kl}^1 + f_{kl}^1 g_{ij}^1}, \tag{28}$$

$$\beta = \frac{g_{ij}^2 g_{kl}^1 - g_{kl}^2 g_{ij}^1}{-f_{ij}^1 g_{kl}^1 + f_{kl}^1 g_{ij}^1}, \tag{29}$$

$$\gamma = \frac{f_{ij}^1 f_{kl}^2 - f_{kl}^1 f_{ij}^2}{-f_{ij}^1 g_{kl}^1 + f_{kl}^1 g_{ij}^1}, \tag{30}$$

and

$$\delta = \frac{-f_{ij}^1 g_{kl}^2 + f_{kl}^1 g_{ij}^2}{-f_{ij}^1 g_{kl}^1 + f_{kl}^1 g_{ij}^1}. \tag{31}$$

(b2) *For* $\delta = \frac{-f_{ij}^1 g_{kl}^2 + f_{kl}^1 g_{ij}^2}{-f_{ij}^1 g_{kl}^1 + f_{kl}^1 g_{ij}^1}$, *then*

$$C_2\left[vec(F_1)\ -vec(G_1)\ vec(F_2)\right] = C_2\left[vec(F_1)\ -vec(G_1)\ -vec(G_2)\right] = \mathbb{O}, \tag{32}$$

And, if $f_{ij}^1 \neq 0$, *then*

$$\alpha = \frac{f_{ij}^2 + g_{ij}^1 \gamma}{f_{ij}^1} \tag{33}$$

and

$$\beta = \frac{-g_{ij}^2 + g_{ij}^1 \delta}{f_{ij}^1}, \tag{34}$$

for arbitrary $\gamma, \delta \in \mathbb{R}$ *and (22) holds. Otherwise, if* $g_{ij}^1 \neq 0$, *then*

$$\gamma = \frac{f_{ij}^1 \alpha - f_{ij}^2}{g_{ij}^1}, \tag{35}$$

and

$$\delta = \frac{f_{ij}^1 \beta + g_{ij}^2}{g_{ij}^1}, \tag{36}$$

for arbitrary $\alpha, \beta \in \mathbb{R}$ *and (23) holds.*

The next examples clarify the importance of Theorem 2.1. Thus, in Example 2.1, the first case provides an example where the pencils are not bilinear equivalent. Alternatively, two pencils are showed to be bilinear equivalent when their ranks of row-vectors are 2. In Example 2.2, the bilinear equivalence when the ranks of row-vectors are 1 for particular pencils is presented.

Example 2.1 a) Consider pencil $(F_1, G_1) = \left(\begin{bmatrix} 0 & 0 \\ 1 & 0 \end{bmatrix}, \begin{bmatrix} 1 & 0 \\ 1 & 1 \end{bmatrix}\right)$.

This is not bilinearly equivalent to pencil $(F_2, G_2) = \left(\begin{bmatrix} 1 & 0 \\ 0 & 0 \end{bmatrix}, \begin{bmatrix} 1 & 1 \\ 0 & 1 \end{bmatrix}\right)$, since $C_3\left[vec(F_1)\ -vec(G_1)\ vec(F_2)\right] = \left[0\ 0\ -1\ 0\right]^T \neq \underline{0} \in \mathbb{R}^4$. (Note that $rank\left[vec(F_1)\ -vec(G_1)\ vec(F_2)\right] = 3$)

b) Consider now pencils $(F_1, G_1) = \left(\begin{bmatrix} 1 & 1 \\ 0 & 0 \end{bmatrix}, \begin{bmatrix} 1 & -1 \\ 0 & 0 \end{bmatrix}\right)$ and $(F_2, G_2) = \left(\begin{bmatrix} 0 & 1 \\ 0 & 0 \end{bmatrix}, \begin{bmatrix} 1 & 0 \\ 0 & 0 \end{bmatrix}\right)$. It is clear that $C_3\left[vec(F_1)\ -vec(G_1)\ vec(F_2)\right] = C_3\left[vec(F_1)\ -vec(G_1)\ -vec(G_2)\right] = \underline{0} \in \mathbb{R}^4$.

Moreover, $rank\left[vec(F_1)\ -vec(G_1)\right] = 2$.

The rows $\begin{bmatrix} 1 & -1 \\ 1 & 1 \end{bmatrix}$ span the matrix $\begin{bmatrix} vec(F_1) & -vec(G_1) \end{bmatrix} = \begin{bmatrix} 1 & 0 \\ 1 & -1 \\ 0 & 0 \\ 0 & 0 \end{bmatrix} \in \mathbb{R}^{4\times 2}$, and the restriction (27) is satisfied since $\det \begin{bmatrix} 0 & 1 \\ 1 & 0 \end{bmatrix} = -1 \neq 0$, then the rows $\begin{bmatrix} 0 & -1 \\ 1 & 0 \end{bmatrix}$ span the matrix $\begin{bmatrix} vec(F_2) & -vec(G_2) \end{bmatrix} = \begin{bmatrix} 0 & -1 \\ 1 & 0 \\ 0 & 0 \\ 0 & 0 \end{bmatrix} \in \mathbb{R}^{4\times 2}$.

Consequently, the pencils above are bilinearly equivalent. Now, by using expression (28)-(31), we obtain

$$\alpha = \frac{1}{2},\ \beta = -\frac{1}{2},\ \gamma = \frac{1}{2}\ and\ \delta = \frac{1}{2}.$$

Finally, we can easily verify that

$$\begin{aligned} \begin{bmatrix} F_2 & -G_2 \end{bmatrix} &= \begin{bmatrix} \begin{bmatrix} 0 & 1 \\ 0 & 0 \end{bmatrix} & \begin{bmatrix} -1 & 0 \\ 0 & 0 \end{bmatrix} \end{bmatrix} \\ &= I_2 \begin{bmatrix} \begin{bmatrix} 1 & 1 \\ 0 & 0 \end{bmatrix} & \begin{bmatrix} -1 & 1 \\ 0 & 0 \end{bmatrix} \end{bmatrix} \begin{bmatrix} \frac{1}{2} I_2 & -\frac{1}{2} I_2 \\ \frac{1}{2} I_2 & \frac{1}{2} I_2 \end{bmatrix} \\ &= \begin{bmatrix} \frac{1}{2} \begin{bmatrix} 1 & 1 \\ 0 & 0 \end{bmatrix} + \frac{1}{2} \begin{bmatrix} -1 & 1 \\ 0 & 0 \end{bmatrix} & -\frac{1}{2} \begin{bmatrix} 1 & 1 \\ 0 & 0 \end{bmatrix} + \frac{1}{2} \begin{bmatrix} -1 & 1 \\ 0 & 0 \end{bmatrix} \end{bmatrix}. \end{aligned}$$

Example 2.2 Consider pencils $(F_1, G_1) = \left(\begin{bmatrix} 1 & 1 \\ 0 & 0 \end{bmatrix}, \begin{bmatrix} 1 & 1 \\ 0 & 0 \end{bmatrix} \right)$ and $(F_2, G_2) = \left(\begin{bmatrix} 2 & 2 \\ 0 & 0 \end{bmatrix}, \begin{bmatrix} 3 & 3 \\ 0 & 0 \end{bmatrix} \right)$. It is clear that $C_3 \begin{bmatrix} vec(F_1) & -vec(G_1) & vec(F_2) \end{bmatrix} = C_3 \begin{bmatrix} vec(F_1) & -vec(G_1) & -vec(G_2) \end{bmatrix} = \underline{0} \in \mathbb{R}^4$.

Moreover, since $rank \begin{bmatrix} vec(F_1) & -vec(G_1) \end{bmatrix} = 1$, then we can verify that $C_2 \begin{bmatrix} vec(F_1) & -vec(G_1) & vec(F_2) \end{bmatrix} = C_2 \begin{bmatrix} vec(F_1) & -vec(G_1) & -vec(G_2) \end{bmatrix} = \mathbb{O} \in \mathbb{R}^{6\times 3}$. And, since $f_{ij}^1 = 1 \neq 0$, then $\alpha = \gamma + 2$ and $\beta = \delta - 3$ for arbitrary $\gamma, \delta \in \mathbb{R}$ and (22) holds. Finally, we can easily verify that

$$\begin{aligned} \begin{bmatrix} F_2 & -G_2 \end{bmatrix} &= \begin{bmatrix} \begin{bmatrix} 2 & 2 \\ 0 & 0 \end{bmatrix} & \begin{bmatrix} -3 & -3 \\ 0 & 0 \end{bmatrix} \end{bmatrix} \\ &= I_2 \begin{bmatrix} \begin{bmatrix} 1 & 1 \\ 0 & 0 \end{bmatrix} & \begin{bmatrix} -1 & -1 \\ 0 & 0 \end{bmatrix} \end{bmatrix} \begin{bmatrix} \alpha I_2 & \beta I_2 \\ (\alpha - 2) I_2 & (\beta + 3) I_2 \end{bmatrix} \\ &= \begin{bmatrix} \alpha \begin{bmatrix} 1 & 1 \\ 0 & 0 \end{bmatrix} + (\alpha - 2) \begin{bmatrix} -1 & -1 \\ 0 & 0 \end{bmatrix} & \beta \begin{bmatrix} 1 & 1 \\ 0 & 0 \end{bmatrix} + (\beta + 3) \begin{bmatrix} -1 & -1 \\ 0 & 0 \end{bmatrix} \end{bmatrix}. \end{aligned}$$

3. Bilinear Equivalence Algorithm (BEA)

In this section, the algorithmic approach for comparing the bilinear equivalence of pencils

$$\left[F_1 \ -G_1 \right] \ and \ \left[F_2 \ -G_2 \right],$$

where $F_i, G_i \in \mathbb{R}^{m\times n}$ for $i = 1, 2$, is analytically presented.

Bilinear Equivalence Algorithm (BEA)

Alg. Step 1: Transform the matrices $F_i, G_i \in \mathbb{R}^{m\times n}$ for $i = 1, 2$ into

$$\bar{F}_i \triangleq vec\left(F_i\right) \triangleq \left[\left(\underline{f}^i_1\right)^t \cdots \left(\underline{f}^i_m\right)^t \right]^t \in \mathbb{R}^{nm}$$

and

$$\bar{G}_i = vec\left(G_i\right) \triangleq \left[\left(\underline{g}^i_1\right)^t \cdots \left(\underline{g}^i_m\right)^t \right]^t \in \mathbb{R}^{nm},$$

for $i = 1, 2$.

Alg. Step 2: Create the matrices $A = \left[\bar{F}_1 \ \bar{G}_1 \right] \in \mathbb{R}^{nm\times 2}$,

$$A_1 = \left[\bar{F}_1 \ -\bar{G}_1 \ \bar{F}_2 \right] \in \mathbb{R}^{nm\times 3}, \ and \ A_2 = \left[\bar{F}_1 \ -\bar{G}_1 \ -\bar{G}_2 \right] \in \mathbb{R}^{nm\times 3}.$$

Alg. Step 3: Compute the

$$C_3\left(A_1\right) = \underline{0} \in \mathbb{R}^{\binom{mn}{3}} \ and \ C_3\left(A_2\right) = \underline{0} \in \mathbb{R}^{\binom{mn}{3}}.$$

In order to calculate these expressions, for instance, the references [20]-[21] may be considered. Note that if one of the two expressions is not satisfied, then the pencils are not bilinear equivalent. Making this test, we can proceed by applying the next step.

Alg. Step 4:

a) If the $rank\left(A\right) = 2$, go to the **Alg. Step 5**.
b) Else the , go to the **Alg. Step 6**

Alg. Step 5: $\left(rank\left(A\right) = 2\right)$. Determine the two independent rows that span matrix A. This is derived by the calculation of the elements of matrix $C_2\left(A\right) \in \mathbb{R}^{\binom{mn}{2}\times 3}$. The whole procedure stops until the first element of the matrix above is non-zero. Consequently, the two independent rows are fully calculated.

Now, since row vectors $\left(f^1_{ij}, \ -g^1_{ij}\right)$ and $\left(f^1_{kl}, \ -g^1_{kl}\right)$ span the matrix A, then

$$\left(f^1_{ij}, \ -g^1_{ij}, f^2_{ij}\right), \ \left(f^1_{kl}, \ -g^1_{kl}, f^2_{kl}\right)$$

and

$$\left(f_{ij}^1, \ -g_{ij}^1, \ -g_{ij}^2\right), \ \left(f_{kl}^1, \ -g_{kl}^1, \ -g_{kl}^2\right)$$

which span the matrices A_1 and A_2 respectively, are determined.
Hence, **if** det $\begin{bmatrix} f_{ij}^2 \ g_{ij}^2 \\ f_{kl}^2 \ g_{kl}^2 \end{bmatrix} \neq 0$, then α, β, γ and are calculated by (28)-(31).
Else, i.e. det $\begin{bmatrix} f_{ij}^2 \ g_{ij}^2 \\ f_{kl}^2 \ g_{kl}^2 \end{bmatrix} = 0$, then the pencils are not bilinear equivalent.
Alg. Step. 6: $(rank\,(A) = 1)$. Calculate the elements of matrix

$$C_2\,(A_1) \ \ and \ C_2\,(A_2) \in \mathbb{R}^{\binom{mn}{2} \times 3}.$$

If there is a non-zero element, then the pencils are not bilinear equivalent.
Else, perform Alg. Step 7.
Alg. Step 7: Determine the independent row that span matrix A. This is derived by the first non zero elements of matrix $C_2\,(A) \in \mathbb{R}^{\binom{mn}{2} \times 3}$.
Now, since row vector $\left(f_{ij}^1, \ -g_{ij}^1\right)$ spans the matrix A, then

$$\left(f_{ij}^1, \ -g_{ij}^1, \, f_{ij}^2\right) and \left(f_{ij}^1, \ -g_{ij}^1, \ -g_{ij}^2\right),$$

which span the matrices A_1 and A_2, respectively, are determined.
Hence, **if** $f_{ij}^1 \neq 0$, then go to the **Alg. Step 8**.
Else, i.e. $g_{ij}^1 \neq 0$, then go to the **Alg. Step** 9.
Alg. Step 8: Since $f_{ij}^1 \neq 0$, then α and β are calculated by (33) and (34) for arbitrary $\gamma, \delta \in \mathbb{R}$. Note that (22) should also hold.
Alg. Step 9: Since $g_{ij}^1 \neq 0$, then γ and δ are calculated by (35) and (36) for arbitrary $\alpha, \beta \in \mathbb{R}$. Note that (23) should also hold.
(end) Algorithm.

4. Conclusions

In this paper, we have presented an algorithmic technique for investigating the necessary and sufficient conditions for two homogeneous matrix pencils $sF_1 - \hat{s}G_1$ and $sF_2 - \hat{s}G_2$ to be bilinear equivalent. The proposed algorithmic technique is implemented in a numerically stable manner, giving efficient results. However, error analysis and the required complexity can be considered as a further extension. Moreover, it should be stressed out that the present paper might be considered as a bird-sight view of bilinear equivalent matrix pencil theory, which is significant for several other extensions and connections with other fields of (numerical) linear algebra and control theory.

Acknowledgments

The project is co-funded by the European Social Fund and National Resources -(EPEAEK II) KAPODISTRIAS (2008-09).

References

1. S.L. Campbell, *Singular systems of differential equations, Pitman*, San Francisco, Vol. 1 (1980); Vol. 2 (1982).
2. R.F. Gantmacher, *The theory of matrices*, vol. I and II, Chelsea, New York, U.S.A. (1977).
3. W. Greub, *Multilinear Algebra*, Second edition, Universitext, Springer - Verlag, New York - Heidelberg (1978).
4. G.I. Kalogeropoulos, *Matrix pencils and linear systems*, Ph.D Thesis, City University, London (1985).
5. G.I. Kalogeropoulos and A.A. Pantelous, *Can be linear difference descriptor systems appeared in insurance?* Proceedings of the 7th International Conference APLIMAT 2008, Bratislava, Slovakia 2008, pp. 467-478.
6. G.I. Kalogeropoulos, A.A. Karageorgos and A.A. Pantelous, A stabilization criterion for matrix pencils under bilinear transformation, *Linear Algebra and its Application* 428 (11-12), 2008, pp. 2852-2862.
7. N. Karcanias and U. Baçer. Exterior algebra and invariant spaces of implicit systems: the Grassmann representative approach, *Kybernetika (Prague)* 30, 1994, pp. 1-22.
8. N. Karcanias and G.E. Hayton, *Generalized autonomous differential systems, algebraic duality and geometry theory*, Proceedings of IFAC VIII, Triennial World Congress, Kyoto, Japan (1981).
9. N. Karcanias and G.E. Hayton, *Generalized autonomous differential systems, algebraic duality and geometry theory*, Research Report, The City University, Systems Science, DSS/N.K./G.E.H./212 (1982).
10. N. Karcanias and G.I. Kalogeropoulos, *Bilinear-strict equivalence of matrix pencils and autonomous singular differential systems*, Proceedings of 4th I.M.A. Intern. Cond. On Control Theory, Robinson College, Cambridge, U.K. (1984).
11. D.A., Kendrick. On the Leontief dynamic inverse. *Quarterly Journal of Economics* 86, 1972, pp. 693-696.
12. D.A. Kendrick, Stochastic control for economic models: past, present and the paths ahead. *Journal of Economic Dynamics and Control* 29, 2005, pp. 3-30.
13. R.G. King and M.W. Watson, The solution of singular linear difference systems under rational expectations. *International Economic Review* 39 (4), 1998, pp. 1015-1026.
14. R.G. Kreijger and H. Neudecker, Kendrick's forward integration method and the dynamic Leontief multisectoral model. *Quarterly Journal of Economics* 90, 1976, pp. 505-507.
15. W. Leontief, Input-Output Economics. 2nd ed., New York: Oxford University

Press (1986).
16. T.P. de Lima, Blending positive matrix pencils with economic models, *Lecture Notes in Control and Information Sciences* (Springer Berlin/Heidelberg) 294, 2004, pp. 289-296.
17. D.A. Livesey, The singularity problem in the dynamic input-output model, *International Journal of Systems Science* 4, 1973, pp. 437-440.
18. D.G. Luenberger and A. Arbel, Singular dynamic Leontief systems, *Econometrica* 45, 1977, pp. 991-995.
19. M. Marcus, and H. Minc, *A survey of matrix theory and matrix inequalities*, Allyn and Bacon, Boston, U.S.A (1964).
20. M. Mitrouli and G. Kalogeropoulos, A compound matrix algorithm for the compu-tation of the Smith form of polynomial matrices, *Numerical Algorithms* 7, 1994, pp. 145-159.
21. M. Mitrouli and C. Koukouvinos, On the computation of the Smith normal form of compound matrices, *Numerical Algorithms* 16 (2), 1997, pp. 95-105.
22. A.A. Pantelous, A.A. Zimbidis and G.I Kalogeropoulos, *A generalized linear discrete time model for managing the solvency interaction and singularities arising from potential regulatory constraints imposed within a portfolio of different insurance products*, Proceedings of ASTIN colloquium 2008, Manchester, U.K.
23. G.J. Semple and T.G. Kneebone, *Algebraic projective Geometry*, Oxford at the Clarendon Press, 1952.
24. G. Sierksma, Non-negative matrices: The open Leontief model, *Linear Algebra and its Applications* 26, 1979, pp. 175-201.
25. M.S. Silva and T.P. de Lima, Looking for nonnegative solutions of a Leontief dynamic model, *Linear Algebra and its Applications* 364, 2003, pp. 281-316.

ELECTRIC FIELD GRADIENT EFFECTS IN ELASTIC DIELECTRICS

A. I. ARVANITAKIS,* V. K. KALPAKIDES

Department of Materials Science and Engineering,
Laboratory of Mathematical Modeling and Scientific Computing,
University of Ioannina, GR-45 110 Ioannina, Greece
E-mail: aarvanit@cc.uoi.gr, vkalpak@cc.uoi.gr

A theory for non-linear elastic dielectrics, assuming higher gradients of the electric field, is presented. The proposed theory can be seen as a theory with quadrupoles because the gradient of electric field is energetically conjugate to the electric quadrupole. The analysis specializes particularly to elastic ferroelectrics generalizing the existing linear electric field gradient theories. Attention is paid to the so called quadrupole piezoelectricity phenomenon, which is present in all crystal symmetry groups even in centrosymmetric materials where the classical piezoelectricity vanishes. Analytical solutions for 1-D simplified problems reveal the role of quadrupole piezoelectricity in elastic dielectrics.

1. Elastic Ferroelectrics with electric quadrupoles

1.1. *Theory*

Consider a ferroelectric body occupying region Ω with boundary $\partial\Omega$. The total potential H of the system is assumed to be

$$H = W\left(e_{ij}, \phi_{,i}, \phi_{,ij}\right) + \Psi(\pi_i^s) + \pi_i^s \phi_{,i} \tag{1}$$

where e_{ij} is the infinitesimal strain, u_i the displacement field, ϕ the electric potential and π_i^s the spontaneous dipole polarization vector. Also, W is the energy of deformation and electric field and Ψ denotes the phase separation energy. The energy Ψ is responsible for the strongly non-linear character of ferroelectrics and for the formation of domain patterns within the material [3]. Notice that the gradient of electric field (or the second gradient of electric potential) is introduced as a new constitutive variable in the total potential H.

*Work partially supported by the State Scholarships Foundation of Greece.

It is assumed that the energy consists of two parts: the energy of the matter and the interaction field–matter energy. Here, the matter energy consists of the first two terms in Eq. (1), i.e.:

$$\Sigma(e_{ij}, \phi_{,i}, \phi_{,ij}, \pi_i^s) = W(e_{ij}, \phi_{,i}, \phi_{,ij}) + \Psi(\pi_i^s). \tag{2}$$

The third term, i.e., $\pi_i^s \phi_{,i}$ concerns the interaction energy between matter and field.

The equilibrium conditions for the mechanical stress σ_{ij} and electric displacement D_i are

$$\sigma_{ij,j} = 0, \quad D_{i,i} = 0. \tag{3}$$

The constitutive equations can be derived by the total potential H

$$\sigma_{ij} = \frac{\partial H}{\partial e_{ij}} = \frac{\partial W}{\partial e_{ij}}, \quad \pi_i = \frac{\partial H}{\partial \phi_{,i}} = \pi_i^s + \frac{\partial W}{\partial \phi_{,i}}, \quad q_{ij} = \frac{\partial H}{\partial \phi_{,ij}} = \frac{\partial W}{\partial \phi_{,ij}} \tag{4}$$

where q_{ij} denotes the quadrupole density moments and π_i the total dipole density moment consisting of two parts: a spontaneous part and a reversible one. Setting $\pi_i^r = \partial W / \partial \phi_{,i}$ as the reversible part, then Eq. $(4)_2$ is written in the simple form

$$\pi_i = \pi_i^s + \pi_i^r. \tag{5}$$

The electric displacement vector is given by the fundamental relation

$$D_i = \varepsilon_0 E_i + P_i, \tag{6}$$

where ε_0 denotes the vacuum permittivity and P_i represents the total polarization vector inside the medium, which in the present analysis is enriched containing quadrupole effects

$$P_i = \pi_i - q_{ij,j}. \tag{7}$$

The equilibrium equations (Eq. (3)) are not adequate enough to describe the behavior of ferroelectrics. The thermodynamical force conjugate to the spontaneous polarization π_i^s is demanded to vanish in order to minimize the total system energy

$$\frac{\delta H}{\delta \pi_i^s} = 0 \Rightarrow \frac{\partial \Sigma}{\partial \pi_i^s} + \phi_{,i} = 0. \tag{8}$$

Defining the first term to be a local electric field, i.e., $E_i^L = \partial \Sigma / \partial \pi_i^s$, Eq. (8) takes the form

$$E_i^L - E_i = 0. \tag{9}$$

The above equation is well known as the intramolecular balance law in general theories of non-linear elastic dielectrics [1,2,8]. It is remarked that Eq. (8) describes the hysterical behavior of ferroelectric materials [10].

For a complete analysis, the above equations have to be accompanied by appropriate boundary conditions. Based on previous work [1,2,6] the boundary conditions are

$$\sigma_{ij} n_j = t_i, \quad \text{or} \quad u_i = \bar{u}_i \quad \text{on} \quad \partial\Omega \tag{10}$$

$$D_i n_i = \bar{\sigma} - \Pi_{i,i}, \quad \text{or} \quad \phi = \bar{\phi} \quad \text{on} \quad \partial\Omega \tag{11}$$

$$q_{ij} n_i n_j = 0, \quad \text{on} \quad \partial\Omega, \tag{12}$$

The values t_i and $\bar{u}_i$ denote prescribed values of mechanical traction and displacement on the boundary. For the electrical part of the problem, $\bar{\sigma}$ is the surface charge and $\bar{\phi}$ the prescribed electric potential. Finally, the vector Π_i is given by the relation

$$\Pi_i = q_{ij} n_j. \tag{13}$$

Concluding the above analysis, it is stated that the fundamental equations describing the behavior of elastic ferroelectrics are Eqs. $(3)_{1,2}$ and (8), accompanied by the boundary conditions (10)–(12).

1.2. *Specific choice of the energy functions*

In this subsection, specific forms of the stored energy function W and the phase separation energy Ψ are chosen, so as to formulate a well-posed boundary value problem. The energy function W is assumed to be

$$\begin{aligned} W(e_{ij}, \phi_{,i}, \phi_{,ij}) = {} & \frac{1}{2} c_{ijkl} \left(e_{ij} - e^0_{ij}\right) \left(e_{kl} - e^0_{kl}\right) + d_{ijk} \phi_{,i} \left(e_{jk} - e^0_{jk}\right) \\ & - D_{ijkl} \phi_{,ij} (e_{kl} - e^0_{kl}) - \frac{1}{2} \varepsilon_0 \varepsilon_{ij} \phi_{,i} \phi_{,j} - \frac{1}{2} \varepsilon_0 a_{ijkl} \phi_{,ij} \phi_{,kl} \\ & - \varepsilon_0 \beta_{ijk} \phi_{,i} \phi_{,jk}, \end{aligned} \tag{14}$$

where c_{ijkl} is the elastic coefficient tensor, ε_{ij} is the dielectric susceptibility tensor and d_{ijk} is the piezoelectric coefficient tensor, which depends on the spontaneous polarization orientation. The coefficient tensors β_{ijk} and a_{ijkl} are new material parameters related with the insertion of the second gradient of the electric potential into the set of the independent constitutive variables. The tensor D_{ijkl} denotes the quadrupole piezoelectric tensor. The tensor e^0_{ij} denotes the spontaneous strain and it is a function of the spontaneous polarization. The electrostriction in ferroelectrics is characterized by a general relation of the form [3]

$$e^0_{ij} = \gamma_{ijkl}\pi^s_k\pi^s_l, \tag{15}$$

where γ_{ijkl} are the electrostrictive coefficients. Combining Eqs. (4),(14) it is possible to determine the dipole and quadrupole density moments as well as the stress tensor

$$\sigma_{ij} = c_{ijkl}(e_{kl} - \gamma_{klmn}\pi^s_m\pi^s_n) + d_{kij}\phi_{,k} - D_{ijkl}\phi_{,kl}, \tag{16}$$

$$\pi_i = \pi^s_i + d_{ijk}(e_{jk} - \gamma_{jklm}\pi^s_l\pi^s_m) - \varepsilon_0\varepsilon_{ij}\phi_{,j} - \varepsilon_0\beta_{ijk}\phi_{,jk}, \tag{17}$$

$$q_{ij} = -\varepsilon_0\beta_{kij}\phi_{,k} - \varepsilon_0\alpha_{ijkl}\phi_{,kl} - D_{ijkl}e_{kl} + D_{ijkl}\gamma_{klmn}\pi^s_m\pi^s_n. \tag{18}$$

It is worth noting that in Eq. (18) the last term is exclusively due to the spontaneous polarization, thus like the dipole polarization, the quadrupole polarization also is separated into two parts: a spontaneous and a reversible one. The spontaneous quadrupole polarization is given

$$q^s_{ij} = D_{ijkl}\gamma_{klmn}\pi^s_m\pi^s_n. \tag{19}$$

It is important to mention that in [6] quadrupole piezoelectricity is ignored and a relation of the form (19) is taken by assumption. Here it is extracted in a natural way. In the same work it has been proved that spontaneous quadrupole is responsible for the existence of domain walls with proper width. Under the above considerations one can write for the total polarization vector

$$P_i = \overbrace{\pi^s_i - q^s_{ij,j}}^{\substack{spontaneous\\ part}} + \overbrace{\pi^r_i - q^r_{ij,j}}^{\substack{reversible\\ part}}. \tag{20}$$

Now, we turn to the phase separation energy Ψ which is taken to be a high order polynomial of the spontaneous polarization vector similar to the Landau-Devonshire energy in polarization based theories [9], i.e.,

$$\Psi(\pi^s_i) = \bar{a}_{ij}\pi^s_i\pi^s_j + \bar{\bar{a}}_{ijkl}\pi^s_i\pi^s_j\pi^s_k\pi^s_l, \tag{21}$$

where higher order terms can be used if necessary. Finally, it is obvious that one has all the necessary information to solve the coupled boundary value problem.

1.3. *Simplification to linear elastic dielectrics*

An important aspect of the proposed theory is that it generalizes the existing theories of linear elastic dielectrics with electric field gradient. Indeed, assuming that the spontaneous parts of dipole and quadrupole polarizations vanish, then Eqs. (16)–(18) become

$$\sigma_{ij} = c_{ijkl}e_{kl} + d_{kij}\phi_{,k} - D_{ijkl}\phi_{,kl}, \tag{22}$$

$$\pi_i = d_{ijk}e_{jk} - \varepsilon_0\varepsilon_{ij}\phi_{,j} - \varepsilon_0\beta_{ijk}\phi_{,jk}, \tag{23}$$

$$q_{ij} = -\varepsilon_0\beta_{kij}\phi_{,k} - \varepsilon_0\alpha_{ijkl}\phi_{,kl} - D_{ijkl}e_{kl} \tag{24}$$

and the set of equations needed to determine the material's behavior is reduced to two equations: the mechanical equilibrium and the Gauss equation. The intramolecular balance law in linear dielectrics becomes an identity. After substitution of Eqs. (22)–(24) into Eqs. $(3)_{1,2}$ one obtains the following equations

$$-\varepsilon_0\epsilon_{ij}\phi_{,ij} + d_{ikl}e_{kl,i} + \varepsilon_0 a_{ijkl}\phi_{,ijkl} + D_{ijkl}e_{kl,ji} = 0, \tag{25}$$

$$c_{ijkl}e_{kl,j} + d_{kij}\phi_{,kj} - D_{ijkl}\phi_{,klj} = \rho\ddot{u}_i, \tag{26}$$

where $\epsilon_{ij} = \varepsilon_0(\varepsilon_{ij} + \delta_{ij})$ and ρ is the mass density necessary to account for dynamical effects. Note that the above set of equations are similar to those extracted in[4,5] with the sole difference that here quadrupole piezoelectricity is taken into account.

2. Quadrupole effects in 1–D isotropic media

Quadrupole piezoelectric phenomenon has been observed very early by Voigt in some natural minerals. Experimental work was performed by Gladkii [7], who proved that the macroscopic quadrupole moment is the main feature of structural changes between non-polar states of dielectric solids.

2.1. *Thin plate capacitor*

Consider an infinite plate capacitor as shown in Fig. 1. It is assumed that the dielectric material under study is isotropic and non-piezoelectric. An electric potential difference between the electrodes and compressive mechanical traction have been applied so as to trigger quadrupole piezoelectricity. For this one dimensional problem, Eqs. (25) and (26) take the form

$$-\varepsilon_0\epsilon\phi'' + Du''' = 0, \tag{27}$$

$$cu'' - D\phi''' = 0. \tag{28}$$

Notice that the coefficient tensors β_{ijk} and d_{ijk} are zero due to the material's isotropy, while α_{ijkl} is assumed to be zero. According to our theoretical analysis the boundary conditions for the capacitor are

$$\phi(-h) = -V,\ \phi(h) = V,\ q_{11}(\pm h) = -D\,u'(\pm h) = 0, \tag{29}$$

$$(cu' - D\phi'')|_{x=-h} = -t_0,\ (cu' - D\phi'')|_{x=h} = -t_0. \tag{30}$$

It is easy to decouple the set of equations in the above problem. Thus combining Eqs. (27), (28) one ends up with the following differential equation

$$\phi''''(x) - (c\varepsilon_0\epsilon/D^2)\phi'' = 0, \tag{31}$$

accompanied by the boundary conditions for the electric potential

$$\phi(-h) = -V,\ \phi(h) = V, \tag{32}$$

$$\phi''(-h) = t_0/D,\ \phi''(h) = t_0/D. \tag{33}$$

It is possible to solve Eq. (31) analytically taking the solution

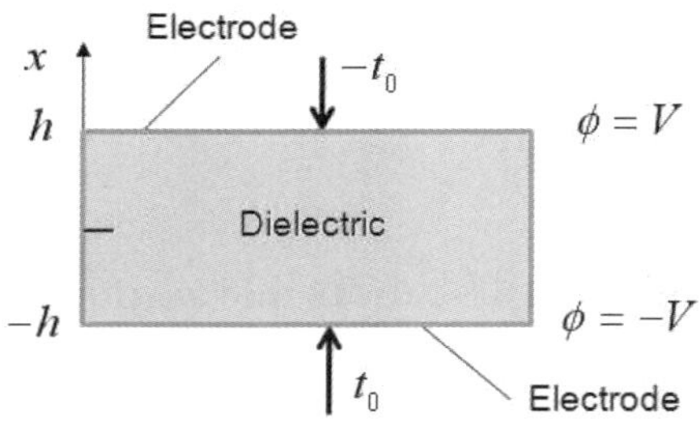

Figure 1. Infinite plate capacitor under potential difference and compressive traction

$$\phi(x) = \frac{V}{h}x + \frac{t_0 D}{c\varepsilon_0\epsilon} \cdot \frac{-1 - e^{2\frac{\sqrt{c\varepsilon_0\epsilon}}{D}h} + e^{\frac{\sqrt{c\varepsilon_0\epsilon}}{D}(h-x)} + e^{\frac{\sqrt{c\varepsilon_0\epsilon}}{D}(h+x)}}{1 + e^{2\frac{\sqrt{c\varepsilon_0\epsilon}}{D}h}}, \tag{34}$$

where the first term represents the classical solution. The electric field is given by the following equation

$$E(x) = -\frac{V}{h} - \frac{t_0}{\sqrt{c\varepsilon_0\epsilon}} \cdot \frac{e^{\frac{\sqrt{c\varepsilon_0\epsilon}}{D}(h+x)} - e^{\frac{\sqrt{c\varepsilon_0\epsilon}}{D}(h-x)}}{1 + e^{2\frac{\sqrt{c\varepsilon_0\epsilon}}{D}h}}. \tag{35}$$

Finally, the capacitance per unit area of the thin dielectric film can also be computed. Thus using Eq. $(11)_1$ one takes

$$C = \frac{\bar{\sigma}}{2V} = \frac{\varepsilon_0\epsilon}{2h} - \sqrt{\frac{\varepsilon_0\epsilon}{c}}\frac{t_0}{2V}\frac{1 - e^{2\frac{\sqrt{c\varepsilon_0\epsilon}}{D}h}}{1 + e^{2\frac{\sqrt{c\varepsilon_0\epsilon}}{D}h}}. \tag{36}$$

The first term in Eq. (36) is the classical thin film capacitance. It is remarked that compression or tension may result in capacitance reduction or increment. This conclusion might be very useful for constructing devices of technological importance.

2.2. *Dispersion of plane elastic waves*

Consider the propagation of a plane wave in an infinite dielectric strip of the form

$$u(x,t) = e^{(ikx-\omega t)}, \tag{37}$$

where k is the wave number and ω is the frequency. For the one dimensional problem under study, the governing equations are given by

$$-\varepsilon_0 \epsilon \phi'' + Du''' = 0, \tag{38}$$

$$cu'' - D\phi''' = \rho \ddot{u}. \tag{39}$$

Following a similar decoupling procedure as in the previous subsection,

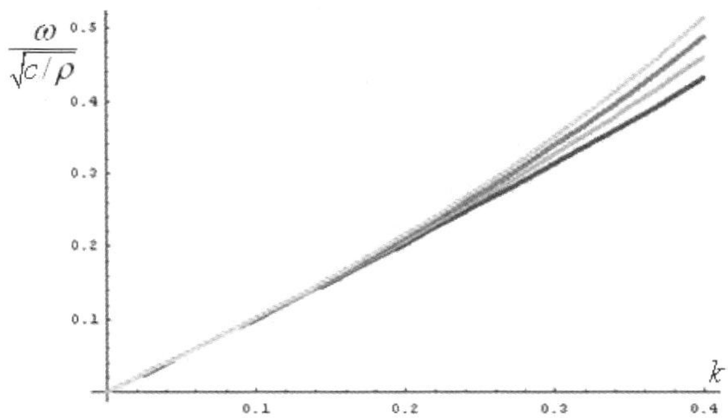

Figure 2. Dispersion of plane elastic waves for different values of the quadrupole coefficient. The values of $D^2/\rho\varepsilon_0\epsilon$ are taken: $1, 2, 3, 4$.

one takes the differential equation

$$cu''(x,t) - (D^2/\varepsilon_0\epsilon)u''''(x,t) = \rho\ddot{u}(x,t). \tag{40}$$

Substitution of Eq.(37) into Eq.(40) yields the following dispersion relation

$$\omega = \sqrt{\frac{c}{\rho}}\, k \sqrt{1 + \frac{D^2}{\rho\varepsilon_0\epsilon} k}, \tag{41}$$

where it is obvious that unlike classical elastic dielectric theories, quadrupole piezoelectricity predicts dispersive waves through the electromechanical coupling of second order. In Fig. 2, Eq. (41) is plotted for different values of the ratio $D^2/\rho\varepsilon_0\epsilon$. Note that in this work the values of D are chosen in an abstract manner with no reference to experimental observations. The major conclusion is that in non-piezoelectric materials, dispersion exists and becomes larger for materials with increased D.

References

1. V.K. Kalpakidis and C.V. Massalas, *Int. J. Engng. Sci.* **31**, 157 (1993).
2. E.P. Hadjigeorgiou,V.K. Kalpakides and C.V. Massalas, *Int. J. Non-linear Mech.* **34**, 831 (1999).
3. M. Kamlah, *Cont. Mech. Thermodyn.* **13**, 219 (2001).
4. J.S. Yang, H.G. Zhou and J.Y. Li, *Proc. R. Soc.* **A462**, 3511 (2006).
5. X.M. Yang, Y.T. Hu and J.S. Yang, *Int. J. Solids Struct.* **41**, 6801 (2004).
6. A.I. Arvanitakis, V.K. Kalpakides and E.P. Hadjigeorgiou, *submitted for publication* (2009).
7. V.V Gladkii, *Phase Transitions.* **6**, 273 (1986).
8. R.A. Toupin, *J. Ration. Mech. Anal.* **5**, 849 (1956)
9. A.F. Devonshire, *Phil. Mag.* **40**, 1040 (1949)
10. E. Bassiouny, A. F. Ghalebs and G. A. Maugin, *Int. J. Engng Sci.* **26**, 1279 (1988)

ON THE ATTENUATION AND VELOCITY DISPERSION OF ULTRASOUND THROUGH TRABECULAR BONE

D. DELIGIANNI, P. KOIKA, T. ASIMAKOPOULOU, T. PHILIPPIDIS

Department of Mechanical Engineering & Aeronautics, University of Patras, Rion 26500, Greece

Many studies report a strong linear correlation between broadband ultrasonic attenuation and apparent density in low density human trabecular bone. In denser bovine bone, however, this correlation is non-linear. Mostly negative and also positive values of dispersion have been measured in trabecular bone *in vitro*. In this work, the experimentally observed non-linear dependence of attenuation on density was explained by attributing it mainly to scattering and the phase velocity was determined by broadband and narrowband pulses.

A standard through-transmission substitution method was used to measure ultrasonic attenuation and velocity. Attenuation coefficient was estimated using a log spectral difference technique. For phase velocity measurements, the phase spectrum of the ratio of the diffraction corrected signals was used. Based on a weak scattering model, the attenuation due to scattering was calculated. The model predicted variations in acoustic parameters with porosity and scatterer size that are consistent with observations, including the nonlinear dependence of attenuation on density.

The phase velocity of cancellous bone is considered strongly frequency-dependent. Negative phase velocity dispersion was measured from the phase spectra whereas negative and positive dispersion was estimated from the transit time velocity of narrowband pulses, depending on the density.

1. Introduction

Many studies report a strong correlation between BUA and apparent density in human cancellous bone with R ranging between 0,84 and 0,97 (Nicholson et al.,1994; Han et al., 1996; Sherpe et al., 1996; Chaffai et al., 2002). In denser bovine cancellous bone, however, the correlation is much weaker or even completely absent and both positive and negative regression slopes have been reported (Töyräs et al., 1999; Han et al., 1996; Alves et al., 1996; Duquette et al., 1997; Töyräs et al., 2002; Hoffmeister et al., 2000; Hakulinen et al., 2004; Lee et al., 2003). The results suggest that the relationship between BUA and density over a large density range of cancellous bone is non-linear (Hodgskinson et al., 1996; Han et al., 1996). The explanation, given in the literature, for this nonlinearity is that between human and bovine trabecular bone there are structural and compositional differences.

Frequency-dependent attenuation and velocity dispersion are acknowledged sources of uncertainty which degrades measurement accuracy when measuring velocity in the time domain. The envelope of received signal may differ considerably from the reference signal, and this makes it difficult to determine the delay time through the specimen.

Several investigators have reported that phase velocity usually -but not always-decreases with frequency in human calcaneus samples *in vitro*

(Nicholson et al., 1996; Droin et al., 1998; Wear, 2000). This is has been considered unusual for biologic tissue and contrary to what one might expect based on models relating frequency-dependent attenuation and frequency-dependent phase velocity (O'Donnell et al., 1981; Waters et al., 2000). A number of theoretical models have been proposed to account for negative dispersion. The so-called "restricted-bandwidth form" of the Kramers-Kronig relations has been shown to accurately predict negative dispersion in bovine cancellous bone (Waters and Hoffmeister, 2005). Negative dispersion can also be explained by interference between fast and slow longitudinal modes in cancellous bone even when each mode is positively dispersive (Marutyan et al., 2006a, b). Negative dispersion in cancellous bone can also be predicted from multilayer models (Wear, 2001, 2007) or a nonlocal version of Biot's theory of poroelasticity (Chakraborty, 2008). Multiple scattering phenomena have been shown that are responsible for negative dispersion values. The resulting velocity dispersion (positive or negative) depends on the relative contribution of multiple scattering and absorption effects (Haiat et al., 2008).

In this work, the experimentally observed attenuation, and its dependence on frequency and density, were explained by attributing it mainly to scattering. The phase velocity and the velocity of pulse propagation were determined by broadband and narrowband pulses. The dispersion determined by narrowband pulse was found to be increasing with frequency.

2. Experimental procedure

Samples of cancellous bone were obtained from ten bovine tibiae of different animals. They were prepared using a band saw to make transverse cuts oriented parallel to the articular surface. These sections were subdivided into approximately cubic specimens with a side length of 22 to 25 mm. The apparent density (the ratio of the dehydrated, defatted tissue mass to the total specimen volume) of the samples was determined from separate measurements of mass and volume.

A number of 25 samples were interrogated. The velocity of the pulse, phase velocity and attenuation of the specimens were measured using both broadband and narrowband pulses. For all measurements unfocused (Panametrics, V303, d=0.5 in., center frequency 1 MHz) immersion transducers were used, connected to an ultrasonic pulser-receiver. RF signals were digitized at 35 MHz.

A standard through-transmission substitution method was used to measure ultrasonic pulse velocity and attenuation. Using two opposite, coaxially–aligned unfocused transducers, the pulse transit time and pulse amplitude spectrum were calculated in the presence and in the absence of the bone sample in the acoustic path. Attenuation coefficient was estimated using a log spectral difference technique. Diffraction related errors were corrected according to the method of Xu et al. (1993). Transmission losses at the interface between the sample and the water were neglected. Attenuation versus frequency was least-squares fit to a linear function over the range from 200 to 900 kHz. The function was characterized by the slope of the resulting line (Broadband Ultrasound Attenuation, BUA).

Narrowband measurements were performed using a pulse/function generator to create a sinusoidal tone burst at selected frequencies from 200 to 1200 kHz in steps of 100 kHz. The velocity of the narrowband pulse was estimated using a time-of-flight method.

2.1. *Attenuation due to scattering*

Bovine cancellous bone possesses a distribution of trabecular thicknesses. The attenuation coefficient due to scattering A_{sc}, with the modified Gaussian correlation model for a continuous distribution of scatterers, will be (Deligianni and Apostolopoulos, 2007):

$$A_{sc}^{T} = 3\int_{a} \frac{A_{sc}\, d\alpha}{a}$$

with

$$A_{sc} = 0.6\langle \gamma^2 \rangle \alpha k^2 \left(1 + \frac{1}{2}\alpha^2 k^2 \right) e^{-\frac{1}{2}\alpha^2 k^2}$$

where: α is the correlation length, k is the wave number and $\langle \gamma^2 \rangle$ is the mean square fluctuation in medium properties. Its nhe numerical value is estimated by taking ρ_b=1800 kg/m^3, ρ_w=1000 kg/m^3, c_b=3300 m/s and c_w=1450 m/s, constant throughout a given specimen as well as for different individuals.

2.2. *Phase velocity calculation*

To obtain the continuous phase spectrum, phase had to be unwrapped by appropriately adding or subtracting multiples of 2π to the principal value until the discontinuities induced by the modulo 2π operation were removed. The unwrapped phase $\varphi_u(f)$ is written:

$\varphi_u(f) = \varphi(f) \pm 2k\pi$, where k is an integer.

The phase velocity could be calculated by the following equation in which the measured phase $\varphi(f)$ has been replaced by the unwrapped phase $\varphi_u(f)$,:

$V(f) = d/(d/V_\omega - \varphi_u(f)/2\pi f d)$

Highly attenuating media like cancellous bone are highly dispersive (Nicholson et al., 1996). The phase velocity is strongly frequency-dependent. For that reason, velocity dispersion was characterized by UBV defined as the value of the phase velocity taken at the center frequency of the initial pulse and by the velocity dispersion magnitude (VDM) defined as:

$VDM = \Delta V = V_{(fmax)} - V_{(fmin)}$,

where f_{min} = 300 kHz and f_{max} = 1100 kHz are the lower and upper limits of the transducer frequency bandwidth.

3. Results

Figure 1 represents the predicted from the densely populated model attenuation due to scattering by a continuous distribution of scatterers, ranging from 100 to 600 μm, versus frequency.

The scattering model predicts that attenuation due to scattering by a distribution of scatterers is linear in the range of 0.3-0.9 MHz. On the contrary, attenuation by identical scatterers is not linear or it is linear in a very limited frequency range, depending on the scatterer size. If two sizes of scatterers were present in the scattering volume, the attenuation versus frequency displays two different slopes. Experimentally, a number of specimens presented two distinct slopes of attenuation function versus frequency. Linear dependence of ultrasonic attenuation from bovine cancellous bone specimens with high density was observed in this frequency range. BUA due to scattering by a distribution of scatterers was predicted to be 45 $dB \cdot cm^{-1} \cdot MHz^{-1}$. This value is of the order of the measured experimentally total attenuation for bovine cancellous bone.

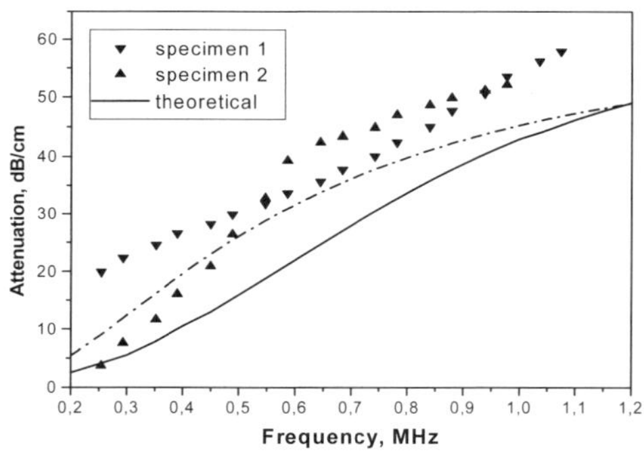

Fig.1. Predicted attenuation due to scattering, with densely populated medium model, for two dominant groups of scatterers sizes, 200 and 600 μm (−·−) and for a continuous distribution of scatterers sizes 100 to 600 μm (—). ▲, ▼: experimental data, which correspond to the attenuation of two representative specimens, one displaying linear behavior with frequency in the examined frequency range and the other two distinct slopes of attenuation in the same frequency range.

Figure 2 illustrates the change with apparent density of the ultrasonic parameter BUA after five successive stages of demineralization of a representative specimen in three perpendicular directions. Specimens demonstrated a non linear relationship of BUA with apparent density. As apparent density decreased at the successive demineralization stages, BUA increased up to apparent density of 0.65 g/cm^3 and then it decreased with apparent density. For specimens with values of apparent densities at the baseline lower than 0.65 g/cm^3, this relationship was approximately linear, as it has been found by various researchers (Alves et al., 1996; Chaffaî et al., 2002; Töuräs et al., 1999). Predicted by the scattering model BUA due to scattering displayed maximum value at the same apparent density. The values of BUA depend on the scatterer size (trabecular thickness). In case of identical scatterers they increase for scatterers size up to 400 μm and, for larger scatterers, they decrease. BUA due to scattering by a spectrum of scatterers was predicted to be approximately 45 dB/cm MHz.

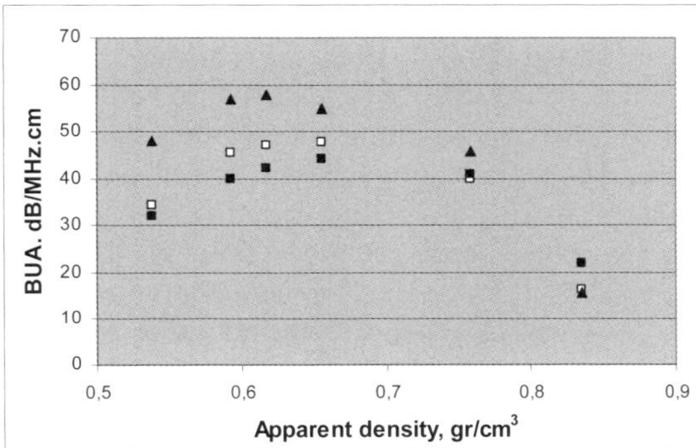

Fig. 2. Change of the ultrasonic property BUA with apparent density after five successive stages of demineralization of a representative specimen in three perpendicular directions.

Figure 3 shows phase velocities derived for three typical specimens. Figure 4 displays the velocity of the narrowband pulse measured by the transit time method for the same specimens. The increase of the pulse velocity with frequency of some specimens, whose phase velocity decreased, was noticeable. Average phase velocity at 700 kHz was 1489 ± 15 m/s (mean ± standard deviation). The VDM was −103 ± 30 m/s. The VMD for a typical specimen was -108 m/s. The corresponding VMD, calculated from the narrow pulse velocity for the same specimen, was 20 ± 36 m/s.

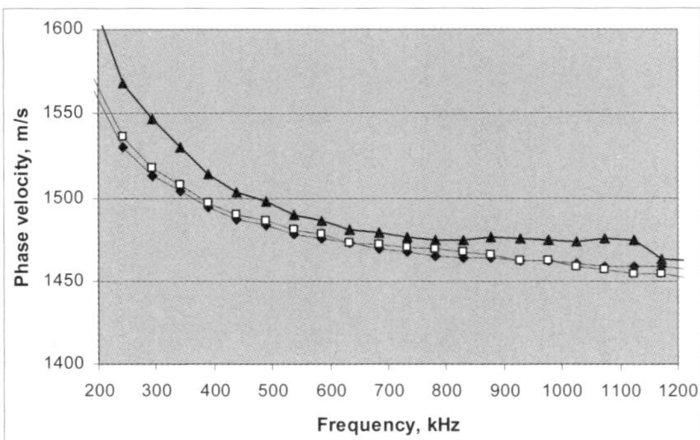

Fig. 3. Phase velocities estimated from the phase spectra for three typical specimens

4. Discussion

The scattering model, proposed for the ultrasonic frequency-dependent backscatter in dense bovine cancellous bone, using a combination of two autocorrelation functions to describe the medium (Deligianni and Apostolopoulos, 2007), was used to estimate the attenuation due to scattering. Attenuation due to scattering (A_{sc}) for bovine cancellous bone was predicted from the model to be strong (70-80% of the measured total attenuation). A_{sc} by a continuous distribution of scatterers was predicted to be linear over a frequency range from 0.3 to 0.9 MHz, suggesting a possibility that scattering may be a significant source of attenuation through bovine trabecular bone..

Indications that scattering may be a significant source of attenuation in human trabecular bone have been reported, although there are also contrary implications (Wear, 2000). Models development (Bossy et al., 2007) has shown that scattering phenomena are a major contribution to the overall attenuation losses for frequencies around 600 kHz.

Kaufman et al. (2003), with computer simulations of ultrasonic propagations concluded that most of the frequency dependent attenuation observed in trabecular bone is due to scattering of the ultrasound wave and not from absorption losses. Much more theoretical analysis and experimentation are necessary to determine the relative roles of scattering and absorption in attenuation of ultrasound through cancellous bone.

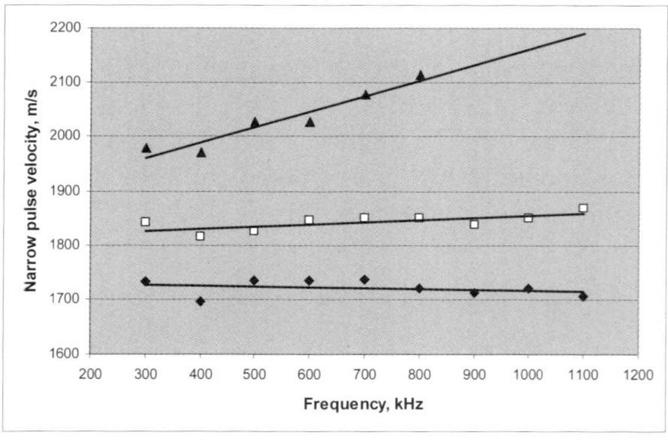

Fig. 4. The transit time velocities of a narrowband pulse through three representative specimens.

Dispersion appears to be an important acoustical property, influencing ultrasound velocity measurements performed clinically. A better understanding of its physical determinant may have important impacts on the clinical relevance of velocity measurements. Possible explanations of the negative values of dispersion measured by phase spectra have been given by developing theoretical models or performing numerical simulations in recent studies in the context of trabecular bone (Haiat et al., 2008; Chakraborty, 2008). The attenuation coefficient in bone being a linearly increasing function of frequency, negative dispersion has been referred to as "anomalous" (Marutyan et al., 2006), due to the apparent contradiction with the Kramers–Krönig relationships. A positive dispersion might be expected in these cases based on consideration of the causality-induced Kramers–Kronig relations.

In this work, for the specimens investigated, broadband and narrowband data in the frequency range 0.3-1.1 MHz do not demonstrate good consistency. Positive and negative dispersion was estimated from the velocity of the narrowband pulses and was depended on the density. The higher the density, the less negative dispersion was observed (Table 1). On the contrary, only negative dispersion was estimated from the phase spectra for the examined specimens. Similar values have been reported for phase velocities by other researchers (Nicholson et al., 1996; Droin et al., 1998; Wear, 2000). Further careful measurements are necessary to resolve the issue of dispersion in cancellous bone.

Table 1. Relation of VMD with density for three representative specimens.

Density (gr/cm^3)	VDM
1269	-15
1324	17
1651	58

References

1. P.H.F. Nicholson, M.J. Haddaway, M.W.J. Davie, *Phys. Med. Biol.* **39**, 1013 (1994).
2. S. Han, J. Rho, J. Medige, I. Ziv, *Osteoporosis Int.* **6**, 291 (1996).
3. L. Serpe and J.Y. Rho, *J. Biomech.* **29**, 963 (1996).
4. S. Chaffaî, F. Peyrin, S. Nuzzo, et al., *Bone* **30**, 229 (2002).
5. J. Töuräs, H. Kröger, J.S. Jurvelin, *Bone* **25**, 725 (1999).
6. J.M. Alves, W. Xu, D. Lin, et al., *IEEE Trans. Biomed. Eng.* **43**, 249 (1996).
7. J. Duquete, T. Honeyman, A. Hoffman et al., *Bone* **21**, 289 (1997).
8. J. Töuräs, M.T. Nieminen, H. Kröger, J.S. Jurvelin, *Bone* **31**, 503 (2002).
9. B.K. Hoffmeister, S.A. Whitten, J.Y. Rho, *Bone* **26**, 635 (2000).
10. M.A. Hakulinen, J. Töyräs, A. Saarakkala, et al. *Ultrasound Med. & Biol.* **30**, 919 (2004).
11. K. Lee, H.-S. Poh, S.W. Yoon, *J. Acoust. Soc. Am.* **113**, 2933 (2003).
12. R. Hodgskinson, C.F. Njeh, M.A. Whitehead, C.M. Langton, *Phys. Med. Biol.* **41**, 2411 (1996).
13. P.H.F. Nicholson, G. Lowet, C.M. Langton, J. Dequeker, and G. Van der Perre, *Phys. Med. Biol.* **41**, 2421 (1996).
14. P. Droin, G. Berger, and P. Laugier, *IEEE Trans. Ultrason. Ferroelectr. Freq. Control* **45**, 581 (1998).
15. K.A. Wear, *J. Acoust. Soc. Am.* **107,** 3474 (2000).
16. M. O'Donell and J.G. Miller, *J. Appl. Phys.* **52**, 1056 (1981).
17. K.R. Waters, M.S. Hughes, J. Mobley, et al., *J. Acoust. Soc. Am.* **108**, 556 (2000).

18. K.R. Waters, and B.K. Hoffmeister, *J. Acoust. Soc. Am.* **118**, 3912 (2005).
19. K.R. Marutyan, G.L. Bretthorst, and J.G. Miller, *J. Acoust. Soc. Am.* **121**, EL8 (2006a).
20. K.R. Marutyan, M.R. Holland, and J.G. Miller, *J. Acoust. Soc. Am.* **120**, EL55 (2006b).
21. K.A. Wear, *IEEE Trans. Ultrason. Ferroelectr. Freq. Control* **48**, 1079 (2001).
22. K.A. Wear, *J. Acoust. Soc. Am.* **121**, 2431 (2007).
23. A. Chakraborty, *J. Acoust. Soc. Am.* **123**, 56 (2008).
24. G. Haiat, A. Lhémery, F. Renaud. et al., *J. Acoust. Soc. Am.* **124,** 4047 (2008).
25. W. Xu and J.J. Kaufman, *IEEE Trans. Biomed. Eng.* **40**, 563 (1993).
26. D. Deligianni and K. Apostolopoulos, *J. Acoust. Soc. Am.* **122**, 1180 (2007).
27. E. Bossy, P. Laugier, F. Peyrin, and F. Padilla, *J. Acoust. Soc. Am.* **122**, 2469 (2007).
28. J.J. Kaufman, G. Luo, R.S. Siffert, *IEEE Ultrasonics Symposium*, 1519 (2003).

ON KELVIN'S TRANSFORMATION OF POTENTIAL PROBLEMS IN AN EQUILATERAL TRIANGLE

G. BAGANIS

School of Science and Technology, Hellenic Open University,

167 R. Feraiou str. GR-26 222 Patras, Greece

E-mail: gbaganis@eap.gr

M. HADJINICOLAOU

School of Science and Technology, Hellenic Open University,

167 R. Feraiou str. GR-26 222 Patras, Greece

E-mail: hadjinicolaou@eap.gr

Abstract

Kelvin's transformation, also known as Kelvin inversion, is a well known mathematical tool used to solve various potential problems. Its significance is due to the fact that the solution of the Laplace equation, in a domain $\Omega \subseteq \mathbb{R}^3$, remains solution of the Laplace equation for the Kelvin image of this domain [1]. In the present work, we study the application of Kelvin's inversion in $\mathbb{R}^2$. In accordance to the 3-D case, we formulate the Kelvin theorem in 2-D. This enables us to obtain solutions of the Laplace equation either in domains that are subsets of $\mathbb{R}^2$ or in those of $\mathbb{R}^3$ that enjoy axial symmetry. Since the Kelvin inversion transforms the boundaries in a nonlinear way, we are able to solve boundary value problems for harmonic functions when one, either the image or the pre-image domain accepts separable solutions. An interesting case is that of the interior of an equilateral triangle. The solution of this problem, for different B.C.'s has been given by Dassios and Fokas [2]. The equilateral triangle is mapped under Kelvin's transformation with respect to the circumscribe circle, into a symmetric non-convex shape, and then, by a straightforward sequence of steps one can construct harmonic functions that fit completely the geometry of the exterior of this particular non-convex domain.

Keywords: Kelvin inversion; equilateral triangle; Dirichlet-to-Neumann map; analytic solution of Laplace equation, non-convex domain.

1. Introduction

Kelvin's transformation (inversion) is a non-linear map, introduced by William Thompson (Lord Kelvin) in 1845. It can be described as a coordinate transformation with respect to a circle, in case of $\mathbb{R}^2$, or to a sphere in case of $\mathbb{R}^3$, and it has the important property to preserve harmonicity. This means that once the solution of the Laplace equation in a domain is known, then the solution of the Laplace equation in the image domain under the Kelvin's transformation can also be obtained. This property provides powerful machinery for solving potential problems in a very effective way and it has been applied successfully in many cases of convex and smooth boundaries, in cases where the method of separation of variables can not be applied. In the present manuscript we employ the Kelvin inversion in order to solve α potential problem for an exterior potential problem, in a non convex domain which is the Kelvin image of an equilateral triangle. As far as we know, there is no general method of deriving analytic solutions of potential problems in non convex domains. Noting that the domain is a subset of $\mathbb{R}^2$, we state the Kelvin's theorem in 2-D and we perform the transformation with respect to the circumscribed circle of the triangle.

The proof of the theorem is given by the authors in [5]. Then we need to invert the solution of the potential problem which holds in the interior of the triangle. As it is well known, there is no curvilinear system that allows separation of variables for the Laplace equation in the equilateral triangle. Only recently, a new approach for obtaining the solution for two-dimensional boundary value problems and for the Laplace equation in a convex polygon was obtained by Dassios and Fokas [2]. The solution is given through an integral representation in terms of the *spectral functions* which are defined by line integrals over each side of the boundary of the convex polygon. These spectral functions satisfy an equation, the so-called *global relation* by Fokas [3] or *functional equation* by Shanin [4]. Depending of the type of problem, the integrand is expressed in terms of the known Dirichlet and the unknown Neumann data or vice versa. In the case of the Dirichlet problem, the missing Neumann data can be evaluated by the *Dirichlet-to-Neumann map* while in the case of the Neumann problem the missing Dirichlet data, are obtained through the *Neumann-to-Dirichlet map*, both are given by Dassios and Fokas in [2].

The combination of the Kelvin's inversion and the Fokas integral representation provide the solution of the Laplace equation in the non-convex domain which is the Kelvin image of an equilateral triangle [5].

The structure of the manuscript is as follows. In section 2, we present the geometry, the properties of the Kelvin inversion, the Kelvin 2-D theorem, as well as the effect of the transformation on some basic differential operators. In the third section we apply the method of the Kelvin inversion to an equilateral triangle with respect to its circumscribed circle, and we describe the way which the image of the equilateral triangle is obtained. In section 4, we indicatively give, as an application, the solution of an exterior Dirichlet problem, subject to the first even component of the Fourier series expansion, in the domain which is the Kelvin image of the equilateral triangle. In section 5 we discuss the outcomes and we comment on some interesting points of the method.

2. Geometry and properties of Kelvin's inversion.

Let V denote the interior of a closed, bounded, smooth surface S in $\mathbb{R}^2$.

We denote the position vector by $\boldsymbol{r}$ and $|\boldsymbol{r}| = r$ (2.1)

We choose the circle of radius $\rho > 0$ and centre (0,0) as the circle of inversion for the Kelvin transformation

$$K : \mathbb{R}^2 \to \mathbb{R}^2, \ \bar{\boldsymbol{r}} := K(\boldsymbol{r}) := \frac{\rho^2}{r^2}\boldsymbol{r} \qquad (2.2)$$

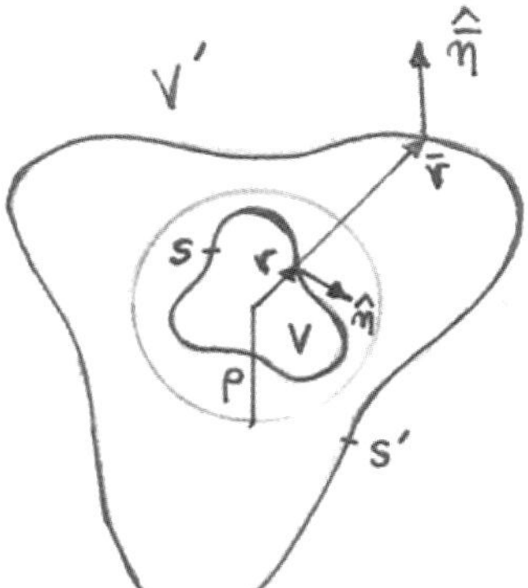

Fig. 1. Kelvin's inversion

An immediate result from the definition is

$$\bar{r} r = \rho^2 \qquad (2.3)$$

which means that the norms of the two position vectors are inversely proportional therefore the point at the origin is mapped to the point at the

infinity and the image of the surface S is a smooth surface S′, lying out of the circle with exterior V′ which is the image of V (*fig. 1*).
Another result that comes right from the definition is

$$\hat{\bar{\boldsymbol{r}}} = \hat{\boldsymbol{r}} \tag{2.4}$$

which indicates the relation between the two unit vectors.
It is also of interest to observe the effect of the Kelvin's inversion on the differential operators [5]:
a) The relation that combines the gradients in the original domain and its image is

$$\nabla_r = \frac{\bar{r}^2}{\rho^2}(\nabla_{\bar{r}} - 2\hat{\bar{r}}\frac{\partial}{\partial \bar{r}}) \tag{2.5}$$

b) The effect of the Kelvin inversion to the Laplace operator is

$$\Delta u(\boldsymbol{r}) = \frac{\bar{r}^4}{\rho^4}\bar{\Delta} u(\bar{\boldsymbol{r}}) \tag{2.6}$$

This relation gives rise to the following fundamental result that we state below as Kelvin's two-dimensional (2-D) theorem, the proof of which is given in [5].

The Kelvin 2-D theorem. *Let V be a bounded domain in* $\mathbb{R}^2$ *and let V′ be its image under the transformation*

$$\bar{\boldsymbol{r}} = \frac{\rho^2}{r^2}\boldsymbol{r}. \tag{2.7}$$

If $u(\boldsymbol{r})$ *is a solution of* $\Delta u(\boldsymbol{r}) = 0$, $\boldsymbol{r} \in V$ then $\bar{u}(\bar{\boldsymbol{r}}) = u(\bar{\boldsymbol{r}})$ *is a solution of* $\bar{\Delta}\bar{u}(\bar{\boldsymbol{r}}) = 0$, $\bar{\boldsymbol{r}} \in V'$.

3. Kelvin Inversion of the equilateral triangle.

Let V be the interior domain, bounded by an equilateral triangle *S*. We assume that the length of each side is *l* and the vertices of this triangle are the images, in the complex plane, of the complex numbers

$$z_1 = \frac{l}{\sqrt{3}}e^{\frac{i\pi}{3}},\ z_2 = \bar{z}_1,\ z_3 = -\frac{l}{\sqrt{3}} \tag{3.1}$$

Then, the sides (z_2, z_1), (z_1, z_3) and (z_3, z_2) of the equilateral triangle are described by the position vectors $\boldsymbol{r}_1$, $\boldsymbol{r}_2$ and $\boldsymbol{r}_3$ respectively.

$$\boldsymbol{r}_1(\theta_1) = (\frac{l}{2\sqrt{3}}, \frac{l}{2\sqrt{3}}\tan\theta_1), \quad -\frac{\pi}{3} \le \theta_1 \le \frac{\pi}{3} \tag{3.2a}$$

$$\boldsymbol{r}_2(\theta_2) = (\frac{l\cos\theta_2}{3\sin\theta_2 - \sqrt{3}\cos\theta_2}, \frac{l\sin\theta_2}{3\sin\theta_2 - \sqrt{3}\cos\theta_2}), \quad \frac{\pi}{3} \le \theta_2 \le \pi \tag{3.2b}$$

$$\boldsymbol{r}_3(\theta_3) = (\frac{l\cos\theta_3}{-3\sin\theta_3 - \sqrt{3}\cos\theta_3}, \frac{l\sin\theta_3}{-3\sin\theta_3 - \sqrt{3}\cos\theta_3}), \quad \pi \le \theta_3 \le \frac{5\pi}{3} \tag{3.2c}$$

Applying the Kelvin 2-D inversion

$$\overline{\boldsymbol{r}}(\theta) = \left(\frac{\rho}{r(\theta)}\right)^2 \boldsymbol{r}(\theta), \tag{3.3}$$

with respect to the circumscribed circle of the equilateral triangle, namely the circle of radius $\rho = \frac{l}{\sqrt{3}}$ and centre (0,0), the above vectors are transformed into the following position vectors $\overline{\boldsymbol{r}}_j$, $j = 1, 2, 3$.

$$\overline{\boldsymbol{r}}_1(\theta_1) = (\frac{2l}{\sqrt{3}}\cos^2\theta_1, \frac{l}{\sqrt{3}}\sin 2\theta_1), \quad -\frac{\pi}{3} \le \theta_1 \le \frac{\pi}{3} \tag{3.4a}$$

$$\overline{\boldsymbol{r}}_2(\theta_2) = (\frac{l}{2}\sin 2\theta_2 - \frac{l\sqrt{3}}{3}\cos^2\theta_2, l\sin^2\theta_2 - \frac{l\sqrt{3}}{6}\sin 2\theta_2), \quad \frac{\pi}{3} \le \theta_2 \le \pi \tag{3.4b}$$

$$\overline{\boldsymbol{r}}_3(\theta_3) = (-\frac{l}{2}\sin 2\theta_3 - \frac{l\sqrt{3}}{3}\cos^2\theta_3, -l\sin^2\theta_3 - \frac{l\sqrt{3}}{6}\sin 2\theta_3), \quad \pi \le \theta_3 \le \frac{5\pi}{3}, \tag{3.4c}$$

The boundary S′ is the image of the perimeter S and the exterior domain V′ is the image of the interior V. (*fig.2*).

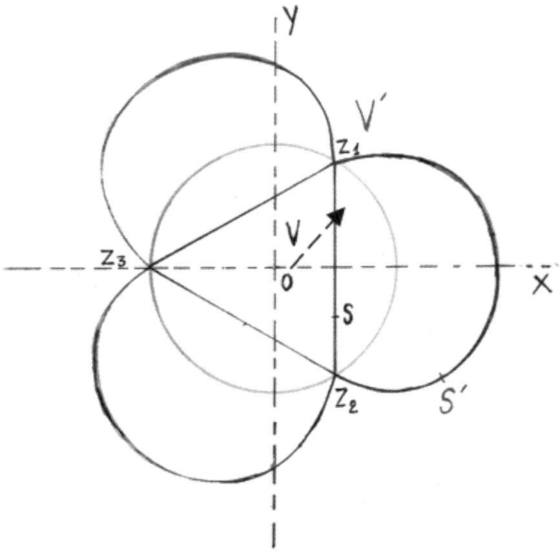

Fig. 2. Kelvin's inversion for an equilateral triangle.

4. Application

We consider the Dirichlet problem:

$$\Delta u(\boldsymbol{r}) = 0 \qquad \boldsymbol{r} \in V', \tag{4.1}$$

$$u(\boldsymbol{r}) = \overline{f}, \qquad \boldsymbol{r} \in S' \tag{4.2}$$

where V′ is the exterior region to a closed and smooth non convex surface in □ ^{2}, bounded by the curve S′, (*Fig. 3*). We further impose that the same function $\overline{f} \in$ □ ^{2}(□) holds on each of the three compact components of the curve S′ and specifically:

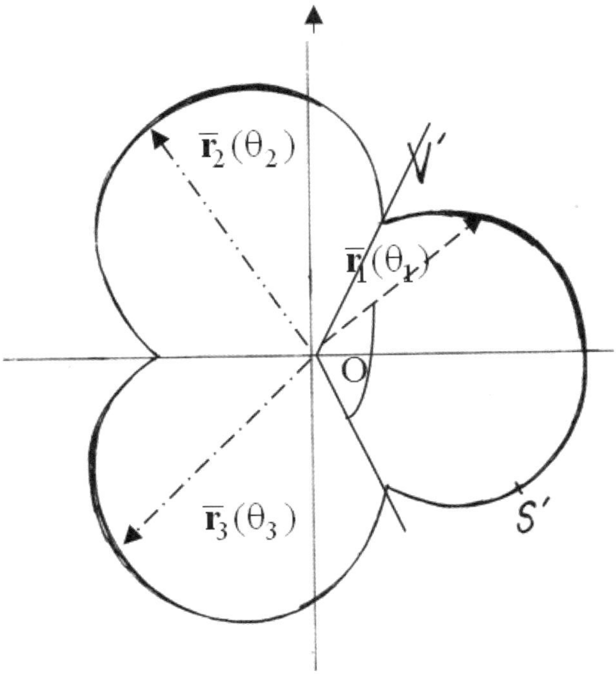

Fig. 3. Geometry of the problem

$$\bar{f}(s_j) = \cos\frac{2\pi s_j}{l}, \quad j=1,2,3 \tag{4.3}$$

where

$$s_1 = \frac{3l\theta_1}{2\pi}, \qquad -\frac{\pi}{3} \le \theta_1 \le \frac{\pi}{3} \tag{4.4a}$$

$$s_2 = \frac{3l\theta_2}{2\pi} - l, \quad \frac{\pi}{3} \le \theta_2 \le \pi \tag{4.4b}$$

$$s_3 = \frac{3l\theta_3}{2\pi} - 2l, \quad \pi \le \theta_3 \le \frac{5\pi}{3} \tag{4.4c}$$

Also, $u(\boldsymbol{r})$ satisfies the asymptotic condition

$|u(\boldsymbol{r})| \le c$, when $\boldsymbol{r}$ tends to infinity, (4.5)

for every $\boldsymbol{r} \in \mathrm{V}'$ with $|\boldsymbol{r}| \ge R$, where R is some positive number and c is a positive constant.

The relation that combines the normal derivatives on the two boundaries is

$$\frac{\partial}{\partial \overline{n'}} = -\frac{3r_i^{'2}}{l^2}\frac{\partial}{\partial n'}, \tag{4.6}$$

where $\frac{\partial}{\partial n'}$ stands for the normal derivative on the surface S and $\frac{\partial}{\partial \overline{n'}}$ for the normal derivative on the surface S′. These quantities are used for expressing the solution of the B.V. problem as an integral representation. The methodology that has to be followed is explicitly given in [5], in a general way. The solution in the above mentioned particular case is then

$$u(\overline{\boldsymbol{r}}) =$$

$$= -\frac{r_1^{'2}\sqrt{3}}{\pi l}\int_{-\pi/3}^{\pi/3}\left[ln\frac{1}{\left|\overline{\boldsymbol{r}}-\overline{\boldsymbol{r}}_1^{'}\right|}\frac{-2\sqrt{3}}{l}\sum_{n=-\infty}^{\infty}e^{\frac{-2in\pi s_1}{l}}\frac{-(-1)^n n^2}{n^4+n^2+1}-\cos\frac{2\pi s_1}{l}\frac{\partial}{\partial n'}ln\frac{1}{\left|\overline{\boldsymbol{r}}-\overline{\boldsymbol{r}}_1^{'}\right|}\right]d\theta_1$$

$$-\frac{r_2^{'2}\sqrt{3}}{\pi l}\int_{\pi/3}^{\pi}\left[ln\frac{1}{\left|\overline{\boldsymbol{r}}-\overline{\boldsymbol{r}}_2^{'}\right|}\frac{-2\sqrt{3}}{l}\sum_{n=-\infty}^{\infty}e^{\frac{-2in\pi s_2}{l}}\frac{-(-1)^n n^2}{n^4+n^2+1}-\cos\frac{2\pi s_2}{l}\frac{\partial}{\partial n'}ln\frac{1}{\left|\overline{\boldsymbol{r}}-\overline{\boldsymbol{r}}_2^{'}\right|}\right]d\theta_2$$

$$-\frac{r_3^{'2}\sqrt{3}}{\pi l}\int_{\pi}^{5\pi/3}\left[ln\frac{1}{\left|\overline{\boldsymbol{r}}-\overline{\boldsymbol{r}}_3^{'}\right|}\frac{-2\sqrt{3}}{l}\sum_{n=-\infty}^{\infty}e^{\frac{-2in\pi s_3}{l}}\frac{-1(-1)^n n^2}{n^4+n^2+1}-\cos\frac{2\pi s_3}{l}\frac{\partial}{\partial n'}ln\frac{1}{\left|\overline{\boldsymbol{r}}-\overline{\boldsymbol{r}}_3^{'}\right|}\right]d\theta_3$$

(4.7)

5. Discussion

We presented an effective way of obtaining solutions of potential problems in analytic form, by employing the Kelvin transformation. This is feasible since the transform has the property to preserve the harmonicity of the solutions of the potential problems. Through the Kelvin inversion, an interior domain is transformed to its image, exterior one. According to the Kelvin theorem in 2-D, which we stated, a solution of an interior potential problem is transformed to the corresponding solution of the equivalent exterior problem and vice versa. The circle of inversion is considered to be the circumscribed circle. It is worth mention here that by varying the radius of the inversion circle, we obtain different, but similar, images of the triangle. This is due to fact that the Kelvin transformation is a conformal map and thus it allows solutions of potential problems in a class of non convex domains.

In the present manuscript we applied the Kelvin inversion to an equilateral triangle, with respect to the circumscribed circle, and we derived the image of it, which is the exterior of a non convex domain described by eqs 3.4, depicted in fig.2. As a case study, we assumed a Dirichlet boundary value problem with a simple boundary condition (cosθ) holding in every compact component of the image boundary. The solution of this exterior problem is given through an integral representation which employs the Kelvin inversion of the solution of the equivalent problem in an equilateral triangle, given by Dassios and Fokas in [2]. Analogous results we obtain in the case where the circle of inversion is assumed to be the inscribed circle. The image of the triangle is a similar non-convex domain, described by eqs. 3.4, divided by the factor 4. An analogous procedure must be followed for obtaining the solution for the case of Neumann problem.

References

[1] W. Thompson, Lord Kelvin, Papers on electrostatics and Magnetism, Mac Millan, London, 1982.

First published in J. Math. Pures Appl.,10 (1845), p.364; 12(1847), p.256.

[2] Dassios, G and Fokas A. S. 2005. The basic elliptic equations in an equilateral triangle. Proc. Soc. A **461**, 2721-2748.

[3] Fokas, A. S. 2001 Two-dimensional linear partial differential equations in a convex polygon. Proc. R. Soc. A **457**, 371-393.

[4] Shanin, A.,V. 1997. Excitation of wave field in a triangle with impedance boundary conditions. Journal of Mathematical Sciences, Springer, New York. Volume 102, Number 4/ December, 2000 pp 4328-4338.

[5] G. Baganis and M. Hadjinicolaou 2009 Analytic solution of an exterior Dirichlet problem in a non-convex domain. IMA Journal of Applied Mathematics 2009 **74**: 668-684

REFLECTION OF THERMOELASTIC PLANE WAVES IN ANISOTROPIC THERMOELASTIC MEDIUM

G. KOSKOVITI, V. KOSTOPOULOS

Department of Mechanical Engineering and Aeronautics,
University of Patras, GR 26500, Greece
E-mail: gkoskov@mech.upatras.gr, kostopoulos@mech.upatras.gr

F. KARIOTOU

School of Science and Technology,
Hellenic Open University, Patras, Greece. E-mail: kariotou@eap.gr

The reflection problem of plane thermoelastic waves on the boundary of a half-space of homogeneous, anisotropic thermoelastic medium has been investigated. The general thermoelastic boundary conditions are presented and their physical interpretation is discussed in details. The problem under investigation considers a plane thermoelastic wave, which propagates in a homogeneous anisotropic medium and is reflected by an anisotropic half space with plane boundary, in the absence of body forces and heat sources. The derivation of the expressions for the reflection coefficients, which are the ratio of the amplitudes of the reflected waves to the amplitude of the incident waves is presented. The variation of reflection coefficients with angle of incident has been also studied. Regenerated case of orthotropic thermoelastic medium has been analyzed and known results have been reproduced.

1. Introduction

Many authors studied problems on reflection of thermoelastic waves in a solid half-space such as Ditri and Rose [2] for transversely isotropic elastic case, Sinha and Elsibai [3] and Abd-Alla, et al. [5] for isotropic thermoelastic case. Sharma, et al. [6] investigated the problem of thermoelastic wave reflection from the insulated and isothermal stress-free as well as rigidly fixed boundaries of homogeneous isotropic solid half-spaces. Singh and Singh [7] studied the reflection of plane waves at the free surface of a fibre-reinforced isotropic elastic half-space. Singh [9] studied the reflection of SV waves from the free surface of an elastic solid in generalized thermoelastic diffusion. Singh, et al. [10] investigated the reflection of generalized thermoelastic waves from an isotropic solid half-space under hydrostatic initial stresses. Finally, Lahiri,et al. [11] studied the reflection of generalized thermoelastic waves from isothermal and insulated boundaries of a half-space.

In this paper we have investigated the problem of reflection of thermoelastic wave from an insulated stress-free boundary of an anisotropic solid half-space. In Sec.2 we give a unified formulation of the thermoelastic problem starting with the complete system of equations of thermoelasticity. We give the representations of the elastothermal and thermoelastic fields and their combinations. In Sec.3 we provide the boundary conditions and the reflection coefficients. Finally, we study the special case of the reflection problem for orthotropic half-space medium and we exhibit the variation of reflected coefficients with angle of incident.

2. Unified formulation

Let V be the medium of propagation, which is assumed to be a homogeneous and thermally conducting anisotropic half-space with respect to all physical constants, both elastic and thermal. The elastic properties of the medium are described by the forth order stiffness tensor $\tilde{\tilde{\mathbf{C}}}$ for the anisotropic body and the mass density ρ. The coefficients of thermal conductivity are included in the second order tensor $\tilde{\mathbf{K}}$. Let $S = \partial V$ be the infinite plane boundary surface of the medium V, which we consider as the coordinate surface Oxy. The medium V is described by $z > 0$ and the boundary S corresponds to $z = 0$.

In the medium V, an incident plane wave propagates in the direction $\mathbf{k}^I$ which, by no loss of generality, is assumed to lie upon the xz-coordinate surface, making an angle θ^I with z-axis. So, the direction of propagation of the incident wave is $\mathbf{k}^I = (\sin\theta^I, 0, \cos\theta^I)$. The incident wave is reflected by the boundary surface S and a reflected wave is produced and also propagates in V, on the xz-plane with direction $\mathbf{k}^R$. The geometry of the problem is shown in Fig.1.

Suppressing the harmonic time dependence $e^{-\imath\omega t}$ we obtain the following time-independent system of linearized thermoelasticity:

$$\begin{aligned}\tilde{\tilde{\mathbf{C}}}\vdots\nabla\nabla\mathbf{u}(\mathbf{r}) + \rho\omega^2\mathbf{u}(\mathbf{r}) &= \tilde{\mathbf{B}}\cdot\nabla\Theta(\mathbf{r}) \\ \tilde{\mathbf{K}} : \nabla\nabla\Theta(\mathbf{r}) + \imath\omega\rho c_\varepsilon\Theta(\mathbf{r}) &= -\imath\omega\mathrm{T}_0\tilde{\mathbf{B}} : \nabla\mathbf{u}(\mathbf{r}).\end{aligned} \tag{1}$$

The system of equations determine the displacement $\mathbf{u}(\mathbf{r})$ and the temperature $\Theta(\mathbf{r})$. Assume that there exist neither body forces nor heat sources. The tensor $\tilde{\mathbf{B}}$ determines the coupling between the elastic and thermal

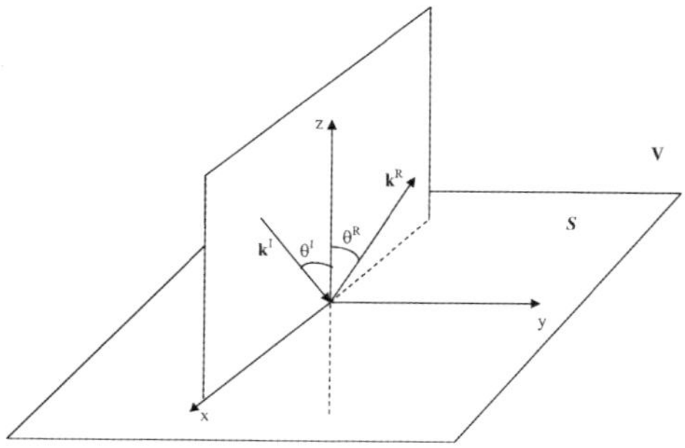

Figure 1. Geometry of the Problem.

effects and the tensor $\tilde{\mathbf{K}}$ denote the heat conduction coefficients. The constant c_ε is called the specific heat at constant strain.

A unified formulation [1] of the thermoelastic problem can be obtained by introducing four-dimensional vector fields, where the first three components are the components of the elastic displacement and the forth component is the temperature field, i.e.,

$$\mathbf{U}(\mathbf{r}) = \left(\mathrm{u}_0^{(1)}, \mathrm{u}_0^{(2)}, \mathrm{u}_0^{(3)}, \Theta_0\right) e^{\imath k \hat{\mathbf{k}} \cdot \mathbf{r}} \tag{2}$$

where $\mathbf{u}_0$ is the polarization vector. Then the system (1) is written as

$$\tilde{\mathbf{L}}(\partial_\mathbf{r}) \cdot \mathbf{U}(\mathbf{r}) = \mathbf{0}, \tag{3}$$

where the time-independent thermoelastic operator

$$\tilde{\mathbf{L}}(\partial_\mathbf{r}) = \left[\begin{array}{c|c} \tilde{\tilde{\mathbf{C}}}\dot{:}\nabla\nabla\tilde{\mathbf{I}}_3 + \rho\omega^2\tilde{\mathbf{I}}_3 & -\tilde{\mathbf{B}}\cdot\nabla \\ \hline \imath\omega \mathrm{T}_0\tilde{\mathbf{B}} : \nabla & \tilde{\mathbf{K}} : \nabla\nabla + \imath\omega\rho c_\varepsilon \end{array}\right] \tag{4}$$

is a 4×4 matrix differential operator. The upper left block of $\tilde{\mathbf{L}}$ is the 3×3 matrix of the corresponding elastic problem, while the lower right block is the scalar differential operator of the corresponding heat conduction equation. The upper right block is a 3×1 matrix representing the effect of the temperature on the elastic behavior of the medium and the lower left block is a 1×3 matrix representing the effect of the elastic field on the thermal behavior of the medium.

The unified total field $\mathbf{\Psi}(\mathbf{r})$ is given by

$$\mathbf{\Psi}(\mathbf{r}) = \mathbf{\Phi}^{\mathrm{I}}(\mathbf{r}) + \mathbf{\Phi}^{\mathrm{R}}(\mathbf{r}) \tag{5}$$

where $\mathbf{\Phi}^{\mathrm{I}}(\mathbf{r})$ is the unified incident and $\mathbf{\Phi}^{\mathrm{R}}(\mathbf{r})$ the unified reflected field. The unified incident and reflected plane waves $\mathbf{\Phi}^{\mathrm{i}}$ have the following form

$$\mathbf{\Phi}^{\mathrm{i}}(\mathbf{r}) = \sum_{j=1}^{4} \mathbf{\Phi}^{\mathrm{i},j}(\mathbf{r}) \tag{6}$$

where

$$\mathbf{\Phi}^{\mathrm{i},j}(\mathbf{r}) = \mathrm{A}^{\mathrm{i},j}(\mathbf{u}_0^{\mathrm{i},j}, \mathrm{c}_j^{\mathrm{i}}) \mathrm{e}^{\imath k_j^{\mathrm{i}} \hat{\mathbf{k}}^{\mathrm{i}} \cdot \mathbf{r}} \tag{7}$$

for $j = 1, 2, 3$ is the unified elastothermal plane wave with amplitude $\mathrm{A}^{\mathrm{i},j}$ and

$$\mathbf{\Phi}^{\mathrm{i},4}(\mathbf{r}) = \mathrm{A}^{\mathrm{i},4}(\mathbf{c}_4^{\mathrm{i}}, \Theta_0^{\mathrm{i},4}) \mathrm{e}^{\imath k_4^{\mathrm{i}} \hat{\mathbf{k}}^{\mathrm{i}} \cdot \mathbf{r}} \tag{8}$$

is the unified thermoelastic plane wave with amplitude $\mathrm{A}^{\mathrm{i},4}$. The constants

$$\mathrm{c}_j^{\mathrm{i}} = -\frac{\omega \mathrm{T}_0 k_j^{\mathrm{I}} \tilde{\mathbf{B}} : \hat{\mathbf{k}}_j^{\mathrm{i}} \otimes \mathbf{u}_0^{\mathrm{I},j}}{\left(\left(k_j^{\mathrm{i}}\right)^2 \tilde{\mathbf{K}} : \hat{\mathbf{k}}_j^{\mathrm{i}} \otimes \hat{\mathbf{k}}_j^{\mathrm{i}} - \imath \omega \rho c_\varepsilon \right)} \tag{9}$$

for $\jmath = 1, 2, 3$ and the vector

$$\mathbf{c}_4^{\mathrm{i}} = \left(\frac{(k_4^{\mathrm{i}})^2}{\omega^2} \frac{\tilde{\tilde{\mathbf{C}}} : \hat{\mathbf{k}}_4^{\mathrm{i}} \otimes \hat{\mathbf{k}}_4^{\mathrm{i}}}{\rho} - \tilde{\mathbf{I}} \right)^{-1} \cdot \left(-\imath \frac{k_4^{\mathrm{i}}}{\omega^2} \left(\frac{\tilde{\mathbf{B}}}{\rho} \cdot \hat{\mathbf{k}}_4^{\mathrm{i}} \right) \Theta_0^{\mathrm{i},4} \right), \tag{10}$$

provide the appropriate factors for $\mathbf{\Phi}^{\mathrm{i},j}$ for $j = 1, .., 4$ and $i = \mathrm{I}, \mathrm{R}$ to satisfy (3).

The wave numbers $k_j \quad (j = 1, .., 4)$ are the complex roots of the following eight degree characteristic equation of the top order symbol of $\tilde{\mathbf{L}}$

$$\zeta_8 k^8 + \zeta_6 k^6 + \zeta_4 k^4 + \zeta_2 k^2 + \zeta_0 = 0. \tag{11}$$

where

$$\zeta_8 = \frac{a}{\rho^3} \mathrm{S}_3 \tag{12}$$

$$\zeta_6 = -\frac{a}{\rho^2} \omega^2 \mathrm{S}_1^* - \frac{\imath \omega \mathrm{T}_0}{\rho^3} \mathbf{b} \cdot \tilde{\mathbf{\Gamma}}^* \cdot \mathbf{b} - \frac{\imath \omega c_\varepsilon}{\rho^2} \mathrm{S}_3 \tag{13}$$

$$\zeta_4 = \frac{a}{\rho} \omega^4 \mathrm{S}_1 + \frac{\imath \omega^3 \mathrm{T}_0}{\rho^2} \left(\mathrm{S}_1 \mathbf{b} \cdot \mathbf{b} - \mathbf{b} \cdot \tilde{\mathbf{\Gamma}} \cdot \mathbf{b} \right) + \frac{\imath \omega^3 c_\varepsilon}{\rho} \mathrm{S}_1^* \tag{14}$$

$$\zeta_2 = -a \omega^6 - \frac{\imath \omega^5 \mathrm{T}_0}{\rho} \mathbf{b} \cdot \mathbf{b} - \imath \omega^5 c_\varepsilon \mathrm{S}_1 \tag{15}$$

$$\zeta_0 = \imath \omega^7 \rho c_\varepsilon \tag{16}$$

with $\tilde{\mathbf{\Gamma}} = \tilde{\tilde{\mathbf{C}}} : \hat{\mathbf{k}} \otimes \hat{\mathbf{k}}$, $\tilde{\mathbf{\Gamma}}^*$ being the adjoint of $\tilde{\mathbf{\Gamma}}$, $a = \tilde{\mathbf{K}} : \hat{\mathbf{k}} \otimes \hat{\mathbf{k}}$ being a positive number and $\mathbf{b} = \tilde{\mathbf{B}} \cdot \hat{\mathbf{k}}$. Also, $\mathrm{S}_1, \mathrm{S}_3, \mathrm{S}_1^*$ represents the trace of $\tilde{\mathbf{\Gamma}}$, the determinant of $\tilde{\mathbf{\Gamma}}$ and the trace of $\tilde{\mathbf{\Gamma}}^*$ respectively. We denote the four complex wave numbers by

$$k_j = \frac{\omega}{v_j} + \imath d_j, \quad v_j > 0, \quad d_j > 0, \quad j = 1, 2, 3, 4 \tag{17}$$

where $v_j (j = 1, 2, 3, 4)$ are the phase velocities of the elastothermal and thermoelastic waves and d_j are the corresponding dissipation coefficients.

3. Boundary Conditions

On the surface of the scatterer S we assume the following boundary condition

$$\tilde{\mathbf{R}}(\partial_{\mathbf{r}}) \cdot \mathbf{\Psi}(\mathbf{r}) = \mathbf{0}, \quad \mathbf{r}\epsilon \mathrm{S}, \tag{18}$$

where

$$\tilde{\mathbf{R}}(\partial_{\mathbf{r}}, \hat{\mathbf{n}}) = \begin{bmatrix} \tilde{\mathbf{T}}(\partial_{\mathbf{r}}, \hat{\mathbf{n}}) & | & -\hat{\mathbf{n}} \cdot \tilde{\mathbf{B}} \\ ----- & | & ----- \\ \mathbf{0} & | & \hat{\mathbf{n}} \cdot \tilde{\mathbf{K}} \cdot \nabla \end{bmatrix} \tag{19}$$

is the boundary operator, $\hat{\mathbf{n}} = (\mathrm{n}_1, \mathrm{n}_2, \mathrm{n}_3)$ is the outward unit normal on S and $\mathbf{R}$ is the unified surface traction-flux operator, with the upper 3×4 block of $\tilde{\mathbf{R}}$ specifying the thermoelastic surface traction and the lower 1×1 block of $\tilde{\mathbf{R}}$ determining the heat flux across S. The operator $\tilde{\mathbf{T}}$ is the surface traction operator of anisotropic elasticity given by

$$\tilde{\mathbf{T}}(\partial_{\mathbf{r}}, \hat{\mathbf{n}}) = \hat{\mathbf{n}} \cdot \tilde{\tilde{\mathbf{C}}} : \nabla. \tag{20}$$

The system (3) of P.D.E's with the boundary conditions (19) form a thermoelastic reflection problem and corresponds to a cavity in thermal insulation.

Finally, on the boundary *Snell's Law* is given as

$$\frac{\hat{\mathbf{k}}^I \cdot \mathbf{r}}{v_j^{\mathrm{I}}} = \frac{\hat{\mathbf{k}}_j^R \cdot \mathbf{r}}{v_j^{\mathrm{R}}}, \; j = 1, .., 4 \tag{21}$$

where $v^{\mathrm{I}}, v^{\mathrm{R}}$, are the phase velocities of incident and reflected plane waves respectively.

Making use of unified plane waves given by Eqs. (7, 8) in the boundary condition, and using Eqs. (9, 10) we get a system of four non-homogeneous equations which can be written as

$$Z_{4\times4}\mathbf{X}_{4\times1} = \mathbf{Y}_{4\times1} \tag{22}$$

where

$$Z = \begin{pmatrix} \left(\mathbf{F}^{\mathrm{R},j}\right)_{3\times 4} \\ ------ \\ \left(\mathrm{G}^{\mathrm{R},j}\right)_{1\times 4} \end{pmatrix}, \ \mathbf{Y} = -\begin{pmatrix} \left(\mathbf{F}^{\mathrm{I},j}\right)_{3\times 4} \\ ------ \\ \left(\mathrm{G}^{\mathrm{I},j}\right)_{1\times 4} \end{pmatrix}, \ \mathbf{X} = \left(\frac{\mathrm{A}^{\mathrm{R},j}}{\mathrm{A}^{\mathrm{I},j}}\right)_{4\times 1} \tag{23}$$

with

$$\mathbf{F}^{i,j} = \left(\imath k_j^i\right) \hat{\mathbf{n}} \cdot \tilde{\tilde{\mathrm{C}}} : \hat{\mathbf{k}}_j^i \otimes \mathbf{u}_0^{i,j} - \left(\hat{\mathbf{n}} \cdot \tilde{\mathbf{B}}\right) \Theta_0^{i,j} \tag{24}$$

$$\mathrm{G}^{i,j} = k_j^i \Theta_0^{i,j} (\hat{\mathbf{n}} \cdot \tilde{\mathbf{K}} \cdot \hat{\mathbf{k}}_j^i) \tag{25}$$

for $i = \mathrm{R}, \mathrm{I}$ and $j = 1, 2, 3$. The reflection coefficients are defined to be the ratio of the absolute values of reflected amplitudes over the absolute values of the incident values, $\left(\frac{\mathrm{A}^{\mathrm{R}}}{\mathrm{A}^{\mathrm{I}}}\right)$, and they are derived from the solution of the system (22).

4. Special Case

We apply the theory on a special case of anisotropic thermoelastic material which is orthotropic. In an orthotropic thermoelastic medium can propagate four waves: quasi-Transverse (qT), elasto-thermal quasi-Longitudinal (elasto-thermal qL), elasto-thermal quasi-Transverse (elasto-thermal qT) and thermo-elastic Tmode. Both qL, qT are coupled with temperature. The elasto-thermal qL, elasto-thermal qT and thermo-elastic Tmode waves are dissipated.

5. Numerical Results and Discussion

Orthotropic Thermoelastic Composite Material: Thermal [4] and mechanical [8] properties are used for 2D woven Silicon Carbide/Silicon Carbide thermoelastic composite (SiC/SiC). We choose for incident wave to be an elasto-thermal qL plane wave at a specific frequency. The reflection coefficients are shown in Figs.(a)-(d). The reflection coefficients of elasto-thermal qT (Fig.(a)) is almost diminished compared to all others, while the qT (Fig.(c)) has the greatest reflection coefficients with almost 2 and 6 orders of magnitude difference from the elasto-thermal qL (Fig.(b)) and thermo-elastic Tmode (Fig.(d)) respectively, partly explained by the fact that it is the only wave thermally uncoupled.

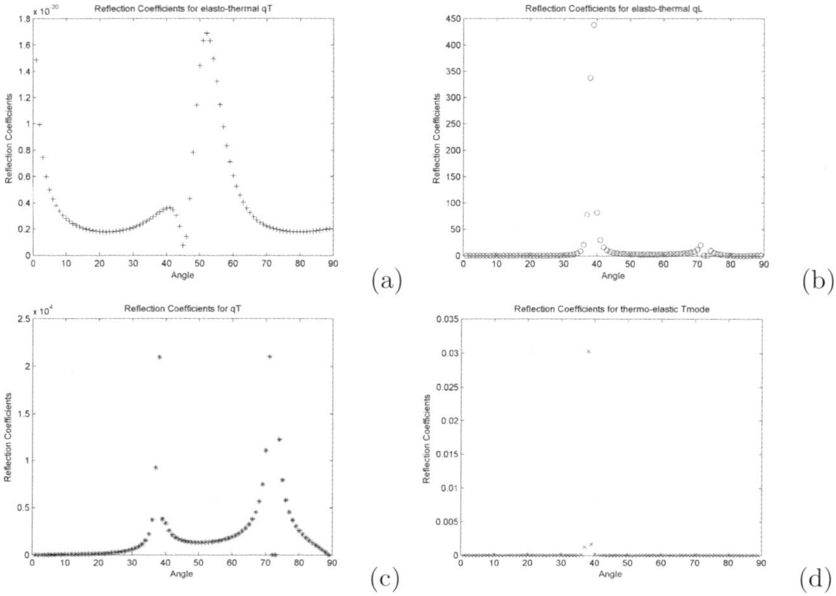

Figure 2. Reflection Coefficients for elasto-thermal qL incident wave

References

1. G. Dassios, and V. Kostopoulos, *SIAM J. APPL. MATH.* **48**, 1 (1988).
2. J. J. Ditri, and J. L. Rose, *J. Acoust. Soc. Am.* **92**, 5 (1992).
3. S. B. Sinha, and K. A. Elsibai, *J. Ther. Str.*, **19**, 8 (1996).
4. S.J. Zinkle, L.L. Snead, *Internal Report, Oak Ridge National Laboratory,* (1998).
5. Abo-El-Nour N. Abd-Alla, and Amira A. S. Al-Dawy, *Int. J. Math. Math. Scien.* **23**, 8 (2000).
6. J. N. Sharma, V. Kumar, and D. Chand, *J. Ther. Str.* **26**, 4 (2003).
7. B. Singh, and S. J. Singh, *Sadhana* **29**, 3 (2004).
8. A. Mouchtachi, R. El. Guerjouma, J. C. Baboux, D. Rouby and D. Bouami, *J. Phys. D: Appl. Phys.* **37**, (2004).
9. B. Singh, *Int. J. Sound and Vabriation* **291**, (2006).
10. B. Singh, A. Kumar, and J. Singh *App. Math. and Comput.* **177**, (2006).
11. N. C. Das, A. Lahiri, S. Sarkar, and S. Basu, *Comp. Math. App.* **56**, (2008).

WHAT PART OF THE NEURONAL CURRENT CAN EEG MEASUREMENTS SEE?

G.DASSIOS AND D. HADJILOIZI
Department of Chemical Engineering, University of Patras and ICE/HT-FORTH, Greece

Electroencephalography is a medical imaging technique where from a finite set of measurements of the electric potential on the surface of the head we want to calculate the position of the neuronal current inside the brain tissue. The fact that this inverse problem has no unique solution is well known. Nevertheless, a characterization of the neuronal current in "visible" and "invisible" parts it is not known. In the present work we provide the answer to this problem. In fact, by utilizing the Hansen decomposition theorem for vector fields, we managed to prove that among the three scalar representation functions of the current, only two are needed to reconstruct the measured electric field. Consequently, we have a complete characterization of that part of the current that is impossible to recover via EEG measurements.

1. INTRODUCTION

Amongst the many advanced imaging modalities in use today, EEG and MEG stand out as the ones that can potentially provide the most comprehensive description of macroscopic brain activity, at the level of cytoarchitectonic areas of the cortex. At the theoretical level, the physics of the problem does not allow the generators of the MEG and EEG to be identified uniquely, even from a complete and perfectly accurate set of measurements of the electrical potential on the scalp and the magnetic field everywhere in space outside the head [**1**]. A unique solution can be extracted from EEG and/or MEG data once constraints are imposed, either at the theoretical level (e.g. mathematical form of the unknown current density) or from anatomical and other information (e.g. confining the generators to the cortical mantle with directions along the local norm to the cortical surface). The first quantitative result for the non-uniqueness of the inverse MEG problem was proved in the fundamental paper by Fokas-Gelfand-Kurylev [**2**]. In fact, it is proved [**2**], for the spherical homogeneous model that from the three scalar functions needed to identify the current only one can be recovered from the knowledge of exterior measurements of the magnetic field it creates. The other two scalar functions needed to represent the vector field in the Hansen [**3**] representation are precisely those that EEG records can recover [**4**].

In the present work, we investigate the non-uniqueness of the solution of the inverse EEG problem and characterize the 'visible' part of the neuronal current that can be retrieved for the case of the spherical-head model.

Section 2 contains the mathematical analysis of the singular support of the exterior electric potential produced by a single dipole. In section 3 we extract the electric potential generated by a continuously distributed current in the form of Green's type integral representation, by utilizing the information obtained in section 2. In section 4, the non uniqueness of the inverse EEG problem is examined and the solution of the relative moment problem that is reduced is also discussed.

2. MATHEMATICAL ANALYSIS OF THE SINGULAR SURRORT OF THE EXTERIOR ELECTRIC

Consider the homogeneous spherical model of the brain which is characterized by the conductivity σ. At the point τ an equivalent dipole current with moment $\mathbf{J}$ is considered, which generate an electromagnetic field. The relative interior electric potential u^- satisfies the Neumann problem [**5**]

$$\Delta u^-(\mathbf{r},\boldsymbol{\tau}) = \frac{1}{\sigma}\mathbf{J}\nabla\cdot\delta(\mathbf{r}-\boldsymbol{\tau}), \qquad r<\alpha,\ \tau<\alpha \tag{2.1}$$

$$\partial_r u^-(\mathbf{r},\boldsymbol{\tau}) = 0, \qquad r=\alpha \tag{2.2}$$

where α is the radius of the sphere and δ is the Dirac measure. The exterior electric potential is the solution of the Dirichlet problem [**5**]

$$\Delta u^+(\mathbf{r},\boldsymbol{\tau}) = 0, \qquad r>\alpha \tag{2.3}$$

$$u^-(\mathbf{r},\boldsymbol{\tau}) = u^+(\mathbf{r},\boldsymbol{\tau}), \qquad r=\alpha \tag{2.4}$$

$$u^+(\mathbf{r},\boldsymbol{\tau}) = O\left(\frac{1}{r}\right), \qquad r\to\infty\,. \tag{2.5}$$

The solution of the Dirichlet problem was first obtained in the form of series [**6**]. The closed form of the solution was given by G.Dassios [5] in the form

$$u^+(\mathbf{r},\tau) = \frac{1}{4\pi\sigma}[w_1(\mathbf{r},\tau)+w_2(\mathbf{r},\tau)], \qquad r>\alpha,\ \tau<\alpha \tag{2.6}$$

where

$$w_1(\mathbf{r},\tau) = 2\mathbf{J}(\tau)\cdot\frac{\mathbf{r}-\tau}{|\mathbf{r}-\tau|^3} = -2\mathbf{J}(\tau)\cdot\nabla_\tau\frac{1}{|\mathbf{r}-\tau|}, \qquad r>\alpha,\ \tau<\alpha \tag{2.7}$$

and

$$w_2(\mathbf{r},\tau) = \mathbf{J}(\tau)\cdot\frac{|\mathbf{r}-\tau|\,\hat{\mathbf{r}}\times(\mathbf{r}-\tau)}{F(\mathbf{r},\tau)}, \qquad r>\alpha,\ \tau<\alpha\,. \tag{2.8}$$

The function F is given by

$$F(\mathbf{r},\tau) = r\,|\mathbf{r}-\tau|^2 + |\mathbf{r}-\tau|\,\mathbf{r}\cdot(\mathbf{r}-\tau)\,. \tag{2.9}$$

and $\hat{\mathbf{r}}$ is the unit vector along $\mathbf{r}$.

Our purpose in this section is to characterize the singular support of the exterior electric potential. Hence, we consider the function Φ which is any compactly supported C^∞-function of $\mathbb{R}^3$ and we will find the value of the integral

$$\Delta u^{+}(\Phi) = \int_{\mathbf{R}^3} (\Delta_{\mathbf{r}} u^{+}(\mathbf{r},\tau))\Phi(\mathbf{r})d\upsilon(\mathbf{r}) . \tag{2.10}$$

The analysis of the forms (2.6)-(2.8) shows that the singularities of u^{+} live in the line segment connecting the position $\boldsymbol{\tau}$ of the dipole with the center of the conductive sphere. Therefore, in order to deal with the integral (2.10) we define the domain

$$\Omega_{\varepsilon}(\tau) = C_{\varepsilon}(\tau) \cup S_{\varepsilon}(\mathbf{0}) \cup S_{\varepsilon}(\tau) \tag{2.11}$$

where $C_{\varepsilon}(\tau)$ denotes the interior of the cylindrical surface

$$\partial C_{\varepsilon}(\tau) = \left\{ \mathbf{r}' \in \mathbb{R}^3 : \rho = \sqrt{x_1^2 + x_2^2} = \varepsilon, 0 \le x_3' \le \tau \right\} , \tag{2.12}$$

$S_{\varepsilon}(\mathbf{0})$ denotes the interior of the semi-sphere

$$\partial S_{\varepsilon}(\mathbf{0}) = \left\{ \mathbf{r}' \in \mathbb{R}^3 : \mathbf{r}' = \varepsilon, \ x_3' \le 0 \right\} , \tag{2.13}$$

and $S_{\varepsilon}(\boldsymbol{\tau})$ denotes the interior of the semi-sphere

$$\partial S_{\varepsilon}(\boldsymbol{\tau}) = \left\{ \mathbf{r}' \in \mathbb{R}^3 : \left| \mathbf{r}' - \boldsymbol{\tau} \right| = \varepsilon, \ x_3' \ge \boldsymbol{\tau} \right\} . \tag{2.14}$$

By using the distributional differentiation, the classical Green's identity, the harmonicity of the electric potential outside of its sources, the fact that both u^{+} and Φ vanish at infinity and taking into account the singularities of functions w_1 and w_2 we arrive

$$4\pi\sigma\Delta u^{+}(\Phi) = I_1(\partial\Omega_{\varepsilon}(\tau)) + I_2(S_{\varepsilon}) \tag{2.15}$$

where

$$I_1(\partial\Omega_{\varepsilon}(\tau)) = -\lim_{\varepsilon\to 0} \int_{\partial\Omega_{\varepsilon}(\tau)} \left[w_1(\mathbf{r},\tau)\frac{\partial}{\partial n}\Phi(\mathbf{r}) - \Phi(\mathbf{r})\frac{\partial}{\partial n}w_1(\mathbf{r},\tau) \right] ds(\mathbf{r}) \tag{2.16}$$

and

$$I_2(S) = -\lim_{\varepsilon\to 0} \int_{S} \left[w_2(\mathbf{r},\tau)\frac{\partial}{\partial n}\Phi(\mathbf{r}) - \Phi(\mathbf{r})\frac{\partial}{\partial n}w_1(\mathbf{r},\tau) \right] ds(r) . \tag{2.17}$$

The singularities of function w_2 impose that the domain S is the cylindrical surface $C_{\varepsilon}(\tau)$, a sphere of radius ε centered at the origin $S = S(\mathbf{0})$ and a sphere of radius ε centered at the point where the dipole is located $S = S(\boldsymbol{\tau})$. Each one of these integrals is separately analyzed.

Lemma 1.
If $I_1(\partial\Omega_{\varepsilon}(\tau))$ is the integral defined in (2.16) then

$$I_1(\partial\Omega_{\varepsilon}(\tau)) = -8\pi\mathbf{J}(\tau)\cdot\nabla_{\tau}\Phi(\tau) . \tag{2.18}$$

Proof.
Since the function w_1 has an isolated singularity at the point $\boldsymbol{\tau}$ the surface of integration is a sphere of radius ε around $\boldsymbol{\tau}$. Then using the equation (2.7), the

Green's second identity and the fundamental solution of the Laplace operator in the definition (2.16) we obtain

$$\mathbf{I}_1(\partial\Omega_\varepsilon(\boldsymbol{\tau})) = -2\mathbf{J}^p(\boldsymbol{\tau})\cdot\nabla_\tau \lim_{\varepsilon\to 0} \int_{|\mathbf{r}-\boldsymbol{\tau}|\le\varepsilon} \left[\frac{1}{|\mathbf{r}-\boldsymbol{\tau}|}\Delta_r\Phi(\mathbf{r}) - \Phi(\mathbf{r})\Delta_r\frac{1}{|\mathbf{r}-\boldsymbol{\tau}|}\right] d\upsilon(\mathbf{r})$$

$$= -2\mathbf{J}(\boldsymbol{\tau})\cdot\nabla_\tau \lim_{\varepsilon\to 0}\int_0^{2\pi}\int_0^{\pi}\int_0^{\tau} \Delta_r\Phi(\mathbf{r})r\sin\theta drd\theta d\varphi - 8\pi\mathbf{J}(\mathbf{r})\cdot\nabla_\tau\Phi(\boldsymbol{\tau})$$

$$= -8\pi\mathbf{J}(\mathbf{r})\cdot\nabla_\tau\Phi(\boldsymbol{\tau}) \qquad (2.19)$$

Lemma 2.
If $I_2(\partial C_\varepsilon(\boldsymbol{\tau}))$ is the integral defined in (2.17) then

$$I_2(\partial C_\varepsilon(\boldsymbol{\tau})) = -\frac{4\pi}{\tau}\mathbf{J}(\boldsymbol{\tau})\cdot(\tilde{\mathbb{I}} - \hat{\boldsymbol{\tau}}\otimes\hat{\boldsymbol{\tau}})\cdot\int_0^{\tau}\nabla\Phi(t\hat{\boldsymbol{\tau}})dt \qquad (2.20)$$

where $\tilde{\mathbb{I}} = \hat{\boldsymbol{\tau}}_1\otimes\hat{\boldsymbol{\tau}}_1 + \hat{\boldsymbol{\tau}}_2 + \hat{\boldsymbol{\tau}}_2 + \hat{\boldsymbol{\tau}}_3 + \hat{\boldsymbol{\tau}}_3$ is the identity dyadic.

Proof.
We introduce the cylindrical coordinates and we assume that the dipole is located on the perpendicular axes. We calculate the asymptotic form of function w_2 and its normal derivative that are given in the forms

$$w_2(\mathbf{r},\boldsymbol{\tau}) = \frac{2}{\tau}(J_1\cos\varphi + J_2\sin\varphi)\frac{1}{\varepsilon} + \frac{J_3}{\tau}\left(\frac{1}{\tau - x_3} - \frac{1}{x_3}\right) + O(\varepsilon) \qquad (2.21)$$

and

$$\partial_n w_2(\mathbf{r},\boldsymbol{\tau}) = -\frac{2}{\tau}(J_1\cos\varphi + J_2\sin\varphi)\frac{1}{\varepsilon^2} + O(1). \qquad (2.22)$$

As far as concerned the test function by utilizing the Taylor expansion we obtain

$$\Phi(\mathbf{r}) = \Phi(0,0,x_3) + \varepsilon\left[\cos\varphi\partial_{x_1}\Phi(0,0,x_3) + \sin\varphi\partial_{x_2}\Phi(0,0,x_3)\right] + O(\varepsilon^2) \quad (2.23)$$

and

$$\partial_n w_2(\mathbf{r},\boldsymbol{\tau}) = -\frac{2}{\tau}(J_1\cos\varphi + J_2\sin\varphi)\frac{1}{\varepsilon^2} + O(1). \qquad (2.24)$$

Then using equations (2.21)-(2.24) the definition (2.17) yields

$$I_2(\partial C_\varepsilon(\boldsymbol{\tau})) = -\frac{4\pi}{\tau}\int_0^{\tau}\left[J_1(\boldsymbol{\tau})\frac{\partial}{\partial x_1}\Phi(0,0,x_3) + J_2(\boldsymbol{\tau})\frac{\partial}{\partial x_2}\Phi(0,0,x_3)\right]dx_3, \qquad (2.25)$$

the invariant form of which is the equation (2.20).
Lemmas 3 and 4 respectively, are the results of the integration of the isolated singularities of function w_2 at the center of the origin and at the point where the dipole is located. We follow a similar approach as Lemma 2 and the surface

integration is a sphere of radius ε centered at the origin and a sphere of radius at the point $\boldsymbol{\tau}$ respectively.

Lemma 3.
If $I_2(S_\varepsilon(\mathbf{0}))$ is the integral defined in (2.17) then

$$I_2(\partial S_\varepsilon(\mathbf{0})) = \frac{4\pi}{\tau}\hat{\tau}\cdot\mathbf{J}(\tau)\Phi(\mathbf{0}) . \tag{2.26}$$

Lemma 4.
If $I_2(S_\varepsilon(\tau))$ is the integral defined in (2.17) then

$$I_2(\partial S_\varepsilon(\tau)) = -\frac{4\pi}{\tau}\hat{\tau}\cdot\mathbf{J}(\tau)\Phi(\tau) . \tag{2.27}$$

Proposition 1.
The action of distribution Δu^+ on the test function Φ yields the formula

$$\Delta u^+(\Phi) = -\frac{2}{\sigma}\mathbf{J}^P(\tau)\cdot\nabla_\tau\Phi(\tau) - \frac{\hat{\tau}\cdot\mathbf{J}^P(\tau)}{\sigma\tau}\left[\Phi(\tau)-\Phi(\mathbf{0})\right]$$

$$-\frac{1}{\sigma\tau}\mathbf{J}^P(\tau)\cdot(\tilde{\mathbb{I}}-\hat{\tau}\otimes\hat{\tau})\cdot\int_0^\tau \nabla_\tau\Phi(t\hat{\tau})dt . \tag{2.28}$$

The first term on the right hand side of (2.28) reflects the contribution of the dipole at the point $\boldsymbol{\tau}$, the second term represents the contribution of the two monopole, one at $\boldsymbol{\tau}$ and one at $\mathbf{0}$, and the third term contains the contribution from the continuous distribution of singularities. On the other hand equation (2.28) gives a first flavor of the way that the current enters. The first term on the right hand side of (2.28) shows the contribution from the full dipole moment, in the second and in the third we have the contribution of the radial and the tangential part of the neuronal current.

3. GREEN'S REPRESENTATION

In this section we consider a continuously distributed neuronal current. The corresponding electric potential that is generated, is obtained by the integration of the electric potential produced by a single dipole

$$u(\mathbf{r}) = \int_{|\tau|\le\alpha} u^+(\mathbf{r},\boldsymbol{\tau})d\upsilon(\mathbf{r}) \tag{3.1}$$

Since we characterize the singular support of the interior electric potential generated by a single dipole we integrate equation (2.28). We analyze each one of the integrals separately and after some long manipulation we arrive at the Green's representation of the electric potential which is describing by the following theorem [**4**].

Theorem 1.
If $u(\mathbf{r})$ is the electric potential defining in (3.1), and the current has the spherical decomposition

$$\mathbf{J}(\mathbf{r}) = J_r(\mathbf{r})\hat{\mathbf{r}} + J_\theta(\mathbf{r})\hat{\boldsymbol{\theta}} + J_\varphi(\mathbf{r})\hat{\boldsymbol{\varphi}}, \tag{3.2}$$

then

$$u(\mathbf{r}) = -\frac{1}{4\pi\sigma r}\int_{|\mathbf{r}'|\le\alpha} \frac{J_{r'}(\mathbf{r}')}{r'} d\upsilon(\mathbf{r}')$$

$$\frac{1}{2\pi\sigma}\int_{|\mathbf{r}'|\le\alpha} \frac{\nabla_r \cdot \mathbf{J}(\mathbf{r}')}{|\mathbf{r}-\mathbf{r}'|} d\upsilon(\mathbf{r}') + \frac{1}{4\pi\sigma}\int_{|\mathbf{r}'|\le\alpha} \frac{\mathbf{J}_r(\mathbf{r}')}{r'|\mathbf{r}-\mathbf{r}'|} d\upsilon(\mathbf{r}')$$

$$-\frac{1}{4\pi\sigma}\int_{|\mathbf{r}'|\le\alpha} \frac{1}{r'^3}\int_{t=r'}^{\alpha} \frac{t}{\sin\theta'}\left[\frac{\partial}{\partial\theta'}\left(\sin\theta' J_{\theta'}(t\hat{\mathbf{r}}')\right) + \frac{\partial}{\partial\varphi'} J_{\varphi'}(t\hat{\mathbf{r}}')\right]\frac{1}{|\mathbf{r}-\mathbf{r}'|} dt\, d\upsilon(\mathbf{r}') \tag{3.3}$$

It is proved [**4**] that, if we represent the current via the Hansen decomposition [5]

$$\mathbf{J}(\mathbf{r}) = J_r(\mathbf{r})\hat{\mathbf{r}} + \hat{\mathbf{r}} \times \nabla F(\mathbf{r}) + \hat{\mathbf{r}} \times (\hat{\mathbf{r}} \times G(\mathbf{r})) \tag{3.4}$$

where F, G are the angular functions, the electric potential yields

$$u(\mathbf{r}) = -\frac{1}{4\pi\sigma r}\int_{|\mathbf{r}'|\le\alpha} \frac{J_{r'}(\mathbf{r}')}{r'} d\upsilon(\mathbf{r}') - \frac{1}{4\pi\sigma}\int_{|\mathbf{r}'|\le\alpha}\left[2\frac{\partial}{\partial r'} J_{r'}(\mathbf{r}') + \frac{3}{r'} J_{r'}(\mathbf{r}')\right]\frac{1}{|\mathbf{r}-\mathbf{r}'|} d\upsilon(\mathbf{r}')$$

$$-\frac{1}{4\pi\sigma}\int_{|\mathbf{r}'|\le\alpha}\left[\frac{2}{r'^2}\mathbb{B}' G(r',\theta',\varphi') + \frac{1}{r'^3}\int_{r'}^{\alpha} \mathbb{B}' G(t,\theta',\varphi')\, dt\right]\frac{1}{|\mathbf{r}-\mathbf{r}'|} d\upsilon(\mathbf{r}'), \tag{3.5}$$

where $\mathbb{B}$ is the Beltrami operator.
This result is crucial since it shows that the electric potential is independent on the scalar function F while is exactly the function that the magnetic potential depends on. On the other hand this result implies that the inverse problem recover the radial component and the irrotational part of the tangential component of the current.

4. THE INVERSE PROBLEM

In this section we investigate the inverse problem of identification the neuronal current from EEG measurements on the surface of the head.
The electric potential has the harmonic expansion

$$u(\mathbf{r}) = \sum_{n=1}^{\infty}\sum_{m=-n}^{n} c_n^m \frac{Y_n^m(\hat{\mathbf{r}})}{r^{n+1}}, \qquad r > \alpha \tag{4.1}$$

and the scalar functions J_τ, G can be expanded as

$$J_r(\boldsymbol{\tau}) = \sum_{n=0}^{\infty}\sum_{m=-n}^{n} J_n^m(t) Y_n^m(\hat{\boldsymbol{\tau}}), \qquad \tau < \alpha \tag{4.2}$$

$$G(\tau) = \sum_{n=0}^{\infty} \sum_{m=-n}^{n} g_n^m(t) Y_n^m(\hat{\tau}), \qquad \tau < \alpha, \tag{4.3}$$

where $Y_n^m(\hat{\tau})$ is the orthonormilized complex form of the spherical harmonics. Then using the fundamental expansion of the Laplace operator in spherical coordinates and the orthogonality of spherical harmonics we arrive at the formula which connect the coefficients c_n^m of the known electric potential with the coefficients j_n^m and g_n^m of the corresponding two representation functions of the current

$$c_n^m = \frac{1}{\sigma} \int_0^{\alpha} J_n^m(t) t^{n+1} dt + \frac{n+1}{\sigma} \int_0^{\alpha} g_n^m(t) t^n dt, \quad n = 1, 2, \ldots, \quad m = -n, \ldots, n \tag{4.4}$$

In order to solve the inverse EEG the moments of the function J_r, G must be determined. We proved the non uniqueness of the inverse problem so a certain assumption must be made. We impose the minimum norm assumption for the current

$$W = \int_{|\tau| \le \alpha} |\mathbf{J}(\mathbf{r})|^2 \, d\upsilon(\mathbf{r}) = \text{minimum}. \tag{4.5}$$

In other words we want to minimize the expression

$$W_{min} = \sum_{n=1}^{\infty} \sum_{m=-n}^{n} \int_0^{\alpha} \left[n(n+1)(g_n^m(r))^2 + (j_n^m(r))^2 \right] dr. \tag{4.6}$$

Under this assumption the inverse problem is uniquely solvable and the scalar function are obtained in the forms [**4**]

$$J_r(\mathbf{r}) = \sigma \sum_{n=1}^{\infty} \sum_{m=-n}^{n} \frac{n(2n+1)(2n+3)}{n(2n+1)\alpha^{2n+3}(n+1)(2n+3)\alpha^{2n+1}} c_n^m r^{n+1} Y_n^m(\hat{\tau}), \qquad \tau < \alpha, \tag{4.7}$$

$$G(\mathbf{r}) = \sigma \sum_{n=1}^{\infty} \sum_{m=-n}^{n} \frac{(2n+1)(2n+3)}{n(2n+1)\alpha^{2n+3} + (n+1)(2n+3)\alpha^{2n+1}} c_n^m r^n Y_n^m(\hat{\tau}), \qquad \tau < \alpha. \tag{4.8}$$

Therefore these formulas answer to the question of what part of the neuronal current can EEG measurements see. The "visible" parts of the current from EEG measurements are the radial component and the irrotational part of the tangential component of the current. The "invisible" part that is the solenoidal part of the tangential component of the current of the EEG measurments is precisely the "visible" part of the current of the MEG problem. This means that for continuously distributed neuronal current information missing in the Electroencephalographic data is exactly information that is available in Magnetoencephalographic data, and vice versa. In the future we will implement numerically the results and explore the solutions using phantom data and real experimental.

REFERENCES

1. Helmholtz, H., Ann. Physiol. Chem., **29** (3), pp. 222, 1853.
2. Fokas, A.S., Gelfand, I.M., Kurylev,Y., "Inversion Method for Magnetoencephalography", *Inverse Problems*, **12**, pp. L9-L11, 1996.
3. Hansen, W.W., "A new type of expansion in radiation problems", *Physical Review*, **47**, pp. 139-143, 1935.
4. Dassios, G and Hadjiloizi, D., "On the non-uniqueness of the inverse problem associate with Electroencephalography", *in press*.
5. Dassios, G., Lecture Notes from the Mini-Course on Mathematics of Emenging Biomedical Imaging, edited by H.Ammary, (Springer-Varleg, Berlin), *in press*.
6. Wilson, F.N., Bayley, R.N., "The electric field of an eccentric dipole in a homogeneous spherical conducting medium" *Circulation* **1**, pp. 84-92, 1950.

EVALUATION OF MASS SENSING CHARACTERISTICS OF SINGLE-WALLED CARBON NANOTUBES

S. K. GEORGANTZINOS AND N. K. ANIFANTIS

Machine Design Laboratory
Mechanical and Aeronautics Engineering Department
University of Patras, GR 26500 GREECE
E-mail: nanif@mech.upatras.gr

This paper utilizes a spring-mass based finite element formulation appropriate for predicting of vibration analysis of single-walled carbon nanotubes in order to investigate its sensing characteristics when a nano-particle is attached on them. Three-dimensional nanoscale elements and corresponding elemental equations are implemented for the simulation of dynamic behaviour of nanotubes. Nanosprings simulate relative translations and rotations between atoms including exclusively physical variables such as bond stretching. The equivalent mass of each atom is defined at the corresponding node location. Thus, molecular mechanics theory is applied directly. The atomistic microstructure of nanotubes is used to assemble the elemental equations and construct the dynamic equilibrium equation. In this way, the resonant frequency shift of a cantilevered or bridged nanotube caused by the addition of a nano-scale particle to the nanotube tip and various intermediate positions can be explored. The simulation results are compared with other numerical data available in the literature demonstrating good agreement. Hence, the current modelling approach can be considered suitable as a coupled-field design tool for the development of carbon nanotube-based NEMS applications.

1. Introduction

Sensors for gaseous molecules play a significant role in monitoring the environment, controlling chemical processes as well as in biomedical applications. Mechanical resonators are widely used as inertial balances to detect small quantities of adsorbed mass through shifts in oscillation frequency. Recently, advances in lithography and materials synthesis have enabled the fabrication of nanoscale mechanical resonators utilizing carbon nanotubes (CNTs) [1,2]. CNTs are molecular-scale tubes of graphitic carbon that present outstanding mechanical properties (i.e. Young's modulus and yield strength near to 1 TPa and 100 GPa, respectively) as well as exceptional structural, electronic, thermal, and optical characteristics. The small size, the light weight and the high stiffness of CNTs enable the increase of resonator sensitivity.

The development of computational tools that are able to simulate the vibration behavior of CNT-based mass sensors is important for two major reasons. Firstly, they can cost-effectively predict the mass sensing characteristics of different nanoresonator types allowing the optimal design of sensors of specific sensing range. Secondly, their cooperation with experimental measurements can improve the detection abilities of the nanodevise. Molecular mechanics atomistic-based methods seem to be the most efficient techniques to approach the vibration analysis problem. They can simulate exactly the atomistic geometry as well as mechanical behavior revealing the overall range of natural frequencies and mode shapes of vibration. Recently, a spring-mass based finite element method [3,4] was developed and efficiently analyzed the vibration problem of single- and multi-walled CNTs revealing conventional and novel modes and natural frequencies.

In the present paper, the previously aforementioned method is utilized in order to evaluate the mass sensing characteristics of single-walled CNTs. Spring-mass based elements are formulated, in such a way, to exclusively include the physical variables provided by molecular theory. The eigenvalue problem is then formulated and solved together with or without an absorved nano-particle mass of variable size. Results available in the literature are directly compared to validate the proposed method. The effect of the added mass in natural frequencies and modes of vibration is investigated for the clamped-free and clamped-clamped support condition cases. The resonant frequency shift of the fundamental and higher order modes is also reported.

2. Modeling Procedure

2.1. *Geometric Specifications*

In order to understand and obtain the CNT geometry, we have to imagine the well known geometry of graphene sheet to be rolled up in the direction of the chiral vector [3]. Knowing the type and length of the tube, the chiral vector and the translational vector that define the ideal rectangular cutting area of the graphene sheet are known [5]. Here, for simplicity, the original coordinate system of the graphene sheet (x', y') is transformed into a new system (x, y, z) of the nanotube such that the translational vector lies along the y-axis. Then, the graphene atomic coordinates are converted to those of the nanotube, according to the equation [6]

$$(x, y, z) = \left(R\cos\left(\frac{x'}{R}\right),\ R\sin\left(\frac{x'}{R}\right),\ y' \right), \tag{1}$$

The diameter D of a single-walled CNT is given by the following equation:

$$D = \frac{a_{\mathrm{c-c}}\sqrt{3(n^2+nm+m^2)}}{\pi}, \tag{2}$$

where the integers (n,m) express the number of steps along the zigzag carbon bonds [3] and $a_{c\text{-}c}$ is the distance between two neighboring carbon atoms and is equal to 0.1421 nm. All the geometric characteristics, such as chirality angles and diameters can be easily evaluated since they are depended only on chiral indexes. A complete list of formulae of all parameters can been found in [5].

2.2. *Interatomic Behavior Modeling*

The total force on each atomic nuclei is the sum of the force generated by the electrons and electrostatic force between the positively charged nuclei themselves. The general formula of potential energy is described as

$$U = \sum U_r + \sum U_\theta + \sum U_\phi + \sum U_\omega + \sum U_{vdw} \tag{3}$$

where U_r represents the energy due to bond stretching, U_θ the energy due to bond angle bending, U_φ the energy due to dihedral angle torsion, U_ω the energy due to out-of-plane torsion, and U_{vdw} the energy due to non-bonded Van der Waals interaction.

Under the assumption of small deformations, the harmonic approximation is adequate for describing potential energy and therefore the force field. By adopting the simplest harmonic forms and combining the dihedral angle torsion with the out-of-plane torsion into a single equivalent term, then the following terms can describe the total potential energy:

$$U_r = \frac{1}{2}k_r(\Delta r)^2, \quad U_\theta = \frac{1}{2}k_\theta(\Delta\theta)^2, \quad U_\tau = U_\phi + U_\omega = \frac{1}{2}k_\tau(\Delta\phi)^2 \tag{4}$$

where k_r, k_θ, and k_τ are the bond stretching, bond angle bending, and torsional resistance force constants, respectively, while Δr, $\Delta\theta$, and $\Delta\phi$ represent the bond length, bond angle, and twisting bond angle variations, respectively. In contrast to the other bonded interactions, the Van der Waals interactions as well as the electrostatic forces may be neglected due to their minor effect on prediction of the dynamic response of CNTs.

The second derivatives of the potential energy terms appearing in Eq. (3) with respect to bond length, bond angle and twisting bond angle variations, respectively, produce the spring stiffness coefficients k_r, k_θ, and k_τ, according to Castigliano's theorem. These are the force field constants provided by molecular mechanics, which are straightforwardly introduced in nanosprings as spring stiffness constants. The angle bending interaction is simulated with an axial nanospring, according to the simplification described in [3]. The stiffness k_b of the specific spring is described by the following equation:

$$k_b = \left(\frac{1}{a_{\text{c-c}} \cos(90^\circ - \gamma)} \right)^2 k_\theta \,, \tag{5}$$

where $\gamma = 30^\circ$ in the hexagonal lattice of the graphene. This angle may vary for each C-C-C microstructure in a CNT according to its type and radius, due to its cylindrical shape. The spring stiffness coefficients of Eq. (3) are taken to be equal to $k_r = 6.52\times10^{-7}$ N nm^{-1}, $k_\theta = 8.76\times10^{-10}$ N nm rad^{-2} and $k_\tau = 2.78\times10^{-10}$ N nm rad^{-2} (see [7,8]), respectively. In order to simulate the inertia effects, a point mass equal to the whole or half carbon atomic nucleus mass ($m = 1.9943\times10^{-26}$ kg) is added onto one node in particular elements only. The masses of electrons are neglected. The analytical description of the spring mass modeling of CNT hexagonal lattice is reported in [3].

2.3. *Computer Implementation*

The elemental equation for the a_i-element, as defined and developed in [3], which represents the bond stretching as well as twisting bond angle interactions is

$$\begin{bmatrix} k_r \mathbf{I_{20}} & \mathbf{0} \\ \mathbf{0} & k_\tau \mathbf{I_{20}} \end{bmatrix} \begin{bmatrix} \mathbf{u} \\ \boldsymbol{\theta} \end{bmatrix} + \begin{bmatrix} m_r \mathbf{I_{21}} & \mathbf{0} \\ \mathbf{0} & \mathbf{0} \end{bmatrix} \begin{bmatrix} \ddot{\mathbf{u}} \\ \ddot{\boldsymbol{\theta}} \end{bmatrix} = \begin{bmatrix} \mathbf{F} \\ \mathbf{T} \end{bmatrix} \quad \text{or} \quad \mathbf{k_a U} + \mathbf{m_a \ddot{U}} = \mathbf{P} \,, \tag{6}$$

where

$$\mathbf{I_{20}} = \begin{bmatrix} 1 & -1 \\ -1 & 1 \end{bmatrix}, \quad \mathbf{I_{21}} = \begin{bmatrix} 0 & -1 \\ 0 & 1 \end{bmatrix}, \tag{7}$$

m_r is the concentrated mass equal to half or whole mass of carbon nuclei [3], $\mathbf{F}$ contains the forces applied on nodes of the element, $\mathbf{u}$ is the vector of nodal displacements ($\mathbf{u} = [u_1 \quad u_2]^{\mathrm{T}}$, $\mathbf{F} = [F_1 \quad F_2]^{\mathrm{T}}$), $\boldsymbol{\theta}$ is the vector of nodal rotations and $\mathbf{T}$ is the vector of the applied torsional moments ($\boldsymbol{\theta} = [\theta_1 \quad \theta_2]^{\mathrm{T}}$, $\mathbf{T} = [T_1 \quad T_2]^{\mathrm{T}}$), $\mathbf{U}$ is the vector of nodal translations and rotations, $\ddot{\mathbf{U}}$ is the vector of nodal accelerations, $\mathbf{P}$ is the column vector of loads, and finally, $\mathbf{k_a}$ and $\mathbf{m_a}$ are the elemental stiffness and mass matrix, respectively. Similarly, the equation for the b_i -element, which describes the bond angle interaction in the hexagonal lattice, is

$$\begin{bmatrix} k_b \mathbf{I}_{20} & \mathbf{0} \\ \mathbf{0} & \mathbf{0} \end{bmatrix} \begin{bmatrix} \mathbf{u} \\ \boldsymbol{\theta} \end{bmatrix} + \begin{bmatrix} \mathbf{0} & \mathbf{0} \\ \mathbf{0} & \mathbf{0} \end{bmatrix} \begin{bmatrix} \ddot{\mathbf{u}} \\ \ddot{\boldsymbol{\theta}} \end{bmatrix} = \begin{bmatrix} \mathbf{F} \\ \mathbf{T} \end{bmatrix} \quad \text{or} \quad \mathbf{k_b U} + \mathbf{m_b \ddot{U}} = \mathbf{P}, \tag{8}$$

where

$$k_b = \begin{cases} k_{b1} & \text{, when } b\text{ - element is straight in respect to the hexagonal cell} \\ k_{b2}\text{,} & \text{when } b\text{ - element is slant in respect to the hexagonal cell} \end{cases}$$

is the stiffness coefficient, as described in [3], and finally, $\mathbf{k_b}$ and $\mathbf{m_b}$ are the corresponding elemental stiffness and mass matrix, respectively.

Hence, the elemental matrices expressed in global coordinates are

$$\mathbf{K_a} = \mathbf{T^T k_a T}, \ \mathbf{M_a} = \mathbf{T^T m_a T}, \ \mathbf{K_b} = \mathbf{T^T k_b T} \tag{9}$$

where $\mathbf{T}$ is the appropriate transformation matrix, as defined in [3]. The displacements and rotations as well as the loads are related by the equation

$$\mathbf{U} = \mathbf{TD}, \ \mathbf{R} = \mathbf{T^T P}, \tag{10}$$

where $\mathbf{D}$ is the vector of displacements and rotations and $\mathbf{R}$ is the vector of loads with respect to the global coordinates. The superscript T in the above equations denotes matrix transposition. Finally, the elemental equations in the global system become

$$\mathbf{K_a D} + \mathbf{M_a \ddot{D}} = \mathbf{R} \tag{11}$$

$$\mathbf{K_b D} + \mathbf{M_b \ddot{D}} = \mathbf{R} \tag{12}$$

After the assembly of global stiffness $\mathbf{K}$ and global mass $\mathbf{M}$ matrices from the above elemental matrices, and considering undamped free vibration for the tubes, the equation of motion becomes

$$\mathbf{M\ddot{X}} + \mathbf{KX} = \mathbf{0} \tag{13}$$

Applying the support conditions of the CNT, the eigenvalue problem can be solved using common finite element procedures. The solution of the eignevalue problem reveals the eigenfrequencies and corresponding eigenmodes.

3. Results and Discussion

Firstly, comparison is performed between results of the present and another theoretical approach [9] (Figure 1), in order to validate the proposed method. The comparison is limited in available fundamental frequencies for clamped-free (Figure 1a) and clamped-clamped (Figure 1b) support conditions, in which attached mass of variable magnitude is added in CNT tip and middle, respectively. The nanotube used for the present comparison is armchair (6,6) of 0.8 nm diameter. Very good agreement between results for different length and support conditions can be reported.

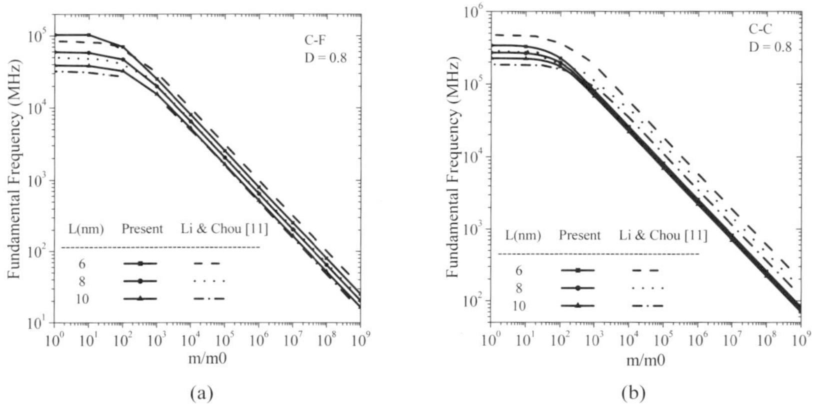

Figure 1. Fundamental frequency variation of CNT due to attached mass of variable magnitude for (a) clamped-free and (b) clamped-clamped support condition.

Regarding the same type of CNT and fundamental frequency, Figure 2(a) illustrates the frequency shift presented due to the added mass in the same position as previously for clamped-free (solid lines) and clamped-clamped (dash lines) support condition. The frequency shift is defined as the difference between the fundamental frequencies of a CNT with and without attached mass. As expected, we observe in both cases the lower the length, the higher the frequency shift and therefore the higher sensitivity. We note remarkable frequency shift ranging 77-350 MHz and 280-1120 MHz when only one carbon atom is attached on cantilever and bridged CNT respectively. Moreover, for very large mass values (higher from 10^7 times of carbon atom nuclei) the frequency shift seems to take a stable value. However, this fact does not influence the sensing ability of the tube whereas these values are much greater than the CNT weight and not need tube of the specific scale in order to be sensed.

Similarly, the effect on the frequency shift of the diameter is presented in Figure 2(a). CNTs of the same length, which is equal to 8 nm, are analyzed. The diameter remarkably influences the sensing performance of the tube. In general, when the diameter increases, the frequency shift increases. However, in the case of clamped-clamped support, we observe different behavior. For diameter equal to 1.08 nm the frequency shift is the lowest. This occurs due to the change of the mode shape corresponds to the fundamental frequency. Here, the fundamental mode is the radial breathing and not the first bending, as in all previous cases. Hence, the limits of fundamental mode changing have to be taken account in the nanosensor design. The atomic resolution (frequency shift when one carbon atom is attached) is in the order of hundreds MHz as previously.

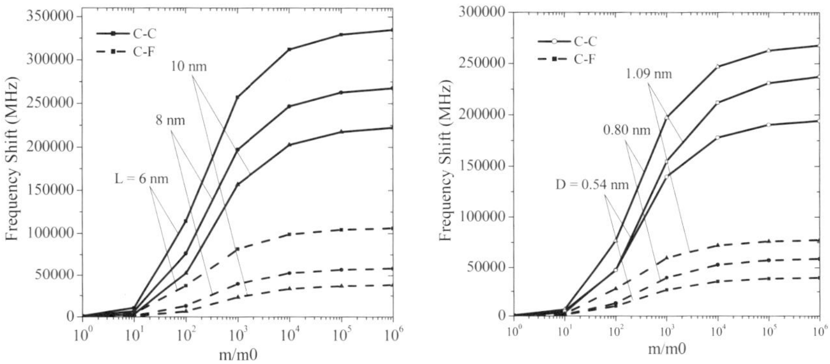

(a) (b)

Figure 2. Frequency shift for CNTs (a) with different length and (b) different diameter.

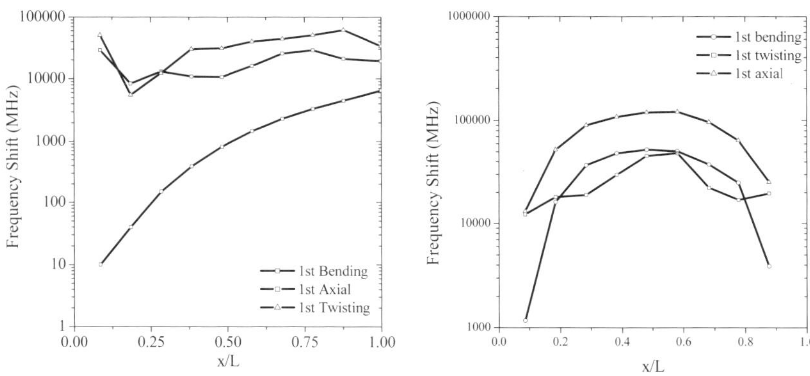

(a) (b)

Figure 3. Frequency shift of different vibration modes for (a) clamped-free and (b) clamped-clamped support condition.

In literature, all the theoretical studies are limited in the shift of the fundamental frequency due to an added mass. However, this is suffice to sense a mass addition on the CNT but not to estimate the weight of this mass because the magnitude of the shift is influenced not only from the mass weight but also from its position on the CNT. A way to sense and measure the mass is to combine the frequency shifts of the fundamental and higher order modes of vibration. Based on this concept, Figure 3(a) depicts the frequency shift of the basic modes of vibration of a cantilever (6,6) CNT with length equal to 8 nm when a mass 10^2 times higher than carbon atom is added in different positions on the tube. Note that the variation of axial and twist mode frequencies is small. This fact may help to find easier the mass weight. Similar results are presented in Figure 3(b). Here, the difference is that the same CNT is bridged.

4. Conclusions

A spring-mass based finite element formulation for vibration analysis of nanotubes was utilized to investigate the mass sensing characteristics of single-walled CNTs. Comparisons demonstrated the acceptable accuracy of the proposed method. The frequency shift, which is the base of mass sensing, was extracted for different geometric parameters. The basic design characteristics, i.e. the diameter and length of CNTs, revealed to considerably influence the mass sensing behavior of CNTs. The CNTs were proved of ultrahigh resolution since the frequency shift could be in the order of GHz when the added mass is just one carbon atom. To the author's best knowledge, the frequency shift of all the basic modes of CNTs is presented for first time in the literature. The position of the attached mass on the nanotube was exposed notably changes the sensing behavior. The combination of frequency shifts of the basic modes could be able not only sense an added mass, but also to find out the weight and its position on CNT.

References

1. K. Jensen, K. Kim, D. A. Zettl, *Nat Nanotechnol* **3**, 533 (2008).
2. R. G. Knobel, *Nat Nanotechnol* **3**, 525 (2008).
3. S. K. Georgantzinos, G. I. Giannopoulos, N. K. Anifantis, *Comput. Mech.* **43**, 731 (2009).
4. S. K. Georgantzinos, N. K. Anifantis, *Comput Mater Sci* **47**, 168 (2009).
5. M.S. Dresselhaus, G. Dresselhaus, R. Saito, *Carbon* **33** 883 (1995).
6. J. Kołoczek, K. Young-Kyun and A. Burian, *J. All. Comp.* **28**, 222 (2001).
7. W. D. Cornell, P. Cieplak, C. I. Bayly, I. R. Gould, K. M. Merz Jr., D. M. Ferguson, D. C. Spellmeyer, T. Fox, J. W. Caldwell, P. A. Kollman, *J. Am. Chem. Soc.* **117**, 5179 (1995).
8. W. L. Jorgensen, D. L. Severance, *J. Am. Chem. Soc.* **112**, 4768 (1990).
9. C. Li, T.-W. Chou, *Appl. Phys. Lett.*, **84**, 5246 (2004).

FINGER VEIN SEGMENTATION AND CENTERLINES EXTRACTION IN INFRARED IMAGES

M. VLACHOS AND E. DERMATAS

Department of Electrical Engineering & Computer Technology

University of Patras, Patras, Greece

{mvlachos,dermatas}@george.wcl2.ee.upatras.gr

In the narrow bandwidth of the near-infrared spectrum, the light is propagated through human tissue with low absorption rates, but strong scattering effects produce extremely low contrast images. In this paper, an algorithm for finger vein segmentation and centerlines extraction in infrared images is presented. Finger veins are detected in pixels with positive divergence of the gradient vector field, while centerlines are extracted in pixels with positive divergence of the normalized gradient vector field estimated at various orientations. The segmentation algorithm has been evaluated on both artificial and real finger infrared images and high segmentation rates are achieved in terms of sensitivity, specificity and accuracy using manual annotation data obtained by human observers.

Introduction

The problem of finger vein extraction arises mainly for biometrics purposes but it is also very important in the biomedical research community, i.e. for diagnosing in vascular pathology. The inspection of angiogenesis in the human body can be used for early diagnosis of several diseases, both in healthy tissue for healing wounds and for restoring blood flow to tissues after injury or insult. Also, inspection of the retinal vasculature may reveal hypertension, diabetes, arteriosclerosis, cardiovascular disease and stroke. In the problem of finger vessel pattern extraction from infrared images, the veins appear darker than the surrounding tissue, due to strong absorption of light from the hemoglobin molecules, and they have a concave shape cross sectional profile. This property can be used to facilitate the segmentation process and several researchers use the concave shape of the vessel to achieve better segmentation accuracy [6-7, 8, 12]. Primitive segmentation methods, based on simple global thresholding techniques, proved inadequate due to the fact that other parts of the finger have similar brightness and the image histogram is not bimodal. Moreover, infrared images suffer from several shading artifacts produced mainly due to uneven illumination, poor acquisition conditions and fingerprint noise. Thus, brightness normalization and shading correction are important pre-processing modules in such applications.

In [1, 2, 4], low pass spatial filtering performs noise removal and high pass spatial filtering is used for enhancing vascular patterns and vascular segmentation. An improved vein pattern extracting algorithm is proposed in [3], which compensates the loss of vein patterns in the edge area, giving better image enhancement, improved vein pattern information, and better performance than the rate reported in [1]. In [5], a direction-based vascular pattern extraction algorithm is developed using two different filters: row and column vascular pattern extraction filter for abscissa and ordinate vascular pattern extraction. The combined output of both filters produces the final hand vascular patterns.

Person identification based on finger-vein patterns is presented and evaluated using line tracking starting from different positions [6-7]. Local dark lines are identified and line tracking is executed by moving along the lines. When a dark line is not detectable, a new tracking operation starts at another position. This iterative procedure detects the same line multiple times, so the line-detection frequency is used to estimate the pixel vein probability.

In the low contrast infrared images, finger veins are extracted in a four step process: the fingerprint lines are removed using the *2D* discrete wavelet filtering and kernel filtering produces multiple images by rotating the kernel in six different directions, focus into the expected directions of the vein patterns. The maximum of all images is transformed into a binary image. Further image enhancement is achieved by a two-level morphological process [8]. A certification system compares near infrared (*NIR*) vein images for low-cost, high speed and high precision vein extraction [9]. In the recognition stage, phase correlation and template matching is applied.

Segmentation and vein-thinning methods are studied in human-hand infrared images and two new methods for threshold-based segmentation, based on end points, crossing points, and conditional thinning are presented in [10]. A method for detecting blood vessels in pathological retina images is proposed [11]. The blood vessel-like objects are extracted using the Laplacian operator and noisy objects are pruned according to the centerlines, which are detected using the normalized vector field. This method has been tested in the STARE database and the experimental results show that it can perform well both on pathological and healthy regions.

In this paper, an algorithm for finger vein segmentation and centerlines extraction in infrared images is presented. Finger veins are detected by estimating the divergence of the gradient vector field or equivalently the Laplacian operator, while centerlines are extracted using the divergence of the normalized gradient vector field. The same methodology has been already used in blood vessel segmentation of retinal images [11]. In this paper several adaptations on the original algorithm are adopted to face the extremely low contrast infrared images of the human finger. In the demanding problem of finger vein segmentation large brightness variations and shading makes the vein segmentation a challenging task. The algorithm has been evaluated on a database consists both of real and artificial infrared images of finger and has achieved high sensitivity, specificity and accuracy rates.

Finger vein segmentation

Finger vein segmentation is a three step procedure: Finger vein-like objects are extracted using the divergence of the gradient vector field, centerlines of veins are detected using the divergence of the normalized gradient vector field and noisy regions are pruned according to the centerlines information.

In infrared finger images the divergence of the gradient vector field can be used to detect vein-like objects. The adopted quantitative attribute of concave structures has great sensitivity in the low contrast images where very small brightness differences are met in neighbor pixels. The vector field does not have the same magnitude of change in every direction and thus the sign of the divergence depends both on magnitude and orientation.

Vein centerlines are located using the normalized gradient vector field. Each pixel has unit magnitude and thus the sign of the divergence depends only on the direction. Pixels belonging to veins are always darker than the immediate background and they have a concave shape. The normalized gradient vector field is expanding. So, a pixel is annotated as vein centerline if its divergence is positive. The divergence of the normalized gradient vector field is used to detect the centerlines. This procedure involves the detection of the high and low contrast centerlines. The detection of low contrast centerlines is applied due to shading.

If $I(x,y)$ is the brightness of the original infrared grayscale image, the image convolution with a Gaussian kernel of variance σ^2 produces a smoothed image:

$$I_G(x,y) = I(x,y) * G_\sigma(x,y)\,. \tag{1}$$

The divergence of the gradient vector field is estimated by:

$$Div(x,y) = \nabla(\nabla I_G(x,y)) = \frac{\partial}{\partial x}\left(\frac{\partial I_G(x,y)}{\partial x}\cdot\vec{i} + \frac{\partial I_G(x,y)}{\partial y}\cdot\vec{j}\right)\cdot\vec{i} + \frac{\partial}{\partial y}\left(\frac{\partial I_G(x,y)}{\partial x}\cdot\vec{i} + \frac{\partial I_G(x,y)}{\partial y}\cdot\vec{j}\right)\cdot\vec{j} = \frac{\partial^2 I_G(x,y)}{\partial x^2} + \frac{\partial^2 I_G(x,y)}{\partial y^2}, \quad (2)$$

which is the Laplacian operator of the image I_G.

The normalized gradient vector field is estimated by:

$$\nabla_n I_G(x,y) = \frac{\nabla I_G(x,y)}{|\nabla I_G(x,y)|} = \frac{\left(\frac{\partial I_G(x,y)}{\partial x}\cdot\vec{i} + \frac{\partial I_G(x,y)}{\partial y}\cdot\vec{j}\right)}{\sqrt{\left(\frac{\partial I_G(x,y)}{\partial x}\right)^2 + \left(\frac{\partial I_G(x,y)}{\partial y}\right)^2}}. \quad (3)$$

Thus, the divergence of the normalized gradient vector field is given by the following equation:

$$NDiv(x,y) = \nabla(\nabla_n I_G(x,y)) = \frac{\left(\frac{\partial^2 I_G(x,y)}{\partial x^2} + \frac{\partial^2 I_G(x,y)}{\partial y^2}\right)}{\sqrt{\left(\frac{\partial I_G(x,y)}{\partial x}\right)^2 + \left(\frac{\partial I_G(x,y)}{\partial y}\right)^2}}. \quad (4)$$

1. Extraction of finger vein-like objects. Finger vein-like objects extraction is performed using the divergence of the gradient vector field *Div(x,y)* of the smoothed image I_G. The original image I is rotated by θ degrees (0, 45) and each rotated version of this image is filtered by a set of Gaussian filter kernels with variance σ^2 (0 to 4 with 0.2 step). A threshold-based annotation process detects the vein-like pixels in the filtered images if $Div(x,y)/(G_\sigma(x,y) * I(x,y)) > \phi_3$, where φ_3 is a predefined threshold, and a connectedness criterion is satisfied. The above procedure delivers the vein-like objects by combining all pixels (logical *OR*) satisfy the above conditions. The connectedness criterion is used for artifact removal. According to this, the skeleton of the vein-like objects or vein centerlines is extracted using a morphological operator [13]. An eight-pixel connectedness is used for the extraction of skeleton. In the sequel, the length of the skeleton is measured and if it exceeds a predefined threshold it is retained, otherwise it is removed. In Fig. 1a the extracted finger vein pattern is shown.

2. Extraction of vein centerlines. This procedure involves the detection of the high and low contrast centerlines. The high contrast centerlines are detected using the divergence of the normalized gradient vector field *NDiv(x,y)*. The original image *I* is rotated by θ degrees (0 to π with $\pi/36$ step) and each rotated version of this image is filtered by a set of Gaussian filter kernels with variance σ^2 (0 to 4 with 0.2 step). Each pixel of all the above filtered images is candidate for belonging to vein centerlines if $NDiv(x,y)>\varphi_1$, where φ_1 is a predefined threshold, and satisfies the connectedness criterion. In a similar procedure the low contrast centerlines are extracted. The rotation parameter θ is set to zero and the smoothness parameter σ^2 varies within a predefined set of values (3 to 5 with 0.2 step). Each pixel of the filtered images is annotated as vein centerlines if $NDiv(x,y)/(G_\sigma(x,y)*I(x,y))>\phi_2$, where φ_2 is a predefined threshold, and the connectedness criterion is satisfied. Finger vein centerlines are obtained by combining (logical *OR*) high and low contrast centerlines. A typical example is shown in Fig. 1b.

3. Pruning according to vein centerlines. In order to remove noisy vein-like objects and to improve the robustness of the finger vein extraction, the finger vein-like objects are pruned according to the vein centerlines. The most distant pixels from the vein centerlines are removed from the vein network.

Therefore, the skeleton of the vein-like objects is estimated. The vein-like objects are classified in two sets. The first set contains pixels closer to the vein centerlines, while the second set contains pixels closer to the skeleton. The connectedness criterion is applied in the second set and the pixels satisfy this criterion are merged to the vein centerlines to construct the finger vein network.

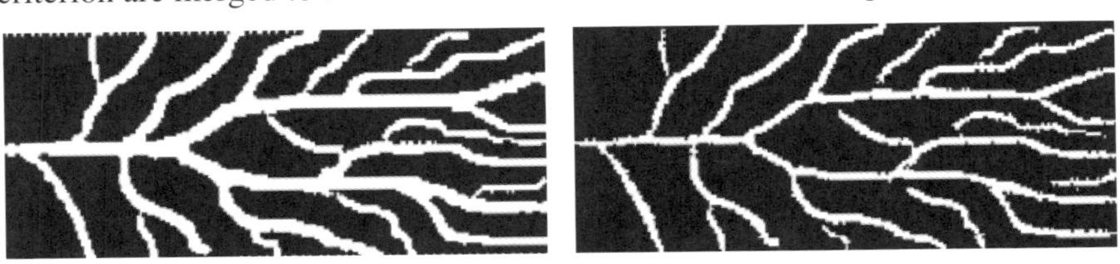

(a) (b)

Figure 1. (a) Finger vein pattern and (b) vein centerlines extracted by the proposed algorithm.

System evaluation

Infrared Image Acquisition. The original infrared images are acquired under infrared light using a low cost CCD camera. The finger is placed between the camera and the light source which consists of a row of five infrared leds with adjustable illumination. In the acquired infrared images strong shading and non-uniform illumination effects are introduced. Thus, vein pixels can be detected by considering the divergence values under different smoothness and orientations. Fig. 2a shows a typical example of a real acquired infrared image. The extracted finger vein network using the proposed algorithm is shown in

Fig.2b.

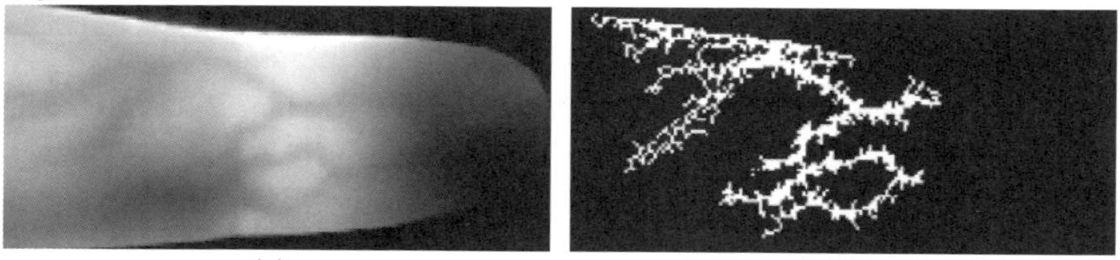

(a) (b)

Figure 2. (a) Real infrared image and (b) the corresponding vein network extracted by the proposed algorithm.

Artificial image database. A quantitative evaluation of the proposed algorithm in real infrared images is difficult due to the absence of manual segmentation data. Even in experts, the extremely low contrast of the infrared images introduces significant difficulties to obtain reliable human annotation of the vein network. Therefore, the proposed method is evaluated using a small set of images each one created by the weighed sum of the two artificial images. The first image is constructed using an artificial vessel-like network. This network consists of connected lines of different widths with junctions and bifurcations and multiple low pass filtering to simulate the blurriness of the edges which is apparent to the real images due to the blood flow and scattering. The second artificial image is used to simulate the non uniform image background, and is created by applying an iterative spatial low pass Gaussian filter with a large window size to the original infrared image. Fig. 3a shows an artificial created infrared image, while Fig. 3b shows an extremely blurred version of this image. Even in this case the segmentation accuracy is satisfactory (comparable with that of Fig. 3a).

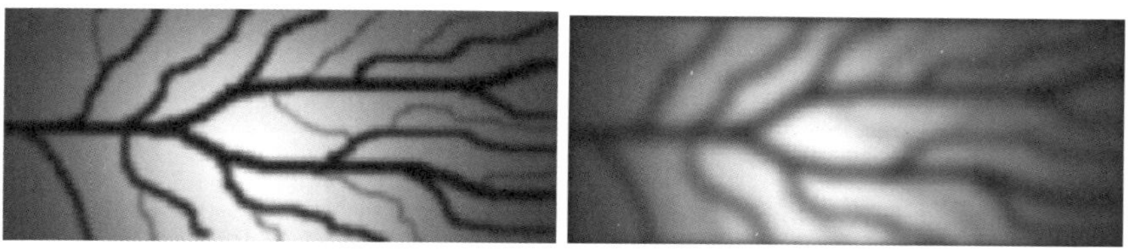

(a) (b)

Figure 3. (a) Artificial infrared image and (b) an extremely blurred version of (a).

Evaluation of the proposed algorithm. In the finger vein segmentation process, each pixel is classified as tissue of vein. Therefore, true positive (*TP*) and true negative (*TN*) appears when a pixel is correctly segmented as a vein or non-vein, and two misclassifications, a false negative *(FN)* appears when a pixel in a vein is segmented in the non-vein area, and a false positive (*FP*) when a non-vein pixel is segmented as a vein-pixel.

Two widely known statistical measures are used for algorithm evaluation:

sensitivity and specificity. The sensitivity is a normalized measure of true positives, while specificity measures the proportion of true negatives. The accuracy is strongly related to the segmentation quality and for this reason is used to evaluate and compare different methods.

Table 1 summarizes the evaluation rates in terms of sensitivity, specificity and accuracy for the artificial image database, while Fig. 4 shows the overall *ROC* curve, obtained by averaging the *ROCs* of each image. Each *ROC* curve is estimated by processing the corresponding segmentation sensitivity and specificity for various values of the threshold parameter φ_3.

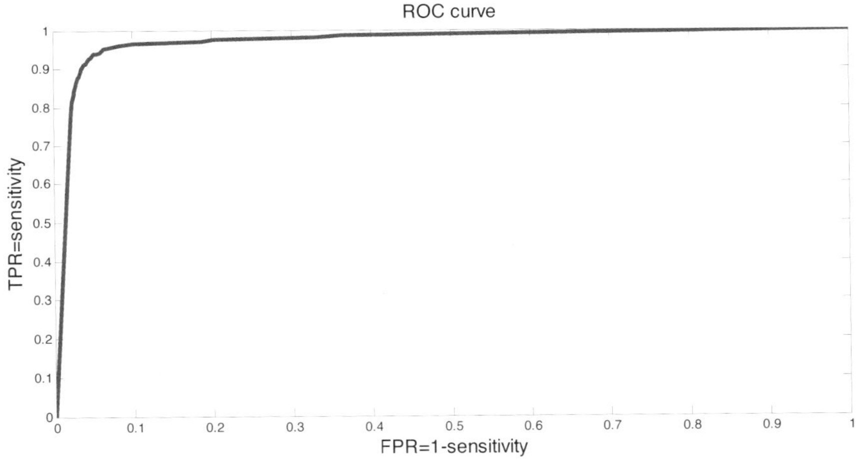

Figure 4. Overall *ROC* curve for the proposed algorithm.

Table 1. Sensitivity, specificity and accuracy for the proposed finger vein segmentation method.

image	sensitivity	specificity	accuracy
im_1	0.990000	0.930805	0.943938
im_2	0.988098	0.942821	0.952722
im_3	0.976388	0.943475	0.951637
im_4	0.973808	0.967901	0.969258
im_5	0.971842	0.95295	0.956842
im_6	0.624453	0.854248	0.805845
im_7	0.980764	0.957437	0.961776
im_8	0.971793	0.957783	0.960583
im_9	0.977645	0.94626	0.953643
im_{10}	0.975414	0.959693	0.963403
average	0.942556	0.941337	0.941965
standard deviation	0.111903	0.032375	0.048354

Acknowledgments. This work is partially supported by grant KARATHEODORIS of the University of Patras.

References

1. G.T. Park, S.K. Im, and H.S. Choi, “A Person Identification Algorithm Utilizing Hand Vein Pattern”, Proc. of Korea Signal Processing Conf., Vol. 10, No. 1, 1997, pp. 1107-1110.
2. D.U. Hong, S.K. Im, and H.S. Choi, “Implementation of Real Time System for Personal Identification Algorithm Utilizing Hand Vein Pattern”, Proc. of IEEK Fall Conf., 1999, Vol. 22, No. 2, pp. 560-563.
3. S.K. Im, H.M. Park, S.W. Kim, C.K. Chung, and H.S. Choi, “Improved Vein Pattern Extracting Algorithm and its Implementation”, Proc. of IEEE ICCE, 2000, pp. 2-3.
4. S.K. Im, H.M. Park, and S.W. Kim, “A Biometric Identification System by Extracting Hand Vein Patterns”, Journal of the Korean Physical Society, Vol. 38, No. 3, 2001, pp. 268-272.
5. S.K. Im, H.S. Choi, and S.W. Kim, “Direction-Based Vascular Pattern Extraction Algorithm for Hand Vascular Pattern Verification”, Korea University, Seoul, Korea. ETRI Journal, Vol. 25, No. 2, April 2003.
6. N. Miura, A. Nagasaka, T. Miyatake, “Feature extraction of finger-vein patterns based on repeated line tracking and its application to personal identification”, Machine Vision and Applications (2004) Vol. 15, pp.194-203.
7. N. Miura, A. Nagasaka, T. Miyatake, “Feature extraction of finger-vein patterns based on iterative line tracking and its application to personal identification”, Systems and Computers in Japan, Vol. 35, No. 7, 2004.
8. M. Vlachos, and E. Dermatas, “A finger vein pattern extraction algorithm based on filtering in multiple directions”, 5th European Symposium on Biomedical Engineering, July 2006.
9. T. Tanaka, N. Kubo, “Biometric Authentication by Hand Vein Patterns”, SICE Annual Conference in Sapporo, August 2004.
10. Y. Ding, D. Zhuang and K. Wang, “A Study of Hand Vein Recognition Method”, Proceedings of the IEEE International Conference on Mechatronics & Automation, Niagara Falls, Canada, July 2005.
11. B.S.Y. Lam, H. Yan, “A Novel Vessel Segmentation Algorithm for Pathological Retina Images Based on the Divergence of Vector Fields”, IEEE Transactions on Medical Imaging, vol. 27(2), pp. 237-246, 2008.
12. C.H. Wu, G. Agam, P. Stanchev, “A general framework for vessel segmentation in retinal images”, International Symposium on Computational Intelligence in Robotics and Automation (CIRA), June 2007 pp. 37 – 421.
13. K.R. Castleman, Digital image processing, Englewood Cliffs, N.J.: Prentice-Hall, 1979.

Signal Processing and Non Destructive Methods

DAMADE CHARACTERIZATION OF AGEING INFRASTRUCTURES BY MEANS OF ELASTIC WAVE TECHNIQUES

T. SHIOTANI
Department of Urban Management, Graduate School of Engineering
Kyoto University
Kyoto University Katsura Campus, Nishikyo-Ku, Kyoto 615-8540, Japan

Infrastructures, one large social stock not only for single state but for the earth, are deteriorating indubitably with elapsed time. Among developed countries recent rapid increase of seriously deteriorated infrastructures calls for urgent repair/ reinforcement; however, sufficient budget is not prepared for such a huge amount of degraded infrastructures. To maintain these with a proper way, paradigm of maintenance shall be shifted from "reactive" to "proactive" since it enables yearly-leveling of maintenance investment with minimal charge. In this idea of proactive maintenance, it will be a mandatory condition that early-stage of damage can be evaluated. This paper demonstrates elastic wave based NDT techniques allowing to evaluate early-stage of damage. Strategic monitoring concept with elastic wave based NDT is specifically drawn. Acoustic emission, providing most likely damaged area in the material/ structures of interest is employed, followed by active NDT as through the thickness/ surface wave measurement aided by computed tomography.

1. Introduction

There is a growing amount of deteriorating infrastructures worldwide that not only affect the productivity of the society but has a great impact on resources, environment and human safety. To maintain such deteriorating infrastructures, reactive maintenance, based on the latest inspection result corresponding to individual structures, has so far been carried out. This concept of the maintenance could be only operative during the era where munificent tax revenue was expected due to rapid economic growth. Scrap and build idea, assuming the structural life span, e.g., 50 years, and replacement after the span was generally assumed, was lay on the base of this. Recent growing concerns on environmental impact [1] as well as governmental revenue shortage, however, resulted that they do not allow such reactive maintenance idea so that we must move on to the other concept, namely proactive maintenance. Several activities can be found on life cycle assessment of maintenance considering environmental impact [2] and management of infra-maintenance investment in a long span of 30 years [3]. The current situation surrounding infrastructures thus requires the paradigm shift of maintenance from reactive to proactive. Proactive maintenance allows early warning for the eventual failures and expansion of service life by executing timely repair/ retrofit program on the basis of LCA (life cycle assessment) [1]. In this proactive maintenance or to establish maintenance program in long-term, there need a requirement: reasonable evaluation of the current health status for the structure can be obtained with some kinds of monitoring techniques. In addition, as the infrastructure is generally large in comparison to others, the health status of the structures shall be investigated roughly for the first stage against the whole area, then for the detail. As stated previously, NDT techniques being essentially desired for the proactive maintenance are ones facilitating both: reasonable status evaluation not only for the whole of the structure but for the local areas; and sensitive enough from the early damage.

2. Role of monitoring

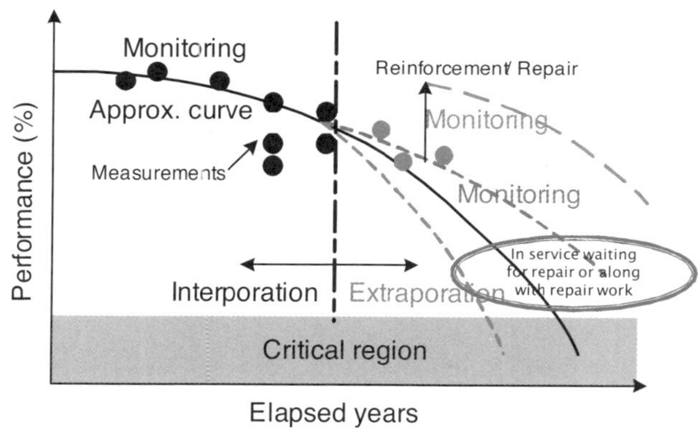

Figure 1. Performance curves approximated by monitoring data

Once established a proper monitoring technique/ procedure, structural performance can be drawn as an approximated performance curve with elapsing time as can be found in Fig. 1 [4]. Based on the curve, the future deterioration process could be estimated, and the performance based risk evaluation in consideration of loss of damage is carried out for individual structure. Prioritization of repair program is then conducted. It is noted that the monitoring has other roles for the infrastructural management. Once repair work has completed, repeated-monitoring could assume the repair effectiveness or recovery rate when comparing to before repair, resulting in updating the performance curve quantitatively [5-7]. And for the seriously deteriorated structures waiting for repair or along with repair work 'in-service,' it could also be monitored in real time to alert eventual failure, which can avoid causalities [8].

The monitoring action, not only for proactive maintenance but through the infrastructural asset management, is consequently very significant.

3. Elastic wave techniques

Due to the complexity of deterioration process as well as various kinds of structures, any decisive investigation technique for proactive maintenance has yet to been established so far. Acoustic emission (AE) and ultrasonic technique (UT), that are nondestructive techniques using stress wave, could be applied for the infrastructures diagnosis since the former allow to cover a wide range with single sensor and the latter visualize the internal damage with velocity structure (namely, tomogram). The covering area of these elastic-wave monitoring and the accuracy of tomogram, both can be complied with various demands with the frequency employed. In general the higher the frequency was set to be excited and detected, the more accurate the evaluation was obtained. A schematic configuration for both AE and UT aided by computed tomography is shown as in Fig. 2. As can be seen in the left, AE uses passive waves generated from the defect/damage of structures, and as same as in the seismic monitoring, less numbers of sensors cover a wider area. When investigating damage status of the structure, AE signals due to friction among existent cracks/ damages, namely secondary AE activity, are expected to be detected, where external force or internal stress variation is mandatory for the monitoring condition. In case that any stress changes could not be expected, UT, as shown in the right, could explore the defect as well as providing the comprehensible distribution of the velocity after the tomographic procedure [9]. As can be found, both techniques of AE and UT complement one another for the damage assessment. Specific cases, AE monitoring in combination with UT for in-situ structure, will be present in the following session.

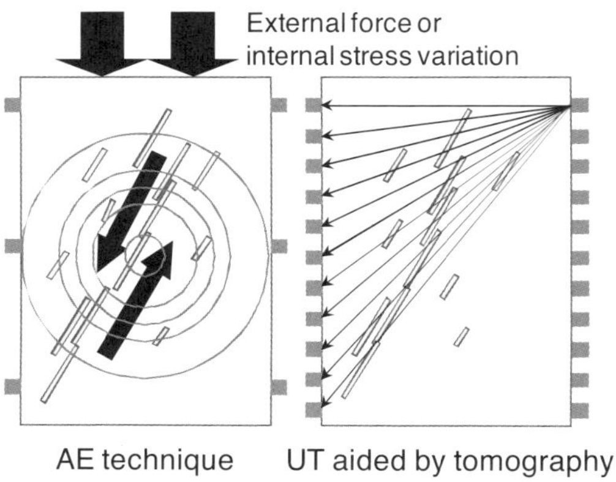

Figure 2. Two elastic wave techniques: acoustic emission, left and ultrasonic testing with tomography, right.

4. Elastic wave activity with materials deterioration processes

As it detects elastic waves continuously, AE is used for real time damage evaluation in infrastructures. AE activity can be classified with its genesis into two types: primary AE activity due to crack-nucleation and -propagation; and secondary AE activity resulted from frictional behavior among existent crack-interfaces. With appropriately arrayed AE sensors on or within materials, the identification of AE source, providing the location, scale and the type of fracture, can be possible. As stated already, additional load application with internal stress variation is necessary to induce the secondary AE activity. On the other hand, active elastic wave techniques can be carried out without load application, instead enormous numbers of elastic wave excitations together with detections of the excited waves in every reference point are required to construct velocity-structure, namely tomogram within the materials. In this session, damage visualization by both AE and UT aided by tomography is demonstrated. Specifically two-layered reinforced concrete slabs were subjected under bending load [10]. AE is used to monitor the deterioration progress of reinforced concrete beams, consisting of two layers of plain and fiber concrete, under four point bending. The specimens are followed to be measured the elastic wave velocities at several loading steps in order to obtain the momentary three dimensional tomographic reconstruction of the interior.

4.1 *Specimens and instrumentations*

Six beam specimens of 150 x 150 x 530 mm were prepared, among which one was plain concrete beam that served as the control specimen; while the other five consisted of plain concrete and a vinyl fiber-reinforced mortar layer of similar thickness, respectively. The concrete and fiber-reinforced mortar have average 28-day compressive strength of 31.9 MPa and 48.6 MPa, respectively. The mortar layer was overlaid by the shotcrete method after seven days of concrete casting. Four-point bending test in line with JSCE-G 552 was conducted on each specimen. The supports and loading configurations are given in Fig. 3. Twelve AE sensors with resonant frequency of 60 kHz were attached to the specimen surfaces. The signals were pre-amplified by 40 dB and recorded by a 16-channel data acquisition system. During the bending tests, loading was terminated at fixed intervals to allow pencil lead break excitations near the location of each AE sensor. The transit time of excited transient waves to the rest of the sensors was used to calculate pulse velocity.

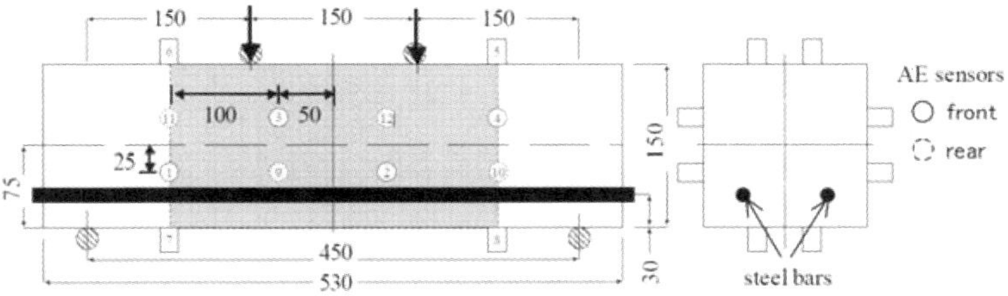

Figure 3. Experimental configuration for specimen as well as AE sensor arrangement.

4.2 Results and discussion

Figure 4 shows example of load vs. displacement curves for both types of specimens. The composite specimen yielded higher maximum load than plain concrete specimen (129.8 MPa and 109.1 MPa respectively). Visual observations suggested no occurrence of debonding between the concrete and the fiber-reinforced mortar layer. Furthermore, the load-displacement curves of the composite specimen exhibited no abrupt change in gradient that is usually associated with significant increase in displacement due to stiffness loss. This implies that fiber-reinforced mortar has enhanced the strength performance of concrete beams without raising concerns of premature debonding.

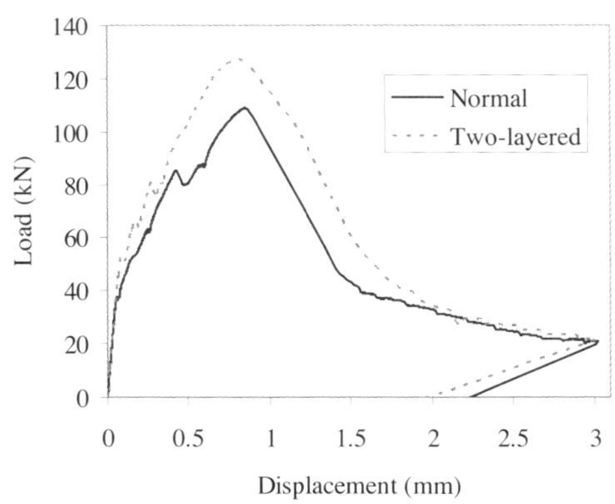

Figure 4. Mechanical behavior of specimens under bending.

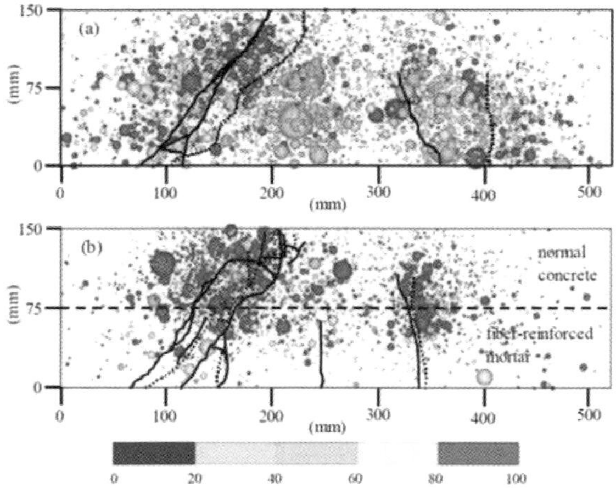

Figure 5. Locations of AE events and actual cracking patterns for a) plain concrete specimen and b) composite specimen (solid line: crack from front view, dash line: rear view).

Figure 5 illustrates typical examples of the AE event locations at specific load steps and cracks observed at ultimate failure. The AE events are represented by circles, where the circle area is proportional to the respective amplitude. The contour color indicates the load in terms of percentage of the ultimate load when an event was detected. For the plain concrete specimen, a great number of sources started to occur from 40% of the ultimate load, with the events concentrated mainly at the center of specimen. Visual observations revealed that this was when concrete fracture started, following the initiation of flexural cracks. At higher load percentages of 60% and onwards, events were found in the vicinity of large diagonal cracks at the left of the specimen, which confirmed that the ultimate failure was caused by shear. It is noted however that the development of cracks hindered accurate acquisition of all signals due to severe attenuation and delay.

In the composite specimen a significant number of events only occurred from 60% of the ultimate load. During 70% to 90% of the ultimate load, an accumulation of events was obtained in the vicinity of diagonal crack developed at the left, and near the tip of smaller cracks at the right of the specimen. Thus, the AE results suggest no premature debonding at the interface, which is in good agreement with the crack observation.

As aforementioned, pulse velocity is indicative of the damage. Data in the form of wave transit times of different paths were analyzed using a program specially developed for the purpose. Tomograms of the central cross-section were constructed, as exemplified in Fig. 6. At the intact condition, the pulse velocity was more than 4000 m/s. Under progressive loading, development of large-scale cracks significantly influenced the velocity structure, resulting in decrease of pulse velocity to 3000 m/s at the zone where the diagonal cracks developed. The tomography results also imply no debonding in the composite specimen, which tally with the results and observations of four-point bending test and AE measurements.

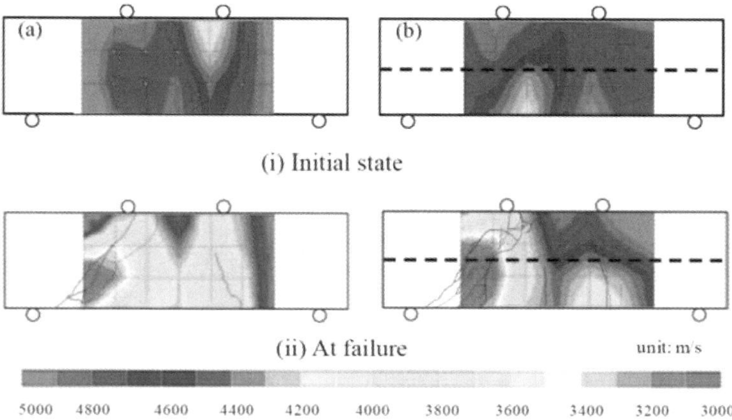

Figure 6. 2-dimensional velocity structures of central cross section for a) plain concrete specimen and b) composite specimen.

In this laboratory test of composite concrete, AE event locations coincide well with the actual crack locations, showing that AE signals can be used to identify the damage in real time. The formation of cracks is also demonstrated by decrease of the velocity in the vicinity of cracks, which is visualized by the pulse velocity tomography. Both AE results and pulse velocity structures imply no premature debonding of fiber-reinforced mortar layer, and are well agreed by results of bending tests and visual observations. It is concluded that the stress wave methods can be used to evaluate the behaviour of composite concrete elements under loading, and provide the general assessment of the internal fracture process.

5. Strategic monitoring for large infrastructures

As aforementioned, NDT technique which can be both applicable for preliminary investigation as well as in-depth measurement is of high demand in large infrastructures with regard to proactive maintenance strategy. Here, such monitoring procedure with elastic wave techniques is demonstrated using in-situ bridge measurement [11]. It is noted that through the preliminary investigations of a 45-m bridge span with surface-crack observation and physical tests of excavated cores, any deterioration did not reveal, however, due to the importance of the bridge, sensitive monitoring of AE was conducted in order to investigate the early damage, followed by UT aided by tomography. In this particular case, the monitoring was planned in the following procedure.

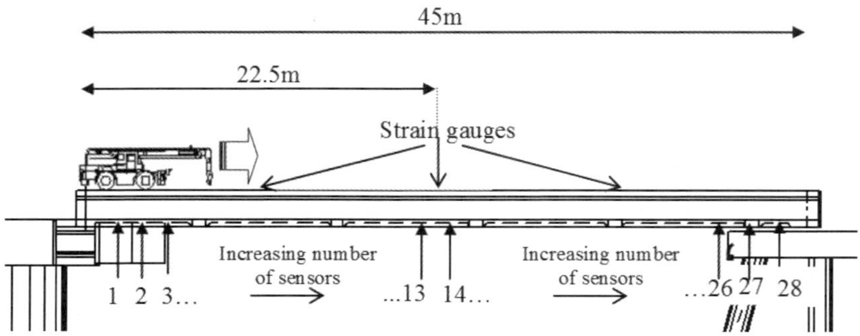

Figure 7. AE monitoring details with crane mobile load. Positions both of AE sensors and strain gauges are also shown.

5.1 *Strategic monitoring procedure*

In order to extract the relatively damaged area from the longitudinal structure, a linear arrangement of AE sensors with a space of 1.5 m over 45 m in a longitudinal direction is carried out. Totally 28 sensors (60 kHz resonance) are placed onto the bottom surface of the bridge with wax. The AE testing was performed during moving a heavy vehicle of 20 tons crane over the bridge as shown in Fig. 7. Based on the AE activity the part of structure being most likely to exhibit higher degree of damage than the other areas was selected for the follow-up detailed investigation, or end up in this first monitoring in case that no active parts of AE were found. In the former case, the selected area is again measured the AE activity with a denser sensor array in 2D. Using the 2D arrayed AE sensors, a surface ultrasonic examination, which is indicative of the concrete quality [12], followed in order to investigate the P-wave velocity of concrete at the area of interest. Pencil lead breaks were performed to excite artificial elastic waves. Detected signals both for AE and UT are amplified by 40 dB at the sensor, and recorded with two synchronized AE monitoring instrumentations.

5.2 *AE data interpretation*

The interpretation of detected AE activity from in-situ measurement is not always easy. This is the reason why AE parametric indices established in experimental studies is not always applicable to in-situ measurement results. Several indices obtained from the AE activity, relating to structural damage, have been suggested and applied for the diagnosis of the infrastructures. As stated earlier, when a material or structure is stressed as to generate cracks, primary AE activity is produced together with crack-nucleation. In addition, the secondary AE activity is intensively generated as damage evolves seriously due to predominant distribution of shear type of crack than of tensile. The number of AE hits during unloading divided by the number of hits during the whole loading cycle is defined as the Calm ratio, and values near zero indicate intact condition of the materials.

Onset of AE activity also suggests notable damage-index when comparison is made of previous one. In case that AE activity is obtained from smaller level of the maximum as such function of load and displacement, this implies unstable i.e, unhealthy state. The ratio of the parameters' level giving onset of AE to that of the past-maximum is referred to as Load ratio [13]. In-situ monitoring, however, since the past-maximum level could not be estimated, RTRI ratio [14], which is based on the maximum level during the monitoring period was used.

Another index comes from the analysis of the amplitude distribution of AE events, or so called the improved *b*-value analysis [15]. Although a large-scale fracture in general corresponds to large AE peak amplitude, the use of the amplitude solely can be misleading. This is because the accumulated damage increases the materials' attenuation rate due to scattering at the cracks. Therefore, even a high amplitude signal will be severely attenuated before being recorded by the sensors. In this respect, the amplitudes are studied through their cumulative distribution that uniquely changes as the damage is accumulated. Specifically, the gradient of the distribution is calculated. With the evolution of damage, this slope decreases, meaning simply that the ratio of the large energy AE events to that of the small relatively increases in the total population of AE events. It has been confirmed that at the moments of extensive cracking, the I*b*-value exhibits severe drops. Other damage-indices can be found in literature [13].

5.3 *Results and discussion*

The load for the AE monitoring was applied by a 20 ton crane vehicle, which passed three times over the bridge with a constant speed of approximately 0.5 m/s, as again shown in Fig. 7. When the crane moved over the bridge, the strain and stress fields changed. The compressive strain measured on the top surface of the bridge at the mid-span can be seen in Fig. 8. The maximum strain was recorded at 88 s, when the vehicle was in the middle of the span, suggesting the highest tensile stress at the bottom layer of the structure. In the same figure, the cumulative number of AE hits recorded by all the sensors is depicted for one passage. It can be seen that the rate of AE hits was more intensive before the crane reached the center of the bridge at 88 s. Up to that moment, more than 70% of the total number of hits was recorded, implying that more active AE sources were located in the first half of the bridge.

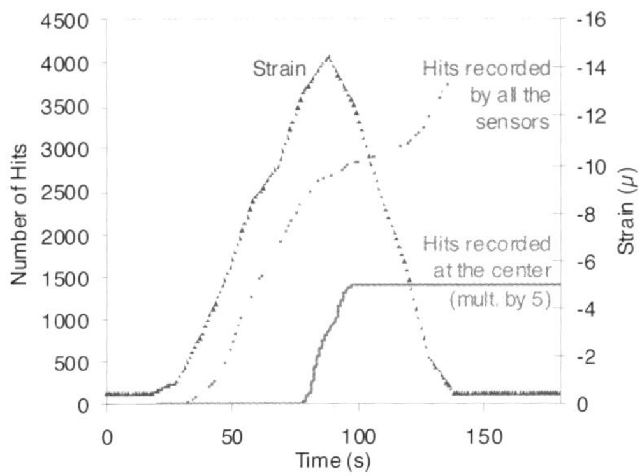

Figure 8. Strain and cumulative AE hits with elapsing time.

As stated already, in order to calculate the Calm ratio, the AE activity should be correlated with a measured mechanical parameter. In this case, this was conducted by the measured strain at the middle point and the AE activity of the center part, recorded by the two sensors placed closest to the center (#13 and #14 of Fig. 7). The activity of these sensors and AE hits of all sensors combined are plotted in Fig. 8. The hits of the middle sensors started at 74.5 s as the crane was approaching, and the last hit was recorded 10 s after the crane had passed over the bridge center, showing the intensive AE activity even during unloading. The number of hits for three different trips of the crane and the resulting Calm ratios ranges between 0.30-0.46, indicating serious damage according to past studies [17].

Using the arrival time difference of AE signals, one dimensional AE sources are determined by linearly placed 28 sensors. The locations of AE sensors and resulted AE events activity by 10 cm are shown in Fig. 9. Intensive AE activity is always emerged about 3 m, irrespective to the trip, and therefore an area at 3 m in the figure was chosen for the subsequent detail monitoring, where the separation distance of sensor is 1.5 m, resulting in the examined area of 3 m by 3 m of three parallel arrays of three. Nine sensors were used for this detail test as shown in Fig. 10.

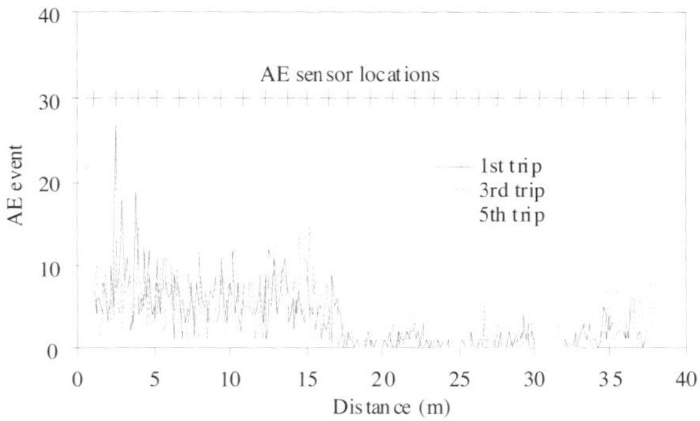

Figure 9. 1D AE sources along with repeated mobile loads.

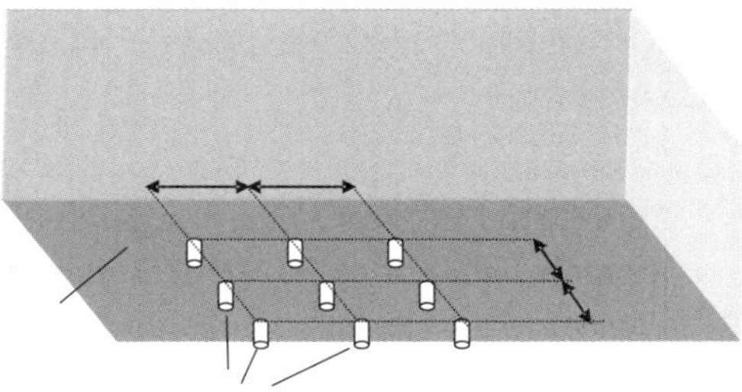

Figure 10. Arrangement of AE sensors both for 2D AE monitoring and UT.

It is again noted that no visual defect was observed on the surface. As can be found in Fig. 11, remarkable numbers of AE events ranging from small to large scale were obtained. Specifically those were appeared along the tires trace of the mobile vehicle employed. Velocity tomogram leads to a characteristic low velocity zone, developed diagonally in the area of interest. The low velocity zone might be attributed to a subsurface defect, although the depth cannot be easily determined. Concerning Rayleigh waves, it is well known that the penetration/ effective depth is approximately similar to the wavelength. In this case, however, the first arrival of waveforms was used to measure velocity, corresponding to the longitudinal wave and always arrives faster than Rayleigh or shear waves. Therefore, it is not straightforward how deep is the surface layer. Considering again the fact that the AE events were intensively generated on this low velocity zone, it can be resulted that concrete cracks, delaminations of different layers (e.g. asphalt on concrete) or friction between the tendon ducts and matrix concrete might be possible origins to induce the AE activity due to the loading. Follow-up investigations focusing on this weak area will clarify the source, while this sequential investigation, which started with AE activity and followed by detailed measurements of ultrasonic P-wave velocity are very useful in characterizing the quality of large-scale concrete structures, leading to contribute to the 'proactive' or 'preventive' maintenance strategy.

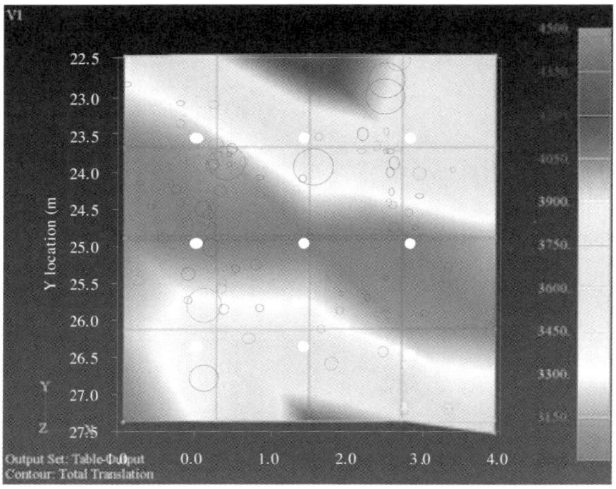

Figure 11. Identified 2D AE sources (see circles) with mobile load and velocity tomogram (see colored contour in m/s) obtained from surface UT measurement.

6. Conclusions

Corresponding to the paradigm shift on infrastructure maintenance from 'reactive' to 'proactive' or 'preventive,' this paper demonstrates elastic wave based NDT techniques allowing to evaluate early-stage of damage, that is necessarily conducted to minimize the investment for maintaining existing infrastructures. As demonstrated, acoustic emission provided most likely damaged area in the material/ structures of interest, and subsequently conducted through the thickness/ surface wave measurement aided by computed tomography, informed the insight of damage within the materials. In elastic wave techniques, the higher the frequency is used, the more accurate the damage visualization becomes and vice versa; however, it is noted that the higher frequency makes detectable/measurement range smaller. Target damage level/scale for the infrastructure determining optimum frequency used will be clarified if more kinds of works for various types of materials/structures were systematically conducted.

Acknowledgement

This series of studies were jointly conducted with a lot of distinguished engineers and researchers. Among them, special thanks should be given to Dr. D. G. Aggelis of Ioannina University for his excellent interpretation of ultrasonic propagation, Mr. S. Momoki and Dr. H. Chai of Tobishima Corporation for their splendid works on elastic wave measurements, and all those whom they involved in this research activity.

References

1. ISO 14040 (1998), 14041 (1998), ISO 14042 (2000), ISO 14043(2000)
2. V. Arskog, S. Fossdal and O. E. Gjorv, *Proc. Int. Workshop on Sustainable Development and Concrete Technology* (2004).
3. Tokyo Metropolitan Government, http:// www.kensetsu.metro.tokyo.jp/asset/index.htm (2009).
4. T. Shiotani, H. Ohtsu, D. G. Aggelis and S. Momoki, *Proc. Tenth Int Conf on Stuctural Safety and Reliability* (ICOSSAR 2009) (2009).
5. T. Shiotani, *Advanced Materials Research*, **13-14** (2006).
6. T. Shiotani, Y. Nakanishi, K. Iwaki, X. Luo and H. Haya, *J. Acoustic Emission*, **23** (2006).
7. D. G. Aggelis and T. Shiotani, *Cement & Concrete Composites*, **29**(9) (2007).
8. T. Shiotani, S. Yuyama, M. Carlos, and S. J. Vahaviolos, *J. Acoustic Emission*, **18** (2000).
9. Y. Kobayashi, T. Shiotani, D. G. Aggelis and H. Shiojiri, *Proc. of Second Int. Operational Analysis Conf.*, **2** (2007).
10. D. G. Aggelis, S. Momoki, T. Shiotani and A. Hirama, *Int. RILEM Conf on Site Assessment of Concrete, Masonry and Timber* (SACoMaTiS 2008) (2008).
11. T. Shiotani, D.G. Aggelis and O. Makishima, *J. Bridge Engineering*, **14**, 3 (2009).
12. T. Gudra and B. Stawinski, *NDT & E Int.*, **33**, (2000).
13. JSNDI, *NDIS 2421*, (2000)
14. T. Shiotani, Y. Nakanishi, X. Luo, H. Haya and T. Inaba, *Key Engineering Materials*, **270-273** (2004)
15. T. Shiotani, K. Fujii, T. Aoki and K. Amou, *Prog. Acoustic Emission* **VII** (1994).
16. C. Grosse and M. Ohtsu, eds, *Acoustic Emission Testing* (2008).
17. M. Ohtsu, M. Uchida, T. Okamoto and S. Yuyama, *ACI Structural Journal*, **99(4)** (2002).

THE TRANSMISSION EIGENVALUE PROBLEM

F. CAKONI AND D. COLTON *
Department of Mathematical Sciences
University of Delaware
Newark, Delaware 19716-2553, USA
E-mail: cakoni@math.udel.edu, colton@math.udel.edu

D. GINTIDES
Department of Mathematics
National Technical University of Athens
Zografou Campus, 15780 Athens, Greece
E-mail: dgindi@math.ntua.gr

This paper provides a survey of recent developments on the transmission eigenvalue problem.

1. Introduction

In recent years the transmission eigenvalue problem has come to play an important role in inverse scattering theory [14]. This is due to the fact that transmission eigenvalues can be determined from the far field data of the scattered wave and used to obtain estimates for the material properties of the scattering object [3], [5], [8]. For the case of scattering of acoustic waves by a bounded simply connected inhomogeneous medium $D \subset \mathbb{R}^3$, the *transmission eigenvalue problem* is to find $k \neq 0$, $w \in L^2(D)$, $v \in L^2(D)$, $w - v \in H^2(D)$ such that

$$\Delta w + k^2 n(x) w = 0 \quad \text{in} \quad D \tag{1}$$

$$\Delta v + k^2 v = 0 \quad \text{in} \quad D \tag{2}$$

$$w = v \quad \text{on} \quad \partial D \tag{3}$$

$$\frac{\partial w}{\partial \nu} = \frac{\partial v}{\partial \nu} \quad \text{on} \quad \partial D \tag{4}$$

*The research of F.C. and D.C. was supported in part by the Air Force of Scientific Research under Grant FA-9550-08-1-0138.

where ν is the unit outward normal to the smooth boundary ∂D and the index of refraction $n(x)$ is positive. Values of $k \neq 0$ such that there exists a nontrivial solution to (1)-(4) are called *transmission eigenvalues*. Research on transmission eigenvalues has focused on two main themes: 1) conditions on $n \in L^\infty(D)$ such that transmission eigenvalues exist and form a discrete set [9], [11], [16], [17], [18] and 2) the determination of lower and upper bounds for $n(x)$ from a knowledge of the transmission eigenvalues corresponding to (1)-(4) [6], [7], [8] [9]. The purpose of this paper is to provide a brief survey of both of these research directions.

2. Scattering by an Inhomogeneous Medium

We begin by considering the scattering problem for a non-absorbing inhomogeneous medium and show how the transmission eigenvalues corresponding to this medium can be determined from the far field pattern of the scattered wave. The scattering problem under consideration is to find a function $u \in H^1_{loc}(\mathbb{R}^3)$ such that

$$\Delta u + k^2 n(x) u = 0 \quad \text{in} \quad \mathbb{R}^3 \tag{5}$$

$$u = u^i + u^s \tag{6}$$

$$\lim_{r\to\infty} r\left(\frac{\partial u^s}{\partial r} - iku^s\right) = 0 \tag{7}$$

where $x \in \mathbb{R}^3$, $r = |x|$, $k > 0$ is the wave number, $u^i(x) = e^{ikx\cdot d}$ with $|d| = 1$ is the incident field, u^s is the scattered field satisfying the Sommerfeld radiation condition (7) uniformly in $\hat{x} = x/|x|$ and $n \in L^\infty(D)$ such that $n(x) > 0$ for $x \in \overline{D}$ and $n(x) = 1$ for $x \in \mathbb{R}^3 \setminus D$. Here D is defined as in the introduction and under our assumptions on D and n the scattering problem (5)-(7) is uniquely solvable in $H^1_{loc}(\mathbb{R}^3)$. It can be shown [12] that u^s has the asymptotic behavior

$$u^s(x) = \frac{e^{ikr}}{r} u_\infty(\hat{x}; d, k) + O(r^{-2}) \tag{8}$$

as $r \to \infty$ uniformly in $\hat{x}$ where u_∞ is the *far field pattern* of the scattered field u^s and we can define the *far field operator* $F : L^2(\Omega) \to L^2(\Omega)$ by

$$(Fg)(\hat{x}) := \int_\Omega u_\infty(\hat{x}; d, k) g(d) ds(d) \tag{9}$$

where Ω is the unit sphere in $\mathbb{R}^3$. We note that by linearity $(Fg)(\hat{x})$ is the far field pattern corresponding to (5)-(7) where the incident field is replaced by the *Herglotz wave function*

$$v_g(x) := \int_{\Omega} e^{ikx\cdot d} g(d)ds(d). \tag{10}$$

Theorem 2.1. [12] *For $k > 0$ the far field operator is injective with dense range if and only if there does not exist a nontrivial solution $v, w \in L^2(D)$, $v - w \in H^2(D)$ of the transmission eigenvalue problem (1)-(4) such that v is a Herglotz wave function.*

Theorem (2.1) suggests the possiblity of being able to determine real transmission eigenvalues from measured far field data. In particular, let F^δ denote the noisy far field operator corresponding to noisy measurements $u^\delta_\infty(\hat{x}, d, k)$ os the far filed data and let $\Phi_\infty(\hat{x}, z, k)$ be the far field pattern of the fundamental solution

$$\Phi(x, z, k) = \frac{e^{ik|x-z|}}{4\pi|x-z|}. \tag{11}$$

We are interested in the Tikhonov regularized solution $g^\delta_{z,\epsilon}$ of the far field equation defined as the unique minimizer of the *Tikhonov functional* [12]

$$\|F^\delta g - \Phi(\cdot, z)\|^2_{L^2(\Omega)} + \epsilon\|g\|^2_{L^2(\Omega)} \tag{12}$$

where the positive number ϵ is known as the *Tikhonov regularization parameter*. Let $\epsilon(\delta)$ be a sequence of regularization parameter such that $\epsilon(\delta)$ be a sequence of regularization parameters such that $\epsilon(\delta) \to 0$ as $\delta \to 0$ and let $g_{z,\delta} := g^\delta_{z,\epsilon(\delta)}$ be the minimizer of (12) with $\epsilon = \epsilon(\delta)$. Then it follows from the results of Arens [1], [2] that if $z \in D$ and $k > 0$ is not a transmission eigenvalue then the Herglotz wave function $v_{g_{z,\delta}}$ converges in the $L^2(D)$ norm as $\delta \to 0$. The following theorem shows that this is not in general true if k is a transmission eigenvalue.

Theorem 2.2. [5] *Let $k > 0$ be a transmission eigenvalue and assume that the far field operator F has dense range. Assume further that either $n(x) > 1$ for $x \in \overline{D}$ or $0 < n(x) < 1$ for $x \in \overline{D}$. Then for almost every $z \in D$, $\|v_{g_{z,\delta}}\|_{L^2(D)}$ can not be bounded as $\delta \to 0$.*

Note that by Theorem 2.1 F has dense range as long as the solution v of the interior transmission problem is not a Herglotz wave function, i.e. v can not be expanded to all of $\mathbb{R}^3$ such that v has the representation (10). Thus in practice real transmission eigenvalues are determined by solving the *far field equation*

$$(Fg)(\hat{x}) = \Phi_\infty(\hat{x}, z, k) \tag{13}$$

for an interval of k values and then locating those values of k for which $\|g\|_{L^2(\Omega)}$ is relatively large [3].

3. The Transmission Eigenvalue Problem: The Case of Spherically Stratified Index of Refraction

In the previous section many questions about transmission eigenvalues were left open. In particular, do transmission eigenvalues exist? Is the set of transmission eigenvalues discrete? What is the relationship between transmission eigenvalues and the index of refraction n? The issue of discreteness is of particular importance in the linear sampling method and the factorization method for solving the inverse scattering problem [4], [15] since in these methods one needs to avoid those frequencies that correspond to transmission eigenvalues and hence is important to know that the set of transmission eigenvalues is discrete. On the other hand, transmission eigenvalues are clearly related to the physical properties of the scattering object and hence it is important to know if they exist and what their connection is to the index of refraction.

In this section we will begin to answer these questions for the case when $n(x) = n(r)$ is spherically stratified, D is the ball $\{x : |x| < a\}$ and $n \in C^2[0, a]$. In this case we can expand v and w in a series of spherical harmonics

$$v(x) = \sum_{l=0}^{\infty} \sum_{m=-l}^{l} a_l^m j_l(kr) Y_l^m(\hat{x}) \tag{14}$$

$$w(x) = \sum_{l=0}^{\infty} \sum_{m=-l}^{l} b_l^m y_l(kr) Y_l^m(\hat{x}) \tag{15}$$

$$\tag{16}$$

where $j_l$is a spherical Bessel function of order l and y_l is a real valued solution of

$$y'' + \frac{2}{r} y' + \left(k^2 n(r) - \frac{l(l+1)}{r^2} \right) y = 0 \tag{17}$$

normalized such that $y_l(r)$ behaves like $j_j(kr)$ as $r \to 0$. From Section 2.3 of [10] we can represent this solution in the form

$$y_l(r) = j_l(kr) + k^2 \int_0^r G(r, s, k) j_l(ks) ds \tag{18}$$

where G is real valued, twice continuously differentiable for $0 \leq s \leq r$ and is an even entire function of k of finite exponential type. It can furthermore be shown that y_l is bounded for k on the positive real axis (c.f. Section 9.4 of [12]).

Theorem 3.1. [13] *Assume that* $n(x) = n(r)$ *is spherically stratified,* D *is the ball* $\{x : |x| < a\}$ *and* $n \in C^2[0, a]$. *Then if* $n(r)$ *is not identically equal to one there exits a countably infinite number of transmission eigenvalues for (1)-(4) with infinity as the only possible accumulation point. If in addition*

$$\frac{1}{a}\int_0^a \sqrt{n(r)}dr \neq 1 \tag{19}$$

then there exist a countably infinite number of positive transmission eigenvalues *for (1)-(4) with infinity as the only possible accumulation point.*

Proof. Assume $n = n(r)$ satisfies the hypothesis of the theorem. A necessary and sufficient condition for k to be a transmission eigenvalue is that k is a zero of the determinant

$$d_l(k) := \det \begin{pmatrix} y_l(a) & -j_l(ka) \\ \\ y_l'(a) & -kj_l'(ka) \end{pmatrix} \tag{20}$$

for some non-negative integer l (c.f. Section 9.4 of [12]). Since the spherical Bessel functions are entire functions of k of finite exponential type and bounded for k on the positive real axis, by the above discussion we see that $d_l(k)$ also has this property. Furthermore, by the series expansion of j_l we see that $d_l(k)$ is an even function of k and $d_l(0) = 0$. Hence if $d_l(k)$ does not have a countably infinite number of zeros, by the Hadamard factorization theorem [19] $d_l(k)$ must be identically zero. We will now show that $d_l(k)$ is not identically zero for every l unless $n(r)$ is identically equal to one.

Assume that $d_l(k)$ is identically zero for every non-negative integer l. Noting that $j_l(kr)Y_l^m(\hat{x})$ is a Herglotz wave function, it follows from the proof of Theorem 8.16 in [12] that

$$\int_0^a j_l(k\rho)y_l(\rho)\rho^2 m(\rho)\, d\rho = 0 \tag{21}$$

for all k where $m(r) := 1 - n(r)$. Hence, using the Taylor series expansion of $j_l(k\rho)$ and (18) we see that

$$\int_0^a \rho^{2l+2} m(\rho)\, d\rho = 0 \tag{22}$$

for all non-negative integers l. By Muntz's theorem [19] we now have that $m(r) = 0$, i.e. $n(r) = 1$.

If (19) is valid then it follows from Theorem 2 of [14] that there exist a countably infinite number of *positive* transmission eigenvalues. □

4. The Transmission Eigenvalue Problem: The Case of General Index of Refraction

We now consider the transmission eigenvalue problem (1)-(4) for general $n \in L^\infty$. We will assume that n satisfies one of the following assumptions: 1) $1+\alpha \leq n_* \leq n(x) \leq n^* < \infty$, 2) $0 < n_* \leq n(x) \leq n^* \leq< 1-\beta$ for some $\alpha > 0$ and $\beta > 0$ where $n_* = \inf_D(n)$ and $n^* = \sup_D(n)$ Analogous to Theorem 3.1, we have the following theorem:

Theorem 4.1. [8] *Assume that $n \in L^\infty(D)$ and $k > 0$. Then under the above assumptions on n there exist infinitely many positive transmission eigenvalues with infinity as the only accumulation point. Furthermore, if $1+\alpha \leq n_* \leq n(x) \leq n^*$ and $\lambda(D)$ is the first transmission eigenvalue for $-\Delta$ in D then the first positive transmission eigenvalue $k_{1,D,n(x)}$ satisfies*

$$k^2_{1,D,n(x)} \geq \frac{\lambda(x)}{n^*}. \tag{23}$$

Outline of Proof: The transmission eigenvalue problem can be written for $u := w - v \in H_0^2(D)$ as an eigenvalue problem for the fourth order equation

$$\left(\Delta + k^2 n\right) \frac{1}{n-1} \left(\Delta + k^2\right) u = 0 \tag{24}$$

which in variational form is formulated as finding a function $u \in H_0^2(D)$ such that

$$\int_D \frac{1}{n-1}(\Delta u + k^2 u)(\Delta\overline{\varphi} + k^2 n \overline{\varphi})\, dx = 0 \tag{25}$$

for all $\varphi \in H_0^2(D)$. To fix our ideas, we assume from now on that $1+\alpha \leq n_* \leq n(x) \leq n^* < \infty$. Letting $\tau := k^2$ we have that

$$\int_D \frac{1}{n-1} \Delta u\, \Delta\overline{\varphi}\, dx - \tau \int_D \frac{n+1}{n-1} \nabla u \cdot \nabla\overline{\varphi}\, dx + \tau^2 \int_D \frac{n}{n-1} u\, \overline{\varphi}\, dx = 0 \tag{26}$$

which by the Riesz representation theorem takes the form

$$\mathbb{K}_1 u - \tau \mathbb{K}_2 u + \tau^2 \mathbb{K}_3 u = 0 \qquad \text{in } H_0^2(D)$$

where $\mathbb{K}_1 : H_0^2(D) \to H_0^2(D)$ is self adjoint and coercive and $\mathbb{K}_2, \mathbb{K}_3 : H_0^2(D) \to H_0^2(D)$ are self adjoint, compact and positive. Hence we can write

$$u - \tau \mathbb{K}_1^{-1}\mathbb{K}_2 u + \tau^2 \mathbb{K}_1^{-1}\mathbb{K}_3 u = 0 \qquad \text{in } H_0^2(D)$$

and setting $\phi = \tau \mathbb{K}_1^{-1/2}\mathbb{K}_3^{1/2}$ we obtain

$$\begin{pmatrix} u \\ \phi \end{pmatrix} - \tau \begin{pmatrix} \mathbb{K}_1^{-1}\mathbb{K}_2 & -\mathbb{K}_1^{-1/2}\mathbb{K}_3^{1/2} \\ \mathbb{K}_1^{-1/2}\mathbb{K}_3^{1/2} & 0 \end{pmatrix} \begin{pmatrix} u \\ \phi \end{pmatrix} = \begin{pmatrix} 0 \\ 0 \end{pmatrix}. \tag{27}$$

This is clearly not an eigenvalue problem for a compact self adjoint operator. However from (27) we see that the transmission eigenvalues (if they exist) form a discrete set with infinity as the only possible accumulation point and that each transmission eigenvalue has finite multiplicity.

We now define the following bounded sesquilinear forms on $H_0^2(D) \times H_0^2(D)$:

$$\mathcal{A}_\tau(u,v) = \int_D \frac{1}{n-1}(\Delta u + \tau u)\,(\Delta\varphi + \tau\varphi)\,ds + \tau^2 \int_D u\,\varphi\,ds, \tag{28}$$

and

$$\mathcal{B}(u,v) = \int_D \nabla u\,\nabla\varphi\,ds \tag{29}$$

and let $\mathbb{A}_\tau : H_0^2(D) \to H_0^2(D)$ and $\mathbb{B} : H_0^2(D) \to H_0^2(D)$ be the corresponding operators defined by the Riesz representation theorem. The interior transmission problem now reads

$$(\mathbb{A}_\tau - \tau\mathbb{B})\,u = 0 \qquad \text{in } H_0^2(D). \tag{30}$$

where $\mathbb{A}_\tau$ is self adjoint, coercive and depends continuously on $\tau \in (0, \infty)$ and $\mathbb{B}$ is a self adjoint compact and non-negative operator.

We now consider the generalized eigenvalue problem

$$(\mathbb{A}_\tau - \lambda(\tau)\mathbb{B})\,u = 0 \qquad \text{in } H_0^2(D) \tag{31}$$

noting that $k^2 = \tau$ defines a transmission eigenvalue $k = \sqrt{\tau}$ if and only if $\tau > 0$ satisfies $\lambda(\tau) = \tau$. Since the generalized eigenvalue problem (31) can be written in the form

$$\left(I - \lambda(\tau)\mathbb{A}_\tau^{-1/2}\mathbb{B}\mathbb{A}_\tau^{-1/2}\right) u = 0 \qquad \text{in } H_0^2(D)$$

we conclude that for a fixed $\tau > 0$ there exists an increasing sequence of eigenvalues $\lambda_j(\tau)$, such that $\lambda_j(\tau) \to \infty$ as $j \to \infty$. These eigenvalues satisfy

$$\lambda_j(\tau) = \min_{W \subset \mathcal{U}_j} \left(\max_{u \in W \setminus \{0\}} \frac{(\mathbb{A}_\tau u, u)}{(\mathbb{B}u, u)} \right) \tag{32}$$

where $\mathcal{U}_j$ denotes the set of all j dimensional subspaces W of $H_0^2(D)$ such that $W \cap \ker(\mathbb{B}) = \{0\}$. We now recall the following theorem from [9]:

Theorem 4.2. *Assume that there exist two positive constants $\tau_0 > 0$ and $\tau_1 > 0$ such that*

(1) $\mathbb{A}_{\tau_0} - \tau_0 \mathbb{B}$ is positive on U,
(2) $\mathbb{A}_{\tau_1} - \tau_1 \mathbb{B}$ is non positive on a m dimensional subspace of U.

Then each of the equations $\lambda_j(\tau) = \tau$ for $j = 1, \ldots, m$, has at least one solution in $[\tau_0, \tau_1]$ where $\lambda_j(\tau)$ is the jth eigenvalue (counting multiplicity) of $\mathbb{A}_\tau$ with respect to $\mathbb{B}$.

To establish the existence of the constant τ_0 we note that

$$(\mathbb{A}_\tau u - \tau \mathbb{B}u, u)_{H^2} = \int_D \frac{1}{n-1} |\Delta u + n\tau u|^2 \, ds - \tau^2 \int_D |u|^2 \, ds + \tau \int_D |\nabla u|^2 \, ds. \tag{33}$$

Since

$$\lambda(D) \leq \inf_{u \in H_0^2(D)} \frac{\int_D |\nabla u|^2 \, ds}{\int_D |u|^2 \, ds}$$

$\mathbb{A}_{\tau_0} - \tau_0 \mathbb{B}$ is a positive operator for all τ_0 such that $0 < \tau_0 < \lambda(D)/n^*$ where $\lambda(D)$ is the first Dirichlet eigenvalue for $-\Delta$ in D (In particular (23) is valid).

To establish the existence of the constant τ_1, let B be a ball contained in D and let $k_{n_*}^B$ and $u_{n_*}^B \in H_0^2(B)$ be a transmission eigenvalue and eigenvector, respectively, corresponding to the ball B and index of refraction n_*. Set $\tau_1 := (k_{n_*}^B)^2$ and let $\tilde{u} \in H_0^2(D)$ be the extension by zero of $u_{n_*}^B$ to the whole of D. Then

$$(\mathbb{A}_{\tau_1}\tilde{u} - \tau_1\mathbb{B}\tilde{u}, \tilde{u})_{H^2} = \int_D \frac{1}{n-1}|\Delta\tilde{u} + \tau_1\tilde{u}|^2\, ds + \tau_1^2 \int_D |\tilde{u}|^2\, ds$$
$$- \tau_1 \int_D |\nabla\tilde{u}|^2\, ds \tag{34}$$
$$\leq \int_B \frac{1}{n_* - 1}|\Delta u^B_{n_*} + \tau_1 u^B_*|^2\, ds + \tau_1^2 \int_B |u^B_{n_*}|^2\, ds$$
$$- \tau_1 \int_B |\nabla u^B_{n_*}|^2\, ds = 0.$$

and hence $\mathbb{A}_{\tau_1} - \tau_1\mathbb{B}$ is a non-positive operator in the subspace spanned by $\tilde{u}$.

The above argument shows that there exists at least one transmission eigenvalue between $\sqrt{\lambda(D)/n^*}$ and $k^B_{n_*}$. To establish the existence of infinitely many transmission eigenvalues take $\epsilon > 0$ small enough such that D contains $m(\epsilon) \geq 1$ disjoint balls $B^1_\epsilon, B^2_\epsilon, \dots B^{m(\epsilon)}_\epsilon$ of radius ϵ. Let k_{ϵ,n_*} be the first transmission eigenvalue for each of these balls with index of refraction n_* and let $u^{B_j}_{n_*} \in H^2_0(B_j)$, $j = 1, \dots, m(\epsilon)$ be the corresponding eigenfunctions and $\tilde{u}^j \in H^2_0(D)$ their extension by zero to the whole of D. Note that $\{\tilde{u}^1, \tilde{u}^2, \dots \tilde{u}^{m(\epsilon)}\}$ are linearly independent in $H^2_0(D)$ since they have disjoint supports. By the above $\mathbb{A}_{\tau_1} - \tau_1\mathbb{B}$ is a non-positive operator in the subspace spanned by $\{\tilde{u}^1, \tilde{u}^2, \dots \tilde{u}^{m(\epsilon)}\}$ where $\tau_1 := (k_{\epsilon,n_*})^2$. Hence there exists $m(\epsilon)$ transmission eigenvalues less then k_{ϵ,n_*}. Letting $\epsilon \to 0$ and noting the finite multiplicity of transmission eigenvalues proves the results.

This completes the outline of the proof of Theorem 4.1.

The inequality (23) can be further improved [7], [8]. To this end let $B_{r_1} \subset D \subset B_{r_2}$ where B_{r_1} is the largest such ball and B_{r_2} is the smallest such ball. For a given $0 < \epsilon \leq r_2$, let $m(\epsilon) \in \mathbb{N}$ be the number of balls B_ϵ of radius ϵ that are contained in D. We denote by k_{1,n_*} and k_{1,n^*} the first transmission eigenvalue for the ball of radius one with index of refraction n_* and n^*, respectively. Finally let $k_{1,D,n(x)}$ denote the first transmission eigenvalue corresponding to D and the given index of refraction $n(x)$.

Corollary 4.1. *Assume that $n \in L^\infty(D)$. Then*

(1) If $1+\alpha \le n_* \le n(x) \le n^* < \infty$, *then*

$$0 < \frac{k_{1,n^*}}{r_2} \le k_{1,D,n(x)} \le \frac{k_{1,n_*}}{r_1}. \tag{35}$$

(2) If $0 < n_* \le n(x) \le n^* < 1-\beta$, *then*

$$0 < \frac{k_{1,n_*}}{r_2} \le k_{1,D,n(x)} \le \frac{k_{1,n^*}}{r_1}. \tag{36}$$

Proof. First we note that from the scaling properties of the spherical Bessel functions, the first transmission eigenvalue for B_{r_2} with index of refraction n^* equals $k_{1,n^*}/r_2$. Similar scaling properties hold for the first transmission eigenvalue for B_{r_2} and index of refraction n_* as well as for B_{r_1} with index of refraction n^* or n_*. The right inequality in (35) is shown in the proof of Theorem 4.1. In the following we show the validity of the left inequality in (35) for the case when $1+\alpha \le n_* \le n(x) \le n^* < \infty$. Obviously for any $u \in H_0^2(D)$ we have

$$\begin{aligned} \frac{\frac{1}{n^*-1}\|\Delta u+\tau u\|_D^2+\tau^2\|u\|_D^2}{\|\nabla u\|_D^2} &\le \frac{\int_D \frac{1}{n(x)-1}|\Delta u+\tau u|^2\,dx+\tau^2\|u\|_D^2}{\|\nabla u\|_D^2} \\ &\le \frac{\frac{1}{n_*-1}\|\Delta u+\tau u\|_D^2+\tau^2\|u\|_D^2}{\|\nabla u\|_D^2}. \end{aligned} \tag{37}$$

Therefore from (32) we have that for an arbitrary $\tau > 0$

$$\begin{aligned} \lambda_1(\tau, B_{r_2}, n^*) \le \lambda_1(\tau, D, n^*) &\le \lambda_1(\tau, D, n(x)) \\ &\le \lambda_1(\tau, D, n_*) \le \lambda_1(\tau, B_{r_1}, n_*) \end{aligned}$$

where $\lambda_1(\tau, D, n^*)$, $\lambda_1(\tau, D, n(x))$ and $\lambda_1(\tau, D, n_*)$ are the first eigenvalues of the auxiliary problem for D and n^*, $n(x)$ and n_*, respectively, and $\lambda_1(\tau, B_{r_2}, n^*)$ and $\lambda_1(\tau, B_{r_1}, n_*)$ are the first eigenvalues of the auxiliary problem for B_{r_2}, n^* and B_{r_1}, n_*, respectively. Now for $\tau_1 := (k_{1,n_*}/r_1)^2$ we have that $\lambda_1(\tau_1, D, n(x)) - \tau_1 \le 0$ since in the subspace spanned by the extension by zero to the whole of D of the eigenfunction $u_{1,n_*}^{B_1}$ the Rayleigh quotient minus τ for $\tau = \tau_1$ becomes negative. On the other hand, for $\tau_0 := (k_{1,n^*}/r_2)^2$, we have $\lambda_1(\tau_0, B_{r_2}, n^*) - \tau_0 = 0$ and hence $\lambda_1(\tau_0, D, n(x)) - \tau_0 \ge 0$. Therefore the first eigenvalue $k_{1,D,n(x)}$ corresponding to D and $n(x)$ is between $k_{1,n^*}/r_2$ and $k_{1,n_*}/r_1$. Note that there is no transmission eigenvalue for D and $n(x)$ that is less then $k_{1,n^*}/r_2$. Indeed if there is a transmission eigenvalue strictly less then $k_{1,n^*}/r_2$ then by the monotonocity of the eigenvalues of the auxiliary problem with respect to the domain and the fact that for τ small enough there are no transmission eigenvalues we would have found an eigenvalue of the ball B_{r_2} and n^* that is strictly smaller then the first eigenvalue. □

Remark 4.1. The inequality (23) was first obtained in [14].

Remark 4.2. The existence of finitely many transmission eigenvalues (under more restrictive assumptions then in Theorem 4.1) was first given in [17].

Remark 4.3. Theorem 4.1 can be generalized to include the case where D has cavities, i.e. there exit regions in D where $n(x) = 1$ [6].

References

1. T. Arens, *Why linear sampling method works*, *Inverse Problems* **20**, 163-173 (2004).
2. T. Arens and A. Lechleiter, *The linear sampling method revisited*, *J. Int. Eqns. Appl.* **21**, 179-202 (2009)
3. F. Cakoni, M. Cayoren and D. Colton, *Transmission eigenvalues and the nondestructive testing of dielectrics*, *Inverse Problems* **24**, 066016 (2008).
4. F. Cakoni and D Colton and H Haddar, *Qualitative Methods in Inverse Scattering Theory*, Springer, Berlin, 2006.
5. F. Cakoni, D Colton and H Haddar, *On the determination of Dirichlet and transmission eigenvalues from far field data*, to appear.
6. F. Cakoni, D. Colton and H. Haddar, *The interior transmission problem for regions with cavities*, to appear.
7. F. Cakoni and D. Gintides, *New results on transmission eigenvalues*, to appear.
8. F. Cakoni, D. Gintides and H. Haddar, *The existence of an infinite discrete set of transmission eigenvalues*, to appear.
9. F. Cakoni and H Haddar, *On the existence of transmission eigenvalues in an inhomogeneous medium*, *Applicable Analysis*, **88**, 475-493 (2009).
10. D. Colton, *Analytic Theory of Partial Differential Equations*, Pitman, Boston, 1980.
11. D. Colton, A. Kirsch and L. Päivärinta, *Far field patterns for acoustic waves in an inhomogeneous medium*, *SIAM Jour. Math. Anal.*, **20**, 1472-1483 (1989).
12. D. Colton and R. Kress, *Inverse Acoustic and Electromagnetic Scattering Theory*, 2nd edition, Springer, New York, 1998.
13. D. Colton, P. Monk and J. Sun, *Analytic and computational methods for transmission eigenvalues*, to appear.
14. D. Colton, L. Päivärinta and J. Sylvester, *The interior transmission problem*, *Inverse Problems and Imaging* **1**, 13-28 (2007).
15. A. Kirsch and N. Grinberg, *Thd Factorization Method for Inverse Problems*, Oxford University Press, Oxford, 2008.

16. A. Kirsch, *On the existence of transmission eigenvalues*, *Inverse Problems and Imaging* **3**, 155-172 (2009).
17. L. Päivärinta and J. Sylvester, *Transmission Eigenvalues*, *SIAM J. Math. Anal.* **40**, 738-753 (2008).
18. B. P. Rynne and B.D. Sleeman, *The interior transmission problem and inverse scattering from inhomogeneous media*, *SIAM J. Math. Anal.* **22**, 1755-1762 (1992).
19. R.M. Young, *An Introduction to Nonharmonic Fourier Series*, Academic Press, San Diego, 2001.

WAVE PROPAGATION IN ANCHOR HEAD FOR ASSESSMENT OF METAL STRAND PRE-STRESS

D. KLEITSA AND D. G. AGGELIS
Dept. Materials Science and Engineering,
University of Ioannina, Ioannina 45110, Greece

K. KAWAI AND T. SHIOTANI
Graduate School of Engineering,
Kyoto University, Kyoto 615-8540, Japan

Pre-stressed steel strands embedded in a structure is a common way to reinforce the weak tensile nature of concrete or soil materials. Due to corrosion, the strands may lose the pre-tension with disastrous consequences. The accessibility to the strands is usually limited to the anchor head from where the assessment should be done. In this paper, through the thickness elastic wave measurements are conducted on the anchor head. The wave amplitude, delay time and other characteristics are influenced by the stress of the strands which affects the bond between strands, wedges and the surrounding anchor head. Numerical simulations are also conducted verifying that the stress of the strands strongly influences the wave parameters. The change of bonding between the strands and anchor head is simulated by modifying the rigidity of a model interphase material. Therefore, a practical non destructive inspection method to assess the pre-stress level of the strands is discussed.

1. Introduction

Millions of ground anchorages have been installed over the past decades with relatively few recorded failures. However, as permanent anchorages in service become older, the subject is of growing importance, particularly for anchorages installed more than 30 years ago, many of which have been designed with corrosion protection considered inadequate by today's standard. Maintenance testing involves visual survey of the physical condition of the structure and the anchorages, measurement of the overall deformation and when applicable measurement of the load of selected tendons. The method to monitor the load of a tendon is the so called "lift-off" method. However, it entails certain difficulties as it requires the anchor head to be raised by some mm using a hydraulic jack and therefore may jeopardize the safety of the structure while only specific tendons can be examined [1]. A routine, fast and easy inspection method which can confirm the satisfactory performance can extend the service life of anchored structures [2].

This study focuses on the ultrasonic method to evaluate the axial stress of the strands. The applicability of the method is examined both experimentally and numerically and suitable wave parameters are sought for in order to yield a robust correlation with the applied stress on the tendons. Due to the wedge geometry, as the pre-stress on the strands increases, the wedges obtain stiffer contact with the matrix material of the anchor. This has an effect on the overall rigidity of the anchor head. Therefore, longitudinal elastic waves are introduced in an anchor head and the time of flight, as well as the amplitude are measured for different levels of pre-stress. The experiment is numerically simulated by finite difference method software modeling the different pre-stress levels with an interphase material of varying elastic constants.

2. Experimental details

A typical example of a pre-stressed structure where the anchor heads are used to constrain the strand wires by means of wedges is shown in Fig. 1. As the pre-stress of the strand wires increases, the wedges obtain better contact with the anchor head, through the toughening of the interface between the wedges and their corresponding holes.

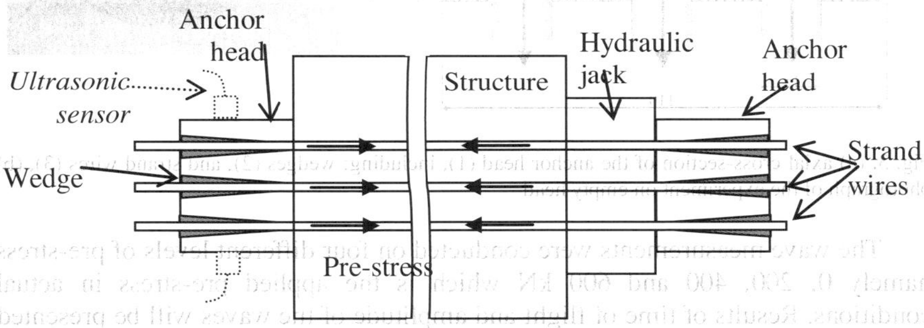

Fig. 1. The basic geometry of pre-stressed structure.

Figure 2(a) shows a close-up of the anchor head where the strand wires and the wedges inside the holes can be seen. Each wedge consists of 2 parts being placed around the strand wire, as shown in Fig. 2(b).

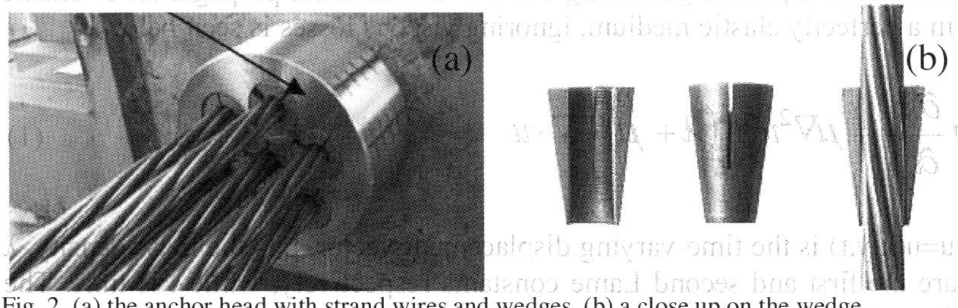

Fig. 2. (a) the anchor head with strand wires and wedges, (b) a close up on the wedge

Since there is no accessibility in the interior of the structure, the assessment should be conducted through the only area located in the surface, the anchor head. Fig. 3(a) shows the cross-section of the anchor head. As the tensile load of the cables increases, the contact between the wedge and the surrounding hole becomes stiffer (see for example the bold dashed line in Fig. 3(a)). At the certain cross-section of the head (e.g. A-A of Fig. 3(a)) the ultrasound sensors can be applied, as seen in Fig. 3(b). In this case the anchor head is empty of wedges and strand wires. The height of the specific measurements is 30 mm below the large opening of the wedge holes (cross-section A-A).

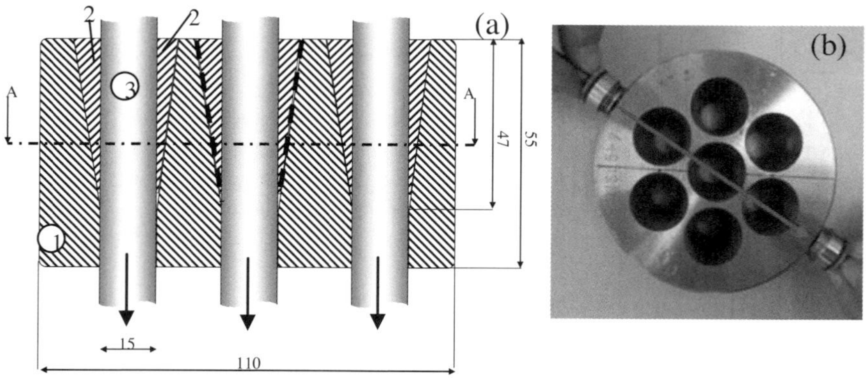

Fig. 3. (a) axial cross-section of the anchor head (1), including: wedges (2), and strand wires (3), (b) photograph of the experiment on empty head

The wave measurements were conducted on four different levels of pre-stress, namely 0, 200, 400 and 600 kN which is the applied pre-stress in actual conditions. Results of time of flight and amplitude of the waves will be presented along with the numerical ones later. More experimental details can be found in [2].

3. Numerical simulations

The fundamental equation governing the two-dimensional propagation of elastic waves in a perfectly elastic medium, ignoring viscous losses is seen below:

$$\rho \frac{\partial^2 u}{\partial t^2} = \mu \nabla^2 u + (\lambda + \mu)\nabla\nabla \cdot u \tag{1}$$

where u=u(x,y,t) is the time-varying displacement vector, ρ is the mass density, λ and μ are the first and second Lame constants respectively, while t is time. The focus, from the engineering point of view is given on simulating the actual cases examined experimentally and not on the numerical method itself. However, certain prerequisites should be followed in order for the analysis to lead to reliable results. The simulations were conducted with commercially available software [3]. It operates by solving the above equation based on a method of finite differences. Equation (1) is solved with respect to the boundary conditions of the object, which include the input source that has pre-defined time-dependent displacements at a given location and a set of initial conditions [4]. For heterogeneous media like the one studied herein, propagation in each distinct homogeneous phase is solved according to eq. (1), while the continuity conditions for stresses and strains must be satisfied on the interfaces [4].

Materials were considered elastic without viscosity components. As the stress increases in the strand, the contact at the interface between the wedge and the matrix material is enhanced (see dashed lines in Fig. 3(b)). This condition is simulated by an imaginary interphase material with varying stiffness, as has been done for the case of surface cracks with faces in partial contact [5]. The minimum value for the stiffness is zero which simply means no contact between the wedge and the matrix while the maximum is the stiffness of steel as if there was no interface at all. The mechanical properties of the steel matrix, the wedge and the strand wire are equal, with Young's modulus 205 GPa, Poisson's ratio of 0.3 and density of 7800 kg/m^3 resulting in a longitudinal wave velocity, C_L, of 5948 m/s. The interphase material's velocity was set to different values starting from 330 m/s which simulates the case of negligible contact. Then, by adjusting the elastic properties of the interphase, the velocity, C_L, was set to the values 1000, 2000, 3000, 4000 and 5000 m/s and 6000 m/s. The last value simulates the case in which the interphase is as rigid as the steel material itself.

The geometry of the model is depicted in Fig. 4. It corresponds to the cross-section A-A, in Fig. 3(a). The thickness of the interphase was 1 mm, being a suggestive value. For simplicity each of the seven strand wires was considered as one solid circle with its surrounding wedge.

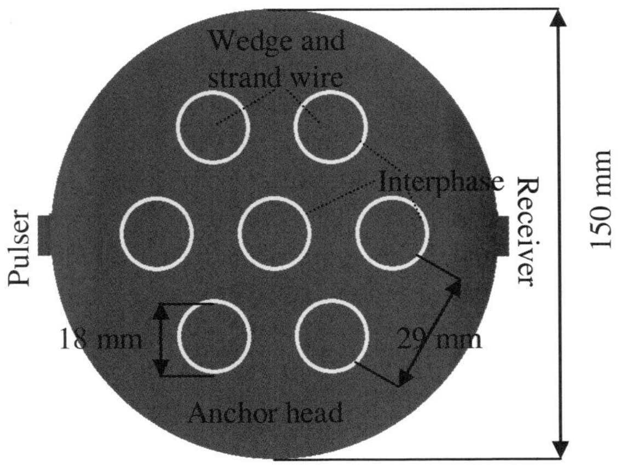

Fig. 4. Numerical geometric model corresponding to cross-section A-A of Fig. 3(a).

The "pulser" introduced an excitation of 1 cycle of 1 MHz vertical to the surface (longitudinal wave). The receiver provided the average vertical displacement on its whole length (10 mm). In Fig. 5(a) and (b) two snapshots are included for the case where the interphase exhibits the stiffness of steel (long. velocity 6000 m/s). The wavefront is obviously not scattered since the mechanical impedance does not change throughout the straight wave path. On the other hand when the interphase is considered similar to air, the wave front is scattered and the straight path cannot be followed by the wave (see Figs. 5(c) and (d)).

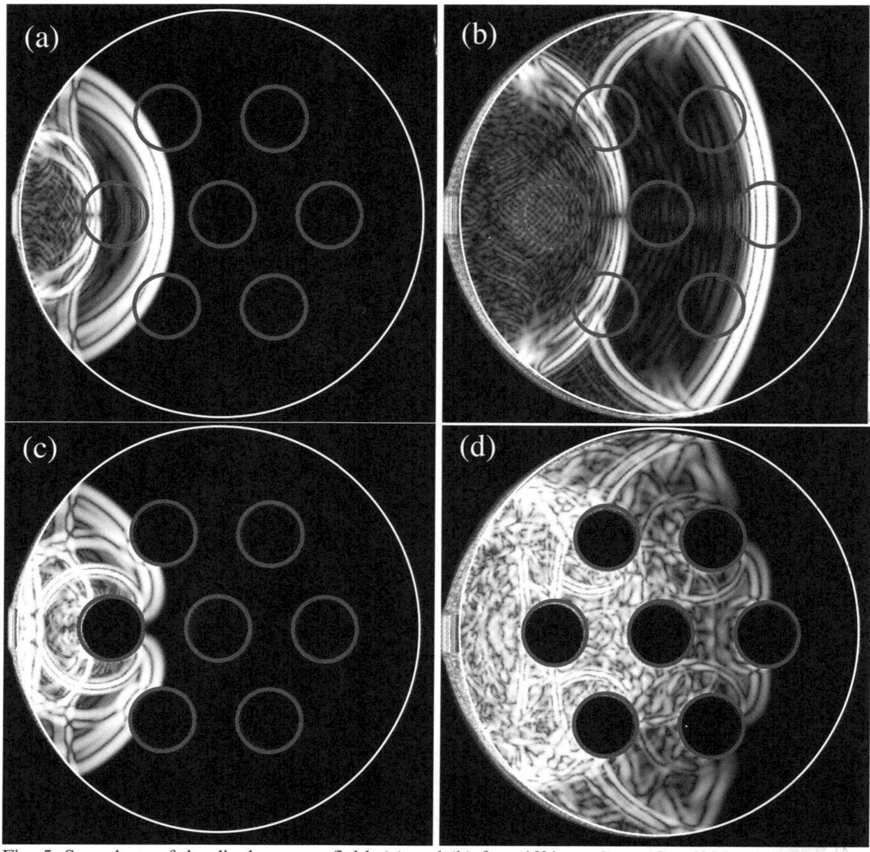

Fig. 5. Snapshots of the displacement field, (a) and (b) for stiff interphase (C_L=6000 m/s). (c) and (d) for weak interphase (C_L=330 m/s)

The element size (mesh) was 0.1 mm, while considering the main frequency of 1 MHz, the wavelength is calculated to about 60 mm. This satisfies the requirement of 10 or even 20 nodes per wavelength and results in reasonable accuracy [6]. The time step was 0.0209 μs, meaning that for the frequency of 1 MHz (period of 1 μs) almost 50 points were used in a cycle ensuring adequate depiction.

4. Results

Below one can see numerical results of transit time for different stiffness of the interphase (see Fig. 6(a)). It is mentioned that the transit time was defined by the first disturbance on the simulated waveform of the receiver.

The transit time is approximately constant until the interphase obtains stiffness corresponding to longitudinal velocity of 4000 m/s. When the interphase rigidity increases, the transit time is shortened. This difference is approximately

0.5 μs and can be measured by any compatible ultrasonic setup, corresponding to 2.7 %. This behavior shows that when the interphase is below some threshold stiffness, the fastest path is around the wedges (seen as holes by the wave front). However, when the interphase obtains stiffness close to the matrix material itself (steel) then the straight path is followed resulting in shorter transit time.

On the same figure one can see the experimental results for the frequency of 1 MHz. When no pre-stress is applied, the transit time is measured at 19.50 μs. As the pre-stress is increased to 200 MPa the transit time is shortened considerably to 18.06 μs and stays approximately constant until the final load of 600 MPa (17.96 μs). The decrease for the experimental measurement is almost 8%, being more sensitive though between zero and low pre-stress level.

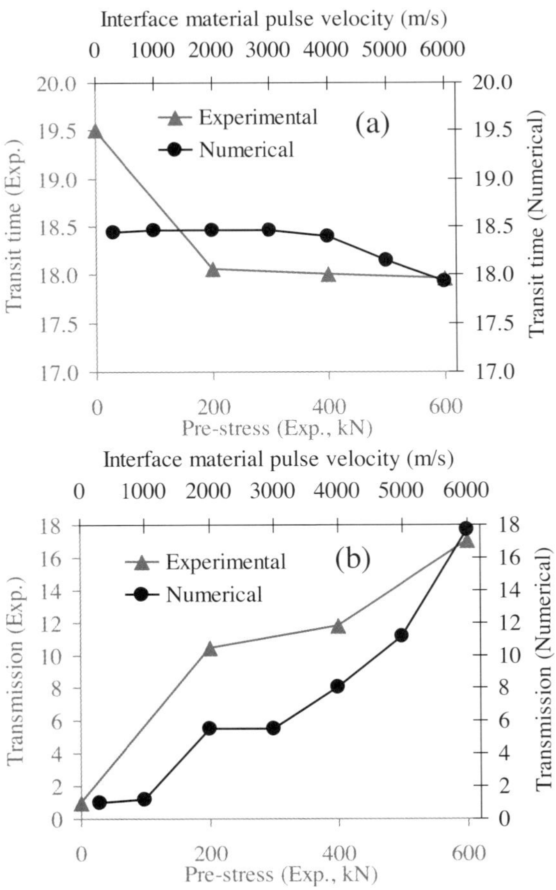

Fig. 6. (a) Experimental transit time vs. pre-stress and numerical transit time vs. interphase C_L. (b) Experimental wavefront amplitude vs. pre-stress and numerical amplitude vs. interphase C_L.

The transit time measurement although is one of the most fundamental ones in non destructive testing, it is not always the most sensitive to the internal condition of the materials. Especially for inhomogeneous materials, where strong scattering interactions are expected, other parameters related to the transmission or energy of the wave are more appropriate [7]. In this case, in order to examine energy-related features, the amplitude of the first cycle of the received wave was measured. Assuming that no viscous components are included, the whole amount of energy excited by the pulser will eventually reach the transducer in full as it is trapped in the metal anchor head surrounded by air. However, the first cycle will contain only the forward scattered components and the rest of the energy will arrive later, especially when strong scattering occurs. Thus, the first cycle is more sensitive to the presence of internal scatterers.

The amplitude of the first cycle for different interphase stiffness (numerical) is depicted in Fig. 6(b) normalized to the amplitude of zero pre-stress. It is evident that as the interphase gets stiffer, the wavefront continuously exhibits higher amplitude until it becomes 18 times higher than the initial value. Again on the same figure the experimental results are depicted. In this case also, the amplitude of the first cycle increases strongly as the loading is building up, reaching approximately the same value as the numerical result.

It is seen that the results for the maximum load almost coincide with the numerical of the highest interphase stiffness (level of steel). This shows that when the maximum load (600 kN) is applied in practice, the whole anchor head behaves as one homogeneous geometry, with a longitudinal velocity of 6000 m/s. In this case, due to the strong contact, the propagating wave will not "see" any interphase. For the case of no pre-stress, the simulation using properties of air for the interphase is also close to the experimental. Concerning the medium level pre-stress, the simulations overestimate the load. However, since the trend is monotonic it seems that amplitude parameters of the received wave are promising features for further study. The transit time, although influenced by the pre-stress changes only by 3 % being therefore a weak parameter to be solely studied.

5. Conclusions

The study presented herein concerns the non destructive evaluation of the stress in metal strand wires. Due to the wedge geometry, increased pre-stress enhances the contact of the metal parts and this has an effect on the propagation of elastic waves through the head, as exhibited by the change of transit time and amplitude. Numerical simulation on the exact geometry exhibit a similar trend for transit time and amplitude of the wavefront, showing that when the actual level of pre-stress is applied (600 kN), the interphase is as stiff as the steel material itself. Further ideas include the use of shear waves which should be more sensitive to the stiffness of the interphase. The shortcoming of this approach is that experimentally, the transmission of shear waves into the specimen is much more difficult and special care should be taken of the coupling conditions. Apart from the indicative wave parameters discussed herein several others should be tested, like the transmitted frequency which should be strongly dependent on the scattering. Also, the time centroid of the rectified signal envelope would be sensitive to the late arrivals due to scattering and thus hopefully sensitive to the stiffening of the interphase by the applied pre-stress. The thickness of the interphase is another parameter that could vary in order to yield results closer to the actual. The thickness of 1 mm was selected as an initial case but it is reasonable to apply also smaller values. Finally, another suggestion includes the excitation of elastic waves on each strand wire directly and receiving by a point on the anchor head so that assessment of the load at each wire separately can be achieved.

References

1. S. Littlejohn (ed.), *Ground Anchorages and Anchored Structures in Service*, 4 (2007).
2. K. Kawai, T. Shiotani, H. Ohtsu, T. Yanase, H. Tanaka, Proc. EIT-JSCE, Geotechnical Infrastructure Management, (in CD, 9) (2009).
3. Wave2000, Cyber-Logic, Inc, NY (http://www.cyberlogic.org)
4. Kaufman, J. J., Luo, G. and Siffert, R. S., *IEEE Transactions on Ultrasonics, Ferroelectrics and Frequency Control*, **55**(6), 1205 (2008)
5. C. Pecorari. *Wave Motion* **33**, 259 (2001).
6. F. Moser, L. J. Jacobs and J. Qu, *NDT&E Int.* **32** 225 (1999).
7. T. Shiotani and D. G. Aggelis, *Mat. Struct.* **42**(3), 377 (2009).

ON-LINE MONITORING OF LOAD INDUCED DEGRADATION OF CROSS PLY LAMINATES

D. G . AGGELIS, N. M. BARKOULA, T. E. MATIKAS, A. S. PAIPETIS
Dept. Materials Science and Engineering,
University of Ioannina, Ioannina 45110, Greece

The scope of this study is to relate the acoustic activity of load induced damage in cross ply composites to the distinct failure mechanisms associated with these material systems. For this purpose cross ply glass fibre reinforced epoxy composites were subjected to incremental tensile loading with concurrent recording of the damage induced acoustic activity. AE parameters were successfully employed to monitor the gradual transition of the damage mechanism from transverse cracking (mode I) to delamination (mode II). Wave propagation studies in between loading steps revealed an increase in the relative amplitude of the propagated wave, which was attributed to the development of delamination that confined the wave to the top longitudinal plies of the composite.

1. Introduction

Multidirectional composites are prone to various forms of damage when they are subjected to thermo-mechanical loading. This damage is different in nature to damage observed in isotropic materials in that fibrous composites inherently possess macroscopic anisotropy which, at the lamina scale is translated in a step property change. Hence, damage is triggered at interfaces where there is a change in reinforcement orientation provided that the equal strain criterion in fibre and/or lamina scale ceases to be valid over a loading threshold.

Typical of the aforementioned behaviour are cross-ply laminates. The premature failure of the plies oriented transversely to the loading axis is known as transverse or intralaminar cracking. This failure creates the shear discontinuity at the lamina scale and the respective stress magnification [1,2] which triggers all other damage mechanisms. Depending on the type of loading, the interlaminar strength and the interfacial strength, this secondary damage may be manifested as interlaminar failure i.e. delamination, interfacial failure i.e fibre matrix debonding, and longitudinal fibre failure i.e. failure of the main load bearing reinforcement. Numerous studies have been performed to study the triggering of the aforementioned mechanisms, their succession and finally their accumulation that leads to the global failure of the composite [3,4].

In general, transverse cracking occurs and creates a stress concentration at the neighbouring 0° plies. Depending on the interlaminar strength, the crack may be deflected to the 0°/90° interlaminar surface causing delamination or proceed in the 0° ply causing longitudinal fibre failure. Transverse cracking is typical of a mode I crack or a tensile crack, whereas delamination is typical of a mode II crack or a shear crack [5]. As is well known, there is a trade of between low interlaminar strength that favours a damage tolerant structure and high interlaminar strength that maximises the reinforcing ability of all ply orientations.

From the above postulations, it is obvious that a good knowledge of the initiation and propagation of the distinct damage mechanisms will provide an insight into the strengthening mechanisms and aid the designer towards the optimization of the interlaminar/ interfacial properties.

The AE technique has been employed in numerous applications for damage characterization on composite materials [6,7]. Suitable sensors are placed on the surface in order to record the transient waves (hits) generated by damage initiation and propagation inside the material. Moreover, Acoustic Emission (AE) has been successfully applied in cross ply composites in order to separate the acoustic activity of various damage modes in a cross ply composite [6]. Unsupervised data clustering of suitable AE descriptors led to the isolation of the acoustic activity of transverse cracking, interlaminar and /or interfacial failure and longitudinal fibre failure during the course of mechanical loading [6].

Further study of the transient waveforms may provide in depth insight of the fracture process. The source of the AE activity is closely connected to the mode of fracture [8]. If the failure sequence is determined, it is possible to tailor the properties of the constituent phases and their interface using proper design or materials so as to optimise the resistance against the specific failure mode. As has already been mentioned, this optimisation does not necessarily coincide with maximum strength.

In general, the nucleation of shear cracks generally follows the propagation of tensile cracks. In case of cross ply laminates, transverse cracking (mode I) of the 90° plies is a prerequisite for the creation of the shear discontinuity that will trigger further failure mechanisms and by definition precedes interlaminar shear or delaminations (mode II). Within the scope of this study is to evaluate qualitatively and quantitatively the signature of mode I and mode II failure in the acoustic activity of the loaded composite.

In the engineering field the shape of the AE waveforms is reported to be characteristic of the fracture mode (Fig. 1a and b). Shear events are characterized by longer Rise Time (RT) and usually higher amplitude (A) than tensile events [9]. This is examined by the RA value which is defined as the ratio of the RT (expressed in μs) to the waveform Amplitude (expressed in V) [10]. It has been shown that lower RA values, indicate more tensile nature of fracture events [8-11]. Another parameter that has been used for the characterization of the cracking mode is the average frequency. It is defined as the number of threshold crossings (counts) divided by the duration of the signal, see Fig. 1b and is expressed in kHz. The shift from higher to lower values indicates the shift of the cracking mode from tensile to shear [12].

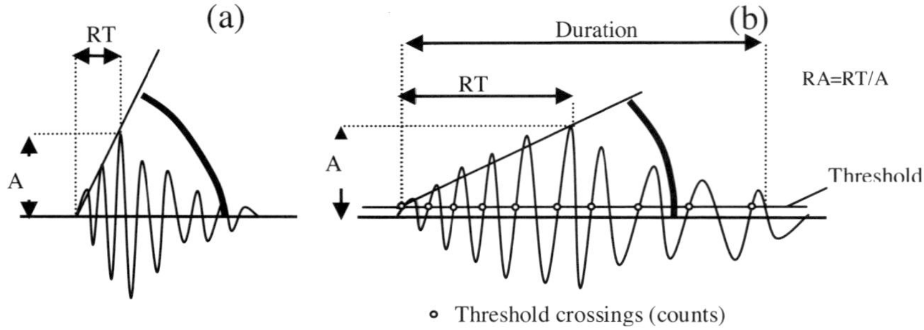

Fig. 1 Typical AE signal of (a) tensile and (b) shear type of crack

2. Experimental details.

$[0°_4/90°_4]_s$ cross-ply laminates were fabricated by hand layup. The UD 220 g/m^2 (Aero) unidirectional glass fibre fabric was impregnated using the HT2 epoxy resin/ hardener matrix system (mixing ratio 100:48) manufactured by R&G Faserverbundwerkstoffe GmbH Composite Technology. A 250x250 mm^2 laminate was manufactured and was allowed to cure for 24h at room temperature. Tensile specimens were subsequently cut according to the ASTM D3039 standard.

The tensile specimens were loaded in load controlled tension, in a step loading mode. The loading spectrum was a saw-tooth spectrum formed from a sequence of triangular loading / unloading steps. A rate 5kN/min was employed for both loading and unloading. The maximum load was incremented by 5 kN at each consecutive step. The step loading continued until the tensile failure of the specimen. All tensile tests were performed using an Instron Universal Testing Machine equipped with hydraulic gripping system, under load control, at controlled environmental conditions of 25 °C and 70% relative humidity.

For the purpose of the AE monitoring, two wide band AE sensors (Pico, Physical Acoustics Corp., PAC) were attached on the same side of the specimen (see Fig. 2a). Electron wax was applied between the sensor and the specimen to enhance acoustic coupling, while it offered the necessary support to the sensors during the experiment. The specific sensors were chosen over other AE transducers mainly due to their spectral response. They are sensitive to frequencies from 100 kHz up to almost 1 MHz, with maximum sensitivity at 500 kHz. Therefore, they can capture a wide range of different sources. The distance between the two receivers was 70 mm. For the purpose of wave propagation measurements another sensor (PAC, R6) was attached to the specimen (Fig 2a). The sensor was connected to a pulse generator which introduced successive signals of 10 cycles of 100, 200, 300, 400 and 500 kHz during the intermittence between the load steps. By capturing the transient response on the two receivers, wave velocity and transmission measurements were performed.

The pre-amplifier gain was set to 40 dB. After performing a pilot test, the threshold was also set to 40 dB in order to avoid the possibility of electronic/environmental noise. The signals were recorded in a two-channel monitoring board PCI-2, PAC. AE hits were recorded with a sampling rate of 5 MHz. A typical Load and AE cumulative hits vs. Time curve is shown in Fig. 2b.

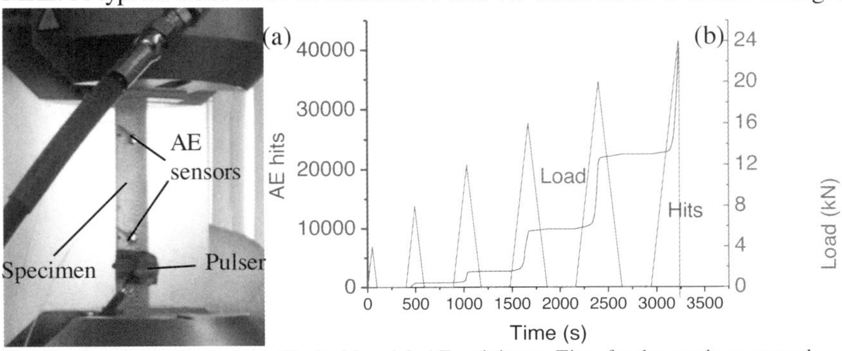

Fig. 2a. Experimental setup, 2b: Typical Load & AE activity vs. Time for the specimens tested

3. AE results

It is essential to present the AE results along with the mechanical data for reference. The most indicative trends concern the RA value shift to higher levels as the load increases. The discussion is focused on later loading steps since the initial loading steps exhibited little AE activity. Some typical cases are presented below.

Fig. 3a depicts the load history of the 5th step corresponding roughly to the time span 2150 s to 2650 s of a single experiment. The load increases linearly with time until the maximum programmed value of 20 kN and afterwards it decreases linearly with the opposite rate. On the same figure the AE cumulative hits are depicted. The first hits are recorded shortly after the application of load. However, significant increase in the hit rate is exhibited after the load passed approximately 12 kN (near 2300 s). Then the hit rate remains high and approximately constant until the maximum load. After the maximum load is reached the hit rate decreases but does not drop to zero. In Fig. 3b the load history of the same specimen together with the transition of the RA value are depicted as a function of time. Until about half the maximum load, the RA exhibits a constant value of approximately 500 μs/V. Despite some fluctuations, the RA value clearly increases with increasing load. After the maximum load is reached, the RA drops immediately to the initial value of 500 μs/V. It should be mentioned that the plotted RA value represents a moving average of the last 500 hits. As explained in the introduction, the RA shift to higher values indicates that more shear events occur in the specimen. This is consistent with the succession of the failure mechanisms that are expected for the tested materials, since transverse i.e. mode I cracking initially takes place in order to trigger delaminations i.e. mode II cracking which dominate later loading steps as transverse cracking asymptotically reaches saturation. This trend is also consistent for the following loading steps until the failure of the specimen (Figs. 3c and d).

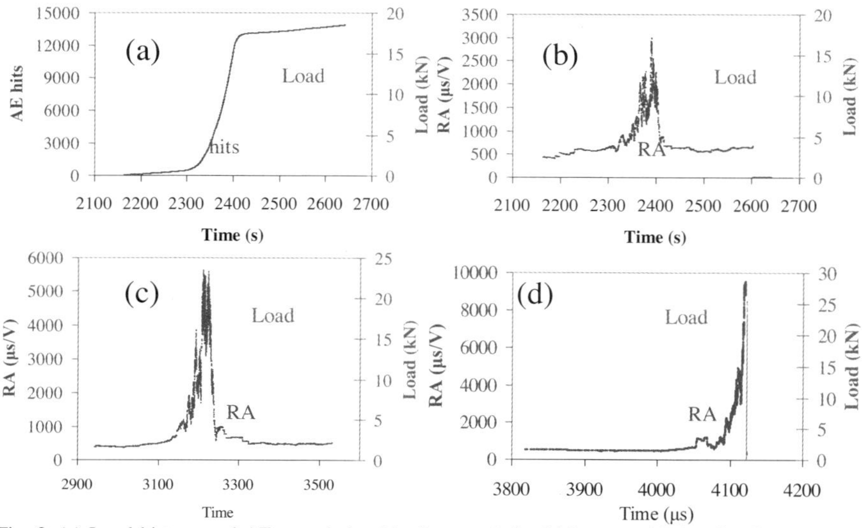

Fig. 3. (a) Load history and AE cumulative hits for step 5, load history and RA value for step 5 (b), step 6 (c), step 7(d).

During step 6 (Fig. 3c), the RA increases to higher levels of almost 6000 μs/V, while during the unloading stage it drops again to less than 1000 μs/V. As for the final stage (step 7 for the specific specimen which failed at 25.5 kN, see Fig. 3d) the RA increases to almost 10000 μs/V before the experiment is terminated due to the failure of the specimen. The increase of the RA value throughout each step suggests that shear failure or delamination is becoming more and more prominent as a failure mode. Additionally, each successive load step results in higher maximum RA values, which is indicates that delaminations become the dominant damage mode at late loading steps, when transverse cracking is near or has reached saturation. Although failure is associated with the failure of the longitudinal plies, the moment of failure the RA exhibits its highest value, suggesting that shear cracking is the dominant mechanism prior to catastrophic failure. This behaviour is consistent for all specimens studied and suggests that the RA value can be potentially used to study the tensile to shear mode conversion in composite systems in order to tailor the interlaminar properties of the material fro maximum performance.

Fig. 4 shows the load histories together with the average frequency of the recorded AE hits for the final step for three different specimens. The average frequency is also indicative of the cracking mode. All three specimens failed at approximately 25 kN. In all cases, the average frequency exhibited a strong drop a few seconds before the final failure of the material to values of around 50 kHz, while earlier it exhibited average values of more than 200 kHz. These changes can be monitored due to the broadband response of the sensors which can record signals even up to 700 kHz. It should be mentioned though that the average frequencies obtained, are certainly influenced by the sensors' response. It is highly likely that the actual frequencies of the sources are different. However, the aim of this study is to identify the moments of failure and characterize the damage status. Therefore, since the same sensors are used throughout the whole experimental series, the relative change with time of either RA or average frequency is related to the damage process associated with the studied material.

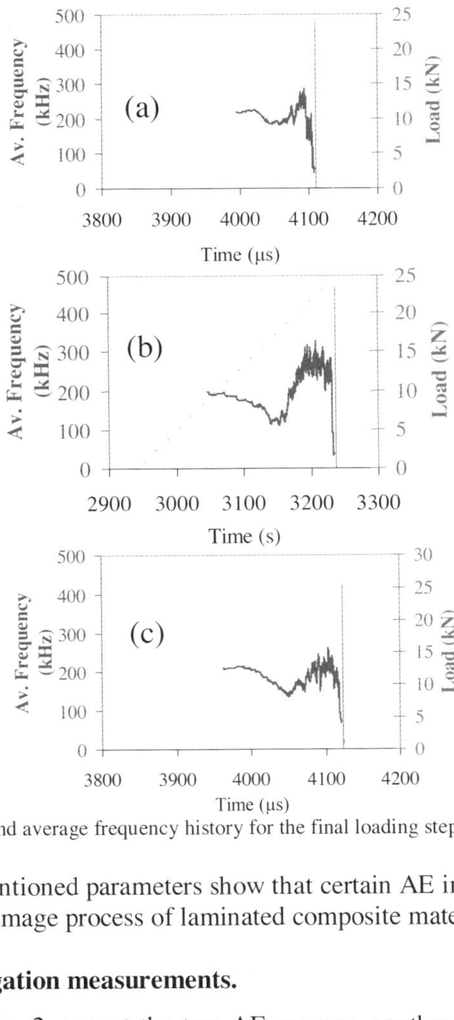

Fig. 4. Load and average frequency history for the final loading step of 3 specimens.

The above mentioned parameters show that certain AE indices can be used to characterize the damage process of laminated composite materials.

4. Wave propagation measurements.

As exhibited in Fig. 2, except the two AE sensors, another transducer acting as pulser was attached to the specimen in order to conduct wave velocity and transmission measurements during the intervals between successive loading steps. Tone bursts of different frequencies were fed to the transducer and the response of the sensors was recorded. In order to increase the signal to noise ratio, 100 signals were recorded and stacked. The wave velocity is measured by the transit time of the first detectable disturbance between the two transducers. This corresponds to what is usually referred to as pulse velocity corresponding in most cases to the longitudinal waves which are the fastest type of waves [13]. Although it cannot be regarded either as phase or group velocity it has been used for decades for structural health monitoring it is indicative of the health status of the material. However, a much more elaborate study can be conducted by identifying the different plate modes and calculate the corresponding velocities.

Pulse velocity does not seem to be sensitive to damage, at least not in the manner associated with damage accumulation. In such cases, pulse velocity in expected to decrease with damage accumulation. In Fig. 5) the pulse velocity for the applied frequencies and for successive load steps is depicted. It is difficult to draw any conclusions because a definite trend is not observed. The pulse velocities of the different frequencies exhibit considerable fluctuations with a slight increasing tendency. The most characteristic trend is that before the final step, the velocities of almost all frequencies exhibit higher values than the initial velocity at the virgin stage. This will be commented with the transmission results which follow. In Fig. 5b the maximum peak to peak amplitude of the 2nd receiver is divided by the amplitude of the 1st. Due to attenuation mechanisms which include geometric spreading, damping as well as scattering on cracks, this ratio is normally less than one. It would be reasonable to expect that as damage in the form of micro-cracks evolves, the amplitude of the 2nd receiver would be decreasing. However, in the present case, there is another strong damage mechanism due to delaminations. As delaminations are accumulated the different layers become separated. The pulser introduces the signal on the top 0° layer. Due to interlaminar failure, the signal propagation on the underlying 90° layers is partially restricted. Therefore, the energy is not spread but concentrated on the top layer. As shown in Fig. 5b, the amplitude is increased for all the frequencies tested. After the first two steps the amplitude ratio is monotonically increasing until failure, more likely being governed by the progressive delaminations which isolate the wave to the top layer. Before the final step, the transmission is approximately double than the initial stage. Thus, this increase in transmitted amplitude can be directly associated to interlaminar failure or the extend of delamination that the specimen can sustain before failure. As should be noted, the longitudinal plies are an order of magnitude stiffer than the transverse plies in the wave propagation direction, which could also account for the increasing tendency of the pulse velocities.

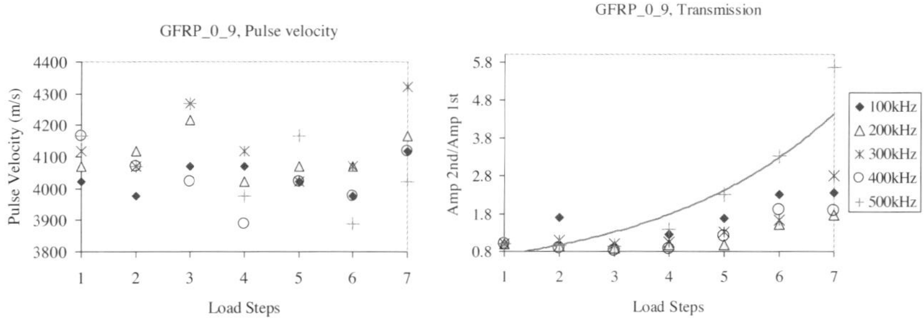

Fig. 5. (a) Pulse velocity for different frequencies and load steps, (b) transmission for different frequencies and loading step.

5. Conclusions

In this study the mechanical and acoustic emission behaviour of GFRP composite laminates is examined under triangular step loading. AE parameters like the RA value and average frequency are found to be sensitive to the failure mode, since they gradually shift as the dominant failure mode changes from transverse cracking to delamination. Therefore, they can act as damage indices associated to the damage conversion in cross ply composites with increasing load. Wave transmission measurements are found to be sensitive to delaminations due to the partial restriction of the wave energy on the top layer. This has an adverse effect on the velocity and amplitude of the transmitted signal compared to the effect of damage in the wave propagation of macroscopically homogeneous and isotropic materials.

References

1. Katerelos, D. G., McCartney, L. N. and Galiotis, C. "Local strain re-distribution and stiffness degradation in cross-ply polymer composites under tension", *Acta Materialia*, 2005; 53(12): 3335-3343.
2. Katerelos, D. T. G., Lundmark, P., Varna, J. and Galiotis, C. "Analysis of matrix cracking in GFRP laminates using Raman spectroscopy", *Composites Science & Technology*, 2007; 67(9): 1946-1954.
3. Talreja, R. "Damage characterization by internal variables". In: Pipes, R.B. and Talreja, R., editors. Damage mechanics of composite materials. *Composite materials series*, vol. 9. Amsterdam: Elsevier; 1994.
4. Nairn, J. and Hu, S. "Matrix microcracking". In: Pipes R.B., Talreja R, editors. Damage mechanics of composite materials. *Composite materials series*, vol. 9. Amsterdam: Elsevier; 1994.
5. D. Callister Jr., *Materials Science and Engineering, An Introduction*, Wiley 2000.
6. D. T. G. Katerelos, A. S. Paipetis, T. Loutas, G. Sotiriadis, V. Kostopoulos, S. L. Orgin, *Plastics, Rubber and Composites* 38(6) 229 (2009).
7. A. Kumar, A.P. Gupta, *Experimental Mechanics* 36(3) 258 (1996).
8. M. Ohtsu, Y. Tomoda, *ACI Materials Journal* 105(2) 194 (2007).
9. T. Shiotani, M. Ohtsu and K. Ikeda. *Construction and Building Materials* 15(5-6) 235 (2001).
10. RILEM TC212-ACD: "*Acoustic Emission and Related NDE Techniques for Crack Detection and Damage Evaluation in Concrete*" – 2009 Recommendation No.3.
11. D. Soulioti, D. G. Aggelis, N. M. Barkoula, A. Paipetis, T. Matikas, T. Shiotani. *Construction and Building Materials*, 23 3532 (2009).
12. T. Shiotani, D. G. Aggelis. J Acoust Emission 2007;25:69–79.
13. T. P. Philippidis, D. G. Aggelis, Ultrasonics, 43, 2005, 584-595.

ACOUSTIC EMISSION MONITORING OF THE FAILURE PATTERNS OF THE ANTERIOR CRUCIATE LIGAMENT UNDER UNIAXIAL TENSION. A CADAVERIC STUDY

N. –M. BARKOULA, D. G. AGGELIS, A. PAIPETIS, T. E. MATIKAS

Department of Materials Science and Engineering, University of Ioannina, Ioannina, Greece

N. K. PASCHOS, D. GKARTZONIKAS, A. D. GEORGOULIS

Orthopaedic Sports Medicine Center of Ioannina, Department of Orthopaedic Surgery, University of Ioannina, Ioannina, Greece

Anterior cruciate ligament (ACL) is not an isometric structure, but it is rather composed of different functional groups of fibers, mainly the anteromendial (AM) and the posterolateral (PL). In a previous study on the biomechanical response of the ACL, it was concluded that ACL failure represents a sequence of interrelated procedures that are related to the biomechanical properties of the distinct fiber bundles (AM and PL). Three different patterns were recognized, where a complete or partial tear of the fiber bundles were preceding while the remaining intact fiber bundles have a potential load resistance. The purpose of the current study was to further investigate the aforementioned failure sequence, and to clarify the events taking place during ACL rupture under uniaxial tension along the ligament. The key objective is to correlate macroscopical response and acoustic emission (AE) activity. Furthermore we aim to use AE as a tool for the confirmation of the distinct role of the AM and PL bundles during rupture of the ACL. For this purpose, AE transducers were placed on both the tibia and the femur, and stress wave signals during tension were monitored. A correlation between AE activity and failure pattern was attempted via the employment of appropriate descriptors.

1. Introduction

Anterior cruciate ligament (ACL) is not an isometric structure. It is composed of different functional groups of fibers, mainly the anteromendial (AM) and the posterolateral (PL) [1-3]. There are indications that each bundle has a different role in knee joint stability and overall knee function [4-8]. Although, the biomechanics of ACL failure have been examined in various cadaveric models, the incidents occurring during ACL rupture with regard to its different fiber bundles have not been emphasized [4, 9-11]. The different macroscopic modes of failure involve ligament disruption and avulsion of bone beneath the ligament insertion site [4, 9-11].

The acoustic emission technique offers the capability of monitoring structural degradation passively and in real time, and can distinguish failure mechanisms, and their location through the analysis of AE parameters [12-16]. Further study of the transient waveforms provides in depth insight of the fracture process. However, the application of AE entails certain difficulties. These mainly concern the accurate interpretation of the results due to the different individual processes that contribute to fracture. In orthopaedics the AE technique has been mainly applied as a predictive tool for the in vitro assessment of cemented implanted orthopaedic constructs [12]. Very limited studies employ the AE technique for the damage monitoring of soft tissues and biomaterials [17, 18] and in all cases on animal specimens.

The purpose of the current study is to employ AE monitoring for better understanding of the different mechanisms that are active during failure of the cadaveric ACL. The key objective is to correlate macroscopical response and AE activity. Furthermore we aim to use AE as a tool for the confirmation of the distinct role of the AM and PL bundles during rupture of the ACL.

2. Experimental details

Four fresh-frozen human cadaveric knee specimens were used in this study. Institutional review board (IRB) approval was obtained by our institution for the use of cadaver specimens. A custom clamping device was designed and manufactured in order to ensure a reproducible placement of the tibial and femoral component of the specimens. With the usage of a digital goniometer the knee angle was measured and each knee was placed in 15° of flexion. It has been shown that in knee flexion of 15° the AM and PL have similar in situ force distribution [7]. Therefore we presume that the rupture occurring would depend on the mechanical properties of the ACL fibers. The axis of the femur and tibia was placed in accordance to the ACL axis to avoid varus/valgus and torsion rotation variations. The biomechanical tests were conducted on a servo-hydraulic machine Instron 8801. For the present study a 5 kN load cell was used. The specimen was loaded from the relaxed position up to failure at a displacement rate of 1.5 mm/s. A non-contact video extensometer, supplied by Instron, was used to monitor the displacement. The video extensometer consisted of a high resolution digital camera and advanced real-time image processing. Two high-resolution high-speed (frame rate of 200Hz) digital cameras, were used, one monitoring the experimental process from the anterior aspect of the knee joint and the other filming the posterior surface.

For the purpose of the AE monitoring, two wide band AE sensors (Pico, Physical Acoustics Corp., PAC) were attached to either side of the ACL. The length of the tissue was too short to allow placing the sensors directly on the ACL, rendering the AE source location troublesome. Therefore, the sensors were attached to the femur and tibia being able to record the events related to the detachment (separation) between the tissue and the bone, as well as the emissions from tissue rupture events which propagate through the tissue into the bone where the sensors are attached. A layer of grease was applied between the sensor and the bone to enhance acoustic coupling, while a tape was used to ensure that the sensors will not be detached during the experiment. The specific sensors were chosen over other AE transducers mainly for two reasons. The first is their small size (diameter 5 mm) which enables to be attached to the bone without the need for specific coupling device (shoe). The other important reason is their spectral response. They are sensitive to frequencies from 100 kHz up to 1 MHz, with maximum sensitivity at 500 kHz. Therefore, they can capture a wide range of different sources. It is mentioned that the point of contact was not the same for each specimen due to variations in each specimen's geometry (length, thickness etc.). The pre-amplifier gain was set to 40 dB. After performing a pilot test, the threshold was also set to 40 dB in order to avoid the possibility of electronic/environmental noise. The signals were recorded in a two-channel monitoring board PCI-2, PAC. AE hits (including whole waveforms) were recorded with a sampling rate of 5 MHz.

3. Results and Discussion

3.1. *Mechanical response and macroscopical observations*

Fig. 1 shows the mechanical response of the ACL specimens tested during this study. All ACL specimens demonstrated a non-monotonic response. If the ACL was behaving as one single fiber then the rupture would translate into complete monotonic failure of the specimen. The fact that load initially drops and then peaks up again leads to the conclusion that ACL acts as a multi-fiber construction. The video analysis revealed three patterns during ACL failure. The first one involved a midsubstance tear of the AM bundle followed by a subsequent tear of the PL bundle (ACL2). The second pattern involved a midsubstance tear of the PL bundle first, which was followed by a tear of the AM bundle (ACL1 and ACL4). The third pattern involved an avulsion of the tibial attachment of part of the fibers of the AM bundle which was followed by a midsubstance tear of the PL bundle. Lastly, there was a midsubstance tear of the remaining fibers of the AM bundle (ACL3). In all cases, the tear of the one of the bundles coincided with the initial decrease in the load-elongation curve. The load carrying ability of the other bundle is depicted by the subsequent plateau/picks in the load-elongation curve. Despite that the AM and PL bundles acted as separate structures, the rupture of the one structure allowed the other to operate under load, therefore we presume that the different bundles act synergistic to each other. Further tensioning of the ACL results in to rupture of the intact fibers and consequent inability to withstand load.

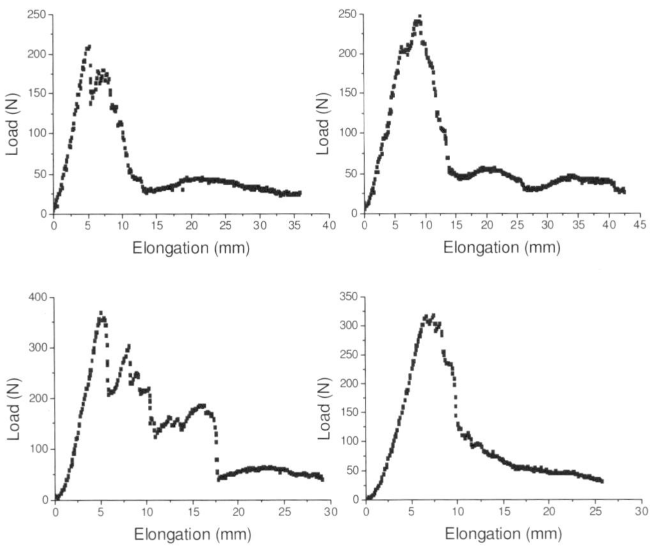

Figure 1: Load – Elongation curves of the 4 ACL specimens tested, a) ACL1, b) ACL2, c) ACL3 and d) ACL4 specimens respectively

3.2. *AE results*

As aforementioned there different patterns were observed during failure of the ACL specimens tested. It is therefore interesting to investigate if there is a correlation between these failure patterns and the AE hits collected during the tensile tests. For that, it is essential to present the AE results along with the mechanical data for reference. Some typical cases are presented below.

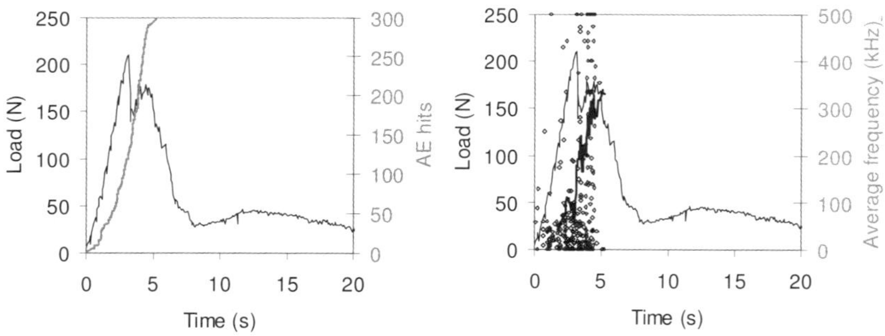

Figure 2: (a) Load and cumulative AE hits history for specimen 1, (b) load and average frequency of AE hits. The solid line is the moving average of the recent 40 points (hits).

In Fig. 2(a) the load time history for ACL1 is depicted. The whole experiment lasted approximately 20 sec, while the failure occurred before 5 sec. After the peak load of 210 N a second smaller peak can be observed. AE generation started quite early after the application of the load. However, during the period between 3 and 5 sec, the AE hit rate (of both transducers combined) was high (100 hits/s) while after 4.9 s when the second load peak was observed, the hit rate decreased to approximately 25 hits/s. Thereafter, the load bearing capacity was critically compromised (the specimen had failed) and no AE was observed. In cases of fracture it is very useful, except for the total hit number to draw information also from the characteristics of the waveforms. Since fracture is a dynamic process, the types of signals emitted may not be the same throughout the experiment. In the field of engineering fracture, descriptors are sought to correlate with the different stages of damage (early – middle – final).

A parameter that has already been used in engineering is the average frequency of the signals [19, 20]. It is calculated as the ratio of the number of threshold crossings of the waveform divided by its duration. This has been shown to change monotonically as the damage is accumulated [19, 20]. It seems that this parameter can also be correlated with damage accumulation for this biological specimen of the present study. In Fig. 2(b) each point is the average frequency of the corresponding AE hit. The solid increasing line is the moving average of the 40 most recent hits. It is clear that the average frequency increases until the second load peak. The average frequency increases considerably from 50 kHz to more than 300 kHz, showing that the type of fracture is gradually shifting. In engineering materials the frequency change is attributed to the shifting of the crack mode from tensile to shear [19, 20]. In the case of the bone specimen the transition cannot be considered a priori similar but it should correspond to activation of different mechanism (fiber rupture or/and separation from the bone). Similar, though not as clear, is the behavior of ACL2 depicted in

Fig. 3(a) and (b). The average frequency through its fluctuations starts at 200 kHz at the load of 50 – 100 N, while after failure it reaches even 400 kHz. It is reminded that this change up to 400 kHz would not be captured if the sensors were of limited bandwidth.

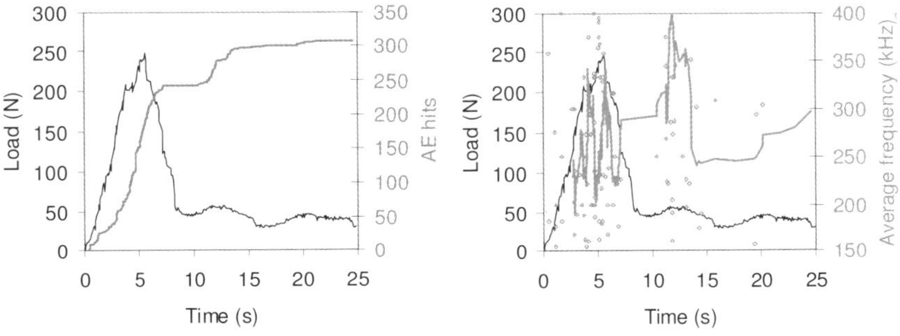

Figure 3: (a) Load and cumulative AE hits history for specimen 2, (b) load and average frequency of AE hits. The solid line is the moving average of the recent 40 points(hits)

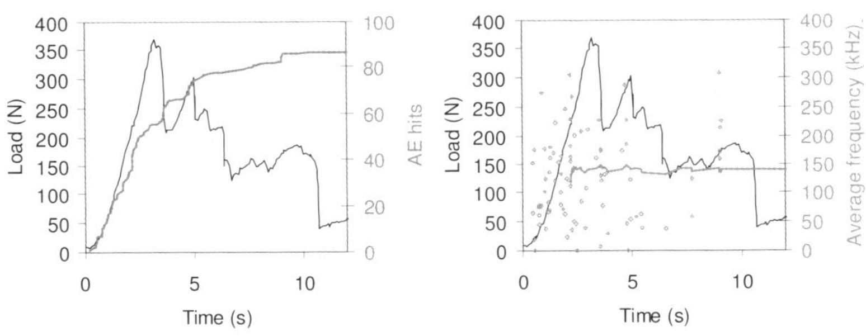

Figure 4: (a) Load and cumulative AE hits history for specimen 3, (b) load and average frequency of AE hits. The solid line is the moving average of the recent 40 points(hits).

Another indicative case is presented in Fig. 4. The load curve exhibited 3 major peaks (at 3.2, 4.9 and a smaller at 9.4 s). These were accompanied by corresponding bursts of incoming AE signals at the exact time of the load peak or slightly earlier, see arrows on the AE hit history line, Fig. 4(a). At these moments, the cumulative hit line exhibited almost vertical increase with a rate of almost 50 hits/s. Between these fracture incidents, the rate was decreased to approximately 4 hits/s corresponding to the stage of more stable deformation of the tissue. As the deformation was accumulated it led to the irreversible fracture accompanied by a strong number of AE hits. For this specimen however, the average frequency demonstrated no serious shift to higher values. This more likely indicates that there was a basic dominant mechanism active throughout the whole fracture procedure.

5. Conclusion

The ACL ligament acts as a multi-fiber construction. In our setting, rupture follows three specific patterns where a complete or partial tear of the fiber bundles are preceding while the remaining intact fiber bundles have a potential load resistance. In general, it can be said that the AE behavior can be used as a precursor of final failure, while study of certain parameters can characterize the process of fracture. The human bone/tissue exhibit high attenuation; however, the mechanisms of AE generation are similar to engineering materials and it seems that specific descriptors obtained from the engineering view point can be applied for these biological materials.

References

1. Girgis FG, Marshall JL, Monajem A., Clin Orthop Relat Res 1975;106:216-231.
2. Amis AA, Dawkins GP., J Bone Joint Surg Br 1991;73:260-267.
3. Petersen W, Zantop T., Clin Orthop Relat Res 2007;454:35-47.
4. Woo SL, Hollis JM, Adams DJ, Lyon RM, Takai S., The effects of specimen age and orientation. Am J Sports Med 1991;19:217-225.
5. Zantop T, Herbort M, Raschke MJ, Fu FH, Petersen W., Am J Sports Med 2007;35:223-227.
6. Lipke JM, Janecki CJ, Nelson CL, et al., J Bone Joint Surg Am 1981;63:954-960.
7. Gabriel MT, Wong EK, Woo SL, Yagi M, Debski RE., J Orthop Res 2004;22:85-89.
8. Lo J, Müller O, Wünschel M, Bauer S, Wülker N., J Biomech 2008;41:1855-1861.
9. Noyes FR, DeLucas JL, Torvik PJ., J Bone Joint Surg Am 1974;56:236-253.
10. Noyes FR, Torvik PJ, Hyde WB, DeLucas JL., J Bone Joint Surg Am 1974;56:1406-1418.
11. Noyes FR, Grood ES., J Bone Joint Surg Am 1976;58:1074-1082.
12. Browne M, Roques A, Taylor A., J Strain Anal 2005; 40(1): 59-78.
13. Kumar A, Gupta AP., Exp Mech 1996; 36(3):258-261.
14. Shiotani T, Ohtsu M, Ikeda K., Constr Build Mater 2001; 15(5-6):235-246.

15. Grosse CU, Finck F., Cement Concr Comp 2006;28:330-336.
16. De Rosa I, Santulli C, Sarasini F., Comp: Part A 2009; 40: 1456–1469.
17. Azangwe G, Fraser K., Mathias K.J., Siddiqui A. M., Med Eng Phys 2000; 22:279-283.
18. Kohn DH., Critical Rev Biomed Eng 1995;23(3-4):221-306.
19. Ohtsu M, Tomoda Y., ACI Mat J 2007;105(2):194-200.
20. Soulioti D, Aggelis DG, Barkoula NM, Paipetis A, Matikas T, Shiotani T., Constr Build Mater 2009; 23: 3532-3536.

DETRENDING OF HRV SIGNALS BASED ON THE BAYESIAN FRAMEWORK AND A NON STATIONARY MODEL FOR THE NON TREND COMPONENT

VANGELIS P. OIKONOMOU

Department of Computer Science, University of Ioannina,
GR 45110 Ioannina, Greece
Email: voikonom@cs.uoi.gr

DIMITRIOS I. FOTIADIS

Unit of Medical Technology and
Intelligent Information Systems
Department of Materials Science and Engineering,
University of Ioannina, GR 45110 Ioannina, Greece
Email: fotiadis@cs.uoi.gr

In this work, a new detrending method for Heart Rate Variability (HRV) signals, based on the Bayesian framework, is proposed. The trend in HRV signals affects the subsequent statistical analysis and must be removed. The HRV signal is modeled as the sum of two components. The first component describes the trend contained in the HRV signal, while the second component describes the rest of the HRV signal. The trend is the slow change in the HRV signal and the smoothness prior is used to describe it. The second component is modeled using a Gaussian distribution with diagonal covariance. This assumption makes the model able to deal with various forms of HRV signal, which are obtained in pathological situations. For the estimation procedure the Bayesian framework is used. However, the true posterior is difficult to be obtained and approximation techniques are sought. In our method the Variational Bayesian (VB) Methodology is employed, where an approximated posterior is used. One advantage of the VB Methodology is that gives closed form solutions. The results compare well with a previous approach where a spherical covariance matrix is used for the non-trend component.

1. Introduction

The Heart Rate Variability (HRV) signal is obtained from the electrocardiogram (ECG) and describes the variations between consecutive cardiac beats. Studies have shown that this signal originates from the Autonomous Nervous System (ANS)[1]. Also, the HRV signal is strongly connected to respiration and blood pressure[2]. Thus the HRV signal can be used as a quantitative marker of the ANS and HRV parameters are used to evaluate the clinical condition of subjects in normal or pathological conditions. The HRV analysis methods can be divided into time-domain, frequency-domain, and nonlinear methods[3]. The analysis of HRV is performed by studying various measures of the signal such as the standard deviation between two normal beats (SDNN) and the power spectral density (PSD)[4,5]. However, the HRV signal contains artefacts which can be originate from other physiological processes, such as the breathing pattern of a human, or technical dysfunctions, such as QRS misdetection[6,7]. These artefacts distort the HRV signal leading in erroneous calculations of various statistical measures.

Thus, it is necessary, before the statistical analysis of an HRV signal, a pre - processing step to remove the various artefacts. In the literature various approaches have been used to remove the trend from the HRV signal. These approaches include linear or high order polynomials[8], wavelet analysis[9] and regularization techniques[10]. In those detrending techniques the trend is described as the slowly changing waveform in the HRV signal. Also, the raw HRV signal is modeled as the sum of two components, a smooth component, corresponding to the trend, and a high pass component, corresponding to the rest of the HRV signal. The approach in[10] proposes the use of the smoothness prior for the trend and a stationary model for the high pass component of the HRV signal. While the assumption of smoothness over the trend seems to be reasonable, due to the definition of the trend, the stationary model for the high pass component of the HRV signal is questionable, particularly the assumption of white Gaussian distribution over this component. In this work we address this problem and propose a non stationary model for the high pass component of the HRV signal.

The scope of this work is to develop a method to remove the trend from the HRV signal. The additive model is used for the modeling of the HRV signal, i.e. the raw HRV signal is composed by the trend and a high pass component. The trend is modeled using the smoothness prior, while for the high pass component a Gaussian distribution with diagonal covariance is used. The use of prior is connected to the Bayesian framework and the posterior distribution. However, the true posterior is not easily calculated and the use of approximation techniques, such as the Variational Bayesian (VB) methodology[11], is required. The VB methodology provides with a closed form solution and a convergence criterion. Our work are compared well to those obtained in[12].

2. Methodology

The observed HRV signal is composed of the trend and the high pass component. The trend describes the slowly varying part of HRV signal and the high pass component mostly describes the variability of RR intervals. The mathematical model which describes such a signal is:

$$\mathbf{y} = \mathbf{s} + \mathbf{n}, \tag{1}$$

where $\mathbf{y}$ is a $N \times 1$ vector which contains the observed HRV signal, $\mathbf{s}$ is a $N \times 1$ vector which contains the trend component and $\mathbf{n}$ is a $N \times 1$ vector which contains the rest of the HRV signal. Our scope is to remove the trend from the HRV. Before removal, the trend must be first estimated. An approach to estimate it is the Maximum Likelihood (ML). However, in our case the ML solution is the observations. To obtain a meaningful solution for the trend we must imposed several constraints. These constraints might be introduced by using the classical regularization theory or the Bayesian framework[13]. In our study the Bayesian framework is chosen because the constraints can be incorporated into our model in a natural way through the prior distribution.

A prior to describe the trend in a signal is the smoothness prior[13]. This type of prior has been used to remove the trend from HRV signal[10] and fMRI data[14]. The smoothness prior is given by:

$$p(\mathbf{s}) \propto \left(\frac{\alpha}{2\pi}\right)^{\frac{N}{2}} \exp\left\{ -\frac{\alpha}{2}\mathbf{s}^{\mathbf{T}}\mathbf{L}^{\mathbf{T}}\mathbf{L}\mathbf{s}\right\}. \tag{2}$$

The matrix $\mathbf{L}$ is a discrete approximation of the second derivative approximation, while the parameter α controls the smoothness of the solution and it is known as the smoothness parameter. We treat this parameter probabilistically and the distribution over this is a gamma distribution given by:

$$p(\alpha) = Ga(\alpha; b_\alpha, c_\alpha). \tag{3}$$

The non trend component is treated as the noise of observations. The non trend component has impulsive nature and in this case a useful distribution to model this component is the student t distribution with zero mean. The model for the non trend component is given by:

$$p(n_i|\nu,\sigma) = \frac{\Gamma((\nu+1)/2)}{\Gamma(\nu/2)\sqrt{\nu\pi\sigma}}\left\{1 + \frac{1}{\nu}\left(\frac{n_i}{\sigma}\right)^2\right\}, \tag{4}$$

where ν is the degree of freedom and σ is the scale parameter. This distribution is characterized for its robustness in the presence of the outliers.

Using this distribution in our model it is difficult to obtain closed form solutions. However, this distribution can be obtained in an alternative way, presented in[15]. Assuming that each sample n_i of the non trend follows a Gaussian distribution with zero mean and precision b_i and each precision b_i follows a gamma distribution, we obtain a hierarchical model for the non trend component. Treating the precision b_i as a nuisance parameter and integrating we obtain the marginal distribution for the non trend component which is a student - t distribution. The marginal distribution is given by:

$$p(n_i|c,d) = \int_0^{\infty} p(n_i|\beta_i)p(\beta_i|c,d)d\beta_i, \tag{5}$$

where

$$p(n_i|\beta_i) = N(0, \beta_i^{-1}), \tag{6}$$
$$p(\beta_i|c,d) = Ga(\beta_i; c, d). \tag{7}$$

This model specifies that the non trend model is a mixture of an infinite number of Gaussians of varying precision β_i. The weights of the mixture are specified by the distributions $p(\beta_i|c,d)$. This decomposition allows the use of an effective inference procedure such as the VB methodology.

The goal in the Bayesian Learning is to obtain the posterior distribution $p(\mathbf{s}, \beta, \alpha|\mathbf{y}, c, d, b_a, c_a)$. However, a simple application of Bayes Theorem is not easy since the dominator $p(\mathbf{y})$ contains intractable integrations. For this reason an approximation scheme is adopted such as the VB methodology. After some algebraic manipulations the $\log p(\mathbf{y})$ can be expressed as:

$$\log p(\mathbf{y}) = F(q(\mathbf{s}, \beta, \alpha)) + KL(q(\mathbf{s}, \beta, \alpha)||p(\mathbf{s}, \beta, \alpha)), \tag{8}$$

where $q(\mathbf{s}, \beta, \alpha)$ is an arbitrary approximating distribution. The first term in the above equation is given by

$$F(q(\mathbf{s}, \beta, \alpha)) = \int q(\mathbf{s}, \beta, \alpha) \log\left\{\frac{p(\mathbf{y}, \mathbf{s}, \beta, \alpha)}{q(\mathbf{s}, \beta, \alpha)}\right\} d\alpha d\mathbf{s} d\beta, \tag{9}$$

and the second term is the KL divergence between the approximating distribution and the true posterior. Since $KL(q(\mathbf{s}, \beta, \alpha)||p(\mathbf{s}, \beta, \alpha)) \geq 0$, $F(q(\mathbf{s}, \beta, \alpha))$ is a lower bound on $\log p(\mathbf{y})$. Maximizing this bound with respect to $q(\mathbf{s}, \beta, \alpha)$ we obtain an approximation to the true posterior. We must also specify $q(\mathbf{s}, \beta, \alpha)$. There are two approaches. In the first approach a parameterized form is adopted for the approximating distribution. The second approach, which we adopted in this work, assumes that the approximating distribution can be written in a separable form, i.e.

$q(\mathbf{s}, \beta, \alpha) = q(\mathbf{s})q(\beta)q(\alpha)$. In that case the bound $F(q(\mathbf{s}, \beta, \alpha))$ is maximized when

$$q(\mathbf{s}) \propto \exp < \log p(\mathbf{y}, \mathbf{s}, \beta, \alpha) >_{q(\beta)q(\alpha)}, \quad (10)$$
$$q(\beta) \propto \exp < \log p(\mathbf{y}, \mathbf{s}, \beta, \alpha) >_{q(\mathbf{s})q(\alpha)}, \quad (11)$$
$$q(\alpha) \propto \exp < \log p(\mathbf{y}, \mathbf{s}, \beta, \alpha) >_{q(\mathbf{s})q(\beta)} . \quad (12)$$

Applying the VB methodology and taking into account the prior distributions of the model parameters $\mathbf{s}$, β and α we obtain the following posteriors:

$$q(\mathbf{s}) = N(\hat{\mathbf{s}}, \mathbf{C}_{\hat{\mathbf{s}}}), \quad (13)$$
$$q(\beta_i) = Ga(\beta_i; c_i^{'}, d_i^{'}), i = 1, 2, \cdots, N, \quad (14)$$
$$q(\alpha) = Ga(\alpha; b_\alpha^{'}, c_\alpha^{'}), \quad (15)$$

where

$$\hat{\mathbf{s}} = (\bar{\alpha}\mathbf{L}^T\mathbf{L} + \bar{\mathbf{B}})^{-1}\bar{\mathbf{B}}\mathbf{y}, \quad (16)$$
$$\mathbf{C}_{\hat{\mathbf{s}}} = (\bar{\alpha}\mathbf{L}^T\mathbf{L} + \bar{\mathbf{B}})^{-1}, \quad (17)$$
$$\frac{1}{b_\alpha^{'}} = \frac{1}{2}(Tr\{\mathbf{C}_{\hat{\mathbf{s}}} + \hat{\mathbf{s}}\hat{\mathbf{s}}^T\}) + \frac{1}{b_\alpha}, \quad (18)$$
$$c_\alpha^{'} = \frac{N}{2} + c_\alpha, \quad (19)$$
$$\bar{\alpha} = b_\alpha^{'}c_\alpha^{'}, \quad (20)$$
$$\frac{1}{d_i^{'}} = (y_i - \hat{s}_i)^2 + \mathbf{C}_{\hat{\mathbf{s}}}(i, i) + \frac{1}{d}, \quad (21)$$
$$c_i^{'} = c + \frac{1}{2}, \quad (22)$$
$$\hat{\beta}_i = d_i^{'}c_i^{'}. \quad (23)$$

3. Results

The proposed approach is applied in simulated and real data and and the results are compared with the method presented in[12]. This method is similar to the Smoothness Prior Approach[10]. The only difference is that the smoothness parameter is estimated from the data and not from visual analysis. This estimator is called VBSPA.

3.1. *Initialization*

We use uniformative priors by setting $b = 10^{+6}$ and $c = 10^{-6}$ for the Gamma prior of the smoothness parameter. The Gamma priors of the non - trend component are initialized by setting $b = 10^{+6}$ and $c = 10^{+2}$. Also, the smoothness parameter α must be initialized. The initialization of the parameter α is very crucial for the obtained results. Letting $\alpha \to 0$ means that both algorithms start from a very noisy solution, while $\alpha \to \infty$ means that the algorithms start from a very smooth solution. In our study we set $\alpha = 10^9$. We made this choice based on the nature of the trend, since we expect the trend to be a smooth signal. The parameters c and d of the prior distribution of the non trend component are optimized in a similar way as it is described in[15]. We note that in this case we obtain point estimates. Typically, the algorithm stops when the Variational Bound converges. However, the computational effort is high. Also, when we calculate this bound we can face numerical problems. In our approach we choose that both algorithms converge when $\frac{\|\mathbf{t}^{(k)} - \mathbf{t}^{(k-1)}\|}{\|\mathbf{t}^{(k)}\|} < \epsilon$ where $\epsilon = 0.0001$ and $\mathbf{t}^{(k)}$ is the estimated trend at step k.

3.2. *Simulated Data*

We have created simulated HRV time series by adding noise, coming from a student - t distribution, to the simulated trend. To study the performance of the algorithm we calculate the normalized mean square error (MSE) between the true and the estimated trend for various SNR configurations for 10 Monte Carlo simulations. Also, measures related to HRV studies such as the standard deviation between normal beats (SDNN) and the root mean square of differences of successive RR intervals (RMSSD) are calculated. The SDNN measure describes the global RR variability, while the RMSSD describes the local RR variability. This interpretation of SDNN and RMSSD is very helpful to the subsequent analysis of the experiments. The trend is an effect that affects the global properties of the HRV time series. In Table 1 the MSE for various SNR is shown. It is obvious that the proposed approach gives better estimates of the trend compared to the VB - SPA. Also, in Tables 2 and 3 we observe that the proposed approach affects the global properties of the time series while at the same time the effect in the local properties is low.

Table 1. Normalized MSE

SNR	VBSPA	Student - t SPA
10	0.0533	0.0324
5	0.0871	0.0511
0	0.1472	0.0916

Table 2. SDNN (in msec)

SNR	Original	VBSPA	Student - t SPA
10	61.2692	59.6565	60.5514
5	192.4104	189.0005	191.0245
0	606.8732	597.9491	603.5701

Table 3. RMSSD (in msec)

SNR	Original	VBSPA	Student - t SPA
10	118.4736	118.4634	118.4721
5	379.7152	379.7100	379.7116
0	1212.2921	1212.2843	1212.2890

3.3. *Real HRV time series*

The VBSPA and Student - t SPA approaches have been applied for the detrending of real short term HRV data (duration < 5 minutes). The HRV data have been obtained from the MIT/BIH Arrhythmia Database[16]. In our experiments the first 5 minutes from the records 100, 101, 102, 104 and 119 have been employed. These records correspond to five different subjects. The choice of these records have been done to demonstrate different situations of the HRV signal and how they influence the estimation of the trend. For the first 5 minutes the record 100 contains 371 beats, from which 367 are normal and 4 are ectopic. The record 101 contains 342 normal beats. The record 102 contains 366 beats, from which 98 are normal, 1 ectopic, and the rest 267 from the pacemaker. The record 104 contains 372 beats, from which 65 are normal, 1 is ectopic, 301 are from the pacemaker and 5 are unclassifiable. Finally, the record 119 contains 326 beats, from which 246 are normal and 80 are ectopic.

Before detrending, each RR time series is interpolated, using cubic interpolation, to obtain a regularly sampled time series. The SDNN and the RMSSD measures have been used to study the effect of detrending in the HRV time series. These two measures have been calculated on the raw and the detrended HRV time series. The obtained results are shown in Tables 4 and 5. We observe that the Student - t SPA approach has small effect in the local properties of the HRV time series, i.e. the RMSSD value after the detrending is similar to the value of RMSSD of the raw HRV time series. This is a valuable feature of the proposed approach since the trend affects only the global properties of the HRV time series. The HRV time series and the estimated trends using the two approaches are shown in Fig. 1. We observe that the two trends are different when an outlier is present in the original HRV time series. This effect can be observed around time=230 sec.

Table 4. SSDN of real HRV time series (in msec)

Record	Original	VBSPA	Student - t SPA
100	33.2834	32.3454	32.5189
101	59.1509	29.1859	34.1285
102	31.3978	28.7050	29.1194
104	36.3418	35.0191	35.3417
119	223.3453	221.5342	222.9595

Table 5. RMSSD of real HRV time series (in msec)

Records	Original	VBSPA	Student - t SPA
100	33.6905	33.6804	33.6873
101	25.6902	25.3115	25.6260
102	25.0943	25.0581	25.0816
104	38.3449	38.3016	38.3255
119	261.8728	261.8261	261.8736

4. Conclusions

In this work a method for the detrending of HRV time series is presented. The method is based on the Variational Bayesian framework and the smoothness prior. The central idea of this work is the decomposition of the HRV time series in two components, the trend and the rest of the HRV signal. The trend is modeled using the smoothness prior, while the rest is modeled by a non stationary model which at the same time has an impulsive distribution. The results indicate that the Student - t SP approach has smaller effect on the local properties of the time series compared to the VBSPA approach. This is a useful feature in the detrending of HRV signal, since the trend is a signal which has small contribution to the RMSSD measure due to its properties, i.e. the neighborhood samples have similar values.

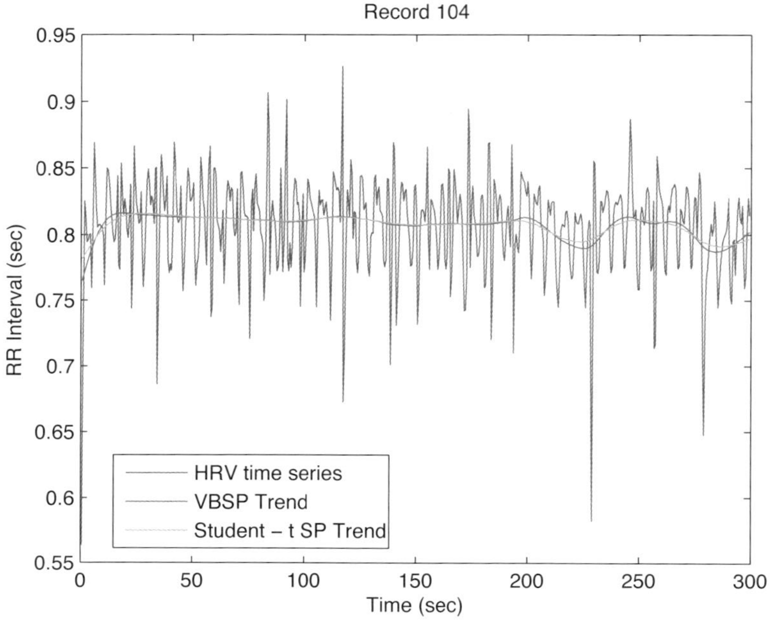

Figure 1. HRV time series and the estimated trends for record 104.

References

1. S. Cerutti, A. Goldberger, and Y. Yamamoto, "Recent advances in heart rate variability signal processing and interpretation," *IEEE Transactions on Biomedical Engineering*, vol. 53, pp. 1–3, January 2006.
2. M. D. Rienzo, G. Mancia, G. Parati, A. Pedotti, and A. Zanchetti, *Frontiers of Blood Pressure and Heart Rate Analysis*. New York, NY, USA: IOS Press, 1997.
3. M. Akay, *Nonlinear Biomedical Signal Processing, Volume 2, Dynamic Analysis and Modeling*. Wiley-IEEE Press, 2000.
4. Electrophysiology, Task Force of the European Society of Cardiology the North American Society of Pacing, "Heart Rate Variability : Standards of Measurement, Physiological Interpretation, and Clinical Use," *Circulation*, vol. 93, no. 5, pp. 1043–1065, 1996.
5. U. R. Acharya, K. P. Joseph, N. Kannathal, C. M. Lim, and J. S. Suri, "Heart Rate Variability: A Review," *Medical and Biological Engineering and Computing*, vol. 44, pp. 1031–1051, December 2006.
6. J. Mateo and P. Laguna, "Analysis of heart rate variability in the presence of ectopic beats using the heart timing signal," *IEEE Transactions of Biomedical Engineering*, vol. 50, pp. 334–343, March 2003.

7. K. Solem, P. Laguna, and L. Sornmo, "An efficient method for handling ectopic beats using the heart timing signal," *IEEE Transactions of Biomedical Engineering*, vol. 53, pp. 13–20, January 2006.
8. I. P. Mitov, "A method for assessment and processing of biomedical signals containing trend and periodic components," *Medical Engineering and Physics*, vol. 20, pp. 660 – 668, 1998.
9. R. Thuraisingham, "Preprocessing rr interval time series for heart rate variability analysis and estimates of standard deviation of rr intervals," *Computer Methods and Programs in Biomedicine*, vol. 83, pp. 78–82, 2006.
10. M. P. Tarvainen, P. O. Ranta-aho, and P. A. Karjalainen, "An advanced detrending method with application to hrv analysis," *IEEE Transactions of Biomedical Engineering*, vol. 49, pp. 172–175, February 2002.
11. M. Beal, *Variational Algorithms for Approximate Bayesian Inference*. PhD thesis, Gatsby Computational Neuroscience Unit,University of London, London, U.K., 2005.
12. V. Oikonomou and D. Fotiadis, "A bayesian approach for biomedical signal denoising," in *Proceedings IEEE ITAB*, 2006.
13. J. Kaipio and E. Somersalo, *Statistical and Computational Inverse Problems*. Springer, 2004.
14. J. Carew, G. Wahba, X. Xie, E. Nordheim, and M. Meyerand, "Optimal spline smoothing of fmri time series by generalized cross-validation," *Neuroimage*, vol. 18, pp. 950–961, April 2003.
15. M. Tipping and N. Lawrence, "Variational inference for student-t models: Robust bayesian interpolation and generalised component analysis," *Neurocomputing*, vol. 69, pp. 123–141, 2005.
16. A. L. Goldberger, L. A. N. Amaral, L. Glass, J. M. Hausdorff, P. C. Ivanov, R. G. Mark, J. E. Mietus, G. B. Moody, C.-K. Peng, and H. E. Stanley, "PhysioBank, PhysioToolkit, and PhysioNet: Components of a new research resource for complex physiologic signals," *Circulation*, vol. 101, no. 23, pp. e215–e220, 2000 (June 13). Circulation Electronic Pages: http://circ.ahajournals.org/cgi/content/full/101/23/e215.

A FAST WAVELET/BEM FOR WAVE SCATTERING SOLUTIONS

D. NTALAPERAS[(A)], S.V. TSINOPOULOS[(B)], D. POLYZOS[(A)]

(a) Department of Mechanical and Aeronautical Engineering, University of Patras, Greece

(b) Department of Mechanical Engineering, Technical Educational Institute of Patras, Greece

Abstract: The conventional Boundary Element Method (BEM) as it is applied to acoustic wave scattering problems, requires the solution of a system of algebraic equations $Ax=b$ with the matrix A being dense and full populated. Thus for large problems, BEM solutions become computationally expensive confined to scattering problems corresponding to low and intermediate frequencies. The present paper demonstrates a new BEM, which in conjunction with a wavelet transform technique reduces the complexity of the matrix A and accelerates the solution process. The Haar wavelet transform, suitable for transforming matrices of arbitrary size, is exploited in the present work. After *thresholding* the matrix A becomes sparse with the well known computational gains in the required solution time. Representative numerical examples dealing with acoustic wave scattering problems are provided in order to assess the efficiency of the proposed methodology.

1. Introduction

The Boundary Element Method (BEM) is a very well known and robust numerical tool successfully used for the solution of acoustic wave scattering problems [$1 \div 4$]. Two remarkable advantages it offers as compared to other numerical methods is the reduction of the dimensionality of the problem by one and its high solution accuracy. This means that acoustic wave scattering problems are accurately solved by descretizing only the boundary of the scatterer and not the exterior domain plus the boundary as the Finite Element Method, the Finite Differences and Meshless Methods do. However, problems associated with the unsymmetric and full-populated nature of the final system of algebraic equations confines the use of the BEM in scattering problems with relatively small number of degrees of freedom (dof), since a direct method of solution, such as Gauss elimination and LU decomposition methods, requires an effort of order $O(dof^3)$.

Acceleration to the solution process can be accomplished with the aid of iterative solvers such as the conjugate gradient solver [5] or the Generalized Minimal Residual (GMRES) solver [6]. Although successful for some types of scattering problems, the required total solution time corresponds to an effort of order $O(dof^2)$. On the other hand the use of Fast Multipole Techniques [7, 8], in spite their very high efficiency ($O(dof \cdot \log(dof))$), requires completely new numerical formulation and reconsideration of the used BEM code.

A very promising alternative to above mentioned techniques is the application of the wavelet transform on the coefficient matrix of the final system of algebraic equations, in a way used for image compression and signal processing. The efficiency of the method is similar to that provided by the Fast Multipole technique as it is illustrated in the representative works of Bucher et al [9] and Ravnik et al [10]. In the present work the Haar wavelet transform technique, used by Ravnik et al [10] for potential problems, is exploited for the solution of acoustic wave scattering problems. The methodology is illustrated in section 3. The final system of algebraic equations is obtained with the aid of an advanced BEM explained in [3, 4] and presented in brief in the next section. Finally, in section 4 a representative numerical example is provided to demonstrate the efficiency and the achieved accuracy of the method.

2. BEM solution of 3D acoustic wave scattering problems

In this section the BEM as it is applied for the solution of acoustic wave scattering problems is presented in brief. More details one can find in [3, 4].

Consider an acoustic scatterer (hard, soft or penetrable) of volume V embedded into a 3D infinitely extended acoustic space surrounded by a surface S characterized by a normal vector $\hat{\mathbf{n}}$ (Figure 1). A harmonic plane wave of frequency ω propagates in $\hat{\mathbf{k}}$ direction and impinges upon the scatterer.

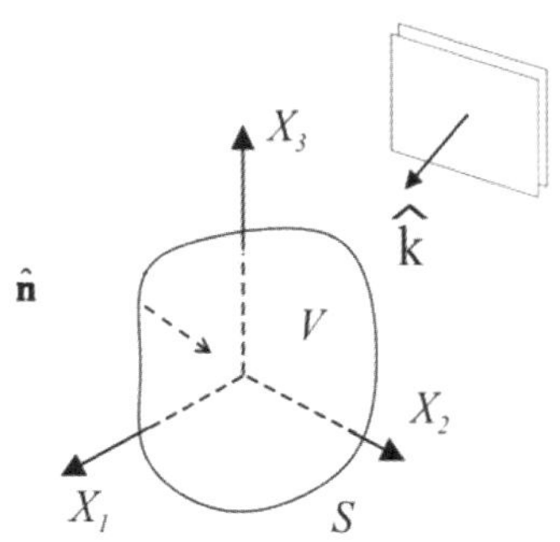

Figure 1: An acoustic plane wave impinges upon a scatterer of volume *V*.

The created scattered field ϕ^s and the incident one ϕ^i form the total field ϕ, which satisfies the following integral equation:

$$c\;\mathbf{x}\;\cdot\phi\;\mathbf{x},\omega\;+\int_S \frac{\partial G\;\mathbf{x},\mathbf{y},\omega}{\partial n_{\mathbf{y}}}\cdot\phi\;\mathbf{x},\omega\;dS_{\mathbf{y}}=\int_S G\;\mathbf{x},\mathbf{y},\omega\;\cdot\frac{\partial\phi\;\mathbf{y},\omega}{\partial n_{\mathbf{y}}}\,dS_{\mathbf{y}}+\phi^i\;\mathbf{x},\omega \tag{1}$$

where G($\mathbf{x}$,$\mathbf{y}$,ω) is the 3D acoustic fundamental solution [3,4] and $\mathbf{x}$, $\mathbf{y}$ are position vectors indicating a field point in the acoustic space and a source point on the boundary S, respectively. $\partial / \partial n$ indicates differentiation with respect the normal direction $\hat{\mathbf{n}}$ at the surface point $\mathbf{y}$, while the coefficient $c(\mathbf{x})$ takes the values 0,1,0.5 when the point $\mathbf{x}$ belongs in $V, R^3 - V, S$ respectively.

Discretizing the surface S into E quadratic line elements, Eq. 1 obtains the form

$$c\ \mathbf{x} \cdot \phi\ \mathbf{x},\omega\ + \sum_{e=1}^{E} \int_{S_e} \frac{\partial G\ \mathbf{x},\mathbf{y}^e,\omega}{\partial n_{\mathbf{y}}} \phi(\mathbf{x},\omega)\, dS_{\mathbf{y}^e} = \sum_{e=1}^{E} \int_{S_e} G\ \mathbf{x},\mathbf{y}^e,\omega\ \cdot \frac{\partial \phi\ \mathbf{y}^e,\omega}{\partial n_{\mathbf{y}}}\, dS_{\mathbf{y}^e} + \phi^i\ \mathbf{x},\omega \tag{2}$$

Expanding in the nodal basis functions and writing (2) for the node k, one obtains

$$c^k \cdot \phi^k + \sum_{e=1}^{E} \sum_{a=1}^{3} \tilde{\mathbf{H}}_{eak} \cdot \phi_a^e = \sum_{e=1}^{E} \sum_{a=1}^{3} \tilde{\mathbf{G}}_{eak} \cdot \frac{\partial \phi_e^a}{\partial n_{\mathbf{y}}} + \phi_k^i \tag{3}$$

where the double summation in Eq 3 runs on all of the nodes of each element.

Adopting a global numbering for all nodes Eq. 3 reads

$$c^k \cdot \phi^k + \sum_{q=1}^{Q} \tilde{\mathbf{H}}_{qk} \cdot \phi^q = \sum_{q=1}^{Q} \tilde{\mathbf{G}}_{qk} \cdot \frac{\partial \phi^{\ q}}{\partial n_{\mathbf{y}}} + \phi_k^i \tag{4}$$

Collocating at all nodes the following system of algebraic equations is obtained

$$\mathbf{H} \cdot \phi = \mathbf{G} \cdot \left\{ \frac{\partial \phi}{\partial n_{\mathbf{y}}} \right\} + \phi^i \tag{5}$$

with $\{q\}$ being a vector containing all the nodal values of the parameter q.

Satisfying the boundary conditions of the problem and rearranging the system (5) takes its final form

$$[\mathbf{A}]\{\mathbf{x}\} = \{\mathbf{b}\} \tag{6}$$

Where the vectors x, b contain all the unknown and known nodal values of the problem, respectively and A is a full populated and dense matrix. As it is mentioned in the introduction, for small dimensions of A the system (6) is solved via a typical LU decomposition algorithm, while for problems with many dof the wavelet transform technique explained below is applied before the final solution of the system.

3. Wavelet transform and compression of matrix [A]

For a given function $f(x)$, the wavelet transform $W_\psi(a,b)$ is defined as

$$W_{\psi}(a,b) = \frac{1}{\sqrt{|a|}} \int_{-\infty}^{\infty} \psi^{*}_{jk}(\frac{x-b}{a}) f(x) dx \tag{7}$$

where ψ_{jk} (with $j,k = 1,\ldots, N$) is a chosen wavelet family, a and b represent scale and position parameters, respectively and the symbol * denotes complex conjugation.

The wavelet family of functions ψ_{jk} is generated from the mother wavelet function ψ by scaling and translation, as follows:

$$\psi_{jk}\ t\ = 2^{1/2}\psi\ 2^{j}t - k \qquad \text{with } j,k = 1,\ldots, N \tag{8}$$

The mother wavelet ψ is defined by a scaling function (father wavelet):

$$\psi(t) = \sum_{k=0}^{N} (-1)^{k} h_{N-k}\, \phi(2t-k) \tag{9}$$

where h_k are the filter coefficients of the wavelet and N must be a power of 2.

In the present work the Haar wavelet is employed with the corresponding father wavelet ϕ and mother wavelet ψ being

$$\phi(t) = \begin{cases} 1 \ , & 0 \le t < 1 \\ 0 \ , & \text{elsewhere} \end{cases} \tag{10}$$

$$\psi(t) = \begin{cases} 1 \ , & 0 \le t < 1/2 \\ -1 \ , & 1/2 \le t < 1 \\ 0 \ , & \text{elsewhere} \end{cases} \tag{11}$$

The compression of matrix [**A**] (Eq 6) is accomplished with the aid of the Haar transformation matrix defined by Eq 8, for $j,k = 1,\ldots, N$. For example, for N=8, the Haar transform matrix, with each row normalized to unity, is given by:

$$H_8 = \begin{bmatrix} 1 & 1 & 1 & 1 & 1 & 1 & 1 & 1 \\ 1 & 1 & 1 & 1 & -1 & -1 & -1 & -1 \\ 1 & 1 & -1 & -1 & 0 & 0 & 0 & 0 \\ 0 & 0 & 0 & 0 & 1 & 1 & -1 & -1 \\ 1 & -1 & 0 & 0 & 0 & 0 & 0 & 0 \\ 0 & 0 & 1 & -1 & 0 & 0 & 0 & 0 \\ 0 & 0 & 0 & 0 & 1 & -1 & 0 & 0 \\ 0 & 0 & 0 & 0 & 0 & 0 & 1 & -1 \end{bmatrix} \tag{12}$$

Since the matrix [**A**] has not always dimensions equal to a power of 2, a suitable "padding" process must be followed in order to apply the wavelet transformation. In this paper, the procedure followed in [5] is adopted. Let [**A**] is a (*dof*)x(*dof*) matrix and k be the smaller number, such that $dof<2^k$. The *modified* (*dof*)x(*dof*) Haar matrix is defined as

$$[\bar{\mathbf{H}}]_{dof\times dof} = [\mathbf{C}]_{dof\times 2^k}[\mathbf{H}]_{2^k\times 2^k}[\mathbf{E}]_{2^k\times dof} \tag{13}$$

where $[\mathbf{C}],[\mathbf{E}]$ are suitable padding matrices, given in [5].

Under these considerations, we can transform the original linear system of equations as follows:

$$[\bar{\mathbf{H}}][\mathbf{A}][\bar{\mathbf{H}}]^{-1}[\bar{\mathbf{H}}]\{\mathbf{x}\} = [\bar{\mathbf{H}}]\{\mathbf{b}\} \tag{14}$$

The system (6) in the wavelet basis becomes:

$$[\hat{\mathbf{A}}]\{\hat{\mathbf{x}}\} = \{\hat{\mathbf{b}}\} \tag{15}$$

with

$$\begin{aligned} [\hat{\mathbf{A}}] &= [\bar{\mathbf{H}}][\mathbf{A}][\bar{\mathbf{H}}]^{-1} \\ \{\hat{\mathbf{x}}\} &= [\bar{\mathbf{H}}]\{\mathbf{x}\} \\ \{\hat{\mathbf{b}}\} &= [\bar{\mathbf{H}}]\{\mathbf{b}\} \end{aligned} \tag{16}$$

The transformed matrix $[\hat{\mathbf{A}}]$ has many elements with small values. Small in this context means, that the value is either very close to zero or is much smaller compared to the large values existing in the matrix. In order to obtain a sparse matrix, a thresholding criterion is applied, setting the values of the elements, satisfying the criterion, equal to zero. In the present work, the thresholding criterion applied, is given by:

$$\hat{a}_{ij} = 0 \quad \left|\hat{a}_{ij}\right| \le \lambda \cdot \min(\hat{a}_{11}, \hat{a}_{22}, \ldots \hat{a}_{NN}) \tag{17}$$

where λ is a parameter. For large λ the matrix becomes sparser and the solution generally becomes less accurate. The optimum value of λ is dependent on the specific problem under consideration.

4. Numerical examples

In order to demonstrate the efficiency of the proposed methodology, a characteristic acoustic wave scattering problem, concerning a rigid spherical scatterer of radius a, is solved. The incidence is a plane wave, having a dimensionless frequency $ka = 4$. The problem is solved utilizing three different meshes with *dof*=290, 1154 and 4610, as shown in figure 2.

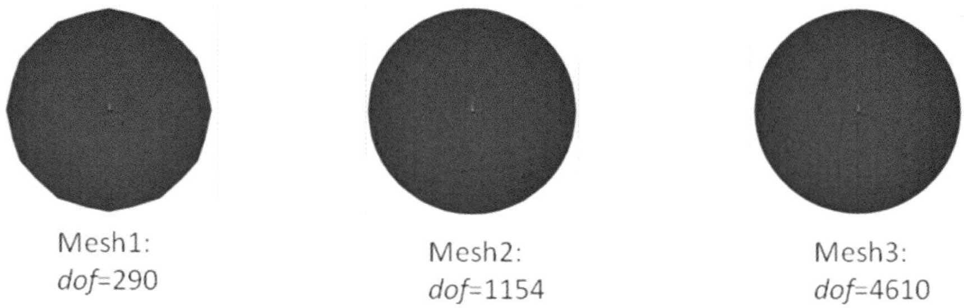

Figure 2. Different meshes of a spherical volume

In figures 3(a) and 3(b), the scattering amplitudes and the corresponding relative errors, respectively, are shown versus the scattering angle, for different values of threshold parameter λ, using Mesh2. The chosen values of λ=0.001, 0.005 and 0.01, correspond to sparsities 51%, 87.2% and 95.2% of matrix A, respectively and as can be observed from figure 3(b), the corresponding maximum percentage relative errors of the computed scattering amplitudes are equal to 0.56%, 7.62% and 13.1%, respectively. In addition, figure 4 depicts both the total computational time for the solution of the linear system and the average percentage relative error of the scattering amplitudes, as a function of λ.

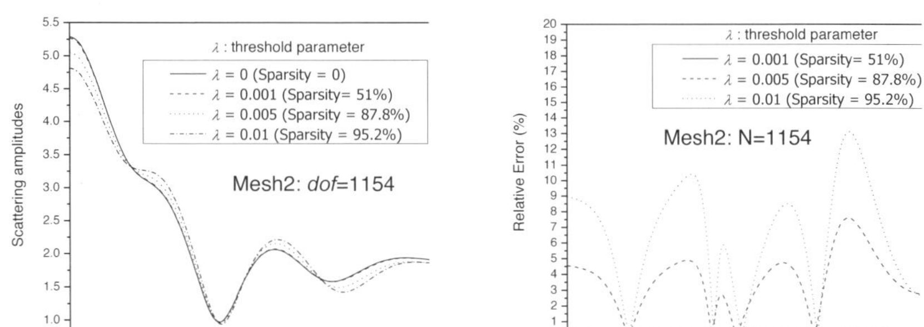

Figure 3: (a) Scattering Amplitudes and (b) their computed relative errors, as functions of the scattering angle, for different values of threshold parameter λ.

Finally, in figure 5, the total system solution time for λ=0.005 as a function of *dof* is shown, in comparison with the corresponding time required for the solution of the fully populated matrix produced by classical BEM (λ=0).

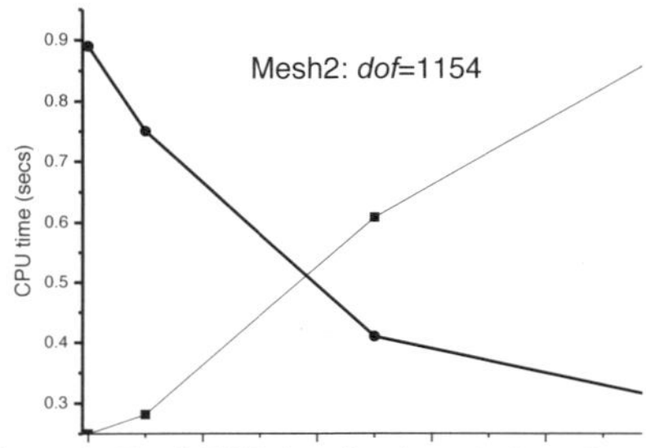

Figure 4. Total CPU time for the solution of [A]{x}={b} and the average relative error of the scattering amplitudes, as a function of λ.

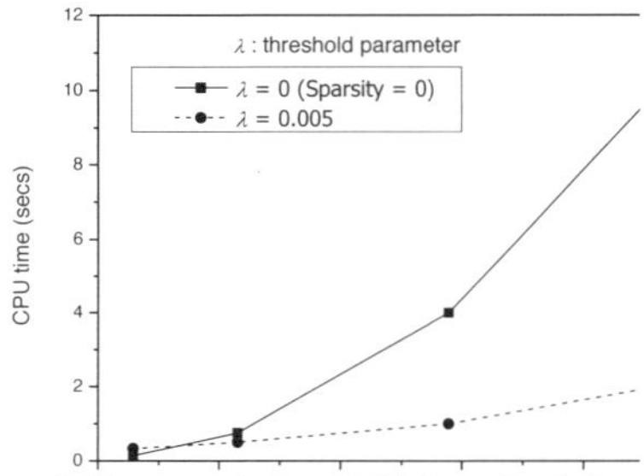

Figure 5. System solution CPU times for λ=0 and λ=0.005 as a function of *dof.*

5. Conclusions

In the present work, the Haar wavelet transform was applied to the fully populated matrix [**A**] generated by the classical BEM, leading to a sparse linear system. The sparsity was controlled by a user defined parameter and it was shown that the needed CPU time for the solution of the sparse system was considerably shorter than that required by classical BEM.

References

[1] Rego Silva JJ. Acoustic and Elastic Wave Scattering using Boundary Elements. Topics in Engineering, Brebbia CA, Connor JJ Eds; Vol. 18, Chap. 2,3, 7-68, Southampton, Boston, Computational Mechanics Publications, 1994.

[2] Ciskowski RD, Brebbia CA. Boundary Element Methods in Acoustics. Southampton, Boston, Computational Mechanics Publications and London, New York, Elsevier Applied Science, 1991.

[3] Agnatiaris JP, Polyzos D, 2003. A boundary element method for acoustc scattering from non-axisymmetric and axisymmetric elastic shells. CMES Comput Model Eng Sci 4:197–212.

[4] Tsinopoulos SV, Agnatiaris JP, Polyzos D, 1999. An advanced boundary element/fast Fourier transform axisymmetric formulation for acoustic radiation and wave scattering problems. Journal of the Acoustical Society of America; 105(3): 1517-1526.

[5] Valente FP, Pina HL, 2006. Conjugate gradient methods for three-dimensional BEM systems of equations. Engineering Analysis with Boundary Elements; 30: 441–449.

[6] Y. Saad and M. H. Schultz, "GMRES:Ageneralized minimal residual algorithm for solving nonsymmetric linear systems," *SIAM J. Sci. Statist. Comput.*, vol. 7, pp. 856–869, 1986.

[7] Gumerov NA, Duraiswami R, 2009. A broadband fast multipole accelerated boundary element method for the three dimensional Helmholtz equation. Journal of the Acoustical Society of America; 125 (1): 192-205.

[8] Liu y, 2009. Fast Multipole Boundary Element Method-Theory and Applications in Engineering. Cambridge University Press.

[9] Bucher HF, Wrobel LC, Mansur WJ, Magluta C, 2003. Fast solution of problems with multiple load cases by using wavelet-compressed boundary element matrices. Commun. Numer. Meth. Engng; 19:387–399.

[10] Ravnik J, Škerget L, Zunic Z, 2009. Comparison between wavelet and fast multipole data sparse approximations for Poisson and kinematics boundary – domain integral equations. Comput. Methods Appl. Mech. Engrg.; 198 ; 1473–1485.

Author Index